Yi-Long Chen and De-Ping Yang

Mössbauer Effect

in Lattice Dynamics

1807–2007 Knowledge for Generations

Each generation has its unique needs and aspirations. When Charles Wiley first opened his small printing shop in lower Manhattan in 1807, it was a generation of boundless potential searching for an identity. And we were there, helping to define a new American literary tradition. Over half a century later, in the midst of the Second Industrial Revolution, it was a generation focused on building the future. Once again, we were there, supplying the critical scientific, technical, and engineering knowledge that helped frame the world. Throughout the 20th Century, and into the new millennium, nations began to reach out beyond their own borders and a new international community was born. Wiley was there, expanding its operations around the world to enable a global exchange of ideas, opinions, and know-how.

For 200 years, Wiley has been an integral part of each generation's journey, enabling the flow of information and understanding necessary to meet their needs and fulfill their aspirations. Today, bold new technologies are changing the way we live and learn. Wiley will be there, providing you the must-have knowledge you need to imagine new worlds, new possibilities, and new opportunities.

Generations come and go, but you can always count on Wiley to provide you the knowledge you need, when and where you need it!

William J. Pesce
President and Chief Executive Officer

Peter Booth Wiley
Chairman of the Board

Yi-Long Chen and De-Ping Yang

Mössbauer Effect in Lattice Dynamics

Experimental Techniques and Applications

WILEY-VCH Verlag GmbH & Co. KGaA

The Authors

Prof. Dr. Yi-Long Chen
Physics Department, Wuhan University
Wuhan, The People's Republic of China
ylchen@whu.edu.cn

Dr. De-Ping Yang
Physics Department
College of the Holy Cross
Worcester, Massachusetts, USA
dyang@holycross.edu

Library of Congress Card No.: applied for

British Library Cataloguing-in-Publication Data
A catalogue record for this book is available from the British Library.

Bibliographic information published by the Deutsche Nationalbibliothek
Die Deutsche Nationalbibliothek lists this publication in the Deutsche Nationalbibliografie; detailed bibliographic data are available in the Internet at ⟨http://dnb.d-nb.de⟩.

Printed in the Federal Republic of Germany
Printed on acid-free paper

Typesetting Asco Typesetters, North Point, Hong Kong
Printing Strauss GmbH, Mörlenbach
Binding Litges & Dopf GmbH, Heppenheim
Wiley Bicentennial Logo Richard J. Pacifico

ISBN 978-3-527-40712-5

Contents

Mössbauer Effect in Lattice Dynamics. Yi-Long Chen and De-Ping Yang

ISBN: 978-3-527-40712-5

Preface

The chief objective of this book is an extensive and updated description of applications of Mössbauer effect in lattice dynamics. As an important component of solid state physics, lattice dynamics is the study of atomic vibrations around their equilibrium positions in a solid and it therefore leads to a better understanding of various properties of the solid. Research in lattice dynamics began in the early 20th century. For perfect crystals, lattice dynamics has been developed very successfully. Recently, the focus has been on dynamics of various imperfect lattice systems.

After the discovery of Mössbauer effect, theorists quickly pointed out the possibility of observing the frequency distribution of atomic vibrations in a solid, $g(\omega)$, using such an effect. Unfortunately, it did not become reality for many years because of technical difficulties. The amount of Doppler shift can, at best, only reach the order of μeV. It is difficult to increase it to cover the phonon energy range (meV) in a stable and reliable manner. Furthermore, the phonon peak is usually broadened and its intensity is at least two orders of magnitude smaller than that for the recoilless γ-ray resonance process, making it nearly impossible to measure $g(\omega)$ with sufficiently good statistics.

Although applications of Mössbauer effect in magnetic materials and chemistry are very extensive, those in lattice dynamics are less straightforward because of the following reasons. To study lattice dynamics, one must adopt a model (often the Debye model) for the phonon frequency distribution and must rely on Mössbauer measurements of recoilless fraction f and second order Doppler shift δ_{SOD}. These allow us to derive parameters such as mean-square displacement $\langle u^2 \rangle$ and mean-square velocity $\langle v^2 \rangle$ of atomic vibrations, Debye temperature θ_{D}, force constants, and the effective vibrating mass M_{eff}. There are many challenges in applying Mössbauer effect in lattice dynamics, which may have been the main reason why there existed very few books on this subject.

In the last 20 years, new progress in the field has been made, especially due to the rapid development of synchrotron radiation. In particular, the long-anticipated direct measurement of α-Fe phonon frequency distribution $g(\omega)$ was achieved for the first time using synchrotron Mössbauer source in 1995, which had motivated us to provide a comprehensive and in-depth description of all aspects of Mössbauer effect and lattice dynamics in this book.

Mössbauer Effect in Lattice Dynamics. Yi-Long Chen and De-Ping Yang

ISBN: 978-3-527-40712-5

There are a total of nine chapters and several appendices. The first three chapters introduce the basics of Mössbauer spectroscopy pertinent to lattice dynamics. Unlike most of the books on Mössbauer effect, we used the theory of coherent states to provide a simple yet rigorous derivation of the recoilless fraction f in Chapter 1. The second chapter deals with an essential part of Mössbauer spectroscopy, i.e., hyperfine interactions and the consequent polarization of γ-rays. Chapter 3 covers the instrumentation and data analysis, with one section especially devoted to describing in detail a method for estimating the optimal thickness of an absorber. The fourth chapter provides the background necessary for interpreting Mössbauer spectra, with a brief mention of the first-principles lattice dynamics because it would help us to understand the experimental results. Chapter 5 focuses on the properties of the two quantities f and δ_{SOD}, their dependence on temperature and pressure, anisotropic behavior of f, as well as the relationship between f and δ_{SOD}. In Chapter 6, scattering of Mössbauer radiation is discussed, with an emphasis on the understanding of coherence phenomena and the applications of Rayleigh scattering of Mössbauer radiation (RSMR) in lattice dynamics. The recent development synchrotron Mössbauer spectroscopy as a scattering method is described in Chapter 7, which contains a great deal of important and updated information, such as the excellent properties of synchrotron radiation (SR), how it makes time-domain Mössbauer spectroscopy possible, how it allows precise measurement of f in addition to hyperfine interactions, and how to measure the $g(\omega)$ of a solid directly. In Chapter 8, the Mannheim model is applied to lattices with very low concentrations of impurity atoms and its success is shown in several examples of experimental work. Chapter 8 also includes how isotopic selectivity of Mössbauer effect permits a unique way of studying the dynamics of impurity atoms. Chapter 9 presents a collection of various experimental results on metals, alloys, amorphous solids, molecular crystals, thin films, and nanocrystals, to show the versatility and applicability Mössbauer effect in lattice dynamics today.

This book may be used as a textbook for an advanced undergraduate course or a graduate course and as a reference book for researchers in Mössbauer spectroscopy, solid state physics, and related fields.

Special thanks go to our colleague and mutual friend Professor Xielong Yang of East China Normal University for introducing the two of us to each other and for initiating a productive collaboration. We would like to thank our respective families for their tremendous support throughout this project. Y.L. Chen is indebted to his wife Yuan Qin, Professor of Neurology, for her immense support and encouragement. D.P. Yang is deeply grateful to his wife Sharon Yang, an English professor at Worcester State College, for her endless love, stimulating discussions, and continued encouragement.

It has been a pleasure to work with Dr. Christoph v. Friedeburg, our Commissioning Editor, and with Ulrike Werner, our Project Editor, at Wiley-VCH. Our sincere appreciation goes to both of them for their professionalism and guidance in so many aspects of the production of this book. We also thank the reviewers for their constructive comments and insightful suggestions.

D.P. Yang would also like to thank Dr. William Hines and Dr. Philip Mannheim, Professors of Physics at University of Connecticut, who were great mentors and remain close friends. Many thanks go to Dr. Janine Shertzer, Professor of Physics at College of the Holy Cross, for her wisdom and insight in the preparation and submission of manuscripts, and to Diane Jepson for her careful reading of the manuscript. The responsibility of all errors lies, of course, with the authors. We welcome comments and corrections.

Yi-Long Chen 陈义龙
Wuhan Univeristy
Wuhan, China

De-Ping Yang 杨德平
College of the Holy Cross
Worcester, Massachusetts

February, 2007

1
The Mössbauer Effect

1.1
Resonant Scattering of γ-Rays

It was at the beginning of the 20th century that resonant scattering of light became experimentally verified. For example, when a beam of yellow light (the D-lines) from a sodium lamp goes through a flask with low-pressure sodium vapor in it, sodium atoms in the ^{2}S ground state will have a relatively large probability of absorbing the incident photons and making a transition to the excited ^{2}P state (as shown in Fig. 1.1). When these atoms return to the ground state, they emit a yellow light of the same wavelength (known as resonance fluorescence) in all spatial directions. In the original direction of the incident beam, the light intensity will be substantially reduced. This phenomenon can be considered as a process of resonant scattering of photons.

In 1929, Kuhn [1] pointed out that a similar γ-ray resonant scattering phenomenon should also exist for the nuclei. However, research during the next twenty plus years failed to produce satisfactory experimental results to support his predictions. The reason was quite clear. Because of the law of momentum conservation, after emitting a γ-ray, the nucleus obtains a velocity in the opposite direction (recoil). Compared to the recoil velocity of an atom when the atom emits a visible photon, the nucleus recoils with a velocity several orders of magnitude larger, takes enough energy away from the emitted γ-ray, and prevents the observation of resonance absorption. We will now discuss this in detail.

Suppose a free nucleus of mass M and initial velocity v is in the excited state E_e, emitting a γ-ray in the x-direction when it returns to the ground state. Figure 1.2 shows the energy levels and recoil of the nucleus, where v_x is the x-component of the initial velocity and v_R its recoil velocity (relative to v_x). According to momentum conservation and energy conservation, we have

$$\begin{cases} Mv_x = \dfrac{E_\gamma}{c} + M(v_x - v_R) \\ E_e + \dfrac{1}{2}M{v_x}^2 = E_g + E_\gamma + \dfrac{1}{2}M(v_x - v_R)^2 \end{cases} \tag{1.1}$$

Mössbauer Effect in Lattice Dynamics. Yi-Long Chen and De-Ping Yang

ISBN: 978-3-527-40712-5

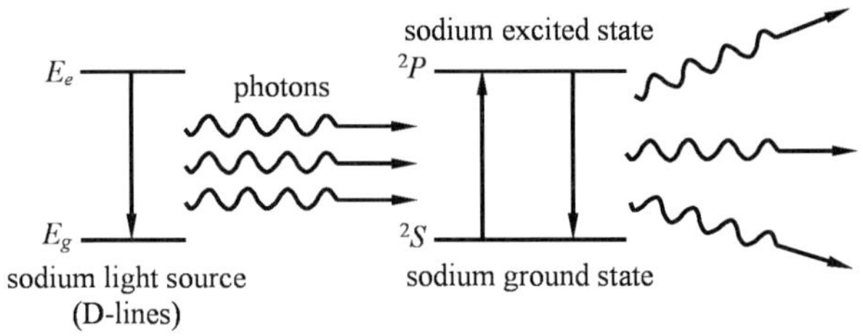

Fig. 1.1 Schematic diagram of resonance scattering of light.

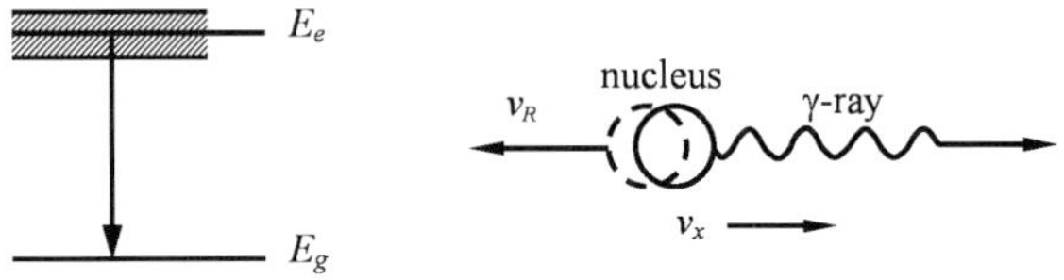

Fig. 1.2 Recoil of a nucleus after emitting a γ-ray.

where E_g is the ground state energy of the nucleus and E_γ is the energy of the emitted γ-ray. From the above equations, we obtain

$$E_\gamma = (E_e - E_g) - \frac{1}{2}Mv_R^2 + Mv_x v_R = E_0 - E_R + E_D, \tag{1.2}$$

where E_0 is the energy difference between the excited state and the ground state

$$E_0 = E_e - E_g, \tag{1.3}$$

E_R is the recoil energy

$$E_R = \frac{1}{2}Mv_R^2 = \frac{E_\gamma{}^2}{2Mc^2}, \tag{1.4}$$

and E_D depends on the initial velocity v_x and is due to the Doppler effect (known as the Doppler energy shift)

$$E_D = Mv_x v_R = \frac{v_x}{c}E_\gamma. \tag{1.5}$$

We will now consider the following two cases ($v_x = 0$ and $v_x \neq 0$) separately.

(a) If $v_x = 0$, then $E_D = 0$. In this case, the excited nucleus is at rest. The energy spectrum of the emitted γ-rays from such nuclei is shown by the dashed line in Fig. 1.3. The spectrum is a sharp peak centered at $E_0 - E_R$, and its width at

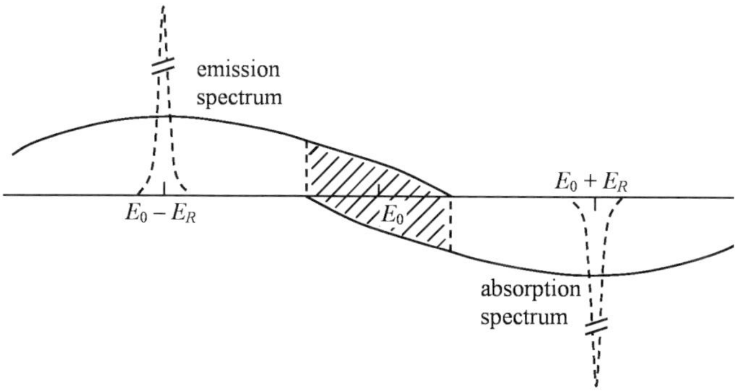

Fig. 1.3 Emission and absorption γ-ray spectra when recoil is present.

the half height is nearly the same as the natural width (Γ_n) of the excited energy level.

The nuclei in the ground state (E_g) may resonantly absorb the incident γ-rays and transit to the excited state (E_e). The energy distribution of these absorbed γ-rays is identical to the emission spectrum, except for a shift of E_R to the right of E_0, as shown in Fig. 1.3. The energy difference between the emitted and the absorbed γ-rays is $2E_R$. Therefore, the fundamental condition necessary for the photon's resonant scattering is

$$\frac{\Gamma_n}{2E_R} > 1, \tag{1.6}$$

that is, the recoil energy must be less than half of the natural width of the excited state. Comparing the data for the ^{57}Fe nucleus and the sodium atom in Table 1.1, we can easily see that for the Na atom, condition (1.6) is completely satisfied, because the emission spectrum and the absorption spectrum are almost overlapping, resulting in very large probability for resonant absorption. For the ^{57}Fe nucleus, however, its $\Gamma_n/2E_R$ value is far from satisfying condition (1.6). Although the natural widths Γ_n of a nucleus and an atom are comparable, the former gives a much more energetic photon than the latter, usually by three orders of magni-

Table 1.1 Comparison between photon emissions from the ^{57}Fe nucleus and the Na atom while each decays from its first excited state to the ground state.

	E_γ (eV)	Γ_n (eV)	E_R (eV)	$\Gamma_n/2E_R$
^{57}Fe nucleus	14.4×10^3	4.65×10^{-9}	1.95×10^{-3}	1.2×10^{-6}
Na atom (D-lines)	2.1	4.39×10^{-8}	1.0×10^{-10}	2.2×10^2

tude. This makes their E_R values differ by more than six orders of magnitude, and this is the reason why resonant absorption of γ-rays is usually not observed.

(b) For $v_x \neq 0$, the situation is more common. Because of the random thermal motions of free atoms, their velocities v_x may have large variations, described by the Maxwell distribution

$$p(v_x)\,dv_x = \left(\frac{M}{2\pi k_B T}\right)^{1/2} \exp\left(-\frac{M}{2k_B T} v_x{}^2\right) dv_x$$

where k_B is Boltzmann's constant and T is the absolute temperature. This distribution will greatly broaden the emission spectrum (or the absorption spectrum), as indicated by the solid line in Fig. 1.3. This broadening is due to the Doppler effect, and hence is known as Doppler broadening. Since the width of the above velocity distribution is $2(2k_B T \ln 2/M)^{1/2}$, the width of the emission spectral line is then

$$\Delta E_D = M v_R \left(2\sqrt{\frac{2k_B T \ln 2}{M}}\right) = 4\sqrt{E_R k_B T \ln 2}. \tag{1.7}$$

For ^{57}Fe at $T = 300$ K, $\Delta E_D = 2.4 \times 10^{-2}$ eV $> 2E_R$. This means that the emission spectrum partially overlaps the absorption spectrum (the shaded region in Fig. 1.3), and it may be possible to observe some effect of resonant absorption.

1.2 The Mössbauer Effect

1.2.1 Compensation for Recoil Energy

As discussed above, if the nucleus is free to move, the lost energy due to recoil must be compensated before substantial resonance absorption of γ-rays can be observed. Several ingenious experiments were devised to achieve this compensation, two of which are briefly explained here.

The first experiment made use of mechanical motion of the source [2]. The radioactive source was mounted on the tip of a high-speed rotor. Due to the Doppler effect, the γ-rays acquired an additional energy ΔE,

$$\Delta E = \frac{v}{c} E_\gamma. \tag{1.8}$$

It was possible to adjust the speed v of the rotor to completely compensate the recoil energy loss, i.e., $(v/c)E_\gamma = 2E_R$ (for ^{57}Fe, $v = 81$ m s^{-1}). This experiment had two problems. First, only during a very short portion of the rotation period could the emitted γ-rays be used in the experiment, and thus the source was largely under-utilized. Second, the experiment was limited by the maximum obtainable speed of the mechanical rotor and especially by the poor stability of the rotor speed.

The second experiment used the fact described in Eq. (1.7) that the Doppler broadening is increased by raising the temperature. As a result, it would cause increases in the overlapping region in Fig. 1.3, and therefore increases in the probability of resonance absorption.

By the above means, the phenomenon of γ-ray resonant absorption had been observed before 1954, but a major shortcoming was that these resonance absorption experiments all involved recoil, which would never be practically significant due to low γ-ray counts and poor energy resolution. A historic discovery by Mössbauer of resonant absorption *without* recoil completely eliminated the need for the above effort to compensate the energy loss. We will now describe this discovery.

1.2.2
The Discovery of the Mössbauer Effect

In 1958, Rudolf L. Mössbauer [3] was investigating the resonant absorption of the 129 keV γ-ray in ^{191}Ir nucleus and discovered that if the source nuclei ^{191}Os and absorber nuclei ^{191}Ir were rigidly bound in crystal lattices, the recoil could be effectively eliminated and the resonant absorption was readily observed.

In a crystal lattice, an atom is held in its equilibrium position by strong chemical bonds corresponding to an energy of typically 10 eV. For the 129 keV transition in free ^{191}Ir nucleus (Fig. 1.4), the recoil energy is 4.7×10^{-2} eV, much smaller than the chemical bond energy. Therefore, from the classical viewpoint, when the γ-ray is emitted by a nucleus bound in a lattice, the nucleus will not recoil alone, but the entire crystal lattice recoils together (a total of about 10^{18} atoms). In this case, the mass M in the denominator of Eq. (1.4) should be the mass of the whole crystal, not the individual nucleus. This reduces the recoil energy to a negligible amount ($\sim 10^{-20}$ eV). Consequently, Eq. (1.6) is satisfied, Eq. (1.2) is simplified to $E_\gamma \approx E_0$, and the entire process becomes a recoilless resonant absorption. A more exact explanation of this phenomenon is given by a quantum mechanical description in Section 1.5.

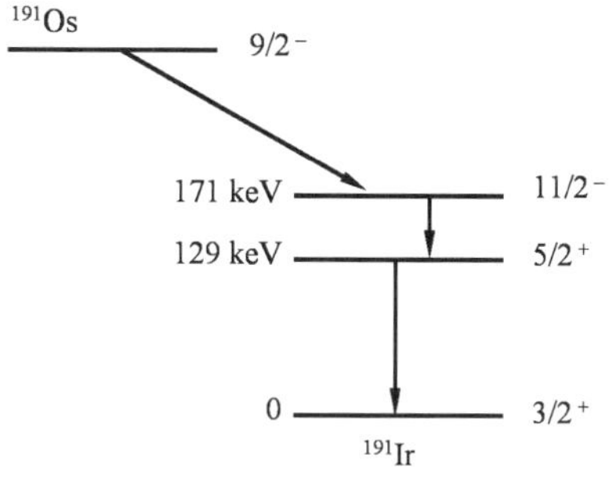

Fig. 1.4 Decay scheme of ^{191}Os.

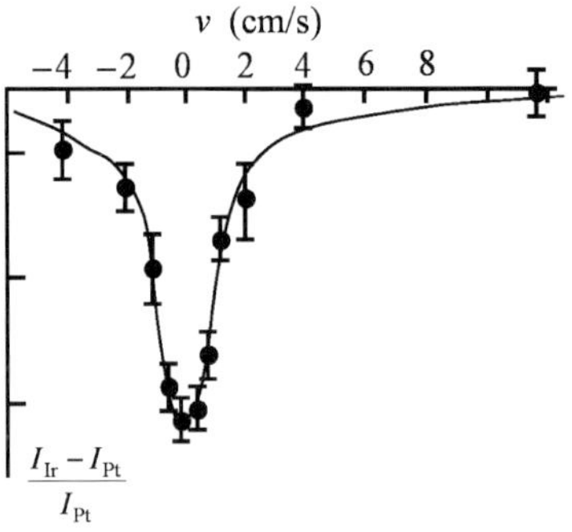

Fig. 1.5 Resonance absorption curve of the 129 keV γ-rays by ^{191}Ir.

In Mössbauer's first experiment where he observed recoilless resonance absorption of γ-rays, the radiation source was a crystal containing ^{191}Os and the absorber was an iridium crystal, both at a temperature of 88 K. A platinum (Pt) comparison absorber of the same thickness was used to measure the background. Because the process was recoilless, the Doppler velocity only needed to be small, about several centimeters per second. The results from that first experiment are reproduced in Fig. 1.5, where the horizontal axis represents the γ-ray energy variation ΔE (or source velocity v). When the source is moving towards the absorber, $v > 0$, and when the source is moving away from the absorber, $v < 0$. The vertical axis represents the relative change in the γ-ray intensity, $(I_{\mathrm{Ir}} - I_{\mathrm{Pt}})/I_{\mathrm{Pt}}$, where I_{Ir} and I_{Pt} are the γ-ray intensities transmitted through the Ir and Pt absorbers, respectively.

As shown in Fig. 1.5 and pointed out by Mössbauer, the width of the spectrum is 4.6×10^{-6} eV, which is just slightly more than twice the natural width of the 129 keV energy level of ^{191}Ir. Never before had such a high resolution in energy $(\Delta E/E \approx 3.5 \times 10^{-11})$ been achieved, and Mössbauer's research results were fundamentally different from what anyone had previously obtained from γ-ray resonant scattering, because he observed γ-ray emission and absorption events in which the recoil was completely absent. Not too long after the discovery of recoilless γ-ray emission and resonant absorption, this effect was named after its discoverer and is now known as the Mössbauer effect.

In reality, the Mössbauer nucleus is not rigidly bound, but is usually free to vibrate about its equilibrium position. Photons may exchange energy with the lattice, resulting in the creation or annihilation of quanta (phonons) of lattice vibrations. Suppose we have an Einstein solid with one vibrational frequency ω, then the lattice can only receive or release energies in integral multiples of $\hbar\omega$ $(0, \pm\hbar\omega, \pm 2\hbar\omega, \ldots)$. So if $E_{\mathrm{R}} < \hbar\omega$, the lattice cannot absorb the recoil energy, i.e., the zero phonon process, and the γ-ray is emitted without recoil. The probability of having such a process is known as the recoilless fraction f, an extremely important parameter in Mössbauer spectroscopy.

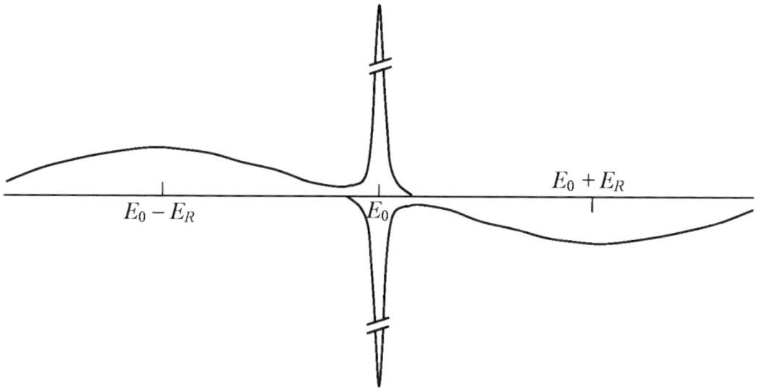

Fig. 1.6 Emission and absorption spectra of γ-rays when $E_R \ll \hbar\omega$.

In a typical lattice, both E_R and $\hbar\omega$ are in the ranges 10^{-3} to 10^{-1} eV. Obviously, the value of f depends on how E_R compares with $\hbar\omega$. Only when $E_R \ll \hbar\omega$ will f be reasonably large (see Fig. 1.6). As we derive it later (see Section 1.5.4), according to Lipkin's sum rule, when a large number of absorption events are considered, the average energy transferred to the lattice must be exactly equal to E_R. Let a total of m γ-photons with E_γ be absorbed among which n of them cause zero phonon creation and the rest $(m-n)$ photons each excites a single phonon (neglecting double phonons), then

$$mE_R = (m-n)\hbar\omega.$$

Based on the Einstein model, we arrive at an approximate expression for the recoilless fraction

$$f = \frac{n}{m} = 1 - \frac{E_R}{\hbar\omega}. \tag{1.9}$$

It can be seen from this expression that, in order to observe the Mössbauer effect, the recoilless fraction f should be sufficiently large, and we would like to have the following condition between E_R and $\hbar\omega$:

$$E_R \ll \hbar\omega. \tag{1.10}$$

A more precise expression for the recoilless fraction is

$$f = e^{-k^2\langle x^2\rangle}$$

where $\langle x^2\rangle$ is the mean square displacement of a nucleus along the direction of the wave vector $\boldsymbol{k}$ of the emitted γ-ray. This expression points out that in a liquid or a gas, the Mössbauer effect is extremely difficult to observe because of the large $\langle x^2 \rangle$ values. Also, a small k value would give a large f value, and therefore γ-rays with lower energies will favor the observation of the Mössbauer effect. At present, the Mössbauer effect has been observed from more than 100 nuclear isotopes (e.g., ^{57}Fe, ^{119}Sn, ^{191}Ir, etc.), among which one of the highest γ-ray energies is 187 keV in ^{190}Os. For a γ-ray energy higher than 100 keV, the source and the absorber are usually kept at low temperatures to reduce their $\langle x^2 \rangle$ values.

1.3
The Mössbauer Spectrum

1.3.1
The Measurement of a Mössbauer Spectrum

To facilitate our discussions in the first two chapters, the basic principles of measuring a Mössbauer spectrum will be given, before the experimental details in Chapter 3.

The shape of a resonance curve is often used to characterize the properties of the resonance system. For example, we can obtain the natural width Γ_n of the excited energy state from the linewidth of the measured γ-ray resonance curve and estimate the life time of the energy state according to the uncertainty relation $\tau\Gamma_n \sim \hbar$. A Mössbauer spectrum is a recoil-free resonance curve. To measure this, we no longer need those high-speed rotors, but it is still necessary to use the Doppler effect for modulating the γ-ray energy E_γ within a small energy range, $E_\gamma(1 \pm v/c)$. A velocity transducer with the mounted source moves with respect to the absorber and the emitted γ-ray energy is therefore modulated, as shown in Fig. 1.7. A Mössbauer absorption spectrum, shown on the right of Fig. 1.7, is a record of transmitted γ-ray counts through the absorber as a function of γ-ray energy, whose linewidth has a minimum value of $\Gamma_s + \Gamma_a$ (the sum of the natural widths of the Mössbauer nuclei in the source and the absorber).

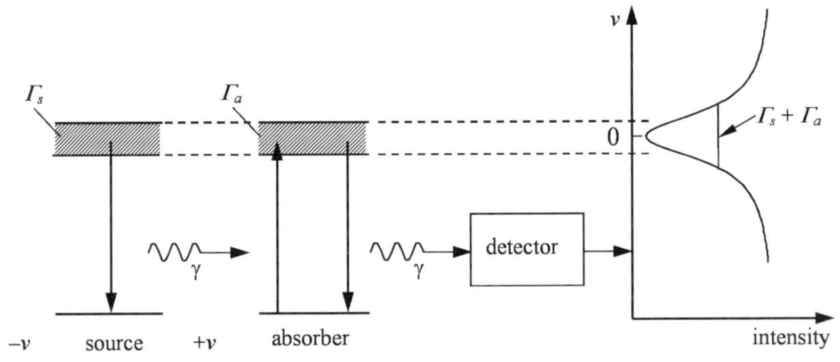

Fig. 1.7 Measuring a Mössbauer spectrum.

For the sake of simplicity, it is customary to use the source velocity (in mm s^{-1}) to label the energy axis. To obtain the energy value, one simply multiplies the velocity by a constant E_γ/c, and for ^{57}Fe, $E_\gamma/c = 4.8075 \times 10^{-8}$ eV mm^{-1} s.

1.3.2
The Shape and Intensity of a Spectral Line

After a Mössbauer resonant absorption, the nuclear excited state is an isomeric state, which can only decay to the ground state through γ-ray emission or internal conversion. The cross-section of resonant absorption of γ-rays (as a function of photon energy E) is described by the Breit–Wigner formula [1, 4]:

$$\sigma_a(E) = \frac{\sigma_0 \Gamma_a^2/4}{(E - E_0)^2 + \Gamma_a^2/4} \tag{1.11}$$

where

$$\sigma_0 = \frac{\lambda^2}{2\pi} \frac{1 + 2I_e}{1 + 2I_g} \frac{1}{1 + \alpha} \tag{1.12}$$

is the maximum resonance cross-section, E_0 and λ are the energy and wavelength of the γ-ray, I_e and I_g are, respectively, the nuclear spins of the excited and the ground states, and α is the internal conversion coefficient.

Because its excited state has a certain width Γ_s, the emitted γ-rays from the source are not completely monochromatic, but follow the Lorentzian distribution around E_0

$$\mathscr{L}(E)\, dE = \frac{\Gamma_s}{2\pi} \frac{1}{(E - E_0)^2 + \Gamma_s^2/4} dE \tag{1.13}$$

where

$$\int \mathscr{L}(E)\, dE = 1. \tag{1.14}$$

Therefore, in a situation where both the source and the absorber are very thin, the observed resonance absorption curve can be calculated by a convolution integral

$$\sigma_a^{exp}(E) \propto \int_{-\infty}^{+\infty} \mathscr{L}(E - x)\sigma(x)\, dx = \frac{\sigma_0 \Gamma_a}{\Gamma_a + \Gamma_s} \frac{\left(\frac{\Gamma_s + \Gamma_a}{2}\right)^2}{(E - E_0)^2 + \left(\frac{\Gamma_s + \Gamma_a}{2}\right)^2} \tag{1.15}$$

and it is clear that the line shape is also Lorentzian, similar to Eq. (1.13) except that the linewidth becomes $\Gamma_s + \Gamma_a$. In reality, because of the finite thicknesses of the source and absorber, the emission and absorption spectral linewidths Γ_s

and Γ_a would be larger than the natural width Γ_n (for ^{57}Fe, $\Gamma_n \approx 0.097$ mm s^{-1}), and the observed resonance line would be broader than $2\Gamma_n$.

We now discuss in detail how the thickness of an absorber influences the shape and intensity of a transmission spectrum. Let the total intensity of the γ-ray emitted by Mössbauer nuclei be I_0, of which only a part I_r is recoil free and distributed according to a Lorentzian shape:

$$I_r(E, v, 0) = f_s I_0 \mathscr{L}\left(E - \frac{v}{c} E_0\right)$$

where f_s and v are the recoilless fraction and the Doppler velocity of the source. Going through the absorber, γ-ray intensity is reduced because of two absorption processes, a non-resonance atomic absorption (mainly the photoelectric effect) with a mass absorption coefficient of μ_a (μ_a values for different elements are tabulated in Appendix H) and a Mössbauer resonance absorption with an absorption coefficient of μ_r:

$$\mu_r(E) = n_a f \sigma_a(E) \tag{1.16}$$

where n_a is the number of Mössbauer nuclei in the absorber per unit mass and f is the recoilless fraction of the absorber. Considering both of these absorption processes, the γ-ray intensity decreases exponentially after transmitting an absorber thickness d (mg cm^{-2}):

$$I_r(E, v, d) = f_s I_0 \mathscr{L}\left(E - \frac{v}{c} E_0\right) e^{-(\mu_a + \mu_r)d}. \tag{1.17}$$

According to this, at a given Doppler velocity of the source, the intensity of the recoil-free γ-ray detected should be an integral over the energy:

$$I_r(v, d) = \int_{-\infty}^{+\infty} I_r(E, v, d)\,\mathrm{d}E = f_s I_0 e^{-\mu_a d} T(v) \tag{1.18}$$

where

$$T(v) = \int_{-\infty}^{+\infty} \mathscr{L}\left(E - \frac{v}{c} E_0\right) A(E)\,\mathrm{d}E, \tag{1.19}$$

$$A(E) = \exp[-\mu_r(E)d] = \exp[-\sigma(E)t_a], \tag{1.20}$$

$$\sigma(E) = \sigma_a(E)/\sigma_0,$$

$$t_a = n_a f \sigma_0 d. \tag{1.21}$$

$T(v)$ is known as the transmission integral. As defined in Eq. (1.21), t_a is called the effective thickness of the absorber, and is temperature dependent in the same manner as f.

The rest of the γ-rays are emitted with recoil, and they are distributed in a rather broad energy range (Fig. 1.6) and absorbed solely due to the non-resonant absorption process. Thus, the intensity after absorption is independent of the Doppler velocity v and can be expressed as

$$I(d) = I_0(1 - f_s)e^{-\mu_a d}. \tag{1.22}$$

Combining Eqs. (1.18) and (1.22), we obtain the total intensity recorded by the detector (whose efficiency is assumed to be 100%) as

$$I(v, d) = I_r(v, d) + I(d) = I(\infty, d)[1 - f_s + f_s T(v)] \tag{1.23}$$

where $I(\infty, d) = I_0 \exp(-\mu_a d)$ is the spectral baseline corresponding to $v = \infty$.

If we neglect hyperfine interactions for the time being, the fractional intensity of the absorbed of γ-rays at a Doppler velocity v can be defined as

$$\varepsilon(v) = \frac{I(\infty, d) - I(v, d)}{I(\infty, d)} = f_s[1 - T(v)] \tag{1.24}$$

which describes the shape of the absorption spectrum. According to Appendix A or Ref. [5], the fractional intensity $\varepsilon(v)$ can be obtained analytically and, at resonance $v = v_r = 0$, $\varepsilon(v)$ reaches its maximum

$$\frac{\varepsilon(v_r)}{f_s} = 1 - e^{-t_a/2} I_0\left(\frac{t_a}{2}\right) - 2e^{-t_a/2} \sum_{n=1}^{\infty} \left(\frac{\xi - 1}{\xi + 1}\right)^n I_n\left(\frac{t_a}{2}\right) \tag{1.25}$$

which is explicitly expressed in terms of $\xi = \Gamma_s/\Gamma_a$. In the above equation, I_n is the modified Bessel function of the first kind of order n. If $\Gamma_s = \Gamma_a$, thus $\xi = 1$, Eq. (1.25) becomes

$$\varepsilon(v_r) = f_s\left[1 - e^{-t_a/2} I_0\left(\frac{t_a}{2}\right)\right] \tag{1.26}$$

which is a well-known result independent of both linewidths.

Next, we discuss the contribution of the third term in $\varepsilon(v_r)$, which will be abbreviated as ε_3:

$$\varepsilon_3 = -2e^{-t_a/2} \sum_{n=1}^{\infty} \left(\frac{\xi - 1}{\xi + 1}\right)^n I_n\left(\frac{t_a}{2}\right). \tag{1.27}$$

The value of $\varepsilon(v_r)/f_s$ in Eq. (1.25) is plotted in Fig. 1.8 as a function of t_a and ξ. Regardless whether n is even or odd, $I_n(t_a/2)$ is always positive. Therefore, the sign of ε_3 is determined by the factor $(\xi - 1)/(\xi + 1)$. When $\xi < 1$, $\varepsilon_3 > 0$, and when $\xi > 1$, $\varepsilon_3 < 0$. The effect of this third term ε_3 is clearly demonstrated in

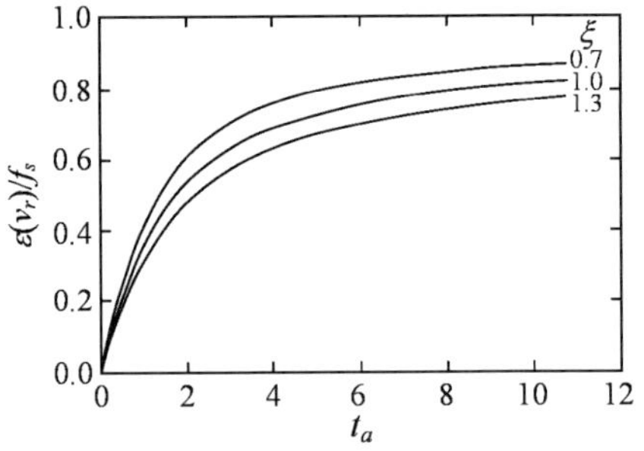

Fig. 1.8 $\varepsilon(\nu_r)/f_s$ as a function of t_a for $\xi = 0.7, 1.0$, and 1.3 [5].

Fig. 1.8. The curve with $\xi = 1$ is completely consistent with those given in Ref. [6]. In practice, cases with $\xi > 1$ are hardly observed and $\xi < 1$ is in the majority. Therefore, the influence of the third term on $\varepsilon(v_r)$ is essentially the addition of a positive contribution. Obviously, when $t_a < 1$, such an influence becomes negligible regardless of the value of ξ.

In fact, the above argument can be understood in the following straightforward way. In the case where $\Gamma_s < \Gamma_a$ (or $\xi < 1$), the absorber in some sense looks like a "black absorber" [7] absorbing the majority of resonant γ-rays. In other words, the resonant γ-rays, as a whole, have a higher probability of becoming absorbed.

Based on the above results, a transmission Mössbauer spectrum is sketched in Fig. 1.9 where I_b represents the background counts. During the above derivation, we assumed that the intensity was corrected by I_b.

As long as the "thin absorber approximation" ($t_a < 1$) is valid, one only need to take the first two terms in the polynomial expansion of $A(E)$ in Eq. (1.20). Then the fractional absorption intensity described in Eq. (1.24) can be easily written as

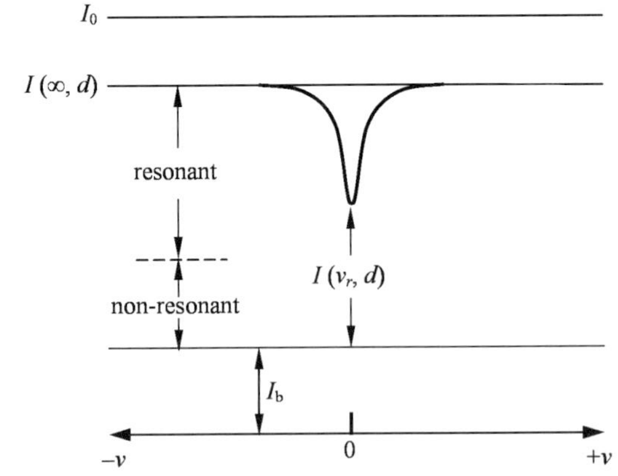

Fig. 1.9 Contributions to the Mössbauer spectrum in transmission geometry.

$$\varepsilon(v) = f_s[1 - T(v)] = f_s \int_{-\infty}^{+\infty} \mathscr{L}\left(E - \frac{v}{c}E_0\right)\left[1 - \exp\left(-\frac{t_a\sigma_a}{\sigma_0}\right)\right] dE$$

$$\approx f_s \int_{-\infty}^{+\infty} \mathscr{L}\left(E - \frac{v}{c}E_0\right) \frac{t_a(\Gamma_a/2)^2}{(E - E_0)^2 + (\Gamma_a/2)^2} dE$$

$$= \frac{\Gamma_a}{\Gamma_s + \Gamma_a} \frac{f_s t_a \left(\frac{\Gamma_s + \Gamma_a}{2}\right)^2}{\left(\frac{v}{c}E_0\right)^2 + \left(\frac{\Gamma_s + \Gamma_a}{2}\right)^2}. \qquad (1.28)$$

This means when $t_a < 1$, the spectral shape is still Lorentzian. At resonance, expression (1.25) becomes identical to (1.28), only if $t_a < 1$ and $\Gamma_s = \Gamma_a$. The area of the absorption spectrum has been accurately calculated [4]:

$$A(t_a) = f_s\Gamma_a\pi\frac{t_a}{2}\exp\left(-\frac{t_a}{2}\right)\left[J_0\left(i\frac{t_a}{2}\right) + J_1\left(i\frac{t_a}{2}\right)\right] \qquad (1.29)$$

where J_0 and J_1 are the zeroth- and first-order Bessel functions. Important parameters of a Mössbauer spectrum are the height, width, area, and position of a spectral line. Because of the constraint in Eq. (1.29), only two of the first three parameters are independent.

As the absorber thickness increases, the area $A(t_a)$, as well as $\varepsilon(v_r)$, deviates considerably from its linearity with t_a and gets saturated (see Figs. 1.10 and 1.8). Interpretation of Mössbauer spectra is often complicated by such a saturation effect due to a finite absorber thickness. A comparison between Figs. 1.10 and 1.8 shows how the area $A(t_a)$ saturates much less rapidly than $\varepsilon(v_r)$. A further analysis reveals that the spectral shape remains Lorentzian for up to $t_a \approx 10$.

Notice that $\varepsilon(v)$ describes the shape of the spectrum and obviously depends on both Γ_s and Γ_a, while the area $A(t_a)$ is an integral of $\varepsilon(v)$ over the Doppler velocity range (see Appendix A) and is only dependent on Γ_a.

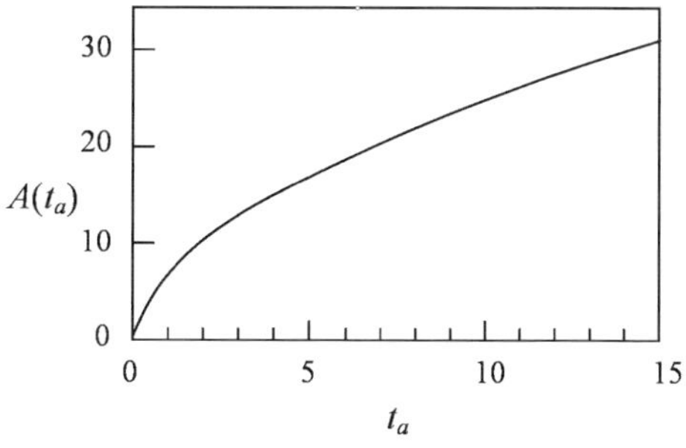

Fig. 1.10 $A(t_a)$ as a function of t_a. In plotting this curve, the proportionality constant $(f_s\Gamma_a\pi)$ in Eq. (1.29) is taken to be 1.

1.4 The Classical Theory

Although the Mössbauer effect is a quantum mechanical effect, its main features can be also derived by the classical theory. The first comprehensive classical description was provided by Shapiro [8]. A radioactive nucleus, as a classical oscillator, does not experience a recoil effect and emits an electromagnetic wave of frequency ω_0. The distribution in frequency is entirely determined by the Doppler effect. Thus, the corresponding vector potential at distance x_0 from the source is then

$$\boldsymbol{A}(t) = \boldsymbol{A}(0)\exp(-\gamma t)\exp[i(\omega_0 t - kx_0)] \tag{1.30}$$

where γ is the damping coefficient, which is half of the natural width of the excited state, $\gamma = \Gamma_n/2$. If thermal motion of the nucleus is neglected, the distance x_0 will be constant. As a result, leaving out the last phase factor in Eq. (1.30) has no effect on the recoilless fraction. Thus the radiation intensity as a function of frequency is

$$I(\omega) = I_0 \frac{(\Gamma_n/2)^2}{(\omega_0 - \omega)^2 + (\Gamma_n/2)^2}. \tag{1.31}$$

In reality, the nucleus in a solid undergoes inevitable thermal motion around its equilibrium position. From the classical point of view, this motion modulates the electromagnetic wave due to the Doppler effect. Let $v(t)$ be the velocity component of the nucleus in the direction of γ-ray propagation. The phase of the wave is modulated and becomes

$$\phi(t) = \int_{-\infty}^{t} \omega_0 \left(1 + \frac{v(t')}{c}\right) \mathrm{d}t' = \omega_0 t + \frac{2\pi x(t)}{\lambda} \tag{1.32}$$

where $x(t)$ is the instantaneous displacement of the nucleus away from its equilibrium position in the direction of the γ-ray propagation, and may be expressed as

$$x(t) = x_0 \sin \Omega t \tag{1.33}$$

where we have used Ω ($\Omega \ll \omega_0$) to represent the frequency of the thermal motion of all Mössbauer nuclei (as is the case with the Einstein model). Incorporating the phase modulation into Eq. (1.30), the vector potential becomes

$$\boldsymbol{A}(t) = \boldsymbol{A}(0)\exp(-i\omega_0 t - \Gamma_n t/2)\exp(ikx_0 \sin \Omega t). \tag{1.34}$$

If we expand the last phase factor into a sum of Bessel functions,

$$\exp(ikx_0 \sin \Omega t) = \sum_{-\infty}^{+\infty} J_n(kx_0) \exp(-in\Omega t),$$

Eq. (1.34) can be written as

$$\boldsymbol{A}(t) = \boldsymbol{A}(0) \sum_{n=-\infty}^{+\infty} J_n(kx_0) \exp(-\Gamma_n t/2) \exp(-i\omega_0 t - in\Omega t).$$

Consequently, the normalized distribution of radiation intensity is

$$I(\omega) = I_0 \sum_{n=-\infty}^{+\infty} [J_n(kx_0)]^2 \frac{(\Gamma_n/2)^2}{(\omega - \omega_0 - n\Omega)^2 + (\Gamma_n/2)^2}. \tag{1.35}$$

It is clear that this radiation includes one spectral line unshifted in frequency (ω_0) as well as a series of satellite lines with frequencies $\omega_0 \pm \Omega$, $\omega_0 \pm 2\Omega$, $\omega_0 \pm 3\Omega$, etc. Each spectral line has a Lorentzian shape with a width of Γ_n, and its intensity is described by the respective coefficient, i.e., the square of the Bessel function value (Fig. 1.11). Therefore, the recoilless fraction is $f = [J_0(kx_0)]^2$.

For a low-energy radiation, we have $kx_0 \ll 1$, and

$$\ln f \approx 2 \ln\left(1 - \frac{k^2 x_0^2}{4}\right) \approx -\frac{k^2 x_0^2}{2}$$

or,

$$f = e^{-k^2 \langle x^2 \rangle} \tag{1.36}$$

where $\langle x^2 \rangle = {x_0}^2/2$ is the mean square of the displacement of the nuclear vibration. This result is identical to the quantum mechanical result to be derived next.

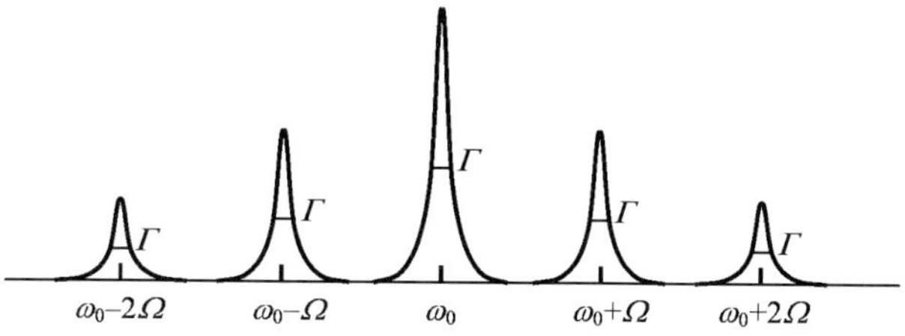

Fig. 1.11 Intensity distribution of γ-ray emission from a classical oscillator.

1.5 The Quantum Theory

Mössbauer [3] and later Visscher [9] derived the f fraction based on the theory of neutron resonance scattering from nuclei bound in a solid [10]. Soon after, Lipkin [11] simplified the derivation of the f fraction. Singwi and Sjölander [12] used a method developed by van Hove [13] to arrive at this result. The reader may find an abundance of relevant references.

In the 1960s, a theoretical method was developed using coherent states [14, 15] (also known as pseudo-classical quantum states). The concept of coherent states has attracted attention from researchers in many areas of physics, and recently found a wide range of applications. The earliest and the most complete studies of the coherent states were those of the harmonic oscillators [16, 17], and these coherent states provide an extremely convenient way of describing certain particular states of vibration. Because harmonic oscillation is an important model for describing the structure and motion of matter on the microscopic scale, the method of coherent states is especially useful in research fields such as studying interactions between radiation and matter. This method not only provides a direct analogy to the classical theory, but also greatly simplifies the calculation. Recently, Bateman et al. [18] calculated the recoilless fraction f for Mössbauer effect using coherent states. Here, we will use this new approach to the derivation of recoilless fraction f.

1.5.1 Coherent States of a Harmonic Oscillator

The Hamiltonian of a one-dimensional harmonic oscillator is

$$\mathscr{H} = \frac{\hat{p}^2}{2m} + \frac{1}{2} m\omega^2 \hat{x}^2. \tag{1.37}$$

Instead of using the position and momentum operators $\hat{x}$ and $\hat{p}$, we will introduce an annihilation operator $\hat{a}$ and a creation operator $\hat{a}^+$:

$$\hat{a} = \sqrt{\frac{m\omega}{2\hbar}}\left(\hat{x} + \frac{i}{m\omega}\hat{p}\right), \tag{1.38}$$

$$\hat{a}^+ = \sqrt{\frac{m\omega}{2\hbar}}\left(\hat{x} - \frac{i}{m\omega}\hat{p}\right). \tag{1.39}$$

Solving for $\hat{x}$ and $\hat{p}$ from the above definitions, we obtain

$$\hat{x} = \sqrt{\frac{\hbar}{2m\omega}}(\hat{a}^+ + \hat{a}), \tag{1.40}$$

$$\hat{p} = i\sqrt{\frac{m\hbar\omega}{2}}(\hat{a}^+ - \hat{a}). \tag{1.41}$$

Substituting into the Hamiltonian, it becomes quite simple

$$\mathscr{H} = \hbar\omega\left(\hat{a}^{+}\hat{a} + \frac{1}{2}\right)$$

and $\hat{a}^{+}\hat{a}$ is known as the number operator $\hat{N}$, and its eigenstates $|n\rangle$ are also eigenstates of the Hamiltonian

$$\mathscr{H}|n\rangle = E_n|n\rangle = \hbar\omega\left(\hat{a}^{+}\hat{a} + \frac{1}{2}\right)|n\rangle = \hbar\omega\left(n + \frac{1}{2}\right) \quad n = 0, 1, 2, \ldots \tag{1.42}$$

$$\hat{N}|n\rangle = \hat{a}^{+}\hat{a}|n\rangle = n|n\rangle. \tag{1.43}$$

This means that each eigenvalue of $\hat{N}$ is the number of energy quanta $\hbar\omega$ in the number state $|n\rangle$. Any excited state $|n\rangle$ can be generated by repeatedly applying the creation operator on the ground state $|0\rangle$:

$$|n\rangle = \frac{1}{(n!)^{1/2}}(\hat{a}^{+})^{n}|0\rangle \tag{1.44}$$

where all possible n values are included, and these states $|n\rangle$ form a complete orthonormal set. We will introduce an important concept, the coherent state, defined as the following linear combination of these states:

$$|\alpha\rangle = \mathrm{e}^{-(1/2)|\alpha|^2}\sum_{n}\frac{\alpha^{n}}{(n!)^{1/2}}|n\rangle \tag{1.45}$$

where α may be any complex number, as is proved later. If we substitute (1.44) into (1.45), a coherent state may also be expressed in terms of $|0\rangle$:

$$|\alpha\rangle = \hat{D}(\alpha)|0\rangle, \tag{1.46}$$

where $\hat{D}(\alpha)$ is called the displacement operator

$$\hat{D}(\alpha) = \exp(\alpha\hat{a}^{+} - \alpha^{*}\hat{a}). \tag{1.47}$$

Among all the coherent states ever developed, harmonic oscillator coherent states were the earliest ones and are now the most widely applied. Interestingly, the coherent states represent those states in which the uncertainty relation takes the minimum value (i.e., they describe situations that best resemble classical systems). The squares of standard deviations of position x and momentum p are

$$\Delta x^2 = \langle\alpha|x^2|\alpha\rangle - \langle\alpha|x|\alpha\rangle^2 = \frac{\hbar}{2m\omega}, \tag{1.48}$$

$$\Delta p^2 = \langle\alpha|p^2|\alpha\rangle - \langle\alpha|p|\alpha\rangle^2 = \frac{\hbar m\omega}{2}. \tag{1.49}$$

The product of the standard deviations is the smallest possible value allowed by the uncertainty principle

$$\Delta x \Delta p = \frac{\hbar}{2}.$$

It is because of this property of satisfying the minimum uncertainty that these quantum states are also known as pseudo-classical coherent states.

For any coherent state, we can show that

$$\hat{a}|\alpha\rangle = \alpha|\alpha\rangle, \tag{1.50}$$

$$\langle\alpha|\hat{a}^{+} = \langle\alpha|\alpha^{*}. \tag{1.51}$$

Since $\hat{a}$ is not a Hermitian operator, α is a complex eigenvalue. It is easily verified that the $|\alpha\rangle$ eigenstates are normalized, but not orthogonal. However, they constitute an overcomplete set, represented by

$$\frac{1}{\pi}\int |\alpha\rangle\langle\alpha|\, d^2a = \hat{I} \tag{1.52}$$

where $\hat{I}$ is the unitary operator. This is a very useful operator, because any other operator, particularly the density operator $\hat{\rho}$, may be expressed in the coherent state basis as

$$\hat{\rho} = \frac{1}{\pi}\int p(\alpha)|\alpha\rangle\langle\alpha|\, \mathrm{d}^2a. \tag{1.53}$$

This is the p-representation of the density operator $\hat{\rho}$. For oscillators at temperature T in thermal equilibrium [19]

$$p(\alpha) = \frac{1}{\langle n\rangle} \exp[-|\alpha|^2/\langle n\rangle] \tag{1.54}$$

where

$$\langle n\rangle = \frac{\exp[-\hbar\omega/k_{\mathrm{B}}T]}{1 - \exp[-\hbar\omega/k_{\mathrm{B}}T]} \tag{1.55}$$

and k_{B} is Boltzmann's constant. Therefore, using $p(\alpha)$ as a weight function, the thermal average of any physical quantity can be evaluated in the coherent state basis.

1.5.2
Gamma Radiation from a Bound Nucleus

Suppose that an atom is not free but moving in the potential of a harmonic oscillator. Although the motion of this atom may not be identical to that in a crystal, this approximation can lead to the basic characteristics of the Mössbauer effect.

Let the initial state of the atom before irradiating a γ-ray be the ground state $|0\rangle$ of the harmonic oscillator. As can be seen in (1.42), this state is not an eigenstate of the momentum operator $\hat{p}$, and it is impossible to immediately write down its final state through momentum conservation. However, the set of eigenstates of the momentum operator constitute a complete orthonormal set $|k'\rangle$, and we may expand $|0\rangle$ in this set as follows:

$$|0\rangle = \sum |k'\rangle\langle k'|0\rangle. \tag{1.56}$$

When an energy transition occurs within a nucleus at $t = 0$, a γ-ray with $k = E_\gamma/c\hbar$ is emitted in the x-direction. The momentum of the atom must change from $\hbar k'$ to $\hbar(k' - k)$ in order to conserve momentum, and the final state of the atom's motion can be written as

$$|f\rangle = \sum_{k'} |k' - k\rangle\langle k'|0\rangle. \tag{1.57}$$

It is obvious that $e^{-ik\hat{x}}$ is a displacement operator of k, thus

$$e^{-ik\hat{x}}|k\rangle = |k' - k\rangle. \tag{1.58}$$

When this is substituted into (1.57), the final state is given by

$$|f\rangle = e^{-ik\hat{x}}|0\rangle. \tag{1.59}$$

Contrary to the case of the free atom, the final state (1.59) is not an eigenstate of the Hamiltonian (1.37) and therefore does not have a well-defined energy. This means that one cannot predict the energy of the γ-ray in advance, but can only provide a probability description. Let us again expand $|f\rangle$ in the complete set $|n\rangle$ of eigenstates (1.44):

$$|f\rangle = \sum_n |n\rangle\langle n|f\rangle = \sum_n |n\rangle\langle n|e^{-ik\hat{x}}|0\rangle \tag{1.60}$$

where we have used Eq. (1.59). The probability that the atom is found in the state $|n\rangle$ with energy $(n + 1/2)\hbar\omega$ is given by the square of the expansion coefficient in

(1.60). Thus, the probability for the atom to remain in the ground state $|0\rangle$ after the γ-emission is none other than the recoilless fraction f of the Mössbauer effect:

$$f = |\langle 0|\mathrm{e}^{-ik\hat{x}}|0\rangle|^2. \tag{1.61}$$

To evaluate f, we express the operator $\mathrm{e}^{-ik\hat{x}}$ in terms of the annihilation operator $\hat{a}$ and the creation operator $\hat{a}^+$ by using (1.40)

$$-ik\hat{x} = \alpha\hat{a}^+ - \alpha^*\hat{a} \quad \alpha = -ik\left(\frac{\hbar}{2M\omega}\right)^{1/2} \tag{1.62}$$

where M is the mass of the nucleus. According to (1.47), the operator e^{-ikx} happens to be a displacement operator $\hat{D}$, and the final state $|f\rangle$ is a coherent state

$$\mathrm{e}^{-ik\hat{x}}|0\rangle = \mathrm{e}^{(\alpha\hat{a}^+ - \alpha^*\hat{a})}|0\rangle = |\alpha\rangle = \mathrm{e}^{-(1/2)|\alpha|^2}\sum_n \frac{\alpha^n}{(n!)^{1/2}}|n\rangle. \tag{1.63}$$

Applying this to (1.61) and taking account of the orthogonal property of the states $|n\rangle$, we obtain

$$f = \mathrm{e}^{-|\alpha|^2}.$$

Substituting the value for α (1.62) into this, we have

$$f = \mathrm{e}^{-k^2(\hbar/2M\omega)}. \tag{1.64}$$

As defined in (1.4), the recoil energy is $E_\mathrm{R} = \dfrac{1}{2}Mv^2 = \dfrac{p^2}{2M} = \dfrac{k^2\hbar^2}{2M}$, and we can express f in terms of E_R:

$$f = \mathrm{e}^{-E_\mathrm{R}/\hbar\omega} \tag{1.65}$$

which is consistent with (1.9).

On the other hand, using a special property of a harmonic oscillator that its average kinetic energy is one half of the total energy (for the ground state, total energy is $\frac{1}{2}\hbar\omega$),

$$\frac{1}{2}M\omega^2\langle x^2\rangle = \frac{1}{2}\left(\frac{1}{2}\hbar\omega\right), \quad \text{or} \quad \langle x^2\rangle = \frac{\hbar}{2M\omega}$$

and substituting into (1.64), we obtain

$$f = \mathrm{e}^{-k^2\langle x^2\rangle} \tag{1.66}$$

which is exactly the same as (1.36).

We know that the harmonic oscillator potential well has a parabolic shape. The larger the ω-value is, the narrower the potential well, and consequently the Mössbauer nucleus is bound more tightly. Based on the above expressions for the recoilless fraction f, if ω is increased such that $\hbar\omega \gg E_R$, $\langle x^2 \rangle$ would be very small, and the f-value could be appreciable.

Finally, it needs to be noted that the recoilless fraction f can also be expressed in terms of the coherent states $|\alpha\rangle$. Since $\hat{D}^*(\alpha)\hat{D}(\alpha) = 1$, we can also write formula (1.61) as

$$f = |\langle 0| \mathrm{e}^{-ik\hat{x}} \hat{D}^*(\alpha)\hat{D}(\alpha)|0\rangle|^2 = |\langle \alpha| \mathrm{e}^{-ik\hat{x}} |\alpha\rangle|^2. \tag{1.67}$$

From the viewpoint of calculating the recoilless fraction f, both the number state basis and the coherent state basis are identical; the latter, however, has an immense advantage shown in the next section.

1.5.3 Mössbauer Effect in a Solid

We will now treat the actual situation in the Mössbauer effect where the γ-source nucleus is bound in a solid. Owing to thermal motion, the lth nucleus is displaced from its equilibrium position $\boldsymbol{l}$ by a distance $\boldsymbol{u}(l)$, and therefore its instantaneous position is $\boldsymbol{R}_l = \boldsymbol{l} + \boldsymbol{u}(l)$. After it emits a γ-ray, the nucleus makes a transition from its initial state $|i\rangle$ to the final state $|f\rangle$. Because of this, the lattice may have a corresponding transition from its initial phonon state $|n_i\rangle$ to $|n_f\rangle$. The nuclear force causing this transition is the strong force, but its range is extremely short, well within the nucleus itself, and it would not perturb the bonding and motion of the atoms in the solid. On the other hand, the bonding forces between the atoms in the solid are relatively weak, and would have negligible effect on the transition process taking place inside the nucleus. Therefore, these two processes can be considered independent of each other and the overall transition matrix element is the product of the matrix element of the phonon state transition and that of the nuclear transition [18, 19]:

$$\langle n_f| \exp(-i\boldsymbol{k} \cdot \boldsymbol{R}_l)|n_i\rangle\langle f|a(\boldsymbol{k})|i\rangle.$$

The nuclear transition $\langle f|a(\boldsymbol{k})|i\rangle$ is solely determined by the nuclear properties, regardless of its lattice position. Here, we are only interested in the matrix element describing a phonon state transition from $|n_i\rangle$ to $|n_f\rangle$ due to the emission of a γ-photon and a transfer of momentum $\hbar\boldsymbol{k}$ from the nucleus to the lattice. The probability of the phonon transition is proportional to

$$p(n_f, n_i) = |\langle n_f| \exp(-i\boldsymbol{k} \cdot \boldsymbol{R}_l)|n_i\rangle|^2. \tag{1.68}$$

After summing $p(n_f, n_i)$ over all possible final states including those in the presence and the absence of recoil, we find the normalization condition:

$$\sum_f |\langle n_f| \exp(-i\mathbf{k}\cdot \mathbf{R}_l)|n_i\rangle|^2$$

$$= \sum_f \langle n_i| \exp(i\mathbf{k}\cdot \mathbf{R}_l)|n_f\rangle\langle n_f| \exp(-i\mathbf{k}\cdot \mathbf{R}_l)|n_i\rangle = 1. \tag{1.69}$$

The relative probability of γ-emission without recoil is the recoilless fraction f written as

$$f = \sum_f |\langle n_f| \exp(-i\mathbf{k}\cdot \mathbf{R}_l)|n_i\rangle|^2 \delta(E_f - E_i). \tag{1.70}$$

At temperature T, the initial phonon states may follow a particular distribution $p_{n_i}(T)$, and Eq. (1.70) is then multiplied by $p_{n_i}(T)$ and summed over all initial states n_i. For the sake of simplicity, we assume that the equilibrium position of the radioactive nucleus to be at the origin, thus $\mathbf{R}_l = \mathbf{u}(l)$, and

$$f = \sum_i \sum_f p_{n_i}(T)|\langle n_f| \exp(-i\mathbf{k}\cdot \mathbf{u}(l))|n_i\rangle|^2 \delta(E_f - E_i)$$

$$= |\langle\langle n_i| \exp(-i\mathbf{k}\cdot \mathbf{u}(l))|n_i\rangle\rangle_T|^2 \tag{1.71}$$

where $\langle\cdots\rangle_T$ represents the thermal average.

In order to evaluate f in the coherent state basis, we start by expressing the component of $\mathbf{u}(l)$ in the $\mathbf{k}$-direction through the normal coordinates q_s ($s = 1, 2, 3, \ldots, 3N$) as

$$u_k(l) = \frac{1}{\sqrt{M}} \sum_{s=1}^{3N} B_k(l, s) q_s, \tag{1.72}$$

with the normalization condition

$$\sum_{s=1}^{3N} |B_k(l, s)|^2 = 1. \tag{1.73}$$

Each q_s may be represented by the operators $\hat{a}_s^+$ and $\hat{a}_s$:

$$q_s = \sqrt{\frac{\hbar}{2\omega_s}}(\hat{a}_s^+ + \hat{a}_s), \tag{1.74}$$

where ω_s is the sth modal angular frequency. For a crystal of $3N$ independent normal mode oscillators, we must use the product of $3N$ individual coherent states $|\{\alpha_s\}\rangle \equiv \prod_s |\alpha_s\rangle$ instead of the number states in (1.71), and

$$f = |\langle\langle\{\alpha_s\}|e^{-i\mathbf{k}\cdot\mathbf{u}(l)}|\{\alpha_s\}\rangle\rangle_T|^2. \tag{1.75}$$

The matrix elements may be written as follows:

$$\langle\{\alpha_s\}|e^{-i\boldsymbol{k}\cdot\boldsymbol{u}(l)}|\{\alpha_s\}\rangle$$
$$= \prod_s \langle\alpha_s| \exp\left[-ik\left(\frac{\hbar}{2M\omega_s}\right)^{1/2} B_k(l,s)(\hat{a}_s^+ + \hat{a}_s)\right]|\alpha_s\rangle. \tag{1.76}$$

Since the operators $\hat{a}_s^+$ and $\hat{a}_s$ do not commute, but $[\hat{a}_s^+, \hat{a}_s] = -1$, we apply Glauber's formula

$$e^{\hat{a}_s^+ + \hat{a}_s} = e^{\hat{a}_s^+} e^{\hat{a}_s} e^{-(1/2)[\hat{a}_s^+, \hat{a}_s]} \tag{1.77}$$

to simplify (1.76). Letting $k(\hbar/2M\omega_s)^{1/2} B_k(l,s) = \rho_s$, we have

$$e^{-i\rho_s(\hat{a}_s^+ + \hat{a}_s)} = e^{-i\rho_s\hat{a}_s^+} e^{-i\rho_s\hat{a}_s} e^{-(1/2)\rho_s^2}. \tag{1.78}$$

Substituting this into (1.76) and using properties of coherent states ((1.50) and (1.51)), each factor in the product of (1.76) becomes

$$\langle\alpha_s| \exp[-i\rho_s(\hat{a}_s^+ + \hat{a}_s)]|\alpha_s\rangle = e^{-(1/2)\rho_s^2} e^{-i\rho_s(\alpha_s^* + \alpha_s)}$$

and the matrix element is

$$\langle\{\alpha_s\}|e^{-i\boldsymbol{k}\cdot\boldsymbol{u}(l)}|\{\alpha_s\}\rangle = \exp\left[-\frac{1}{2}\sum_s \rho_s^2\right] \exp\left[-2i\sum_s \rho_s \operatorname{Re}(\alpha_s)\right]. \tag{1.79}$$

The next step is to take a thermal average $\langle\cdots\rangle_T$ over the probability of a particular distribution $p(\alpha_s)$ as described in (1.54), and we have

$$\langle\!\langle\cdots\rangle\!\rangle_T = \int \langle\cdots\rangle p(\alpha_s)\, d^2\alpha_s$$
$$= \exp\left[-\frac{1}{2}\sum_s \rho_s^2\right] \prod_s \int \frac{d^2\alpha_s}{\pi\langle n_s\rangle} \exp\left[-\frac{|\alpha_s|^2}{\langle n_s\rangle}\right] \exp[-2i\rho_s \operatorname{Re}(\alpha_s)].$$

Letting $\alpha_s = \xi + i\eta$,

$$\langle\!\langle\cdots\rangle\!\rangle_T = \exp\left[-\frac{1}{2}\sum_s \rho_s^2\right] \exp\left[-\sum_s \rho_s\langle n_s\rangle\right]$$
$$\times \prod_s \int_{-\infty}^{+\infty} \frac{1}{(\pi\langle n_s\rangle)^{1/2}} \exp\left[-\frac{1}{\langle n_s\rangle}(\zeta + i\rho_s\langle n_s\rangle)^2\right] d\zeta$$
$$\times \prod_s \int_{-\infty}^{+\infty} \frac{1}{(\pi\langle n_s\rangle)^{1/2}} \exp\left[-\frac{1}{\langle n_s\rangle}\eta^2\right] d\eta$$

$$= \exp\left[-\sum_s \rho_s^2\left(\langle n_s\rangle + \frac{1}{2}\right)\right]$$

$$= \exp\left[-\frac{E_R}{2}\sum_s \frac{|B_k(l,s)|^2}{\hbar\omega_s}\coth\left(\frac{\hbar\omega_s}{2k_B T}\right)\right]. \tag{1.80}$$

The two integrals in the above calculations are Gaussian integrals: $\langle n_s\rangle$ is the average phonon number at T defined in (1.55) and $E_R = \hbar^2k^2/(2M)$ corresponds to the recoil energy of the free nucleus. Using (1.80), the recoilless fraction in (1.75) is reduced to

$$f = \exp\left[-E_R\sum_s \frac{|B_k(l,s)|^2}{\hbar\omega_s}\coth\left(\frac{\hbar\omega_s}{2k_B T}\right)\right]. \tag{1.81}$$

This still contains a summation over different normal modes, but it may be replaced by a frequency integral over density of states $g(\omega)$. For a cubic crystal, it is only necessary to consider one displacement component, and the corresponding coefficient $|B_k(l,s)|^2$ is equal to $1/(3N)$ (see Eq. (8.63)). Therefore, we have the final result for f:

$$f = \exp\left[-E_R\int \frac{g(\omega)}{\hbar\omega}\coth\left(\frac{\hbar\omega}{2k_B T}\right)d\omega\right] \tag{1.82}$$

where $g(\omega)$ is normalized to unity.

For an Einstein lattice and for $T \to 0$, Eq. (1.81) becomes (1.65). Further evaluation of f in a general case requires the knowledge of $g(\omega)$. A more realistic model is the Debye model whose density of states is

$$g(\omega) = \frac{3\omega^2}{\omega_D^3} \quad (\omega < \omega_D), \tag{1.83}$$

and in this case

$$f = \exp\left\{-\frac{3E_R}{2k_B\theta_D}\left[1 + 4\left(\frac{T}{\theta_D}\right)^2\int_0^{\theta_D/T}\frac{x\,dx}{(e^x - 1)}\right]\right\} \tag{1.84}$$

where $x = \hbar\omega/k_B T$ and $\theta_D = \hbar\omega_D/k_B$ is the Debye temperature. This is an approximate formula of the recoilless fraction f that is often used in practice.

Here again we have demonstrated the equivalency of phonon number states $|n\rangle$ and the coherent states $|\alpha\rangle$, when they are used in calculating f. In principle, other basis functions, if possible, may also be used, provided they satisfy the requirement that the energy state of the crystal is not changed after the γ-ray emission. However, one can see from Sections 1.5.2 and 1.5.3 that the derivation using coherent states is not only simpler (Gaussian integrals are all one needs to use), but also more rigorous (because no approximation was made for the derivation of (1.81)). In addition, coherent states are those quantum states that are most similar to classical situations, and their applications in the derivation of f indicates that there is a classical correspondence in the Mössbauer effect. It is then not surprising that in Section 1.4 the classical radiation theory was able to give a recoilless fraction f that is identical to Eq. (1.66).

1.5.4
Average Energy Transferred

In the source, a large number of excited Mössbauer nuclei (e.g., ^{57}Fe) are imbedded in a crystal lattice. During the γ-emission, the average energy transferred to the lattice is exactly equal to the recoil energy for a free nucleus E_R [20, 21]. This was first proved by Lipkin [11], and it is known as Lipkin's sum rule, which we discuss again in Chapter 7.

Suppose that the interactions between the atoms in the lattice are dependent only on their positions, but not on their velocities. The only term in the lattice Hamiltonian that does not commute with $\exp(i\boldsymbol{k}\cdot\boldsymbol{u}(l))$ is the kinetic energy operator $\hat{p}^2/(2M)$ of the emitting nucleus. Accordingly,

$$[\mathscr{H}, \exp(i\boldsymbol{k}\cdot\boldsymbol{u}(l))] = \left[\frac{\hat{p}^2}{2M}, \exp(i\boldsymbol{k}\cdot\boldsymbol{u}(l))\right]$$
$$= \exp(i\boldsymbol{k}\cdot\boldsymbol{u}(l))\left(\frac{\hbar^2k^2}{2M} + \frac{\hbar\boldsymbol{k}\cdot\hat{\boldsymbol{p}}}{M}\right). \tag{1.85}$$

Utilizing

$$\mathrm{e}^{\pm i\boldsymbol{k}\cdot\boldsymbol{u}(l)}\hat{p}\mathrm{e}^{\mp i\boldsymbol{k}\cdot\boldsymbol{u}(l)} = \hat{p} \pm \hbar\boldsymbol{k},$$

we can calculate the double commutator

$$[[\mathscr{H}, \exp(i\boldsymbol{k}\cdot\boldsymbol{u}(l))], \exp(-i\boldsymbol{k}\cdot\boldsymbol{u}(l))] = -\frac{\hbar^2k^2}{M} = -2E_R. \tag{1.86}$$

On the other hand, this commutator can also be written as

$$[[\mathscr{H}, \exp(i\boldsymbol{k}\cdot\boldsymbol{u}(l))], \exp(-i\boldsymbol{k}\cdot\boldsymbol{u}(l))]$$
$$= 2\mathscr{H} - \exp(i\boldsymbol{k}\cdot\boldsymbol{u}(l))\mathscr{H}\exp(-i\boldsymbol{k}\cdot\boldsymbol{u}(l))$$
$$- \exp(-i\boldsymbol{k}\cdot\boldsymbol{u}(l))\mathscr{H}\exp(i\boldsymbol{k}\cdot\boldsymbol{u}(l)). \tag{1.87}$$

If we calculate the expectation value of this commutator when the system is at its initial state $|n_i\rangle$ with energy E_i, we get

$$\begin{aligned}
&\langle n_i|[[\mathscr{H}, \exp(i\mathbf{k}\cdot\mathbf{u}(l))], \exp(-i\mathbf{k}\cdot\mathbf{u}(l))]|n_i\rangle \\
&= 2E_i - \sum_f \langle n_i|\exp(i\mathbf{k}\cdot\mathbf{u}(l))|n_f\rangle\langle n_f|\mathscr{H}\exp(-i\mathbf{k}\cdot\mathbf{u}(l))|n_i\rangle \\
&\quad - \sum_f \langle n_i|\exp(-i\mathbf{k}\cdot\mathbf{u}(l))|n_f\rangle\langle n_f|\mathscr{H}\exp(i\mathbf{k}\cdot\mathbf{u}(l))|n_i\rangle \\
&= 2E_i - 2\sum_f E_f|\langle n_f|\exp(-i\mathbf{k}\cdot\mathbf{u}(l))|n_i\rangle|^2 \qquad (1.88)
\end{aligned}$$

where a complete set of final states $|n_f\rangle$ was inserted. Taking into account Eqs. (1.68), (1.69), (1.86), and (1.88), we arrive at Lipkin's sum rule

$$\sum_f (E_f - E_i)p(n_f, n_i) = E_R. \qquad (1.89)$$

When $i = f$, $E_R = 0$, $p(n_i, n_i)$ is none other than the recoilless fraction f, i.e., the portion of the γ-ray emission process that has no energy exchange with the lattice. The rest of the emission process will cause recoil, whose recoil energy will have to be sufficiently large so that the average energy transferred to the lattice is E_R.

In order to obtain a relatively large f value, the Mössbauer nucleus should be tightly bound in a localized potential well to form a localized state, and the Debye temperature θ_D should be as high as possible. One of the best examples is the ^{57}Fe impurities in diamond [22]. Diamond has the highest known Debye temperature $\theta_D = 2230$ K, and $f(295\ \text{K}) = 0.94 \pm 0.06$ [23], which is probably the highest recoilless fraction at room temperature ever detected thus far.

References

1 W. Kuhn. Scattering of thorium C″ γ-radiation by radium G and ordinary lead. *Phil. Mag.* 8, 625–636 (1929).

2 P.B. Moon and A. Storruste. Resonant nuclear scattering of ^{198}Hg gamma-rays. *Proc. Phys. Soc. (London)* 66, 585–589 (1953).

3 R.L. Mössbauer. Kernresonanzfluoreszenz von Gammastrahlung in Ir^{191}. *Z. Phys.* 151, 124–143 (1958); R.L. Mössbauer. Kernresonanzabsorption von Gammastrahlung in Ir^{191}. *Naturwiss.* 45, 538–539 (1958).

4 J.M. Williams and J.S. Brooks. The thickness dependence of Mössbauer absorption line areas in unpolarized and polarized absorbers. *Nucl. Instrum. Methods* 128, 363–372 (1975).

5 Chen Yi-long, Lu Ning, Peng Li-li, and Zhao Lei. An exact expression for fractional absorption in Mössbauer spectroscopy. *Wuhan University J. Natur. Sci.* 6, 784–786 (2001).

6 S.L. Ruby and J.M. Hicks. Line shape in Mössbauer spectroscopy. *Rev. Sci. Instrum.* 33, 27–30 (1962).

7 R.M. Housley, N.E. Erickson, and J.G. Dash. Measurement of recoil-free fractions in studies of the Mössbauer effect. *Nucl. Instrum. Methods* 27, 29–37 (1964).

8 F.L. Shapiro. The Mössbauer effect. *Soviet Physics Uspekhi* 4, 881–887 (1961) [Russian original: *Uspekhi Fiz. Nauk* 72, 685–696 (1960)].
9 W.M. Visscher. Study of lattice vibrations by resonance absorption of nuclear gamma rays. *Ann. Phys.* 9, 194–210 (1960).
10 W.E. Lamb. Capture of neutrons by atoms in a crystal. *Phys. Rev.* 55, 190–197 (1939).
11 H.J. Lipkin. Some simple features of the Mössbauer effect. *Ann. Phys.* 9, 332–339 (1960).
12 K.S. Singwi and A. Sjölander. Resonance absorption of nuclear gamma rays and the dynamics of atomic motions. *Phys. Rev.* 120, 1093–1102 (1960).
13 L. Van Hove. Correlations in space and time and Born approximation scattering in systems of interacting particles. *Phys. Rev.* 95, 249–262 (1954).
14 R.J. Glauber. Coherent and incoherent states of the radiation field. *Phys. Rev.* 131, 2766–2788 (1963).
15 C. Cohen-Tannoudji, B. Diu, and F. Laloë. *Quantum Mechanics*, vol. 1, p. 295 (John Wiley, New York, 1977).
16 S. Howard and S.K. Roy. Coherent states of a harmonic oscillator. *Am. J. Phys.* 55, 1109–1117 (1987).
17 R. Loudon. *The Quantum Theory of Light*, pp. 148–153 (Oxford University Press, Oxford, 1973).
18 D.S. Bateman, S.K. Bose, B. Dutta-Roy, and M. Bhattacharyya. The harmonic lattice, recoilless transitions, and the coherent state. *Am. J. Phys.* 60, 829–832 (1992).
19 R.J. Glauber. *Quantum Optics*, pp. 15–56 (Academic Press, New York, 1969).
20 J. Callaway. *Quantum Theory of the Solid State* (Academic Press, New York, 1974).
21 C. Kittel. *Quantum Theory of Solids*, p. 389 (John Wiley, New York, 1963).
22 J.A. Sawicki and B.D. Sawicka. Properties of ^{57}Fe hot-implanted into diamond crystals studied by Mössbauer emission spectroscopy between 4 and 300 K. *Nucl. Instrum. Methods B* 46, 38–45 (1990).
23 T.W. Sinor, J.D. Standifird, F. Davanloo, K.N. Taylor, C. Hong, J.J. Carroll, and C.B. Collins. Mössbauer effect measurement of the recoil-free fraction for ^{57}Fe implanted in a nanophase diamond film. *Appl. Phys. Lett.* 64, 1221–1223 (1994).

2
Hyperfine Interactions

For a free atom in a gas, the interactions between the nucleus and the electromagnetic fields produced by the surrounding electrons are called the hyperfine interactions. In a solid, we need to also include the electromagnetic fields produced by the neighboring atoms or ions. Hyperfine interactions have several different types, and they are usually quite weak. A relatively prominent type was first observed in the atomic spectra where extremely small splittings of spectral lines are produced by the coupling between nuclear spin and the total angular momentum of a valence electron [1]. Many years after the initial observation, the only means for studying hyperfine interactions was free atom optical spectroscopy, which had its historic importance in determining ground state nuclear spins, nuclear magnetic dipole moments, and nuclear quadrupole moments.

The advent of nuclear magnetic resonance (NMR) in condensed matter [2] marked the beginning of using bound atoms to study hyperfine interactions. NMR is a method of observing a resonance spectrum, which is very different from an optical spectrum. Soon after the initial discovery of NMR, many other nuclear methods for studying hyperfine interactions were discovered or developed, such as nuclear quadrupole resonance (NQR) [3], nuclear spin orientation (NO) [4], perturbed angular correlation (PAC) [5], perturbed angular distribution (PAD) [6], and muon spin resonance (μSR) [7]. Gradually, there emerged a new research field – hyperfine interactions – linking together atomic physics, nuclear physics, and solid-state physics. But the fastest development in this field was after the discovery of the Mössbauer effect, because the energy resolution of the Mössbauer effect is much better than that of the above methods, and even higher than that of NMR by an order of magnitude.

Mössbauer spectroscopy is simply the science of using the Mössbauer effect to observe hyperfine interactions for studying the microscopic environment surrounding a nucleus. Therefore, we need to have a detailed description of hyperfine interactions. In the Mössbauer effect, there are mainly the following three types of hyperfine interactions:

1. Electric monopole interaction, which causes isomer shift δ, a shift of the entire resonance spectrum.
2. Electric quadrupole interaction, which causes quadrupole splittings of the spectral lines.

Mössbauer Effect in Lattice Dynamics. Yi-Long Chen and De-Ping Yang

ISBN: 978-3-527-40712-5

3. Magnetic dipole interaction, which causes Zeeman splittings of the spectral lines – magnetic hyperfine splittings.

2.1 Electric Monopole Interaction

2.1.1 A General Description

Both the isomer shift and the quadrupole splitting are due to electric hyperfine interactions. We give a general description of the origins of these electric hyperfine interactions.

The nucleus may not be considered as a point charge, since it has a certain finite volume and a charge distribution within it. We choose the center of the nucleus as the origin of the coordinate system, i.e., $\mathbf{r}' = 0$. Let $\rho_{\mathrm{n}}(\mathbf{r}')$ be the nuclear charge density at $\mathbf{r}'$, and $V(\mathbf{r}')$ be the electric potential at $\mathbf{r}'$ due to all the electric charge outside of the nucleus. Their Coulomb interaction energy is

$$E_{\mathrm{e}} = \int \rho_{\mathrm{n}}(\mathbf{r}') V(\mathbf{r}')\, \mathrm{d}\tau' \tag{2.1}$$

where the integral is over the entire volume of the nucleus. Because the nuclear diameter is small compared with the distance of the outside electric charge producing $V(\mathbf{r}')$, we can approximate the potential by its Taylor expansion near the origin:

$$V(\mathbf{r}') = V(0) + \sum_{i=1}^{3} \left(\frac{\partial V}{\partial x_i'}\right)_0 x_i' + \frac{1}{2} \sum_{i,j=1}^{3} \left(\frac{\partial^2 V}{\partial x_i' \partial x_j'}\right)_0 x_i' x_j' + \cdots \tag{2.2}$$

Substituting Eq. (2.2) into Eq. (2.1),

$$E_{\mathrm{e}} = V(0) \int \rho_{\mathrm{n}}(\mathbf{r}')\, \mathrm{d}\tau' + \sum_{i=1}^{3} \left(\frac{\partial V}{\partial x_i'}\right)_0 \int \rho_{\mathrm{n}}(\mathbf{r}') x_i'\, \mathrm{d}\tau'$$
$$+ \frac{1}{2} \sum_{i,j=1}^{3} \left(\frac{\partial^2 V}{\partial x_i' \partial x_j'}\right)_0 \int \rho_{\mathrm{n}}(\mathbf{r}') x_i' x_j'\, d\tau' + \cdots \tag{2.3}$$

In this expansion, the first term is the interaction energy with the potential if the nucleus is treated as a point charge. This energy is a constant, and therefore will have no effect on what we are studying. The second term is zero, because the nucleus has no electric dipole moment. The third term (denoted by E_3) is not zero, and its physical meaning may be understood more easily if we rewrite it in terms of a sum of two contributions:

$$E_3 = \frac{1}{2}\left[\sum_{i=1}^{3} V_{ii}\right] \int \frac{1}{3} r'^2 \rho_n(\mathbf{r}') \, d\tau' + \frac{1}{6} \sum_{i,j=1}^{3} V_{ij} Q_{ij} \tag{2.4}$$

where

$$Q_{ij} = \int (3x_i' x_j' - \delta_{ij} r'^2) \rho_n(\mathbf{r}') \, d\tau' \tag{2.5}$$

is known as the nuclear quadrupole moment tensor, and

$$V_{ij} = \left(\frac{\partial^2 V}{\partial x_i' \partial x_j'} \right)_0 \tag{2.6}$$

is the electric field gradient (EFG) tensor evaluated at the nucleus.

In Eq. (2.4), the first contribution is the monopole interaction energy, which is due to the finite volume of the nucleus. The second contribution is the quadrupole interaction energy because of the existence of a quadrupole moment in some of the nuclear states.

We now discuss the specific features of each of these electric hyperfine interactions.

2.1.2 The Isomer Shift

To simplify the calculations, we may choose a new coordinate system x, y, z as the EFG tensor principal axis system where the tensor V_{ij} is diagonal and its trace is given by Poisson's equation

$$V_{xx} + V_{yy} + V_{zz} = -4\pi \rho_e(0) \tag{2.7}$$

where $\rho_e(0) = -e|\psi(0)|^2$ is the s electron charge density at the origin. If we use δE to represent the first term in Eq. (2.4), and carry out the sum using (2.7), it becomes

$$\delta E = \frac{2\pi}{3} z e^2 |\psi(0)|^2 \langle r^2 \rangle \tag{2.8}$$

where

$$\langle r^2 \rangle = \frac{\int r^2 \rho_n(\mathbf{r}) \, d\tau}{\int \rho_n(\mathbf{r}) \, d\tau} = \frac{\int r^2 \rho_n(\mathbf{r}) \, d\tau}{ze} \tag{2.9}$$

is the mean-square radius of the nuclear charge distribution.

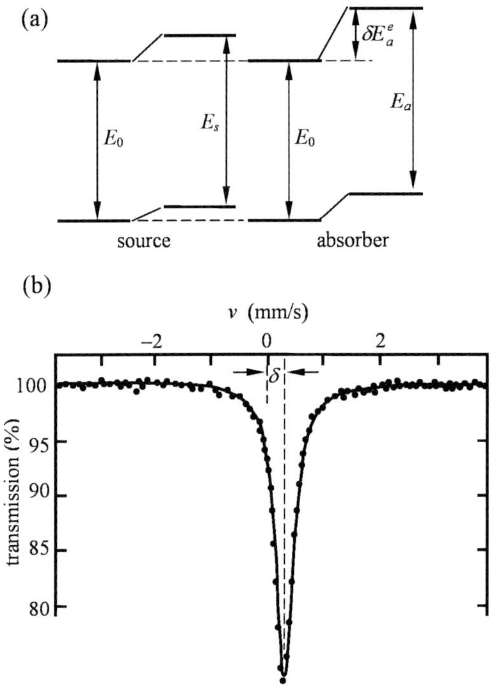

Fig. 2.1 (a) Shift of nuclear energy levels due to electric monopole interaction. (b) A typical Mössbauer spectrum in the presence of an isomer shift.

From Eq. (2.8), we see that, because of the finite volume of the nucleus, the energy level will change by an amount of δE with respect to a point charge nucleus. This happens regardless whether the nucleus is at its ground state or at an excited state. However, the nuclear radius at an excited state may be different from that at the ground state, and the corresponding energy changes δE^{g} and δE^{e} are therefore different. Furthermore, in the radiation source and in the absorber, the same Mössbauer isotope may be in different chemical environments, resulting in $|\psi_{\mathrm{s}}(0)|^2 \neq |\psi_{\mathrm{a}}(0)|^2$. So in a general case, $\delta E_{\mathrm{s}}^{\mathrm{g}} \neq \delta E_{\mathrm{a}}^{\mathrm{g}}$ and $\delta E_{\mathrm{s}}^{\mathrm{e}} \neq \delta E_{\mathrm{a}}^{\mathrm{e}}$, as shown in Fig. 2.1(a). The energy of the emitted γ-ray by a source is

$$E_{\mathrm{s}} = E_0 + \delta E_{\mathrm{s}}^{\mathrm{e}} - \delta E_{\mathrm{s}}^{\mathrm{g}} \tag{2.10}$$

and resonant absorption can occur in an absorber only if the γ-ray energy is

$$E_{\mathrm{a}} = E_0 + \delta E_{\mathrm{a}}^{\mathrm{e}} - \delta E_{\mathrm{a}}^{\mathrm{g}}. \tag{2.11}$$

Figure 2.1(b) shows a Mössbauer spectrum, where the peak position has a shift of δ with respect to the zero velocity, i.e., the resonance occurs at $v \neq 0$. This δ is known as the isomer shift, which can be calculated by taking the difference between Eqs. (2.11) and (2.10):

$$\delta = E_a - E_s = \frac{2\pi}{3} zS'(z)e^2(|\psi_a(0)|^2 - |\psi_s(0)|^2)(\langle r^2\rangle_e - \langle r^2\rangle_g)$$

$$= \frac{2\pi}{3} zS'(z)e\Delta\rho(0)\Delta\langle r^2\rangle. \tag{2.12}$$

Here, $\Delta\langle r^2\rangle = \langle r^2\rangle_e - \langle r^2\rangle_g$ is the difference between the mean squares of the charge radii of the excited state and the ground state, one of the parameters of a nucleus. Also, $\Delta\rho(0) = e(|\psi_a(0)|^2 - |\psi_s(0)|^2)$ is the difference between the s electron charge densities at the nuclei in the absorber and the source. $S'(z)$ is called the relativistic factor, and is introduced due to relativistic effects in heavy elements. This factor takes different values for different nuclei, e.g., $S'(z) = 1.32$ for ^{57}Fe but $S'(z) = 19.4$ for ^{237}Np.

To calculate δ one step further, it is often assumed that a nucleus is a uniformly charged sphere when it is either in the ground state (with radius R_g) or in the excited state (with radius R_e). Therefore, we have

$$\rho_n^g = \frac{ze}{\frac{4}{3}\pi R_g^3} \quad \text{and} \quad \rho_n^e = \frac{ze}{\frac{4}{3}\pi R_e^3}.$$

Using Eq. (2.9), we can evaluate both $\langle r^2\rangle_e$ and $\langle r^2\rangle_g$, and Eq. (2.12) becomes

$$\delta = \frac{4\pi}{5} zS'(z)eR^2\left(\frac{\Delta R}{R}\right)\Delta\rho(0) = \alpha\Delta\rho(0) \tag{2.13}$$

where $\Delta R = R_e - R_g$, $R = (R_e + R_g)/2$. It can be seen that δ is directly proportional to $\Delta\rho(0)$ with a proportionality constant α (known as the calibration constant). Any of the above quantities ($\langle r^2\rangle$, $\Delta R/R$, or α) is a parameter characterizing the nucleus. Equation (2.12) clearly shows that the isomer shift is essentially a measure of the difference in the s electron charge densities at the nuclei in the source and in the absorber. If the same source is used for a series of absorbers, then $\rho_s(0)$ is a constant ($\rho_s(0) = c$) and the isomer shift is a linear function of the s electron density at the nuclear site in the absorber:

$$\delta = \alpha(\rho(0) - c). \tag{2.14}$$

The sign of α may be positive or negative, and it happens to be negative for ^{57}Fe. A positive isomer shift indicates that the electron density at the nucleus in the absorber is less than that in the source.

It should be mentioned that δ is usually very small. For example, in Fig. 2.1(b), $\delta = 0.3$ mm s^{-1} which corresponds to an energy shift of $\sim 10^{-8}$ eV. Only in the Mössbauer effect can this minute amount of energy change be detected.

2.1.3 Calibration of Isomer Shift

As sources of a particular isotope (e.g., ^{57}Co) imbedded into different host matrices (e.g., Rh, Pd) are used to obtain Mössbauer spectra, the same absorber will give different δ values, because the sources have different $|\psi_s(0)|$ values. In order to be able to compare and discuss results from experiments using sources with different hosts, the isomer shift of an absorber is customarily given relative to that of a reference absorber. If the isomer shift of the sample is δ_1 and that of the reference absorber is δ_{ref}, we have, according to Eq. (2.14),

$$\delta_1 = \alpha[\rho(0) - \rho_s(0)],$$
$$\delta_{ref} = \alpha[\rho_{ref}(0) - \rho_s(0)].$$

Therefore, the isomer shift of the absorber relative to the reference absorber is

$$\delta = \delta_1 - \delta_{ref} = \alpha[\rho(0) - \rho_{ref}(0)]. \tag{2.15}$$

It is clear that this is independent of $\rho_s(0)$. In other words, even when different sources are used (e.g., ^{57}Co/Rh or ^{57}Co/Pd), isomer shifts obtained in this way should all be the same for a particular absorber, $\delta = \delta_1(\text{Rh}) - \delta_{ref}(\text{Rh}) = \delta_1(\text{Pd}) - \delta_{ref}(\text{Pd}) = \ldots$ For example, the δ_1 values of sodium nitroprusside $Na_2Fe(CN)_5NO{\cdot}2H_2O$ obtained from experiments using ^{57}Co/Rh and ^{57}Co/Pd are -0.366 and -0.437 mm s^{-1}, respectively. However, relative to α-Fe, isomer shifts of sodium nitroprusside in both cases are -0.260 mm s^{-1}. The values of δ_{ref} of several reference absorbers for ^{57}Fe are indicated in Fig. 2.2 [8]. When re-

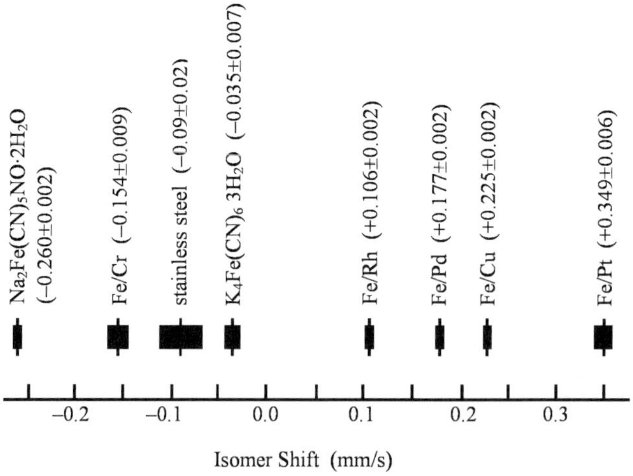

Fig. 2.2 Isomer shift scale for ^{57}Fe (14.4 keV) reference materials (relative to α-Fe) at $T = 300$ K.

porting isomer shifts, one must clearly indicate which type of the reference absorber was used.

There are two major factors that determine the electron density at the nuclear site. The first is due to the inner s electrons of the Mössbauer atom, and the second is due to valence electrons in the outer shells and valence electrons of ligands. The first contribution is not sensitive to changes in the chemical environment, and therefore it is not uncommon to consider it as a constant. The second contribution exerts its effect through the following two mechanisms:

1. A direct interaction, which involves a change in *s* electrons of the valence shell, thus influencing $\rho(0)$.
2. An indirect interaction, which involves a change in the shielding of the s electrons through the increase or decrease of p, d, and f valence electrons. For example, $\rho(0)$ at an Fe^{3+} ($3d^5$) nucleus is larger than that at an Fe^{2+} ($3d^6$) nucleus. For metallic iron ($3d^6 4s^2$), $\rho(0)$ is even larger. Since $\alpha < 0$, we have $\delta(Fe^{2+}) > \delta(Fe^{3+})$, both of which are positive with respect to metallic iron.

In addition, indirect interactions also arise from covalent bonds affecting electronic distribution, from electronegative ligands influencing σ-electrons, and from d_π backbonding which reduces the shielding effect.

Chemical bonds are formed by valence electrons in the outer shell and valence electrons of ligands. These electrons contribute directly or indirectly to $\Delta\rho(0)$ in Eq. (2.12). The isomer shift δ is therefore closely related to the properties of the electronic structure and the chemical bonds.

2.1.4
Isomer Shift and Electronic Structure

Since the observation of δ in Fe_2O_3 in 1960 [9], the isomer shift has been intensely and extensively investigated in many areas of chemistry and in materials science. Using isomer shift for studying the electronic structure in solids has been considered an extremely useful experimental method [10].

The isomer shift δ can provide important information on the character of a chemical bond, as well as on oxidation state, spin state, electronegativity of a ligand, coordination number, etc. So far, large amounts of experimental isomer shift data have been accumulated, but the interpretation of these results is not an easy task. The main difficulty is that there is still lacking a unified model for the chemical bonds that could satisfactorily explain the isomer shift data. For practical applications, the critical problem is to obtain an accurate value for the calibration constant α (alternatively, $\Delta R/R$ or $\langle r^2 \rangle$) in Eq. (2.13). But it cannot be determined without a good understanding of the chemical environment; therefore it is difficult to measure α and $\rho(0)$ separately. Although isomer shift has been widely utilized in research, many conclusions are still qualitative. In addition, since isomer shift is a relative quantity, measurements from a series of sam-

Table 2.1 Measured isomer shift δ in ^{121}Sb compounds (relative to InSb) and calculated s electron density $\rho(0)$. Here a_0 is the Bohr radius.

	δ (mm s^{-1})	$\rho(0)$ (a_0^{-3})
AlSb	0.78(5)	81.31
GaSb	0.22(3)	83.43
InSb	0.00	84.37
SnSb	−1.98(7)	90.12
Sb	−3.10(2)	91.42

ples of similar properties are usually required, and then through a comparative study one can extract information on the electronic structure. The following are several specific examples.

1. *Chemical bond character.* In this example, five covalent compounds of Sb were studied (Table 2.1), and the δ values were obtained from their ^{121}Sb Mössbauer spectra. We will discuss how δ is related to the nature of the local chemical bond [11].

The first three compounds (AlSb, GaSb, and InSb) form a "vertical sequence" of Sb-based binary compounds in the zincblende crystal structure with lattice constants 6.14, 6.12, and 6.48 Å, respectively. Arranged in the same vertical column of the periodic table with one p electron in their outer shell, Al, Ga, and In would have rather similar physical and chemical properties. Nevertheless their effects on the s electron densities at the Sb nuclei of above three compounds are different. Analysis showed that δ is related to the number of p electrons around Sb, and so the δ value of AlSb is the largest because its bond ionicity is definitely higher than that of GaSb or InSb. It was not certain whether or not GaSb is more ionic than InSb, but it was speculated that the difference in δ is caused by the change in the crystal volume. Because InSb has a larger lattice constant than GaSb, the Sb atoms in InSb would be less compressed than in GaSb, and the looser p electron clouds around the Sb nucleus would reduce the shielding of the s electrons.

The last three samples (InSb, SnSb, and Sb) form a "horizontal sequence" from which δ was observed to vary significantly, reflecting the changes in the lattice structure and in the chemical bond character from sp^3 hybridization to $(s^2)p^3$.

2. *Ligand electronegativity.* The difference between the electronegativities of the atom and the surrounding ligands is obviously related to the isomer shift δ.

1. In ferrous halides, δ has a linear relationship with the electronegativity of the halogen atoms [12], as shown in Fig. 2.3. This is direct evidence for the participation of 4s electrons in the formation of the chemical bonds. The Fe electronic configuration is $3d^6 4s^x$, where x can be regarded

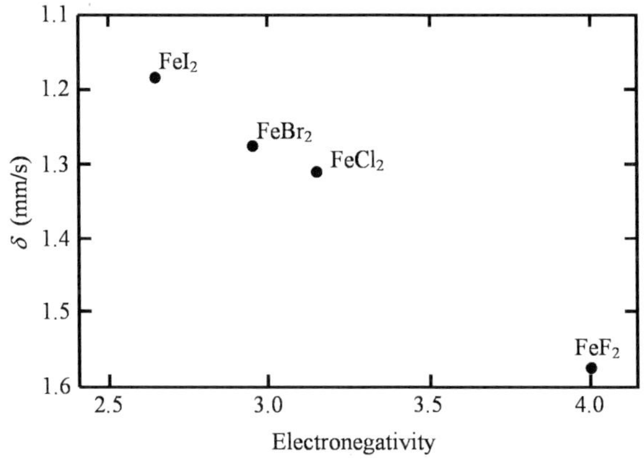

Fig. 2.3 Isomer shift (relative to α-Fe) versus Pauling electronegativity for ferrous halides.

as a measure of the ionicity of the compound (ionicity increases as x decreases). For nuclei of the 4d, 5d, and 5f elements, such a linear relation also exists.

2. In intermetallic compounds of iron, the relationship between δ and electronegativity may be expressed in the following empirical formula:

$$\delta = 1.08(\Phi_A - \Phi_{Fe}) - 2.51\frac{(n_{WS}^{A} - n_{WS}^{Fe})}{n_{WS}^{Fe}} \tag{2.16}$$

where Φ and n_{WS} are the electronegativity and electron density on the Wigner–Seitz cell surface. A relationship similar to Eq. (2.16) can be found in various amorphous Fe-based alloys ($A_{1-x}Fe_x$) [13]. A linear relation between δ and electronegativity also exists in some compounds of ^{119}Sn, ^{121}Sb, and ^{181}Ta [14].

3. *Volume effect.* A different lattice constant can cause a change in $\rho(0)$, resulting in a volume effect. External pressure may be used to change the lattice constant of a compound; Fig. 2.4 shows the results from one such investigation [15]. Isomer shift δ is related to the unit cell volume V by

$$\frac{d(\delta)}{d\ln V} = \frac{\alpha\rho(0)}{\ln V}. \tag{2.17}$$

The linear relation in Fig. 2.4 gives a slope of $d(\delta)/d(\ln V) = -1.42(3)$ mm s^{-1}. In another investigation, an empirical formula for the molecular iodine I_2 was obtained [16], relating δ to the number of holes h_p in the 5p orbital:

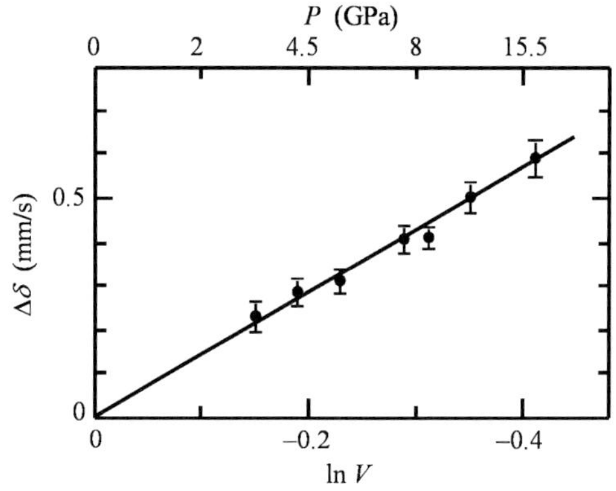

Fig. 2.4 Isomer shift of iodine versus natural logarithm of unit cell volume.

$$\delta = 1.50h_{\mathrm{p}} + A\ (\mathrm{mm\ s^{-1}}) \tag{2.18}$$

where A is a constant. Under atmospheric pressure, $h_{\mathrm{p}} = 1$, and when Eq. (2.18) is substituted into (2.17),

$$\frac{\mathrm{d}(h_{\mathrm{p}})}{\mathrm{d}\ln V} = -0.95. \tag{2.19}$$

Here, $\mathrm{d}h_{\mathrm{p}}$ is the number of p electrons that moved into the conduction band. When the pressure is 16 GPa, $\mathrm{d}h_{\mathrm{p}} = 0.38$, and I_2 becomes a conductor with an electronic configuration of $5s^2 5p^{4.62}$. Still one more example worth mentioning is the volume effect in iron borides as well as α-Fe and γ-Fe, and their isomer shifts can be described by the following relation [17]:

$$\delta\ (\mathrm{mm\ s^{-1}}) = 0.02917 n_{\mathrm{B}} + \frac{\bar{d}_{\mathrm{Fe}} - 2.64}{11} n_{\mathrm{Fe}} \tag{2.20}$$

where n_{B} and n_{Fe} are the numbers of boron and iron nearest neighbors surrounding the Mössbauer nucleus and $\bar{d}_{\mathrm{Fe}}$ is the distance (in Å) between the adjacent Fe atoms.

4. *Evaluation of the calibration constant α.* As mentioned above, it is important to determine the calibration constant α accurately. The basic procedure includes calculating the values of $\rho(0)$ for several series of compounds and measuring the corresponding δ values from their Mössbauer spectra. A least-squares fitting according to Eq. (2.14) will allow the determination of α. The difficult part is that $\Delta\rho(0)$ could be about four to six orders of magnitude smaller than $\rho(0)$ [18, 19], requiring precise measurement of $\rho(0)$ from each of the compounds in the series. In some cases, $\Delta\rho(0)$ may be as large as a few percent. As an example, the isomer shifts of a series of Sb compounds are plotted in Fig. 2.5 against the calculated $\rho(0)$ values [11], and from the approximate linear relation we obtain

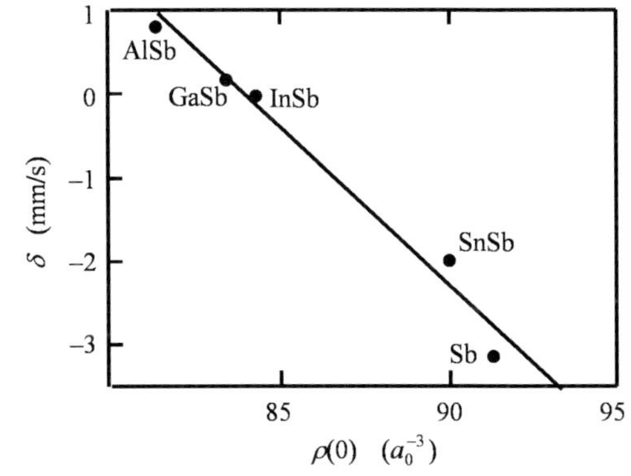

Fig. 2.5 Relationship between measured isomer shifts of Sb compounds and calculated $\rho(0)$ values.

$$\alpha\,(\mathrm{Sb}) = (-0.368 \pm 0.035)a_0^3\ \mathrm{mm\ s^{-1}} \tag{2.21}$$

or

$$\frac{\Delta R}{R} = (-10.4 \pm 1.0) \times 10^{-4}. \tag{2.22}$$

There are many results for the calibration constant α of ^{57}Fe in the literature [19–25], the most recent result being $\alpha(^{57}\mathrm{Fe}) = (-0.22 \pm 0.01)a_0^3$ [26], which agrees with the previous results quite well.

Isomer shift is one of the hyperfine interaction parameters that can only be measured through the Mössbauer effect, and the information on the electronic structure provided by isomer shift is not available from any other methodologies. For example, Sb atoms can be doped into a semiconductor substitutionally and used to monitor the electron density around the atoms that are replaced by Sb. Information about the nature of local chemical bonds is uniquely provided by the isomer shift of ^{121}Sb [11]. Another example is that δ measurements can uniquely determine how electron density varies as alloys are formed [18].

2.2 Electric Quadrupole Interaction

2.2.1 Electric Quadrupole Splitting

In the case of an axially symmetric nucleus, we can choose its symmetry axis (i.e., its quantization axis) as the principal z' axis of the nuclear quadrupole moment tensor defined in Eq. (2.5). In such a coordinate system, only the diagonal elements Q_{11}, Q_{22}, and Q_{33} are nonzero. We also have $Q_{11} = Q_{22}$ because of the axial symmetry and $Q_{11} + Q_{22} + Q_{33} = 0$ because the tensor is traceless. Therefore, only one independent quantity Q is needed to describe the nuclear quadrupole moment for this case:

$$eQ = Q_{33} \tag{2.23}$$

or

$$Q = \frac{1}{e}\int (3z'^2 - r'^2)\rho_{\rm n}(\mathbf{r}')\,d\tau'. \tag{2.24}$$

If a nucleus has a prolate spheroid shape (longer along the z' axis, and shorter along the x' or y' axis), then $Q > 0$; if it has a oblate shape, then $Q < 0$. When a nucleus spin $I = 0$ or $1/2$, the nucleus has spherical symmetry, $Q = 0$. Only when $I > 1/2$ will there be electric quadrupole interaction.

To study quadrupole interactions in a solid, the principal axis system of the EFG tensor, as defined in Section 2.1.2, must be chosen such that $|V_{zz}| \geq |V_{xx}| \geq |V_{yy}|$. Since EFG at the nucleus can only arise from electrons other than s electrons and from ligand charges, both of which have zero electron density at the nucleus, Eq. (2.7) for this case becomes the Laplace equation:

$$V_{xx} + V_{yy} + V_{zz} = 0. \tag{2.25}$$

As a result, only two independent parameters are needed to describe EFG. These two parameters are usually taken as V_{zz} and an asymmetry parameter η, defined by $\eta = (V_{xx} - V_{yy})/V_{zz}$. It is evident that $0 \leq \eta \leq 1$.

The Hamiltonian for quadrupole interaction is

$$\mathscr{H}_{\rm Q} = \frac{1}{2}\sum_{i,j=1}^{3} V_{ij}Q_{ij} \tag{2.26}$$

which can be eventually expressed as

$$\mathscr{H}_{\rm Q} = \frac{eQV_{zz}}{4I(2I-1)}\left[3\hat{I}_z^2 - \hat{I}^2 + \frac{1}{2}\eta(\hat{I}_+^2 + \hat{I}_-^2)\right] \tag{2.27}$$

where $\hat{I}_+ = \hat{I}_x + i\hat{I}_y$ and $\hat{I}_- = \hat{I}_x - i\hat{I}_y$ are the raising and lowering operators, respectively. The eigenvalues of the Hamiltonian are

$$E_{\rm Q} = \frac{eQV_{zz}}{4I(2I-1)}[3m^2 - I(I+1)]\left(1 + \frac{\eta^2}{3}\right)^{1/2} \tag{2.28}$$

where $m = I, I-1, \ldots -|I|$.

The 14.4-keV energy level of ^{57}Fe has a nuclear spin of $I = 3/2$, and this energy level splits into two sublevels ($m = \pm 3/2$ and $m = \pm 1/2$) due to quadrupole interaction. Because the quantum number m in Eq. (2.28) appears only as its square, each sublevel is doubly degenerate. The energy eigenvalues of the two sublevels and the corresponding eigenvectors are

$$E_0 + E_Q\left(\pm\frac{3}{2}\right) = E_0 + \frac{eQV_{zz}}{4}\left(1+\frac{\eta^2}{3}\right)^{1/2},$$

$$\begin{cases} \left|+\frac{3}{2}\right\rangle' = \cos\zeta\left|+\frac{3}{2}\right\rangle + \sin\zeta\left|-\frac{1}{2}\right\rangle \\ \left|-\frac{3}{2}\right\rangle' = \cos\zeta\left|-\frac{3}{2}\right\rangle + \sin\zeta\left|+\frac{1}{2}\right\rangle \end{cases}, \tag{2.29}$$

$$E_0 + E_Q\left(\pm\frac{1}{2}\right) = E_0 - \frac{eQV_{zz}}{4}\left(1+\frac{\eta^2}{3}\right)^{1/2},$$

$$\begin{cases} \left|-\frac{1}{2}\right\rangle' = \cos\zeta\left|-\frac{1}{2}\right\rangle - \sin\zeta\left|+\frac{3}{2}\right\rangle \\ \left|+\frac{1}{2}\right\rangle' = \cos\zeta\left|+\frac{1}{2}\right\rangle - \sin\zeta\left|-\frac{3}{2}\right\rangle \end{cases}, \tag{2.30}$$

where

$$\cos\zeta = [1+\sqrt{3}(3+\eta^2)^{-1/2}]^{1/2}/\sqrt{2},$$
$$\sin\zeta = [1-\sqrt{3}(3+\eta^2)^{-1/2}]^{1/2}/\sqrt{2}.$$

The ground state of ^{57}Fe has $I = 1/2$, so $Q = 0$ and the energy level does not split, as shown in Fig. 2.6(a).

When ^{57}Fe is in a crystal of non-axial symmetry ($\eta \neq 0$), $\mathscr{H}_Q$ and $\hat{I}_z$ do not commute, and each of the four new eigenvectors is a linear combination of the original eigenvectors as in Eqs. (2.29) and (2.30). However, since $0 \leq \eta \leq 1$, $\sin^2\zeta$ is practically very small, and can be neglected as a first-order approximation. For example, if $\eta = 0.5$, $\sin^2\zeta = 0.02$ while $\cos^2\zeta = 0.98$, and therefore the four new states are essentially identical to the original four pure states $|\pm 3/2\rangle$ and $|\pm 1/2\rangle$.

The energy difference between the two sublevels in (2.29) and (2.30) is

$$\Delta E_Q = \frac{eQV_{zz}}{2}\left(1+\frac{\eta^2}{3}\right)^{1/2} \tag{2.31}$$

where V_{zz} is in V cm^{-2}, Q is in cm^2, and therefore ΔE_Q is in eV.

Now that the ^{57}Fe excited state is split into two sublevels, the singlet Mössbauer spectrum becomes a doublet as shown in Fig. 2.6(b). The separation ΔE_Q between the two resonance lines is known as quadrupole splitting, which is another important parameter in Mössbauer spectroscopy.

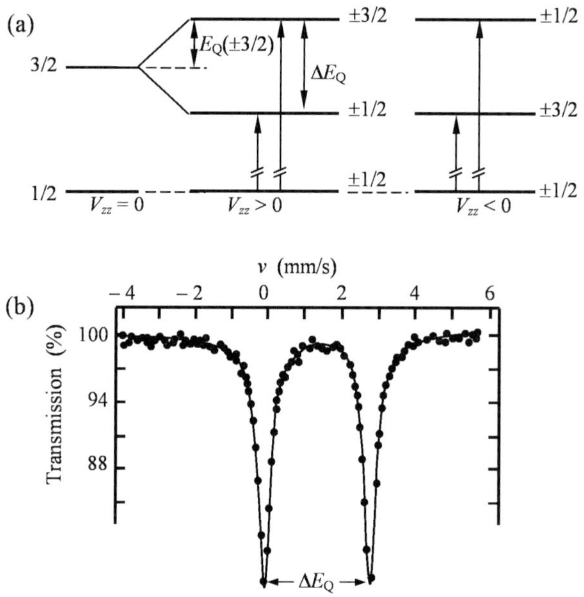

Fig. 2.6 (a) Quadrupole interaction splits the ^{57}Fe energy levels. (b) A quadrupole splitting Mössbauer spectrum.

Electric quadrupole interaction is essentially an electric interaction, which can be understood by considering the following simplified physical picture. Suppose that two point charges are located at a and $-a$ on the z-axis as shown in Fig. 2.7(a), and we want to calculate the potential $V(z_0)$ at a point z_0 far from the point charges ($z_0 \gg a$). It is easy to show that

$$V(z_0) = \left(\frac{2e}{z_0} + \frac{2ea^2}{z_0^3}\right). \tag{2.32}$$

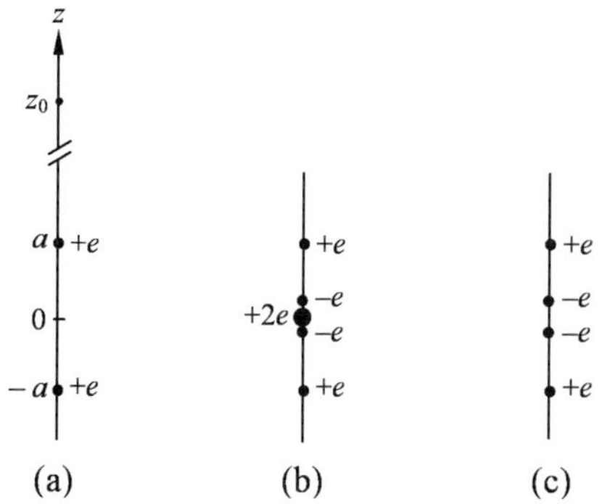

Fig. 2.7 (a) Two positive point charges. (b) The equivalent charge distribution. (c) The quadrupole component.

This expression is the sum of the potential due to a point charge of $2e$ at the origin and the potential due to a quadrupole, as shown in Fig. 2.7(b), which is equivalent to Fig. 2.7(a). When Eq. (2.24) is applied to calculate the quadrupole moment of the point charges shown in Fig. 2.7(c), the result is $Q = 4a^2$. Since the potential due to a quadrupole is

$$\frac{1}{2}\frac{eQ}{z_0^3},$$

the second term in Eq. (2.32) is indeed a contribution from the quadrupole shown in Fig. 2.7. We also see that in this case $Q = 4a^2 > 0$. If the charges were arranged on the x-axis instead (the oblate situation), a similar calculation would show $Q = -2a^2 < 0$.

In order to understand the quadrupole interaction, we can treat the quadrupole in Fig. 2.7(c) as two back-to-back dipoles along the z-axis [27]. A dipole in an electric field would experience a torque, having the tendency of rotating itself until parallel to the external electric field. However, the net torque on the quadrupole in a uniform electric field would be zero, resulting in no such rotational tendency. If the electric field is not uniform and has a gradient (EFG), the situation is entirely different. One dipole tends to move towards a direction where the electric field is more positive, and the other tends to move towards the opposite direction. A net torque is then applied on the quadrupole and it will rotate to a position where the system's energy is the lowest. This is the quadrupole interaction. If the EFG varies in space from one point to the next, there will also be a net force on the quadrupole, causing translational motion. But for a nucleus, this force is extremely small and may be neglected.

Let us consider the case when $V_{zz} > 0$, which means that there are excess negative charges in the xy-plane around the nucleus. In addition, for the excited state of ^{57}Fe, $Q > 0$. The nuclear quadrupole would therefore tend to lie near the xy-plane to minimize the interaction energy. Consequently, the $m = \pm 1/2$ states would have lower energy than the $m = \pm 3/2$ states. These can also be easily verified by calculating the energy eigenvalues in Eqs. (2.29) and (2.30). The energy levels are shown in Fig. 2.6(a).

2.2.2 The Electric Field Gradient (EFG)

2.2.2.1 Sources of EFG

The electric charges distributed around a Mössbauer nucleus can contribute to the EFG tensor only when they have a symmetry lower than cubic. In general, there are two fundamental sources for EFG:

1. The charges on the neighboring ions or ligands surrounding the Mössbauer atom, known as the lattice/ligand contribution $(V_{zz})_{\text{Lat}}$.

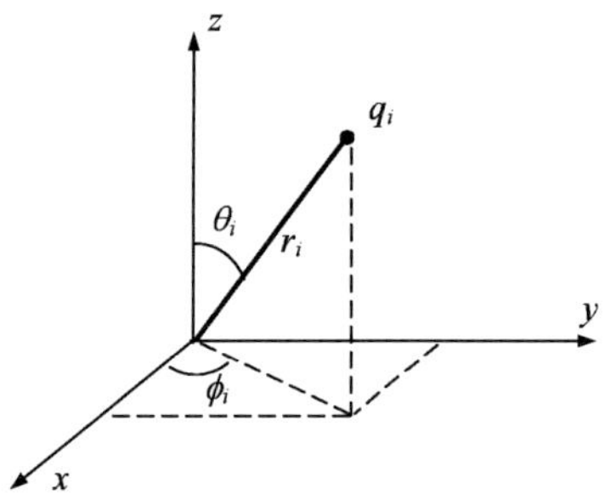

Fig. 2.8 Position of a point charge in the spherical coordinate system.

2. The charges in partially filled valence orbitals of the Mössbauer atom, known as the valence electron contribution $(V_{zz})_{\text{Val}}$.

In a solid, every lattice point around the Mössbauer atom may be regarded as a point charge q_i. The position of each lattice point may be described in the spherical coordinate system as shown in Fig. 2.8. An axially symmetric EFG can then be calculated by

$$(V_{zz})_{\text{Lat}} = \sum_i q_i \frac{3z_i^2 - r_i^2}{r_i^5} = \sum_i q_i \frac{3\cos^2\theta_i - 1}{r_i^3}$$

and

$$\eta_{\text{Lat}} = \frac{3}{V_{zz}} \sum_i q_i \frac{\sin^2\theta_i \cos 2\phi_i}{r_i^3}. \tag{2.33}$$

However, what the Mössbauer nucleus feels is not just this EFG, because the core electrons that are polarized and distorted by $(V_{zz})_{\text{Lat}}$ also contribute to the EFG tensor. The net contribution from the lattice at the nucleus is $(1 - \gamma_\infty)(V_{zz})_{\text{Lat}}$, where γ_∞ is called the Sternheimer antishielding factor. The inner core electron antishielding may cause an appreciable enhancement of EFG, so γ_∞ is negative and usually quite large. Some typical values are $\gamma_\infty = -9.14$ for $^{57}Fe^{3+}$ and $\gamma_\infty = -105$ for $^{141}Pr^{3+}$.

In cases where the crystal structure is complicated and the principal axes of the EFG cannot be found *a priori*, one is forced to choose an arbitrary coordinate system for calculating its elements V_{ij} and then diagonalize the tensor matrix.

For the contribution from the valence electrons which may be considered as orbiting around the Mössbauer nucleus at high speeds, V_{zz} is evaluated by its expectation value with $-e(3\cos^2\theta - 1)/r^3$ for each valence electron and adding all the contributions together,

$$(V_{zz})_{\text{Val}} = -e\sum_i \langle l_i m_i | 3\cos^2\theta - 1 | l_i m_i\rangle \langle r_i^{-3}\rangle. \tag{2.34}$$

For calculating the asymmetry parameter, we evaluate V_{xx} and V_{yy} in the similar way,

$$(V_{yy})_{\text{Val}} = -e\sum_i \langle l_i m_i | 3\sin^2\theta\sin^2\phi - 1 | l_i m_i\rangle \langle r_i^{-3}\rangle,$$

$$(V_{xx})_{\text{Val}} = -e\sum_i \langle l_i m_i | 3\sin^2\theta\cos^2\phi - 1 | l_i m_i\rangle \langle r_i^{-3}\rangle.$$

The core electrons shield the effect of the valence electrons, only allowing the nucleus to see a smaller EFG than $(V_{zz})_{\text{Val}}$ as calculated above. This effect is accounted for by multiplying $(V_{zz})_{\text{Val}}$ by a quantity $(1 - R)$, which is less than 1. Here R is called the Sternheimer shielding factor, which is positive and rather small. For $^{57}Fe^{2+}$, $R \approx 0.32$.

To summarize, the z component of total EFG at the nucleus is then written as

$$V_{zz} = (1 - R)(V_{zz})_{\text{Val}} + (1 - \gamma_\infty)(V_{zz})_{\text{Lat}} \tag{2.35}$$

where both $(V_{zz})_{\text{Val}}$ and $(V_{zz})_{\text{Lat}}$ are inversely proportional to r^3. Since the valence electrons are closer to the nucleus than the lattice charges, the former is much larger than the latter. For $^{57}Fe^{3+}$, we have a half-filled $3d^5$, and $(V_{zz})_{\text{Val}} = 0$ because of symmetry. For $^{57}Fe^{2+}$, the sixth electron in $3d^6$ is the main source producing a significant contribution to the EFG. The quadrupole splitting for Fe^{2+} ($\Delta E_Q \approx 3.0$ mm s^{-1}) has been found to be much larger than that for Fe^{3+} ($\Delta E_Q \approx 0.5$ mm s^{-1}) [28]. For the low spin states of $^{57}Fe^{2+}$ and $^{57}Fe^{3+}$, the former has a smaller ΔE_Q (less than 0.8 mm s^{-1}) while the latter has a larger ΔE_Q (0.7 to 1.7 mm s^{-1}) [29], again reflecting the one electron difference in their d orbitals.

When studying $(V_{zz})_{\text{Val}}$, we usually consider only the p and d electrons, because the f electrons rarely participate in chemical bonding. Table 2.2 lists the contributions to $(V_{zz})_{\text{Val}}$ from various d and p orbitals.

2.2.2.2 Temperature Effect on EFG

$(V_{zz})_{\text{Val}}$ can be further divided into two contributions: $(V_{zz})_{\text{CF}}$ due to the aspherical population of the d-orbital electrons in the valence orbital caused by a crystal field (see Fig. 2.9) and $(V_{zz})_{\text{MO}}$ due to anisotropic molecular bonding [30]. The $(V_{zz})_{\text{CF}}$ term dominates when there is little overlapping of orbitals, while the $(V_{zz})_{\text{MO}}$ term dominates for ions having symmetric electron ground states (e.g., Fe^{3+} high spin or Fe(II) low spin). We will focus on $(V_{zz})_{\text{CF}}$. Because of a crystal field, the Fe ^{5}D orbital may be split into five orbitals in which the lower energy orbitals will be populated according to the Boltzmann distribution. Hence $(V_{zz})_{\text{CF}}$ is strongly temperature dependent. On the other hand, $(V_{zz})_{\text{MO}}$ and $(V_{zz})_{\text{Lat}}$ are hardly affected by temperature. Figure 2.9 shows how Fe^{2+} high

Table 2.2 Contributions to $(V_{zz})_{\mathrm{Val}}$ from various d and p orbitals.

	Orbital	**Magnetic quantum number, m**	$\frac{1}{e}(V_{zz})_{\mathrm{Val}}$
d	$\lvert xy\rangle$	-2	$+\frac{4}{7}\langle r^{-3}\rangle$
	$\lvert yz\rangle$	-1	$-\frac{2}{7}\langle r^{-3}\rangle$
	$\lvert 3z^2 - r^2\rangle$	0	$-\frac{4}{7}\langle r^{-3}\rangle$
	$\lvert xz\rangle$	$+1$	$-\frac{2}{7}\langle r^{-3}\rangle$
	$\lvert x^2 - y^2\rangle$	$+2$	$+\frac{4}{7}\langle r^{-3}\rangle$
p	$\lvert y\rangle$	-1	$+\frac{2}{5}\langle r^{-3}\rangle$
	$\lvert z\rangle$	0	$-\frac{4}{5}\langle r^{-3}\rangle$
	$\lvert x\rangle$	$+1$	$+\frac{2}{5}\langle r^{-3}\rangle$

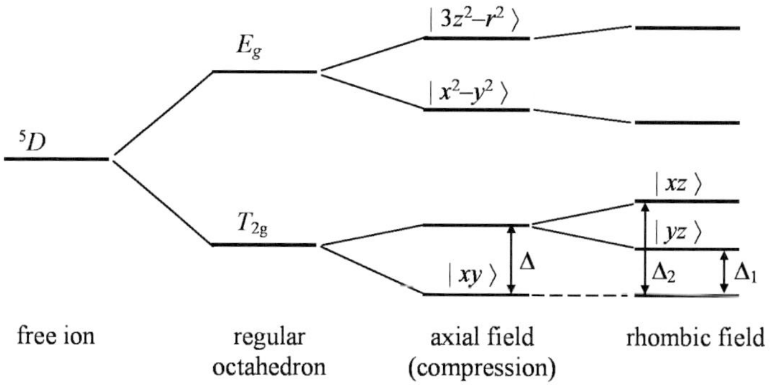

Fig. 2.9 ^{5}D orbitals of high-spin Fe^{2+} are split by crystal fields of various symmetries.

spin ^{5}D orbitals will split in crystal fields, and Fig. 2.10 shows how electrons populate the T_{2g} and E_g orbitals. For Fe^{2+} or Fe^{3+} ions in a regular octahedron, $(V_{zz})_{\mathrm{CF}} = 0$. But the octahedral symmetry is usually distorted (such as Jahn–Teller distortion). When the octahedron is compressed (or elongated) in the z-direction, the two-fold degeneracy of E_g will be removed and the three-fold degeneracy of T_{2g} will be partially lifted. When additional compression or elongation is

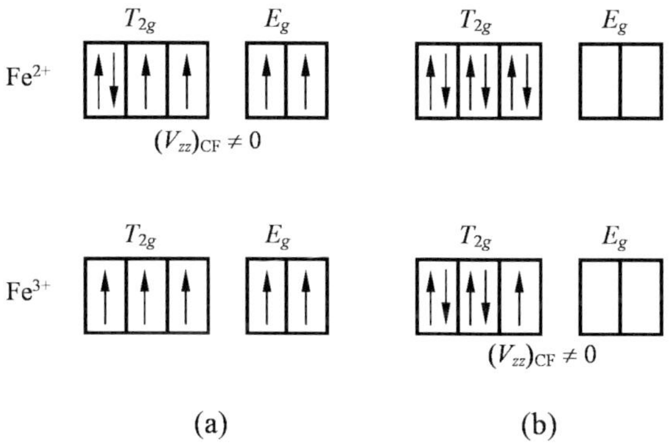

Fig. 2.10 Electronic configurations of (a) high-spin states and (b) low-spin states of Fe ions.

present in the direction perpendicular to the triangular faces, degeneracy will be completely removed. At low temperatures, the sixth electron in Fe^{2+} occupies the $|xy\rangle$ state, but as T increases, it has a higher probability of occupying a high-energy state, and the symmetry of electron population will increase, causing ΔE_Q to decline. However, in Fe^{3+} and Fe(III) compounds, removing the orbital degeneracy will not affect ΔE_Q, except for very high temperatures.

Ingalls [31] first studied in detail the temperature dependence of ΔE_Q in Fe^{2+} compounds. Suppose $(V_{zz})^i_{CF}$ is the contribution to EFG from an electron in orbital i, then the total EFG is the thermal average of contributions from all orbitals:

$$(V_{zz})_{CF} = \frac{\sum_i (V_{zz})^i_{CF} \exp[-\Delta_i/(k_B T)]}{\sum_i \exp[-\Delta_i/(k_B T)]} \tag{2.36}$$

where Δ_i is the energy of the ith orbital with respect to the ground state.

If we assume $\Delta_1 \approx \Delta_2 \approx \Delta$ in Fig. 2.9 and neglect spin–orbital coupling, then

$$(V_{zz})_{CF} = \frac{4e}{7}\langle r^3\rangle \frac{1 - e^{-\Delta/(k_B T)}}{1 + 2e^{-\Delta/(k_B T)}}. \tag{2.37}$$

For Fe^{2+} in tetrahedral compounds, the E_g orbital has a lower energy than T_{2g}, and $(V_{zz})_{CF}$ has a similar expression

$$(V_{zz})_{CF} = \frac{4e}{7}\langle r^3\rangle \frac{1 - e^{-\Delta/(k_B T)}}{1 + \frac{1}{2}e^{-\Delta/(k_B T)}}. \tag{2.38}$$

An experimental example of how ΔE_Q varies with temperature is given in Fig. 2.11 [32], where we can also see that ΔE_Q of the Fe^{3+} site is essentially independent of temperature.

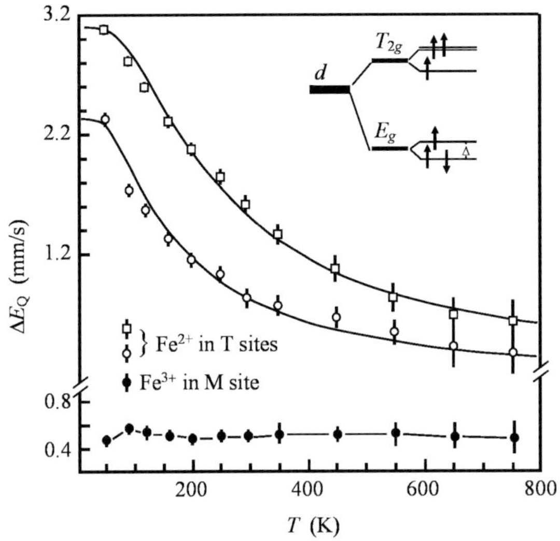

Fig. 2.11 Temperature dependence of ΔE_Q for Fe^{2+} and Fe^{3+} in chromite spinels.

2.2.3
Intensities of the Spectral Lines

Based on the calculations by Karyagin [33], Zory [34], and Alimuddin et al. [35], the relative intensities of the spectral lines of the quadrupole doublet in a single crystal can be written as

$$I_{3/2} = \frac{c}{\sqrt{1+(\eta^2/3)}}\left(4\sqrt{1+(\eta^2/3)} + 3\cos^2\theta - 1 + \eta\sin^2\theta\cos 2\phi\right),$$
$$I_{1/2} = \frac{c}{\sqrt{1+(\eta^2/3)}}\left(4\sqrt{1+(\eta^2/3)} - 3\cos^2\theta + 1 - \eta\sin^2\theta\cos 2\phi\right) \tag{2.39}$$

where c is a constant, and θ and ϕ are the polar and azimuthal angles of the incident γ-ray in the EFG principal axis system.

In the case where the EFG is axially symmetric, $\eta = 0$, and Eq. (2.39) gives the following ratio of the intensities of the two absorption lines:

$$\frac{I_{3/2}}{I_{1/2}} = \frac{1+\cos^2\theta}{\frac{5}{3} - \cos^2\theta}. \tag{2.40}$$

For a polycrytalline sample or a powder sample, the crystal axes are randomly oriented. Spatial average of the two intensities in Eq. (2.39) gives

$$\frac{\langle I_{3/2}\rangle}{\langle I_{1/2}\rangle}=\frac{\dfrac{1}{4\pi}\displaystyle\int I_{3/2}\sin\theta\,\mathrm{d}\theta\,\mathrm{d}\phi}{\dfrac{1}{4\pi}\displaystyle\int I_{1/2}\sin\theta\,\mathrm{d}\theta\,\mathrm{d}\phi}=1, \tag{2.41}$$

which means that the two lines have the same intensity.

2.2.4 The Sign of EFG

Since $Q>0$ for ^{57}Fe, the sign of the quadrupole splitting is determined by the sign of the EFG component V_{zz}. From Fig. 2.6, we see that whether V_{zz} is positive or negative determines whether the energy level of $|\pm 3/2\rangle$ is higher or lower than the energy level of $|\pm 1/2\rangle$.

1. Consider first a single crystal. If $\eta=0$, the two spectral lines will in general have different intensities. For example, $I_{3/2}/I_{1/2}$ is 3.0 when $\theta=0$, and it is 0.6 when $\theta=\pi/2$. Therefore, a single measured spectrum can unambiguously determine the sign of V_{zz} provided the crystal axes are known. If $\eta\neq 0$, this method fails. Signs of V_{zz} in single crystals have also been determined by using polarized γ-rays [36].
2. For polycrystalline or powder samples, an external magnetic field is required for the determination of the sign of V_{zz}. Collins [37, 38] extensively investigated this problem, and the results are illustrated in Fig. 2.12. The external magnetic field further splits the energy levels. In cases where these splittings are smaller than the quadrupole splittings,

$$\varepsilon_1 \ll \Delta E_Q, \tag{2.42}$$

the admixture between $|\pm 3/2\rangle$ and $|\pm 1/2\rangle$ can be neglected and we may obtain approximate analytical solutions. In the presence of an applied weak external magnetic field B, the original doublet in the spectrum transforms into a quartet and a doublet. The quartet appears to be a triplet due to two lines not being completely resolved. If the doublet is on the positive velocity side, then $V_{zz}>0$; if the quartet is on the positive velocity side, $V_{zz}<0$. This method may also be applied to single-crystal samples. Figure 2.13 shows a concrete example of determining the sign of V_{zz} in a powder sample using this method [39].

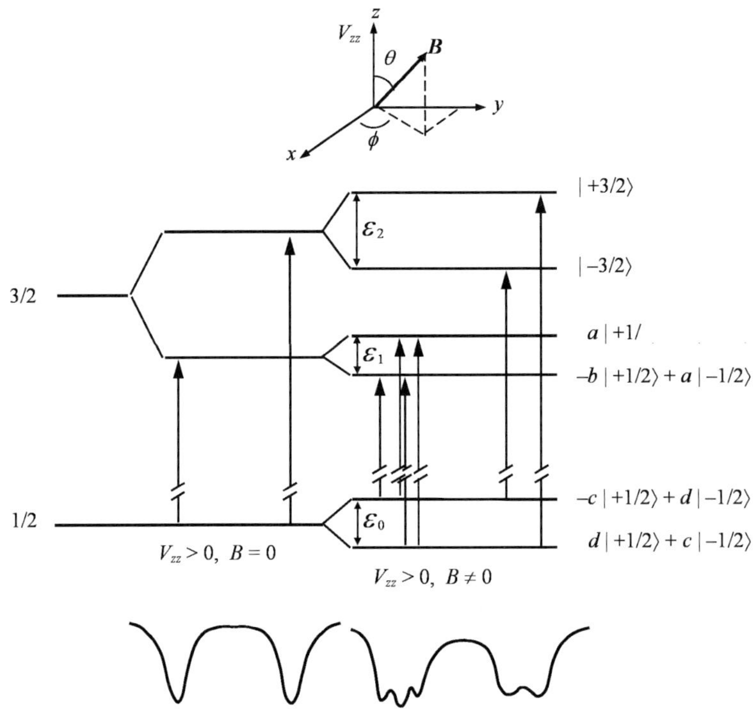

Fig. 2.12 ^{57}Fe nuclear energy level splittings and absorption spectra due to γ-ray transitions under an axially symmetric EFG ($\eta = 0$) and a weak external magnetic field.

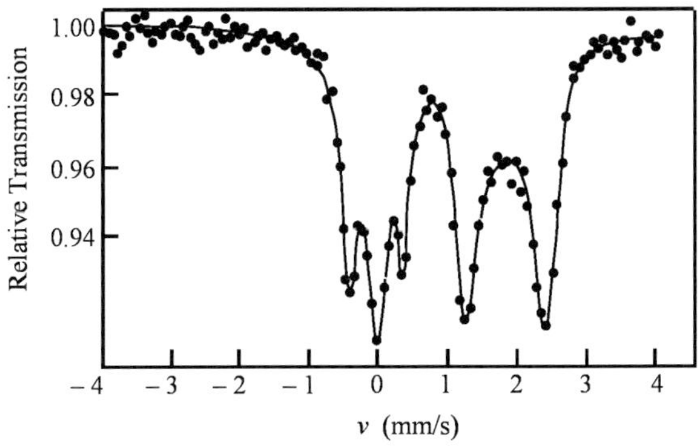

Fig. 2.13 Mössbauer spectrum of $FeCO_3$ ($V_{zz} > 0$) with an external magnetic field $B_{ext} = 3.5$ T.

It should be noted that if the external magnetic field is too large or ΔE_Q is too small, condition (2.42) may not be satisfied. Consequently, after the $|\pm 3/2\rangle$ and $|\pm 1/2\rangle$ levels split, they also mix with one another, making it very difficult to determine the sign of V_{zz} [40].

2.3
Magnetic Dipole Interaction

The interaction between the nuclear magnetic dipole moment μ and the magnetic field $\boldsymbol{B}$ at the nucleus produced by the surrounding electrons or ions is called the magnetic hyperfine interaction. This interaction lifts the degeneracy of the energy level of a nucleus of spin I and splits it into $(2I + 1)$ sublevels. This type of splitting in the nuclear ground state had been observed in nuclear magnetic resonance and paramagnetic resonance. In optical spectroscopy, the Zeeman effect in atoms was observed a long time ago. But the nuclear Zeeman effect, a similar effect in principle, was impossible to observe before the discovery of the Mössbauer effect, because the splittings between the nuclear sublevels are too small to resolve. Using the Mössbauer effect, Hanna [41, 42] first observed the magnetic hyperfine interaction in the nucleus, i.e., the nuclear Zeeman effect.

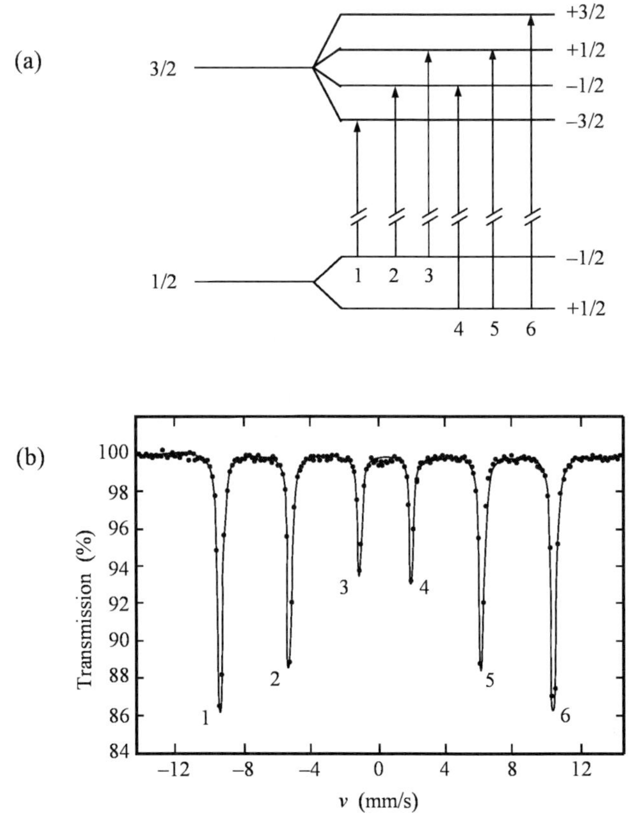

Fig. 2.14 (a) Magnetic splittings of the ^{57}Fe nuclear energy levels. (b) A Mössbauer spectrum of FeF_3 at 4.2 K showing a sextet due to magnetic splittings [44].

2.3.1
Magnetic Splitting

The Hamiltonian of the interaction between the nuclear magnetic dipole moment $\boldsymbol{\mu}$ and the magnetic field $\boldsymbol{B}$ is

$$\mathscr{H}_{\mathrm{M}} = -\boldsymbol{\mu}\cdot\boldsymbol{B} = -g\mu_{\mathrm{N}}\boldsymbol{I}\cdot\boldsymbol{B} \tag{2.43}$$

and the corresponding sublevel energy is

$$E_{\mathrm{M}} = -gmB\mu_{\mathrm{N}} \tag{2.44}$$

where g is the nuclear g-factor, $m = I, I-1, \ldots, -I$, and μ_{N} is the nuclear magneton.

The ^{57}Fe first excited state with spin 3/2 splits into four sublevels equally separated by $g_{\mathrm{e}}B\mu_{\mathrm{N}}$, while the ground state with spin 1/2 splits into two sublevels, as shown in Fig. 2.14(a). The g_{g}-factor of the ground state, in general, is different from g_{e} of the excited state; therefore the separation between the sublevels in the ground state is different from that in the excited state. Since the γ transition in ^{57}Fe is of the magnetic dipole type (M1), it can take place provided the selection rule $\Delta m = \pm 1$ or 0 is obeyed. Thus the six allowed transitions give six absorption lines as shown in Fig 2.14(b). The transitions with $\Delta m = \pm 2$ are forbidden.

The position of each line in the characteristic sextet can be easily calculated according to Eq. (2.44). As can be seen from Fig. 2.14(b), the successive separations between adjacent lines are in the ratio of 1:1:x:1:1, where for ^{57}Fe $x = (g_{\mathrm{g}} - |g_{\mathrm{e}}|)/|g_{\mathrm{e}}| \approx 3/4$, $g_{\mathrm{g}} = 0.1808$, and $g_{\mathrm{e}} = -0.1031$.

2.3.2
Relative Line Intensities

The excited substates $|I_{\mathrm{e}}m_{\mathrm{e}}\rangle$, the ground substates $|I_{\mathrm{g}}m_{\mathrm{g}}\rangle$, and the quantum state of the incident photon $|\chi_L^m\rangle$ are related by the Clebsch–Gordan (C-G) coefficients:

$$|I_{\mathrm{e}}m_{\mathrm{e}}\rangle = \sum_{m=m_{\mathrm{e}}-m_{\mathrm{g}}} \langle I_{\mathrm{g}}m_{\mathrm{g}}Lm|I_{\mathrm{e}}m_{\mathrm{e}}\rangle|I_{\mathrm{g}}m_{\mathrm{g}}\rangle|\chi_L^m\rangle. \tag{2.45}$$

The relative absorption probability W for each of the transitions is

$$W(\theta) = |\langle I_{\mathrm{g}}m_{\mathrm{g}}Lm|I_{\mathrm{e}}m_{\mathrm{e}}\rangle\chi_L^m(\theta\phi)|^2. \tag{2.46}$$

The angular dependence of the probability is contained in χ_L^m, and for M1-type radiation [43]

Table 2.3 Angular distribution of the relative intensities for the six allowed transitions in ^{57}Fe. $C = \langle I_g m_g L m | I_e m_e \rangle$ are the C-G coefficients.

Subspectral line	Transition	Δm	C^2	$W(\theta)$	$W(0)$	$W(90°)$	W
1 (6)	$\pm\frac{1}{2} \rightarrow \pm\frac{3}{2}$	± 1	1	$\frac{3}{4}(1+\cos^2\theta)$	3/2	3/4	1
2 (5)	$\pm\frac{1}{2} \rightarrow \pm\frac{1}{2}$	0	2/3	$\sin^2\theta$	0	1	2/3
3 (4)	$\pm\frac{1}{2} \rightarrow \mp\frac{1}{2}$	∓ 1	1/3	$\frac{1}{4}(1+\cos^2\theta)$	1/2	1/4	1/3

$$\begin{aligned} \chi_1^{\pm 1} &= \sqrt{\frac{3}{4}}(-\mathbf{z}\sin\theta\cos\phi + \mathbf{x}\cos\theta) \pm i(-\mathbf{z}\sin\theta\sin\phi + \mathbf{y}\cos\theta) \\ \chi_1^0 &= i\sqrt{\frac{3}{2}}\sin\theta(-\mathbf{x}\sin\phi + \mathbf{y}\cos\phi) \end{aligned} \tag{2.47}$$

where **x**, **y**, **z** are the unit vectors, and θ, ϕ are the polar and azimuthal angles describing the direction of the incident γ-ray with respect to the magnetic field direction. The ϕ dependence disappears in $W(\theta)$ because in this case $|\chi_L^m|^2$ depends only on θ. The function χ_L^m is normalized as follows:

$$\int_0^{2\pi}\int_0^{\pi} |\chi_L^m|^2 \sin\theta \, d\theta \, d\phi = 4\pi. \tag{2.48}$$

The angular distribution of relative intensities for the six allowed transitions is listed in Table 2.3. For a thin absorber, the Mössbauer fraction f is isotropic, and when the angle between $\boldsymbol{B}$ and the γ-ray is $\theta = 0$, the relative intensities of the six subspectral lines are in the ratio of 3:0:1:1:0:3. When $\theta = 90°$, the ratio is 3:4:1:1:4:3. If the magnetic field vectors at the nuclei are randomly oriented, then an integral over θ gives an intensity ratio of 3:2:1:1:2:3.

2.3.3 Effective Magnetic Field

A convenient way to describe the magnetic hyperfine interaction is using the effective magnetic field, which is the sum of the local magnetic field $\boldsymbol{B}_{\text{loc}}$ at the Mössbauer nucleus by the lattice and the hyperfine magnetic field $\boldsymbol{B}_{\text{hf}}$ by the Mössbauer atom's own electrons:

$$\boldsymbol{B}_{\text{eff}} = \boldsymbol{B}_{\text{loc}} + \boldsymbol{B}_{\text{hf}}. \tag{2.49}$$

The local field by the lattice may be due to the material's magnetic ordering, or may be applied externally, or both. It may have the following contributions:

$$\boldsymbol{B}_{\mathrm{loc}} = \boldsymbol{B}_{\mathrm{ext}} - D\boldsymbol{M} + \frac{4\pi}{3}\boldsymbol{M} \tag{2.50}$$

where $\boldsymbol{B}_{\mathrm{ext}}$ is an external field, $\boldsymbol{M}$ is magnetization, $D\boldsymbol{M}$ represents the demagnetization field, and $4\pi\boldsymbol{M}/3$ represents the Lorentz field. In general, the local field is much smaller than the hyperfine field.

The hyperfine field $\boldsymbol{B}_{\mathrm{hf}}$ has three contributions:

$$\boldsymbol{B}_{\mathrm{hf}} = \boldsymbol{B}_{\mathrm{s}} + \boldsymbol{B}_{\mathrm{L}} + \boldsymbol{B}_{\mathrm{D}} \tag{2.51}$$

where $\boldsymbol{B}_{\mathrm{s}}$ is called Fermi contact field produced by the s electron spin density at the nucleus, and may be expressed as

$$\boldsymbol{B}_{\mathrm{s}} = -\frac{2\mu_0}{3}\mu_{\mathrm{B}}\sum_{n}[|\psi_{ns\uparrow}(0)|^2 - |\psi_{ns\downarrow}(0)|^2] \tag{2.52}$$

where μ_{B} is the Bohr magneton and $|\psi_{\mathrm{ns}\uparrow}(0)|^2$ and $|\psi_{\mathrm{ns}\downarrow}(0)|^2$ represent the ns spin-up and spin-down electron densities at the nucleus, respectively. Actually, such a difference in the spin densities is caused by unpaired d electrons. To understand this mechanism, let us consider ^{57}Fe as an example. The basic reason is that there exists an exchange interaction which would slightly reduce the repulsion between electrons having the same spin quantum number m_{s}. The unpaired d electrons all have the same spin (say, spin-up, or $m_{\mathrm{s}} = +1/2$), and they tend to repel the spin-down s electrons more than the spin-up electrons, therefore polarizing the spins in the otherwise perfectly balanced s shells. The s electrons having relatively large densities at the nucleus are polarized into two groups that interact differentially with the nucleus. This net interaction is equivalent to a magnetic field, called the Fermi contact field, in the opposite direction to the field generated by the spins in the d orbitals. In iron compounds, $\boldsymbol{B}_{\mathrm{s}}$ is the largest among the three terms in Eq. (2.51). In high-spin Fe^{3+} ($3d^5$), the number of unpaired d electrons is the largest, resulting in a maximum spin polarization of the s electrons, and the Fermi contact field becomes as high as 50 to 60 T, while high-spin Fe^{2+}($3d^6$) has four unpaired d electrons, so $\boldsymbol{B}_{\mathrm{s}}$ is lower and varies in the range 20 to 50 T.

B_{L} is called the orbital field, due to the orbital motions of the unpaired electrons around the nucleus. This motion constitutes a circular current, which in turn produces a magnetic field at the nucleus:

$$B_{\mathrm{L}} = -\frac{\mu_0}{2\pi}\mu_{\mathrm{B}}\langle r^{-3}\rangle\langle L_z\rangle \tag{2.53}$$

where L_z is the z-component of the orbital angular momentum. For Fe^{3+} ($3d^5$) in a solid, orbital angular momentum is quenched, $L_z \approx 0$. When the spin–orbit coupling mixes the non-degenerate states with the excited state, thus introducing an angular momentum in the ground state, B_{L} has a finite value. Because the excited state mixing may cause the g-factor of the ion to deviate from the spin-only value of 2, B_{L} has been expressed as [45]

$$B_L = -\frac{\mu_0}{2\pi}\mu_B\langle r^{-3}\rangle(g-2)\langle S_z\rangle. \tag{2.54}$$

For Fe^{3+}, $g \approx 2$, and consequently $B_L \approx 0$. For high-spin Fe^{2+}, $B_L \approx 20$ T and opposite to $\boldsymbol{B}_S$. The B_L values for rare earth compounds are relatively large.

The third term in Eq. (2.51) is the dipole field at the nucleus, produced by the total spin magnetic moment of the valence electrons. It can be written as

$$B_D = \frac{\mu_0}{8\pi}\mu_B\langle r^{-3}\rangle\langle 3\cos^2\theta - 1\rangle\langle S_z\rangle. \tag{2.55}$$

Obviously, B_D is zero for a charge distribution with cubic symmetry. For Fe group ions, B_D is small even in non-cubic systems, ranging only from 0 to 8 T. However, in rare earth compounds where the orbital momentum is not quenched, B_D can be quite large.

A necessary condition for a magnetic hyperfine field is that the atom has a magnetic moment due to unpaired electrons. There are two main characteristics associated with this field: it is very strong ($B_{hf} \approx -33$ T in α-Fe) and it is local (not over the entire lattice). There are several ways to measure the sign of the magnetic hyperfine field. One is to apply an external magnetic field of 2 to 5 T and detect whether B_{eff} increases or decreases. When it increases, B_{hf} is positive; otherwise it is negative.

So far we have limited ourselves to isolated magnetic hyperfine interactions. Now we will discuss B_{eff} in ferromagnetic, antiferromagnetic, and paramagnetic materials. In ferromagnetic materials, the coupling between spins of different atoms is very strong (spontaneous magnetization), forming areas known as domains. Within each domain, the magnetic moments of all atoms are parallel to one another. Even at room temperature, the nucleus feels the interaction of a stable B_{eff}, although the domains have random orientations. In paramagnetic compounds, the coupling between atomic moments is weak, and thermal excitation causes random fluctuations of the spins. As a result, the hyperfine field at the nucleus also fluctuates rapidly, so that the nucleus may not catch the instantaneous values of B_{eff}, but the average value $B_{eff} = 0$. The fluctuation process has a characteristic time called relaxation time τ_R. It is possible to observe hyperfine field only when the following condition is met, $\tau_R > \tau_L$ (τ_L is the Larmor precession period of the nucleus). For many paramagnetic compounds, τ_R can be increased by lowering both the temperature and the density of the Mössbauer nuclei in the material, thus allowing the observation of the magnetic hyperfine splitting.

The Mössbauer effect has found extensive applications in magnetism and in research of magnetic materials. Without requiring an applied external magnetic field, Mössbauer spectroscopy can be used to study the temperature dependence of spontaneous magnetization, the magnitude and orientation of hyperfine fields, and the magnetic structure of new materials. It can also be used to measure the ordering temperatures (T_C, T_N) and spin reorientation temperature, to detect phase transitions and determine phase compositions, to study magnetic lattice anisotropy and relaxation phenomena, etc. From 1960 when Hanna first observed a magnetic hyperfine spectrum to 1980 when $Nd_2Fe_{14}B$ was investigated until the recent discovery of new types of rare earth permanent magnets $R_2Fe_{17}N_x$ and $R_2Fe_{17}C_x$, the field of magnetism would not be this successful without Mössbauer spectroscopy. There is a tremendous amount of literature on this subject, and there are many excellent books and review articles [46–50].

Magnetism arises mainly from the atomic magnetic moments. Transition metals (3d, 4d, 5d), the lanthanides (4f), and the actinides (5f) all have unfilled valence electrons and have atomic magnetic moments. Fortunately, many isotopes of these elements are Mössbauer nuclei, e.g., ^{57}Fe, ^{61}Ni, ^{99}Ru, ^{193}Ir, ^{149}Sm, ^{151}Eu, ^{155}Gd, ^{159}Tb, ^{161}Dy, ^{165}Ho, ^{166}Er, ^{169}Tm, ^{170}Yb, and ^{237}Np. Obviously, ^{57}Fe is the best one because Fe is the most important element in magnetism. Not surprisingly, most of the Mössbauer effect studies in magnetism involve ^{57}Fe.

2.4
Combined Quadrupole and Magnetic Interactions

It is often the case when both a magnetic field and an electric field gradient are present. The shape of a Mössbauer spectrum depends on not only the relative strengths of these two interactions but also the relative orientations of the EFG principal axis, the magnetic field, and the incident γ-ray. In the principal axis system of the EFG (see Fig. 2.15), the total Hamiltonian as the sum of Eqs. (2.27) and (2.43) is

$$\begin{aligned}\mathscr{H}_{QM} &= \mathscr{H}_Q + \mathscr{H}_M \\ &= \frac{eQV_{zz}}{4I(2I-1)}\left[3\hat{I}_z^2 - \hat{I}^2 + \frac{1}{2}\eta(\hat{I}_+^2 + \hat{I}_-^2)\right] \\ &\quad - g\mu_N B\left\{\left[\frac{1}{2}(\hat{I}_+ + \hat{I}_-)\cos\phi + \frac{1}{2}(\hat{I}_+ - \hat{I}_-)\sin\phi\right]\sin\theta + \hat{I}_z\cos\theta\right\}. \qquad (2.56)\end{aligned}$$

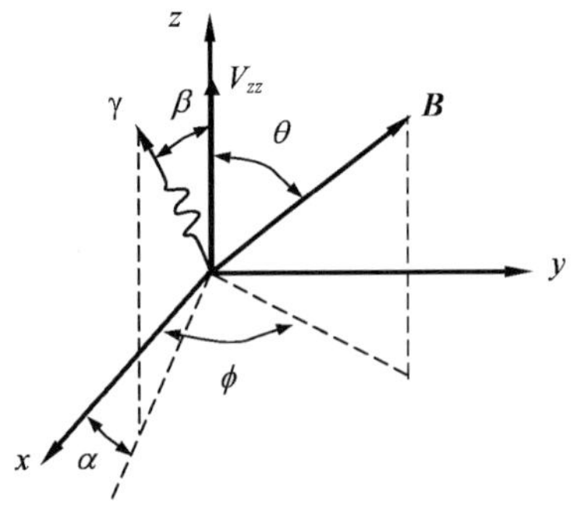

Fig. 2.15 Relative orientations of ***B***, V_{zz}, and the γ-ray direction.

It is quite easy to obtain the eigenvalues and eigenstates of this Hamiltonian for $I = 1/2$, but somewhat more complicated for $I > 1/2$. In the early days, such eigenvalue problems were solved essentially by numerical methods [51–56]. Various computer programs developed for this purpose are widely used. Analytical solutions for the eigenvalues were only available for the case where the quadrupole interaction is much weaker than the magnetic interaction, i.e., $V_{zz}eQ/2 \ll g_e\mu_N B$, or η and θ are both zero.

An important step was made when analytical solutions for eigenvalues of Hamiltonian (2.56) were obtained. Häggström [57] first presented an analytical solution of the secular equation. By an improved method for spherical angular averages [58], his results allow a faster and more precise calculation of measured spectra. Later, Blaes et al. [59] gave an analytical expression for the line intensities by the superoperator technique. Even though their expressions are complicated, they allow angular averages to be performed analytically, not numerically as done in Häggström's procedure.

Although the numerical method is always necessary and the analytical method still has some imperfections, the idea has fundamental significance. Therefore, we choose Häggström's method for discussing the combined hyperfine interactions.

For ^{57}Fe (or ^{119}Sn), the ground state and the excited state have $I = 1/2$ and $3/2$, respectively. The matrix elements of the Hamiltonian (2.56) are

$$\left\langle \frac{1}{2}, m' | \mathcal{H}_{QM} | \frac{1}{2}, m \right\rangle = -\varepsilon_g \begin{bmatrix} \cos\theta & \cos\theta e^{-i\phi} \\ \sin\theta e^{i\phi} & -\cos\theta \end{bmatrix} \tag{2.57}$$

$$\left\langle \frac{3}{2}, m' | \mathcal{H}_{QM} | \frac{3}{2}, m \right\rangle = \varepsilon_e \boldsymbol{H}', \tag{2.58}$$

where

$$\varepsilon_g = \frac{1}{2} g_{1/2}\mu_N B,$$

$$\varepsilon_e = \frac{1}{2} g_{3/2}\mu_N B,$$

$$\boldsymbol{H}' = \begin{bmatrix} R - 3\cos\theta & -\sqrt{3}\sin\theta e^{-i\phi} & \dfrac{\eta R}{\sqrt{3}} & 0 \\ -\sqrt{3}\sin\theta e^{i\phi} & -R - \cos\theta & 2\sin\theta e^{-i\phi} & \dfrac{\eta R}{\sqrt{3}} \\ \dfrac{\eta R}{\sqrt{3}} & -2\sin\theta e^{i\phi} & -R + \cos\theta & -\sqrt{3}\sin\theta e^{-i\phi} \\ 0 & \dfrac{\eta R}{\sqrt{3}} & -\sqrt{3}\sin\theta e^{i\phi} & R + 3\cos\theta \end{bmatrix}, \tag{2.59}$$

and

$$R = \frac{eQV_{zz}}{4\varepsilon_e}.$$

The above matrix is not diagonal in the angular momentum states $|Im\rangle$ except when η and θ are both zero. As can be seen from (2.29) and (2.30), the electric quadrupole interaction will mix states differing in m-value by ± 2 units. Similarly, the magnetic dipole interaction will mix states differing in m-value by ± 1 unit. Accordingly, for $I = 3/2$ the four eigenvectors can then be represented by linear combinations of the states $|Im\rangle$:

$$\left|\frac{3}{2}, j\right\rangle = b_{j,3/2}\left|\frac{3}{2}, +\frac{3}{2}\right\rangle + b_{j,1/2}\left|\frac{3}{2}, +\frac{1}{2}\right\rangle + b_{j,-1/2}\left|\frac{3}{2}, -\frac{1}{2}\right\rangle + b_{j,-3/2}\left|\frac{3}{2}, -\frac{3}{2}\right\rangle \tag{2.60}$$

where the coefficients b_{j,m_e} are normalized to unity:

$$\sum_{m_e} |b_{j,m_e}|^2 = 1, \quad \text{for } j = 1, 2, 3, 4.$$

Analogously, for $I = 1/2$

$$\left|\frac{1}{2}, i\right\rangle = a_{i,1/2}\left|\frac{1}{2}, +\frac{1}{2}\right\rangle + a_{i,-1/2}\left|\frac{1}{2}, -\frac{1}{2}\right\rangle \tag{2.61}$$

where the coefficients a_{i,m_g} are also normalized to unity:

$$\sum_{m_g} |a_{i,m_g}|^2 = 1, \quad \text{for } i = 1, 2.$$

Let λ be the unknown eigenvalue of the matrix $\boldsymbol{H}'$. The secular determinant equation can be then expressed as

$$\lambda^4 + p\lambda^2 + q\lambda + r = 0 \tag{2.62}$$

where

$$\begin{aligned} p &= -10 - 2R^2(1 + \eta^2/3), \\ q &= -8R(3\cos^2\theta - 1 + \eta \sin^2\theta \cos 2\phi), \\ r &= 9 + 2R^2(6\sin^2\theta + \eta^2 \cos 2\theta + 4\eta \sin^2\theta \cos 2\phi - 5) \\ &\quad + R^4(1 + \eta^2/3)^2. \end{aligned} \tag{2.63}$$

The real roots of such an equation of the fourth order for $q \geq 0$ are [60]

$$\begin{aligned}
\lambda_1 &= \sqrt{y_1} - \sqrt{y_2} - \sqrt{y_3} \\
\lambda_2 &= -\sqrt{y_1} - \sqrt{y_2} + \sqrt{y_3} \\
\lambda_3 &= -\sqrt{y_1} + \sqrt{y_2} - \sqrt{y_3} \\
\lambda_4 &= \sqrt{y_1} + \sqrt{y_2} + \sqrt{y_3}
\end{aligned} \tag{2.64}$$

where

$$y_k = \frac{1}{6}\sqrt{p^2 + 12r}\cos[\varphi/3 + (k-1)2\pi/3] - \frac{p}{6} \quad k = 1, 2, 3, \tag{2.65}$$

$$\cos\varphi = \frac{2p^3 + 27q - 72pr}{2(p^2 + 12r)^{3/2}}.$$

For $q < 0$, all λ_i given above should be multiplied by -1.

The energy eigenvalues of the Hamiltonian with $I = 3/2$ are then given as a product

$$E(3/2, j) = \varepsilon_e \times \lambda_j. \tag{2.66}$$

Substituting these values into the eigenvector equation, the coefficients b_{j,m_e} can be calculated (see Appendix B).

The transition energies, E_T, are the differences between the eigenvalues of the excited and ground states

$$E_T(3/2, j; 1/2, i) = E(3/2, j) - E(1/2, i). \tag{2.67}$$

When γ-rays in a direction (β, α) are absorbed, the relative transition probabilities are then [51]

$$W\left(\beta, \alpha; \frac{3}{2}, j; \frac{1}{2}, i\right) = \left|\sum_{m=-1}^{+1} c_m \chi_L^m\right|^2 \tag{2.68}$$

where

$$c_m = \sum_{m_g=-1/2}^{+1/2} a_{m_g}^* b_{(m_g+m)} \langle I_g m_g 1 m | I_e(m_g + m)\rangle, \quad \text{for } m = -1, 0, +1. \tag{2.69}$$

Using the values of appropriate Clebsch–Gordan coefficients, the coefficients in (2.69) are reduced to

$$
\begin{aligned}
c_{\pm 1} &= a^*_{i,\pm 1/2} b_{j,\pm 3/2} + \sqrt{\frac{1}{3}} a^*_{i,\mp 1/2} b_{j,\pm 1/2}, \\
c_0 &= \sqrt{\frac{2}{3}} (a^*_{i,+1/2} b_{j,+1/2} + a^*_{i,-1/2} b_{j,-1/2}).
\end{aligned}
\tag{2.70}
$$

Inserting the vector spherical harmonics χ_1^m from (2.47) into (2.68), one obtains

$$
\begin{aligned}
W\left(\beta,\alpha;\frac{3}{2},j;\frac{1}{2},i\right) = \frac{3}{4}[(|c_{+1}|^2 + |c_{-1}|^2)(1+\cos^2\beta) + |c_0|^2 2\sin^2\beta \\
- \sqrt{2}\sin 2\chi\, \mathrm{Re}(c_{+1}c_0^* \mathrm{e}^{i\alpha}) \\
+ 2\sin^2\beta\, \mathrm{Re}(c_{+1}c_{-1}^* \mathrm{e}^{2i\alpha}) \\
- \sqrt{2}\sin 2\beta\, \mathrm{Re}(c_0 c_{-1}^* \mathrm{e}^{i\alpha})].
\end{aligned}
\tag{2.71}
$$

This formula should be used when the sample under investigation is a single crystal. For a powder sample, Eq. (2.71) must be averaged over β and α, giving the following simple result:

$$
W\left(\frac{3}{2},j;\frac{1}{2},i\right) = |c_{+1}|^2 + |c_0|^2 + |c_{-1}|^2. \tag{2.72}
$$

In order to understand the physical meanings of $|c_{\pm 1}|^2$ and $|c_0|^2$, let us consider the pure magnetic hyperfine interaction just discussed above. In this case, $V_{zz} = 0$ (or $R = 0$). The coordinate system is chosen so that $\theta = 0$, and the matrix (2.59) will be diagonal. This means the absence of mixture between substates. Indeed, since $R = 0$, we have $p = -10$, $r = 9$, and $q = 0$. Solving Eq. (2.58) gives four eigenvalues: $\pm 3\varepsilon_e$ and $\pm\varepsilon_e$ which are the same as Eq. (2.44). For the $\Delta m = +1$ transitions (corresponding to line 6 in Fig. 2.14), we have $b_{4,3/2} = 1$, $b_{4,1/2} = b_{4,-1/2} = b_{4,-3/2} = 0$, $a_{1,1/2} = 1$, and $a_{1,-1/2} = 0$. Using Eq. (2.70), we can calculate $c_{+1} = 1$, $c_{-1} = c_0 = 0$, where c_{+1} is none other than the C-G coefficient for this transition whose relative transition probability is $W_{+1} = |c_{+1}|^2 = 1$. Analogously for the $\Delta m = 0$ and $\Delta m = -1$ transitions, the relative probabilities are $W_0 = |c_0|^2 = 2/3$ and $W_{-1} = |c_{-1}|^2 = 1/3$, respectively. The values of W are exactly the same as earlier results listed in Table 2.3.

A subroutine program has been written which calculates the transition energies E_T, and the corresponding intensities, using the above analytical formulas. This subroutine has been incorporated into a least-square fitting program to provide a practical way of analyzing Mössbauer spectra in the presence of combined interactions. Here we will show the results from a hexagonal FeGe powder as an example [57]. The parameters used are B, ΔE_Q, η, θ, ϕ, together with ordinary parameters such as background, area, linewidth, etc. In Table 2.4 are given experimental results of ^{57}Fe Mössbauer spectra at four temperatures, and one of the spectra is presented in Fig. 2.16.

Table 2.4 Results from the fitting of ^{57}Fe Mössbauer spectra of FeGe hexagonal compound at four temperatures.

T (K)	B (T)	ΔE_Q (mm s^{-1})	δ (mm s^{-1})	η	θ (°)	ϕ (°)
10	15.7(1)	−0.56(3)	0.39(1)	0.3(1)	82(2)	70(25)
295	11.8(1)	−0.60(2)	0.28(1)	0.3(1)	89(1)	70(10)
360	8.7(1)	−0.58(1)	0.23(1)	0.4(1)	89(1)	80(15)
390	5.2(1)	−0.55(2)	0.22(1)	0.3(1)	90(1)	70(10)

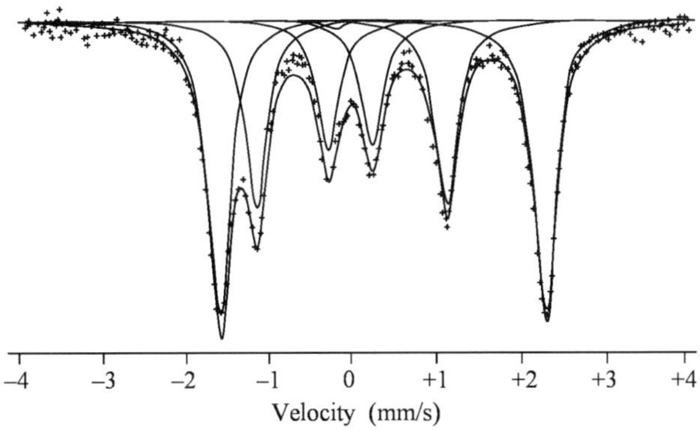

Fig. 2.16 Fitted ^{57}Fe Mössbauer spectrum of hexagonal FeGe at room temperature.

It should be noted that the exact expressions, such as (2.63), (2.68), (2.69), etc., are indeed very tedious for practical purposes. It is often the case where either the magnetic hyperfine interaction or electric quadrupole interaction is dominant while the other occurs as a perturbation, i.e., $R \gg 1$ or $R \ll 1$. We now briefly consider these two extreme cases.

When the magnetic interaction is dominant, $R \ll 1$, expanding λ_j into a Taylor series at $R = 0$ (see Appendix B) can give the approximate energy eigenvalues for the four sublevels as follows:

$$
\begin{aligned}
E\left(\frac{3}{2}, 1\right) &= \varepsilon_e(-3 + Rk_1), \\
E\left(\frac{3}{2}, 2\right) &= \varepsilon_e(-1 - Rk_1), \\
E\left(\frac{3}{2}, 3\right) &= \varepsilon_e(+1 - Rk_1), \\
E\left(\frac{3}{2}, 4\right) &= \varepsilon_e(+3 + Rk_1)
\end{aligned}
\tag{2.73}
$$

where

$$k_1 = \frac{1}{2}(2\cos^2\theta - 1 + \eta \sin^2\theta \cos 2\phi).$$

This means that each of four sublevels in Fig. 2.14(a) shifts an amount of $+Rk_1$ or $-Rk_1$ further.

If the electric quadrupole interaction is dominant as shown in Fig. 2.12, $R \gg 1$, to a first-order approximation we may neglect the four elements in matrix (2.59) that mix the $\pm 3/2$ states with $\pm 1/2$ states. Suppose also $\eta = 0$, then only the inner 2×2 matrix is required to diagonalize. A similar calculation gives the approximate eigenvalues and eigenstates for the excited state with:

$$\begin{aligned}
&E\left(\frac{3}{2},2\right) = -\frac{eQV_{zz}}{4} - \varepsilon_e\sqrt{4-3\cos^2\theta},\\
&\left|\frac{3}{2},2\right\rangle = -\sin\eta\left|\frac{3}{2},\frac{1}{2}\right\rangle + \cos\eta\left|\frac{3}{2},-\frac{1}{2}\right\rangle\\
&E\left(\frac{3}{2},3\right) = -\frac{eQV_{zz}}{4} + \varepsilon_e\sqrt{4-3\cos^2\theta},\\
&\left|\frac{3}{2},3\right\rangle = \cos\eta\left|\frac{3}{2},\frac{1}{2}\right\rangle + \sin\eta\left|\frac{3}{2},-\frac{1}{2}\right\rangle\\
&E\left(\frac{3}{2},1\right) = \frac{eQV_{zz}}{4} - 3\varepsilon_e\cos\theta, \quad \left|\frac{3}{2},1\right\rangle = \left|\frac{3}{2},-\frac{3}{2}\right\rangle\\
&E\left(\frac{3}{2},4\right) = \frac{eQV_{zz}}{4} + 3\varepsilon_e\cos\theta, \quad \left|\frac{3}{2},4\right\rangle = \left|\frac{3}{2},+\frac{3}{2}\right\rangle
\end{aligned} \tag{2.74}$$

where

$$\tan^2\eta = \frac{(4-3\cos^2\theta)^{1/2} - \cos\theta}{(4-3\cos^2\theta)^{1/2} + \cos\theta}.$$

2.5
Polarization of γ-Radiation

From the viewpoint of electromagnetic waves, polarized γ-rays behave basically the same as polarized visible light, except that γ-rays have more of a particle character. In optics, polarization is described by the vibration of the electric field vector $\boldsymbol{E}$. Suppose an electromagnetic wave travels in the z-direction, and all $\boldsymbol{E}$ vectors are in one particular plane, then this wave has a plane polarization or linearly polarization. If the $\boldsymbol{E}$ vectors at different locations form a perfect helix around the z-axis, then the wave is circularly polarized, with either a left-handed or a right-handed helicity. For a particle, polarization is described by its spin. The polarization of a γ-ray can be defined by its helicity h, which is the projection of the photon's angular momentum along the z-axis. Therefore, it can have two values, $h = \pm 1$, corresponding to right- and left-circularly polarized γ-rays.

When $h = +1$, we use a polarization vector $\mathbf{e}_+$ and the state $|+1\rangle$ to describe the right-circularly polarized γ-ray. Similarly, when $h = -1$, we use $\mathbf{e}_-$ and $|-1\rangle$ to describe the left-circularly polarized γ-ray. The two vectors $\mathbf{e}_+$ and $\mathbf{e}_-$ form an orthogonal basis for circular polarization. Two other vectors $\mathbf{e}_x$ and $\mathbf{e}_y$ can also be taken as an orthogonal basis for linear polarization and $\mathbf{e}_\pm$ can be written in terms of $\mathbf{e}_x$ and $\mathbf{e}_y$,

$$\mathbf{e}_\pm = \mp \frac{1}{\sqrt{2}}(\mathbf{e}_x \pm i\mathbf{e}_y) \tag{2.75}$$

and

$$|\pm 1\rangle = \mp \frac{1}{\sqrt{2}}(|e_x\rangle \pm i|e_y\rangle). \tag{2.76}$$

The γ-rays as well as light are electromagnetic waves of very high frequencies; however, the above definition of helicity for right- and left-circularly polarized γ-rays is opposite to that in optics where $h = +1$ and -1 correspond to left- and right-circularly polarized light. As is known in optical spectroscopy, an external magnetic field $\boldsymbol{B}_0$ makes the emitted light linearly π- or σ-polarized, depending on whether $\boldsymbol{E} \,/\!/\, \boldsymbol{B}_0$ or $\boldsymbol{E} \perp \boldsymbol{B}_0$. These conventions are also followed in Mössbauer spectroscopy, but it is more often to use the polarization plane or the polarization vector to describe linear polarization. For instance, the polarization plane of the γ-rays emitted from the $\Delta m = \pm 1$ transitions is perpendicular to that of the γ-rays emitted from the $\Delta m = 0$ transitions. The polarization vector of the former is parallel to $\boldsymbol{B}_0$, while that of the latter is perpendicular to $\boldsymbol{B}_0$.

Mössbauer experiments with polarized resonance radiation yield a considerable amount of information not otherwise easily available. In particular, it can be used to determine the signs of both the quadrupole coupling (eQV_{zz}) [36] and the magnetic hyperfine field B_{eff} [61], and to determine the magnetic structure [62, 63]. It is also this method that allows one to observe the resonant γ-ray Faraday effect [64] and to decompose poorly resolved spectra [65]. Investigation of polarization effects in resonant γ-ray diffraction, interference, and refraction leads to development of the γ-ray optics theory [66].

In this section, we will first describe the methods for producing multi-line and single-line polarized sources, followed by a detailed analysis of Mössbauer spectra measured with these two kinds of γ-ray radiations.

2.5.1 Polarized Mössbauer Sources

The methods to produce polarized Mössbauer γ-rays from radioactive sources are almost invariably based on hyperfine interactions that split nuclear energy levels into sublevels. The γ-rays emitted from these excited sublevels are all or partially polarized. One has to use the hyperfine fields because external magnetic fields in excess of 33 T are not easily obtainable today, and it is extremely difficult to produce an external EFG that would result in discernible ΔE_Q in the spectrum.

1. *Multi-line sources.* The effective fields B_{eff} at all of the ^{57}Co nuclei are aligned by a weak external magnetic field applied to a $^{57}Co/\alpha$-Fe source. The γ-rays emitted in a direction perpendicular to the magnetic field ($\theta_s = 90°$) will give a spectrum of six lines, all linearly polarized. The $\Delta m = \pm 1$ transitions have a π-polarization and the $\Delta m = 0$ transitions have a σ-polarization; the polarization plane of the former being perpendicular to that of the latter. If we look at the γ-rays emitted in the direction of the applied magnetic field ($\theta_s = 0°$), then the system has an axial symmetry. Angular momentum along the magnetic field direction should be conserved, and thus those spectral lines due to the $\Delta m = 0$ transitions will disappear. Consequently, the spectrum has only four lines, among which two are left-circularly polarized and two are right-circularly polarized. In an arbitrary observation direction ($0° < \theta_s < 90°$), the emitted γ-rays are elliptically polarized.

2. *Single-line sources.* Using a magnetized filter to produce monochromatic polarized γ-rays is the most promising approach. The filter is placed between an unpolarized single line source and the absorber, and is moving at a constant Doppler velocity with respect to the source such that γ-rays with one polarization are resonantly absorbed and the remaining γ-rays with the other polarization are allowed to pass and reach the sample absorber. Thus, the filter magnetized by an external magnetic field serves as a polarizer. The first successful construction of such a source was reported in 1969 [67]. However, the required additional velocity transducer makes the apparatus quite complicated. Several source-polarizer systems in which relative motion is not required have since been found. This means the resonance absorption in the polarizer occurs at zero Doppler velocity. An example of such a system used for producing linear polarization is a $^{57}Co/CoO$ source with an Fe–Rh–Ni polarizer [68]. The ^{57}Fe nucleus in the CoO matrix has a very large isomer shift so that its γ-rays can be in resonance with the fourth line of the sextet of the Fe–Rh–Ni polarizer. The polarizer is glued to the source and both are mounted on the standard Mössbauer transducer. In general, the radiation transmitted through the polarizing filter contains the complementary or orthogonal polarization to that absorbed by the filter. When an external magnetic field applied to this polarizer is perpendicular to the γ-ray direction, a beam of monochromatic radiation of 80% linear polarization is obtained. Another system used for producing circular polarized radiation has a $^{57}Co/Cr$ source and an Fe–Si polarizer, which seems to be a better combination [69]. Such a source is shown schematically in Fig. 2.17. It is found that the mismatch between the third absorption line of $^{57}Fe_{2.85}Si_{1.15}$ and the emission line of $^{57}Co/Cr$ does not exceed 0.01 mm s^{-1}. This system has many advantages. For example, Fe in Cr matrix has a relatively narrow single line, so Cr is used as a standard matrix; in Fe–Si, the separation between the absorption line and the nearest line with a different polarization is large. The polarizer of thickness 34 mg cm^{-2} is magnetized by an

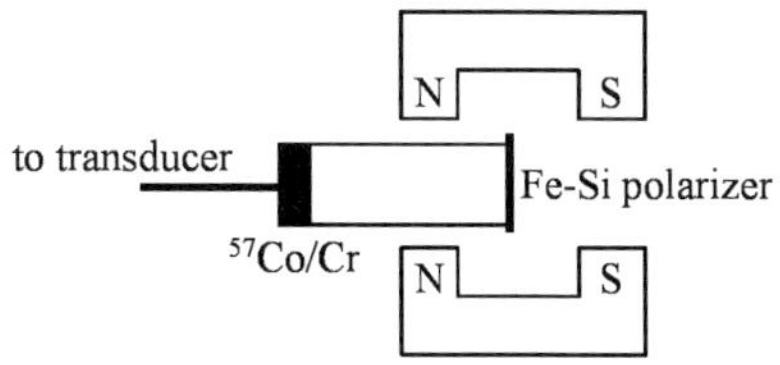

Fig. 2.17 Layout of the circularly polarized source.

applied longitudinal magnetic field of about 1 T. Using this system, monochromatic circularly polarized γ-rays with a $(80 \pm 2)\%$ polarization was obtained.

In addition, one can construct a polarized γ-ray source by using quadrupole interaction. In a case where the principal axis of the EFG is perpendicular to the direction of observation, the γ-rays from the $\pm 3/2 \to \pm 1/2$ transitions $(\Delta m = \pm 1)$ are linearly polarized, and the γ-rays from the $\pm 1/2 \to \pm 1/2$ transitions $(\Delta m = \pm 1, 0)$ are only partially linearly polarized.

In all these methods, a polarized source is obtained at the expense of the radiation intensity. Today, Mössbauer sources from synchrotron radiation (SR) are becoming more accessible, which usually have high intensity and high degree of polarization.

2.5.2
Absorption of Polarized γ-Rays

Now we focus our attention on a comparison of ^{57}Fe absorption spectra taken using a multi-line polarized source with those using a single-line polarized source. In each case, the absorber is magnetized. In Fig. 2.18, the γ-ray is traveling in the z-direction, and $\boldsymbol{B}_s$ and $\boldsymbol{B}_a$ represent the magnetic fields applied to the

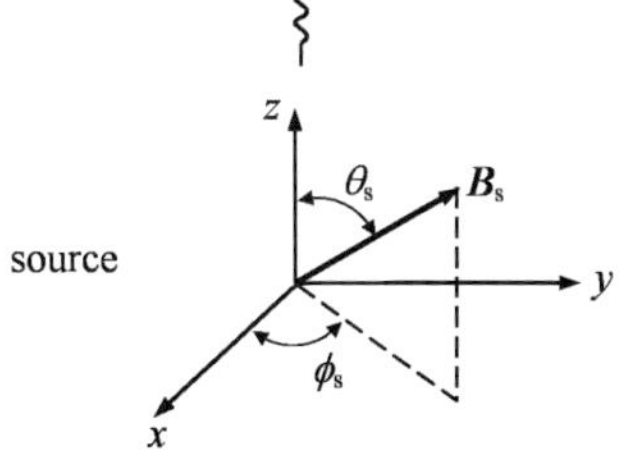

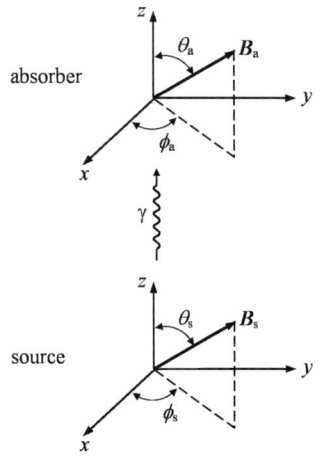

Fig. 2.18 Relative arrangement of the source and the absorber in external magnetic fields $\boldsymbol{B}_s$ and $\boldsymbol{B}_a$, respectively.

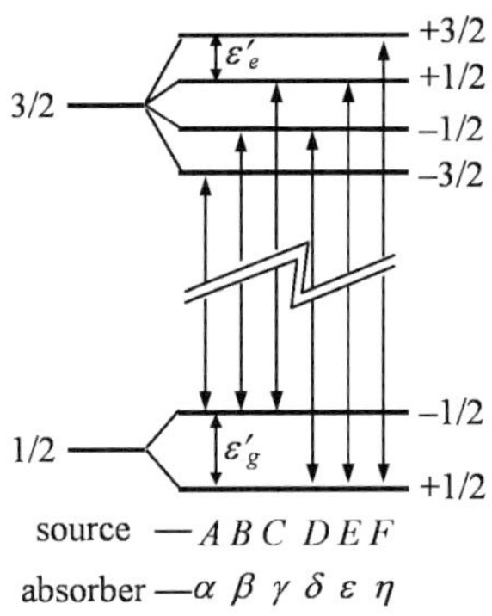

Fig. 2.19 Nuclear Zeeman sublevels and allowed transitions in ^{57}Fe.

source (or the polarizer) and absorber, respectively. We will designate the emission spectral lines as A, B, C, D, E, and F, and the absorption spectral lines as α, β, γ, δ, ε, and η (Fig. 2.19) in addition to the number notation in Fig. 2.14(a).

1. *Circularly polarized γ-rays.* In Fig. 2.18, if $\theta_s = \theta_a = 0$, the source (^{57}Co/α-Fe) and the absorber (α-Fe) are in collinear longitudinal magnetic fields. This multiline source emits only four spectral lines corresponding to the $\Delta m = \pm 1$ transitions. In the absorption process, angular momentum conservation in the magnetic field direction is also required; therefore, the helicity of the γ-ray ($h = +1$ or -1) will have to match the angular momentum change of the transition in the absorber. When the source is driven with the appropriate Doppler velocities, the total number of absorption spectral lines observed is not 16 (4×4), but 8 because of the constraints of angular momentum conservation. The positions and intensities of the 8 lines depend on the ratio B_a/B_s as indicated in a nomograph at the bottom of Fig. 2.20 (see also Table 2.5). The spectrum in the upper portion of Fig. 2.20 is observed when $\boldsymbol{B}_a = \boldsymbol{B}_s$ [61]. Let us analyze the origin of the spectral line on the right. When the source velocity corresponds to an energy ($|\varepsilon'_e| + |\varepsilon'_g|$), where ε'_e and ε'_g are defined in Fig. 2.19, emission line A has the same energy and helicity as the absorption line δ, and resonance absorption takes place. In the meantime, C is absorbed by η. Both these absorptions, $A_\delta + C_\eta$, occur at the same velocity. Similarly, the rest of the six absorptions superimpose to give two spectral lines, one in the center and one on the left, with an intensity ratio of 10:3. We can see from the nomograph that, when $B_a/B_s = 0$, there are four expected lines because of the absence of nuclear level splitting in the absorber. When $\boldsymbol{B}_a = -\boldsymbol{B}_s$, six absorption lines are expected.

If the single line source shown in Fig. 2.17 is used, the spectra become simpler. Four spectra taken from α-Fe and $HoFe_2$ absorbers are illustrated in Fig. 2.21, each of which consists of two lines despite that the fields $\boldsymbol{B}_s$ (acting only on the polarizer) and $\boldsymbol{B}_a$ are parallel ($\theta_s = \theta_a = 0$) or antiparallel ($\theta_s = 0$, $\theta_a = 180°$) [69].

Based on these spectra, we can discuss an important problem, namely to determine the sign of hyperfine magnetic field $\boldsymbol{B}_{hf}$. As we know, the sign of $\boldsymbol{B}_{hf}$ is reflected in the sense of the circularly polarized hyperfine transition γ-ray (i.e.,

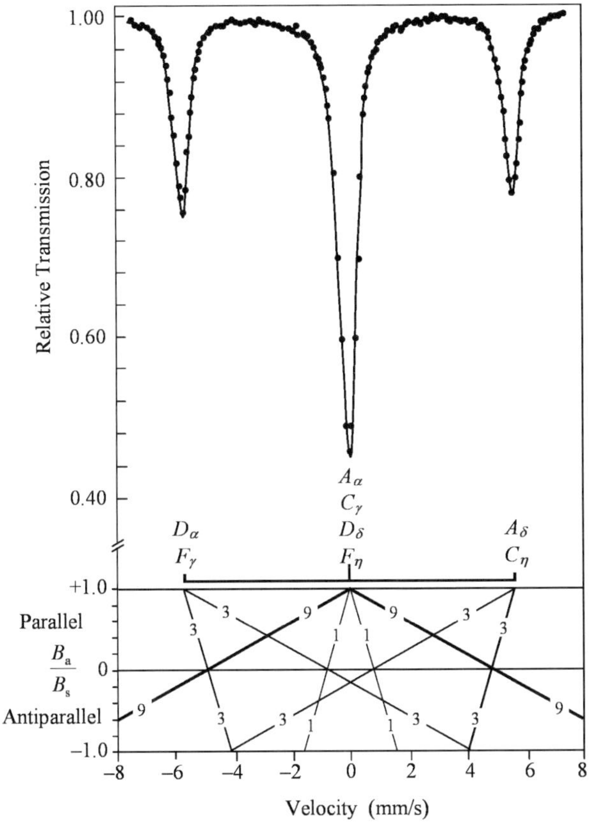

Fig. 2.20 Top: a Mössbauer spectrum using a $^{57}Co/\alpha\text{-}Fe$ source and an α-Fe absorber, both equally magnetized by an external longitudinal magnetic field of 5.2 T. Bottom: a nomogram showing the intensities of the spectral lines and their positions as functions of the ratio B_a/B_s (see Table 2.5) [70].

γ-ray helicity) along the field direction. In other words, if the helicity of emission or absorption γ-ray is found, the sign of $\boldsymbol{B}_{hf}$ in a source or an absorber can be unambiguously determined. The helicity of the emitted γ-ray can be analyzed by a longitudinally magnetized absorber whose $\boldsymbol{B}_{hf}$ direction is known. When the fields $\boldsymbol{B}_s$ and $\boldsymbol{B}_a$ are parallel, only lines 1 and 4 have large intensities. Traces of other lines come from incomplete polarization and incomplete alignment of the moments in the absorber. Since $\boldsymbol{B}_{hf}$ in α-Fe is negative, lines 1 and 4 have the helicity $h = +1$, thus the emitted γ-rays must also have $h = +1$. Consequently, the sign of $\boldsymbol{B}_{hf}$ in Fe–Si polarizer is determined to be negative. After reversing the direction of $\boldsymbol{B}_a$ ($\theta_a = 180°$), lines 3 and 6, as expected, become dominant. Similarly, the sign of $\boldsymbol{B}_{hf}$ in the ferrimagnetic $HoFe_2$ has been determined to be positive as evidently shown in Fig. 2.21. As far as unambiguously determining

Table 2.5 Relative intensities in Mössbauer spectra using polarized sources and thin absorbers, either of which may be a magnetized ferromagnet with the Mössbauer nuclei in unique sites with an axially symmetric crystal field [70].

Circularly polarized γ-ray source ($\theta_s = 0°$)

$\theta_a = 0°$, $\Delta\phi$ indeterminate					$\theta_a = 180°$, $\Delta\phi$ indeterminate				
	α	γ	δ	η		α	γ	δ	η
A	9	0	3	0	A	0	3	0	9
C	0	1	0	3	C	3	0	1	0
D	3	0	1	0	D	0	1	0	3
F	0	3	0	9	F	9	0	3	0

linearly polarized γ-ray source ($\theta_s = 90°$)

θ_a and $\Delta\phi$ variable	α, η	β, ε	γ, δ
A, F	$9(1 - \sin^2\theta_a \sin^2\Delta\phi)$	$12\sin^2\theta_a \sin^2\Delta\phi$	$3(1 - \sin^2\theta_a \sin^2\Delta\phi)$
B, E	$12(1 - \sin^2\theta_a \cos^2\Delta\phi)$	$16\sin^2\theta_a \cos^2\Delta\phi$	$4(1 - \sin^2\theta_a \cos^2\Delta\phi)$
C, D	$3(1 - \sin^2\theta_a \sin^2\Delta\phi)$	$4\sin^2\theta_a \sin^2\Delta\phi$	$(1 - \sin^2\theta_a \sin^2\Delta\phi)$

	$\theta_a = 90°$, $\Delta\phi = 0°$			$\theta_a = 90°$, $\Delta\phi = 90°$			$\theta_a = 0°$, $\Delta\phi$ ind.		
	α, η	β, ε	γ, δ	α, η	β, ε	γ, δ	α, η	β, ε	γ, δ
A, F	9	0	3	0	12	0	9	0	3
B, E	0	16	0	12	0	4	12	0	4
C, D	3	0	1	0	4	0	3	0	1

the sign of $\boldsymbol{B}_{hf}$ is concerned, a partially circularly polarized γ-radiation may be sufficient, provided the spectrum changes appreciably when the applied $\boldsymbol{B}_a$ (or $\boldsymbol{B}_s$) is reversed [71].

2. *Linearly polarized γ-rays.* When $\theta_s = 90°$, the six emission lines from a magnetized ^{57}Co/α-Fe are all linearly polarized. For simplicity, we will assume $\phi_s = 90°$. In this case, four of six lines are of the π-type ($\Delta m = \pm 1$), with the electric field vector $\boldsymbol{E}$ parallel to $\boldsymbol{B}_s$, i.e., the polarization plane is in the yz-plane. The other two lines are of the σ-type ($\Delta m = 0$), $\boldsymbol{E} \perp \boldsymbol{B}_s$, with the polarization plane in the xz-plane. In general, using such a source for a magnetically ordered absorber would result in a total of 36 absorption lines. But if we limit ourselves to the cases

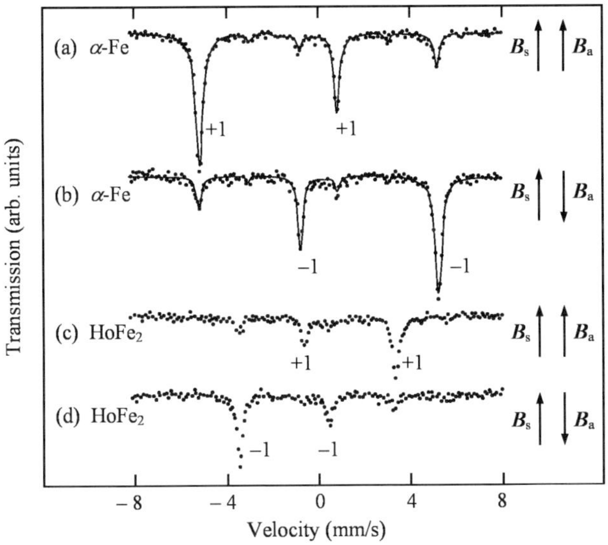

Fig. 2.21 Spectra of α-Fe and $HoFe_2$ measured with a single-line circularly polarized source shown in Fig 2.17 when the fields $\boldsymbol{B}_s$ and $\boldsymbol{B}_a$ are parallel (a, c) and antiparallel (b, d). The helicity of each line is indicated by +1 or −1.

where the magnetic hyperfine field in the absorber is also perpendicular to the γ-ray direction ($\theta_a = 90°$), the spectrum will be greatly simplified, especially when $\boldsymbol{B}_s \,/\!/\, \boldsymbol{B}_a$ or $\boldsymbol{B}_s \perp \boldsymbol{B}_a$. Figure 2.22 shows Mössbauer spectra obtained in these two special arrangements [70].

Now that we have linearly polarized γ-rays perpendicular to both $\boldsymbol{B}_s$ and $\boldsymbol{B}_a$, it is impossible to use simply angular momentum conservation to determine whether an emitted line would be absorbed or transmitted. We may use the intensity of each absorption line calculated in Table 2.5 to determine how many lines are expected in a spectrum. But here we describe a more intuitive graphical method to analyze them. Take the emission lines A and B as examples. Line A has a π-polarization with its polarization plane parallel to $\boldsymbol{B}_s$, while line B has a σ-polarization with its polarization plane perpendicular to $\boldsymbol{B}_s$. The squares in Fig. 2.23 represent the respective polarization planes. There are two types of absorption lines, one with its polarization plane parallel to $\boldsymbol{B}_a$ and the other perpendicular to $\boldsymbol{B}_a$. When $\boldsymbol{B}_s \,/\!/\, \boldsymbol{B}_a$, A will cause four absorption lines: A_η, A_δ, A_γ, and A_α. Although the other two lines A_β and A_ε meet the energy requirement, they are not absorbed. Now we change the magnetization of the absorber so that $\boldsymbol{B}_s \perp \boldsymbol{B}_a$, and we expect that the originally transmitted lines (A_β and A_ε) be absorbed and the originally absorbed lines will go through. For emission line B, similar diagrams in Fig. 2.23 will help us determine which lines are absorbed when $\boldsymbol{B}_s \,/\!/\, \boldsymbol{B}_a$ and $\boldsymbol{B}_s \perp \boldsymbol{B}_a$. All told, 20 out of the possible 36 spectral lines will

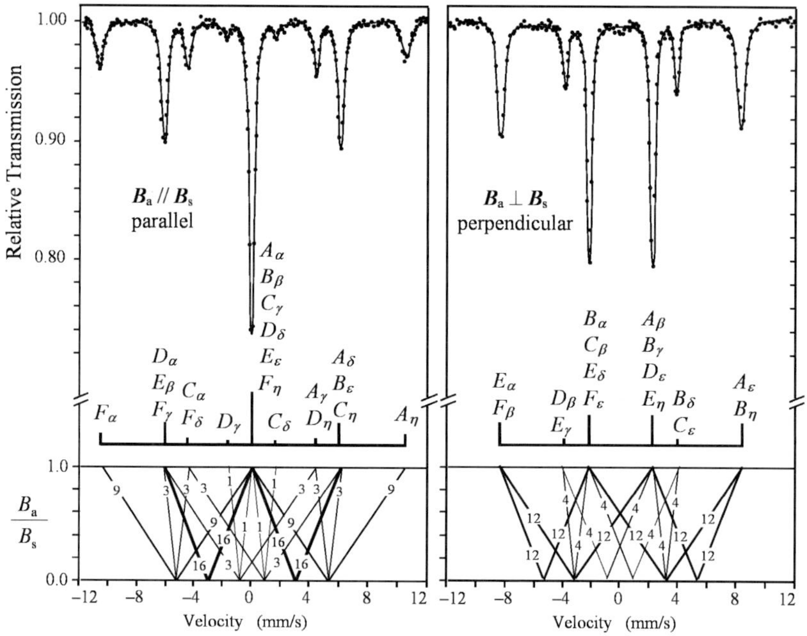

Fig. 2.22 Mössbauer spectra obtained with a $^{57}Co/\alpha$-Fe source and an α-Fe absorber, both at room temperature and equally magnetized perpendicular to the γ-ray direction ($B_a/B_s = 1$, $\theta_s = \theta_a = 90°$). Left: $\boldsymbol{B}_s$ and $\boldsymbol{B}_a$ are parallel ($\phi_s = \phi_a$). Right: $\boldsymbol{B}_s$ and $\boldsymbol{B}_a$ are perpendicular ($\phi_s - \phi_a = 90°$). The letters next to the stick diagrams indicate the origins of the lines. The nomograms at the bottom show the positions and intensities of the lines for any ratio of B_a/B_s.

appear if $\boldsymbol{B}_s \mathbin{/\!/} \boldsymbol{B}_a$, and 16 lines will appear if $\boldsymbol{B}_s \perp \boldsymbol{B}_a$. In cases where $\boldsymbol{B}_s$ and $\boldsymbol{B}_a$ have equal magnitudes, some lines are superimposed, resulting in spectra as shown in Fig. 2.22. We can clearly see a trend that when the source and absorber are both magnetized, an emission line resonates with an absorption line if their polarization planes are parallel, and the emission line will transmit if the these polarization planes are perpendicular. This selective absorption based on polarization is identical to the phenomenon of dichromism in optics.

If the above multi-line source is replaced by the single-line source consisting of $^{57}Co/CoO$ and the Fe–Rh–Ni polarizer, the spectra in Fig. 2.22 will change into that shown in Fig. 2.24 [68]. As can be seen from these comparisons, using a monochromatic source significantly simplifies the complex spectra, and this is of the greatest importance in practice. In fact, the intensity and position of a line in Fig. 2.22 are strongly dependent on the ratio B_a/B_s. Moreover, if the absorber is not simply α-Fe, but $SrFe_{12}O_{19}$ or $Nd_2Fe_{14}B$, the measured spectra can hardly be decomposed. In other words, the practical application of a multi-line source may be limited, and a single-line source has a unique advantage.

22 E.V. Mielczarek and D.A. Papaconstantopoulos. Isomer shift and charge density in FeAl and the ^{57}Fe isomer shift. *Phys. Rev. B* 17, 4223–4227 (1978).

23 W.C. Nieuwport, D. Post, and P.Th. van Duijnen. Calibration constant for ^{57}Fe Mössbauer isomer shifts derived from *ab initio* self-consistent-field calculations on octahedral FeF_6 and $Fe(CN)_6$ clusters. *Phys. Rev. B* 17, 91–98 (1978).

24 J.P. Sanchez, J.M. Friedt, A. Trautwein, and R. Reschke. Electronic charge and spin distribution in some iron halides from the interpretation of the ^{57}Fe and ^{129}I hyperfine interactions. *Phys. Rev. B* 19, 365–375 (1979).

25 Q.M. Zhang, Y.L. Zhang, and D.S. Wang. Isomer shift calibration of Fe nucleus by self-consistent all-electron LAPW band calculation. *Commun. Theor. Phys.* 8, 139–151 (1987).

26 O. Eriksson and A. Svane. Isomer shift and hyperfine fields in iron compounds. *J. Phys.: Condens. Matter* 1, 1589–1599 (1989).

27 T.E. Cranshaw, B.W. Dale, G.O. Longworth, and C.E. Johnson. *Mössbauer Spectroscopy and its Applications* (Cambridge University Press, Cambridge, 1985).

28 B.V. Thosar, P.K. Iyengar, J.K. Srivastava, and S.C. Bhargava (Eds.). *Advances in Mössbauer Spectroscopy: Applications to Physics, Chemistry and Biology*, p. 32 (Elsevier, Amsterdam, 1983).

29 U. Gonser (Ed.). *Mössbauer Spectroscopy* (Springer-Verlag, New York, 1975).

30 G.M. Bancroft. *Mössbauer Spectroscopy: An Introduction for Inorganic Chemists and Geochemists* (Wiley, New York, 1973).

31 R. Ingalls. Electric-field gradient tensor in ferrous compounds. *Phys. Rev.* 133, A787–A795 (1964).

32 Y.L. Chen, B.F. Xu, J.G. Chen, and Y.Y. Ge. Fe^{2+}–Fe^{3+} ordered distribution in chromite spinels. *Phys. Chem. Minerals* 19, 255–259 (1992).

33 S.V. Karyagin. A possible cause for the doublet component asymmetry in the Mössbauer absorption spectrum of some powdered tin compounds. *Proc. Acad. Sci. USSR Phys. Chem. Sect.* 148, 110–112 (1963) [Russian original: *Doklady Akad. Nauk* 148, 1102–1105 (1963)].

34 P. Zory. Nuclear electric-field gradient determination utilizing the Mössbauer effect (Fe^{57}). *Phys. Rev.* 140, A1401–A1407 (1965).

35 A.L. Alimuddin and K.R. Reddy. On the polarization and angular distribution of *M*1 and *E*2 radiations in the Mössbauer resonance. *Nuovo Cimento B* 32, 389–406 (1976).

36 C.E. Johnson, W. Marshall, and G.J. Perlow. Electric quadrupole moment of the 14.4-keV state of Fe^{57}. *Phys. Rev.* 126, 1503–1506 (1962).

37 R.L. Collins. Mössbauer studies of iron organometallic complexes. IV. Sign of the electric-field gradient in ferrocene. *J. Chem. Phys.* 42, 1072–1080 (1965).

38 R.L. Collins and J.C. Travis. The electric field gradient tensor. In *Mössbauer Effect Methodology*, vol. 3, I.J. Gruverman (Ed.), pp. 123–161 (Plenum Press, New York, 1967).

39 R.W. Grant, H. Wiedersich, A.H. Muir Jr., U. Gonser, and W.N. Delgass. Sign of the nuclear quadrupole coupling constants in some ionic ferrous compounds. *J. Chem. Phys.* 45, 1015–1019 (1966).

40 L. Häggström, A. Narayanasamy, T. Sundqvist, and A. Yousif. The sign of the electric field gradient in FeGe and FeSn systems. *Solid State Commun.* 44, 1265–1267 (1982).

41 S.S. Hanna. The discovery of the magnetic hyperfine interaction in the Mössbauer effect of ^{57}Fe. In *Mössbauer Spectroscopy II: The Exotic Side of the Method*, U. Gonser (Ed.), pp. 185–190 (Springer-Verlag, Berlin, 1981).

42 S.S. Hanna, J. Heberle, C. Littlejohn, G.J. Perlow, R.S. Preston, and D.H. Vincent. Polarized spectra and hyperfine structure in Fe^{57}. *Phys. Rev. Lett.* 4, 177–180 (1960).

43 S. DeBenedetti. *Nuclear Interactions* (John Wiley, New York, 1964).

44 G.K. Wertheim, H.J. Guggenheim, and D.N.E. Buchanan. Sublattice magnetization in FeF_3 near the critical point. *Phys. Rev.* 169, 465–470 (1968).

45 G.J. Perlow, C.E. Johnson, and W. Marshall. Mössbauer effect of Fe^{57} in a cobalt single crystal. *Phys. Rev.* 140, A875–A879 (1965).

46 V.G. Bhide. *Mössbauer Effect and Its Applications*, p. 179 (Tata McGraw-Hill, New Delhi, 1973).

47 R.W. Grant. Mössbauer spectroscopy in magnetism: characterization of magnetically-ordered compounds. In *Mössbauer Spectroscopy*, U. Gonser (Ed.), pp. 97–137 (Springer-Verlag, New York, 1975).

48 F. van der Woude and I. Vincze. Magnetism from Mössbauer spectroscopy. *J. de Physique (Colloque)* 41, C1-151–154 (1980).

49 J. Chappert. Mössbauer spectroscopy and magnetism. *Hyperfine Interactions* 13, 25–43 (1983).

50 W. Marshall. Orientation of nuclei in ferromagnets. *Phys. Rev.* 110, 1280–1285 (1958).

51 W. Kündig. Evaluation of Mössbauer spectra for ^{57}Fe. *Nucl. Instrum. Methods* 48, 219–228 (1967).

52 J.R. Gabriel and S.L. Ruby. Computation of Mössbauer spectra. *Nucl. Instrum. Methods* 36, 23–28 (1965).

53 K.A. Hardy, D.C. Russell, R.M. Wilenzick, and R.D. Purrington. A computer code for evaluation of complex Mössbauer spectra. *Nucl. Instrum. Methods* 82, 72–76 (1970).

54 G.R. Hoy and S. Chandra. Effective field parameters in iron Mössbauer spectroscopy. *J. Chem. Phys.* 47, 961–965 (1967).

55 J.R. Gabriel and D. Olson. The computation of Mössbauer spectra, II. *Nucl. Instrum. Methods* 70, 209–212 (1969).

56 E. Kreber and U. Gonser. Evaluation of relative line intensities in Mössbauer spectroscopy. *Nucl. Instrum. Methods* 121, 17–23 (1974).

57 L. Häggström. Determination of hyperfine parameters from $1/2 \rightarrow 3/2$ transitions in Mössbauer spectroscopy. Report UUIP–851, Uppsala, Sweden (1974). L. Häggström, T. Ericsson, R. Wäppling, and E. Karlsson. Mössbauer study of hexagonal FeGe. *Physica Scripta* 11, 55–59 (1975).

58 K.M. Hasselbach and H. Spiering. The average over a sphere. *Nucl. Instrum. Methods* 176, 537–541 (1980).

59 N. Blaes, H. Fischer, and U. Gonser. Analytical expression for the Mössbauer line shape of ^{57}Fe in the presence of mixed hyperfine interactions. *Nucl. Instrum. Methods B* 9, 201–208 (1985).

60 G.A. Korn and T.M. Korn. *Mathematical Handbook for Scientists and Engineers* (McGraw-Hill, New York, 1968).

61 N. Blum and L. Grodzins. Sign of the magnetic hyperfine field in dilute iron alloys using the Mössbauer effect. *Phys. Rev.* 136, A133–A137 (1964).

62 U. Gonser, R.W. Grant, H. Wiedersich, and S. Geller. Spin orientation determination by linearly polarized, recoil-free γ-rays. *Appl. Phys. Lett.* 9, 18–21 (1966).

63 K. Szymanski. Magnetic texture determination by means of the monochromatic circularly polarized Mössbauer spectroscopy. *Nucl. Instrum. Methods B* 134, 405–412 (1998).

64 P. Imbert. Etude des phénomènes de dispersion associés aux raies d'absorption Mössbauer de ^{57}Fe. *J. de Physique* 27, 429–432 (1966).

65 I. Vincze. Evaluation of complex Mössbauer spectra in amorphous and crystalline ferromagnets. *Solid State Commun.* 25, 689–693 (1978).

66 J.P. Hannon, N.J. Carron, and G.T. Trammell. Mössbauer diffraction. III. Emission of Mössbauer γ rays from crystals. *Phys. Rev. B* 9, 2791–2831 (1974).

67 S. Shtrikman and S. Somekh. Mössbauer spectroscopy with monochromatic circularly polarized

radiation. *Rev. Sci. Instrum.* 40, 1151–1153 (1969).

68 J.P. Stampfel and P.A. Flinn. A simple monochromatic, polarized Fe^{57} Mössbauer source. In *Mössbauer Effect Methodology*, vol. 6, I.J. Gruverman (Ed.), pp. 95–107 (Plenum Press, New York, 1971).

69 K. Szymanski, L. Dobrzynski, B. Prus, and M.J. Cooper. A single line circularly polarized source for Mössbauer spectroscopy. *Nucl. Instrum. Methods B* 119, 438–441 (1996).

70 U. Gonser and H. Fischer. Resonance γ-ray polarimetry. In *Mössbauer Spectroscopy II: The Exotic Side of the Method*, U. Gonser (Ed.), pp. 99–137 (Springer-Verlag, New York, 1981).

71 K. Szymanski. ^{57}Co in Cr matrix as a single line polarised source. In *International Conference on the Applications of the Mössbauer Effect, ICAME-95* (Italian Physical Society Conference Proceedings No. 50), I. Ortalli (Ed.), pp. 891–894 (Editrice Compositori, Bologna, 1996).

72 B. Kolk, A.L. Bleloch, and D.B. Hall. Recoilless fraction studies of iron near the Curie temperature. *Hyperfine Interactions* 29, 1377–1380 (1986).

3
Experimental Techniques

The development and advancement of Mössbauer spectroscopy represents one of the great achievements in experimental physics. To someone new to Mössbauer spectroscopy, this ingenious experimental method often seems mysterious as to why it can offer an energy resolution of the same order of magnitude as the "natural width" of the energy level. In this chapter, we first describe the principle of energy modulation using Doppler velocity, which is a key step in observing a Mössbauer spectrum. This technique is well developed and well documented in the literature [1, 2]. Next, we describe the Mössbauer radiation sources and the γ-ray detectors. These sources and detectors must possess certain particular properties and are specially prepared. The data acquisition system is relatively simple, which we briefly deal with.

3.1
The Mössbauer Spectrometer

To measure the characteristics of any resonance phenomenon, e.g., the resonance curve of an LC circuit, one must have a signal generator whose frequency can be continuously adjusted. To obtain the resonance curve of a nucleus absorbing γ-rays, the energy of the incoming γ-ray must also be modulated. This is achieved using the Doppler effect, in which the perceived frequency of a wave is different from the emitted frequency if the source is moving relative to the receiver. Suppose the source and the receiver have a relative velocity of v, then the perceived frequency of the γ-radiation is

$$f = f_0\left(1 + \frac{v}{c}\cos\theta\right)\left(1 - \frac{v^2}{c^2}\right)^{-1/2} \tag{3.1}$$

where f_0 is the frequency of the radiation when the source is at rest, c is the speed of light, and θ is a small angle between the relative velocity and the γ-ray direction. To obtain a typical Mössbauer spectrum, $v_{\max} < 1$ m s^{-1}, thus $v/c \ll 1$, and a very good approximation of the above equation is

Mössbauer Effect in Lattice Dynamics. Yi-Long Chen and De-Ping Yang

ISBN: 978-3-527-40712-5

$$\Delta f = f - f_0 = f_0 \frac{v}{c} \cos\theta,$$

or

$$\Delta E = E_0 \frac{v}{c} \cos\theta. \tag{3.2}$$

In principle, the energy of the emitted γ-rays may be changed by either raising the source temperature or applying an external magnetic field to cause Zeeman splitting of the nuclear energy levels in the source [3]. But both have serious limitations and have never become widely adopted. At the present time, almost every Mössbauer spectrometer has a velocity transducer based on Eq. (3.2), modulating the γ-ray energies in order to observe the resonance curve. In most cases, the source undergoes a mechanical motion, whereas the absorber is at rest so that it is easier to change its temperature or to apply an external magnetic field to the absorber.

The velocity transducers are generally operated in two modes: constant velocity and velocity scan. The first is the simplest, developed in the early 1960s. In this case the spectrum is recorded "point by point" throughout the selected velocities provided that the measurement time interval at each velocity is fixed. The Mössbauer spectrometers used at the present time are almost exclusively constructed using the second mode, in which the source scans periodically through the velocity range of interest. If every increment/decrement in velocity between adjacent points is the same, the source motion must have a constant acceleration, and the velocity-scanning spectrometer becomes a constant-acceleration one. For recording the transmitted γ-rays, each velocity has its own register (usually called a channel) which is sequentially held open for a fixed, short time interval synchronized with the velocity scan. The number of channels, i.e., the number of velocity points, is usually chosen to be 256, 512, or sometimes 1024, etc.

Figure 3.1 shows a block diagram of a velocity-scanning spectrometer in transmission geometry. It consists of a radiation source, an absorber, a detector with its electronic recording system, a clock signal and a function generator, a drive circuit, and a transducer.

The radiation source is not monochromatic. For example, in addition to emitting the 14.4-keV γ-rays, a ^{57}Co source also emits γ-rays and x-rays of other energies (see Section 3.3). In order to pick out the signal due to the 14.4-keV radiation, a single-channel analyzer (SCA) is placed behind the amplifier. Figure 3.2 shows various control signals and an observed spectrum.

The clock generates a synchronizing signal, which sets the starting moment ($t = 0$) for velocity scanning. A triangular wave from the waveform generator begins to increase (decrease) from its minimum (maximum), and the first channel also begins to open. After that, each channel is opened in turn by an advance pulse alone. The velocity of the transducer is scanned linearly from $-v_m$ to $+v_m$,

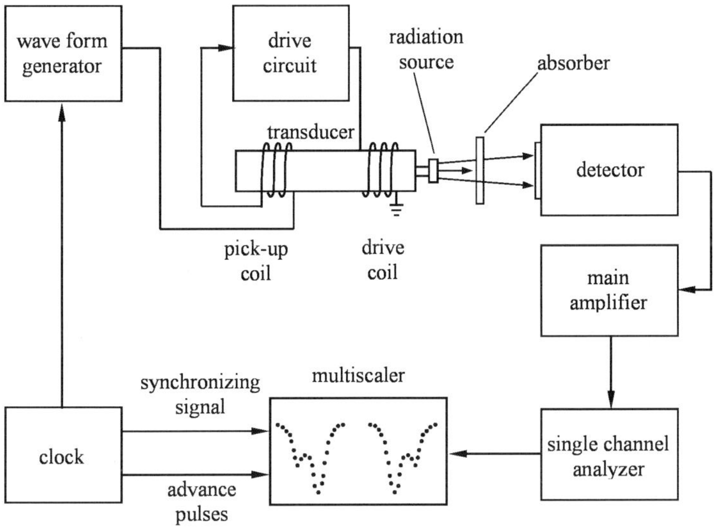

Fig. 3.1 Block diagram of a Mössbauer spectrometer in transmission geometry.

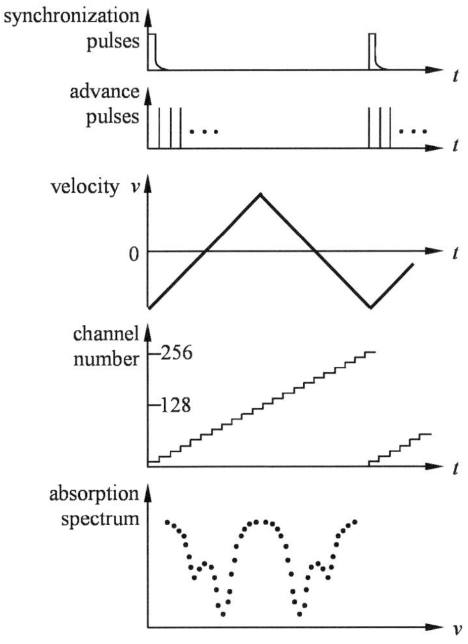

Fig. 3.2 Various control signals in a constant-acceleration spectrometer and an absorption spectrum.

and a spectrum taken during the linear ramp is stored in one half of the total channels. Then, the velocity decreases from $+v_m$ back to $-v_m$, completing a backward scan, during which the measured data are stored in the other half of the available channels. Therefore, in one scan period, a multiscaler or a computer will record two spectra, which are mirror images of each other. In order to obtain a spectrum with a good signal-to-noise ratio, hundreds of thousands of scans are usually necessary. An occasional synchronization problem would have no impact, because it recovers at the next scan period.

One obvious advantage of using a constant-acceleration spectrometer is that the stability requirement is not as strict as in a constant-velocity spectrometer. If instability, such as a discrimination voltage drift at SCA, should cause a decrease in the absorbed line intensities during one scan or several scans, it has a small effect on the absolute intensities but no effect on the positions and the shape of the spectral lines because this process is equivalent to shortening the experiment duration slightly. Another advantage is that this mode can make full use of digital technology, improving the properties of the spectrometer and allowing automatic data acquisition.

3.2 Radiation Sources

Among the isotopes in which the Mössbauer effect has been observed, ^{40}K is the lightest one. It is a pity that there exist no lighter Mössbauer isotopes. The Mössbauer isotopes are not distributed evenly, with three-quarters of them concentrated in elements with atomic numbers between 50 and 80. There are only a little over 20 Mössbauer isotopes that are in practical use, which amount to about one-quarter of the total number of Mössbauer isotopes. ^{57}Fe and ^{119}Sn are the most popular, whose decay schemes are shown in Fig. 3.3. ^{57}Fe is by far the most important one, for more than 69% of research work involves ^{57}Fe. The

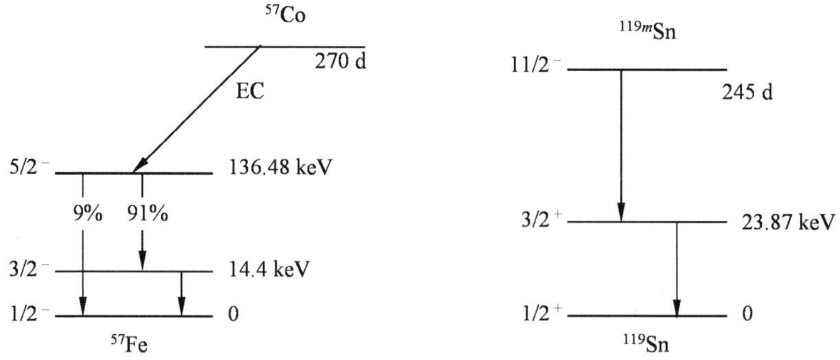

Fig. 3.3 Nuclear decay schemes of ^{57}Co and ^{119}Sn.

next few frequently used isotopes are ^{151}Eu, ^{197}Au, ^{129}I, ^{121}Sb, and ^{125}Te. Increased attention has been paid to ^{237}Np, ^{155}Gd, ^{161}Dy, and especially to ^{67}Zn and ^{181}Ta, which are used to obtain high-resolution spectra.

The quality of a Mössbauer source depends on the properties of the isotope and the host (matrix) material. The Mössbauer isotope and the host should satisfy the following criteria:

1. E_γ should be limited within 5 to 150 keV, preferably less than 50 keV. This is because both f and σ_0 decrease as E_γ increases, and especially f decreases more severely. For γ-rays of energies less than 5 keV, too much self-absorption makes Mössbauer radiation very weak.
2. It is desired that the half-life $T_{1/2}$ of the excited state should lie between 1 and 100 ns. If $T_{1/2}$ is too long, the natural width Γ_n of the excited state is very narrow and a slight mechanical vibration may destroy the resonance condition. Conversely, if $T_{1/2}$ is too short, Γ_n would be too large such that a spectrum with hyperfine structure may not be resolved.
3. The internal conversion coefficient α should be small (<10), to ensure a relatively large probability for γ-ray emission, which is especially important for the transmission geometry.
4. It is preferred that the parent nucleus has a long half-life, and allows for easy production of a high-activity source.
5. The Mössbauer isotope should not have a high spin, which would produce complicated spectra and make analysis more difficult.
6. The Mössbauer isotope should have a reasonably large natural abundance, so that isotopic enrichment in the absorber can be avoided. Except for its low natural abundance, ^{57}Fe satisfies the above criteria the best, followed by ^{119}Sn. A good Mössbauer source also requires an appropriate host, a suitable fabrication process, etc.
7. The radiation from the source should be monochromatic with an energy width as close to the natural width as possible ($\Gamma_s \approx \Gamma_n$). This requires that the host matrix material be a nonmagnetic crystal of cubic symmetry with a very low impurity content.
8. In order to have the highest possible f value, the host material should have a high Debye temperature. This is why metals or ionic crystals of high melting points are usually chosen as the matrix materials. Also, any nonequilibrium charge distribution in a metal would only last for less than 10^{-12} s, shorter than the life time of a typical Mössbauer excited state (between 10^{-6} and 10^{-10} s). Because of this, metals are better than ionic crystals.

9. The number of stable Mössbauer nuclei in the host material should be minimized; otherwise, the resonant self-absorption would broaden the emitted γ-rays.
10. The host material should be made very thin, to reduce the photoelectric effect and Compton scattering caused by the Mössbauer γ-rays.
11. The host should be chemically stable, so that it does not change its chemical composition or structure due to oxidation, hydrolysis, etc.

Satisfying the above conditions, a Mössbauer source would provide a high recoilless fraction f, a small line width ($\approx \Gamma_n$), and intense radiation. For ^{57}Fe, Rh and Pd are good hosts, giving an admirable f-value of about 0.784 at room temperature.

A Mössbauer source is usually custom-made with a particular radioactive isotope. A nuclear reaction in an accelerator or in a reactor produces a very highly excited state. It quickly decays to a relatively long-lived parent nucleus (still an excited state), which is then isolated and diffused into the host material. Other methods include the "in-beam" implantation, which could provide some of the difficult-to-produce isotopes such as ^{40}K or Mössbauer isotopes with short life times. In the recent years, researchers have also been attracted to synchrotron radiation as a new Mössbauer source.

3.3 The Absorber

In Mössbauer spectroscopy, the absorber is usually the sample to be investigated. In transmission geometry, the thickness of the absorber significantly affects the quality of the spectrum and must be carefully chosen. In this section, we mainly discuss this effect and the methods of correcting for it.

3.3.1 Estimation of the Optimal Thickness

The optimal sample thickness d_{opt} means such a thickness that would produce a minimum statistical uncertainty in the Mössbauer spectrum measured in a given time duration, and therefore provide the most accurate values for the spectral parameters. When the sample is too thin, it would contain too few Mössbauer nuclei. Consequently, the spectral intensity would be weak, with too much background and a large statistical uncertainty, and accurately extracting spectral parameters (especially absorption intensities) would be difficult. When the sample is too thick, significant atomic absorption would occur, causing resonant absorption to decrease. Also, as the sample thickness increases, the Lorentzian shape of the spectral lines would be gradually distorted. As the sample thickens to a certain extent, the Mössbauer effect can no longer be observed. Obviously, there should be an optimal thickness between these two extremes.

In studies of subjects such as lattice dynamics, phase analysis, distribution of anions, etc., it is necessary to measure the absorption intensity accurately. This demands a high-quality Mössbauer spectrum. Therefore, it is important to prepare a sample with an optimal thickness. Within three years of the discovery of the Mössbauer effect, there were several reports on studies of optimal sample thickness [3–5]. But at the present time, it is still very difficult to calculate the exact optimal thickness. It would require an understanding of all interactions between the γ-radiation and the sample, as well as the details of sample composition and structure. Therefore, we would be content with an approximate estimation of d_{opt}.

In order to do that, we need to select a physical quantity for judging whether the sample has the optimal thickness. Some use the height [6] or the area [7] of the spectral lines; others use the signal-to-noise ratio (S/N) [8–12]. When one of these quantities reaches a maximum, it is deemed to correspond to the optimal sample thickness. It seems that using S/N is a good method [8]. For convenience, Q (quality factor) is often used to represent S/N, with the following definition [11]:

$$Q = \frac{S}{N} = \frac{I(\infty, d) - I(v_{\mathrm{r}}, d)}{[(\Delta I(\infty, d))^2 + (\Delta I(v_{\mathrm{r}}, d))^2]^{1/2}} = \frac{I(\infty, d) - I(v_{\mathrm{r}}, d)}{[I(\infty, d) + I(v_{\mathrm{r}}, d)]^{1/2}}$$

where $I(\infty, d)$ and $I(v_{\mathrm{r}}, d)$ are the total γ counts off- and on-resonance, respectively. ΔI is the corresponding statistical uncertainty in I, and $\Delta I = \sqrt{I}$ because it is a random process.

For a very thin absorber, the difference between $I(\infty, d)$ and $I(v_{\mathrm{r}}, d)$ is very small, often less than 10% of $I(\infty, d)$, which means $I(\infty, d) + I(v_{\mathrm{r}}, d) \cong 2I(\infty, d)$. Therefore, we may redefine Q (within a constant factor) in a simpler manner [9, 13]:

$$\begin{aligned} Q &= \frac{S}{N} = \frac{I(\infty, d) - I(v_{\mathrm{r}}, d)}{[I(\infty, d)]^{1/2}} \\ &= f_{\mathrm{s}} {I_0}^{1/2} \mathrm{e}^{-\mu_{\mathrm{a}} d/2} [1 - \mathrm{e}^{-\mu_{\mathrm{r}} d/2} J_0(i\mu_{\mathrm{r}} d/2)] \\ &= f_{\mathrm{s}} {I_0}^{1/2} F(\mu_{\mathrm{a}}, \mu_{\mathrm{r}}, d) \end{aligned} \tag{3.3}$$

where I_0 is the total intensity of incident γ-rays, f_{s} is the recoilless fraction of the source, d is the sample thickness in mg cm^{-2} (note that $\mu_{\mathrm{r}} d = t_{\mathrm{a}}$ is the effective thickness), J_0 is the zeroth-order Bessel function, μ_{a} (in cm^2 mg^{-1}) is the atomic mass absorption coefficient, and μ_{r} (in cm^2 mg^{-1}) is the γ resonance absorption coefficient. Both μ_{a} and μ_{r} are defined in Chapter 1 [see Eq. (1.17)]:

$$\mu_{\mathrm{a}} = \sum_i p_i \mu_{\mathrm{a}}^i, \tag{3.4}$$

$$\mu_{\mathrm{r}} = r\sigma_0 f a_{\mathrm{m}} \frac{N_{\mathrm{A}}}{A_{\mathrm{m}}} n_0 \tag{3.5}$$

where p_i is the mass fraction of the ith element in the sample, σ_0 is the maximum cross-section of γ resonance absorption, a_m is the natural abundance of the Mössbauer isotope, N_A is Avogadro's number, A_m is the molecular mass (in mg) per mole, n_0 is the number of Mössbauer atoms in a molecule of the absorber compound (e.g., $n_0 = 3$ for Fe_3BO_6), and f is the recoilless fraction of the absorber. The factor r in Eq. (3.5) needs to be further explained. It is the weight factor to account for the intensity distribution in the case of hyperfine field splitting. For a single line absorber, $r = 1$. When there is a quadrupole splitting and the sample is polycrystalline, resulting in a doublet of equal intensity, one of the lines, which is used for calculating $I(v_r, d)$ in Eq. (3.3), corresponds to an effective thickness of $t_a/2$, so $r = 1/2$. If there is a magnetic hyperfine field in the sample (e.g., α-Fe_2O_3), we calculate $I(v_r, d)$ of line 3 or line 4 where the saturation effect is the weakest according to the discussion in Section 2.6. If the sample is polycrystalline, the saturation effect is negligible, thus the effective thickness of line 3 or line 4 is $t_a/12$, so $r = 1/12$. When the sample is a single crystal, r can be determined in a similar way.

One can rewrite Eq. (3.3) as

$$Q = \frac{I(\infty, d) - I(v_r, d)}{I(\infty, d)} \sqrt{I(\infty, d)} = \varepsilon(v_r) \Big/ \frac{1}{\sqrt{I(\infty, d)}} \tag{3.6}$$

where $1/\sqrt{I(\infty, d)}$ is the fractional standard deviation of the baseline counts, a most commonly used precision index in "counting" experiments, and $\varepsilon(v_r)$ is given by Eq. (1.24). Equation (3.3) is a general definition of the signal-to-noise ratio, but once it is rewritten as Eq. (3.6), we can see its physical significance more clearly. A better Q requires a combination of a large Mössbauer effect and a high precision in the measurement. Suppose we let $f_s[I_0]^{1/2} = c$ (a constant), $\mu_a d = x$, and $\mu_r/\mu_a = b$, then Eq. (3.3) becomes

$$Q(b, x) = c\mathrm{e}^{-x/2} \left[1 - \mathrm{e}^{-bx/2} \sum_{k=0}^{\infty} \frac{1}{(k!)^2} \left(\frac{bx}{4} \right)^{2k} \right] = cF(b, x) \tag{3.7}$$

where

$$F(b, x) = \mathrm{e}^{-x/2} \left[1 - \mathrm{e}^{-bx/2} \sum_{k=0}^{\infty} \frac{1}{(k!)^2} \left(\frac{bx}{4} \right)^{2k} \right]. \tag{3.8}$$

Now we treat b as a parameter, and plot $F(b, x)$, which is proportional to $Q(b, x)$, in Fig. 3.4. For every b, the curve has a maximum, and its corresponding x-value gives the product $\mu_a d_{opt}$. From given values of μ_a and μ_r, a computer program based on Eq. (3.7) may be used to calculate such a curve, find its maximum, and obtain the optimal thickness d_{opt}. To calculate the maximum of $Q(b, x)$ analytically is somewhat more complicated, if not impossible. Table 3.1 lists numerically calculated results for potassium ferrocyanide, sodium nitroprusside, and fer-

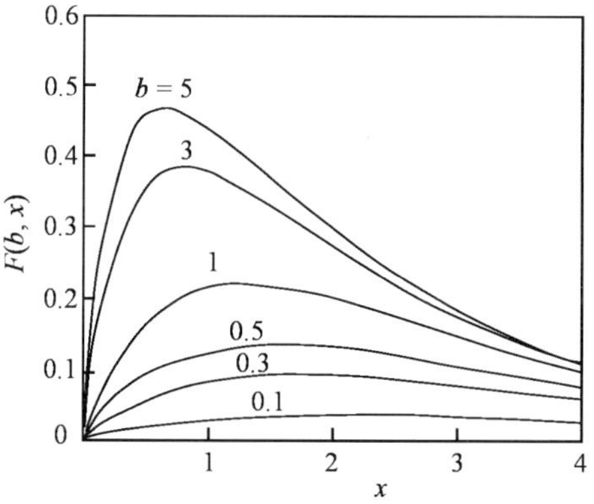

Fig. 3.4 Theoretical curves of $F(b,x)$ versus x for various b-values as indicated.

Table 3.1 Calculated d_{opt} values when f takes two different values for each of the three materials.

Absorber	μ_a (10^{-3} cm^2 mg^{-1})	μ_r (10^{-3} cm^2 mg^{-1})	f	d_{opt} (mg cm^{-2})
$K_4Fe(CN)_6{\cdot}3H_2O$	19.14	15.9	0.2	68
		23.8	0.3	63
$Na_2Fe(CN)_5NO{\cdot}2H_2O$	13.73	33.8	0.3	65
		56.25	0.5	51
$Fe(C_5H_5)_2$	19.79	9	0.05	76
		18	0.1	66

rocene, assuming f (or μ_r) is known and neglecting hyperfine splittings. In this approach, it is imperative to know the precise values of μ_a and μ_r. The μ_a values have been tabulated for all the elements (see Appendix H). A μ_r value may be calculated using Eq. (3.5), since we can easily get precise values for all the factors in Eq. (3.5) except for f. For a new sample under investigation, f is unknown. The f value is not precisely known even for many common materials. But f can be determined experimentally, and once f is known, d_{opt} can be readily calculated. To do this, we obtain a set of spectra from samples of the same material but with different thickness values d, to deduce the experimental Q_{exp} as a function of sample thickness. Now, Eq. (3.7) is used to fit the Q_{exp} values with f as the only adjustable parameter. The fit would give the f value, and the maximum of the fitted curve would give a value for d_{opt}. The results of this procedure applied to seven samples are listed in Table 3.2 and the fitted curves for six samples are shown in Fig. 3.5.

Table 3.2 Parameters obtained from fitting the $Q(b,x)$ curves for seven samples.

Sample	d_{opt} (mg cm^{-2})	μ_a (10^{-3} cm^2 mg^{-1})		f (295 K)	$\frac{1}{\mu_a} \sim \frac{2}{\mu_a}$	Ref.
	exp.	cal.	exp.			
$Fe(C_5H_5)_2$	81	19.79	19.36	0.08	50∼100	13
Cu_5FeS_4	25	63.39	59.82	0.77	15.8∼31.8	13
$FeSO_4{\cdot}7H_2O$	89	16.23	15.98	0.16	61.6∼123	13
SNP	79	13.73	12.64	0.37	73∼146	13
$K_3Fe(CN)_6$	72	21.02	20.25	0.10	47.6∼95	13
$K_4Fe(CN)_6{\cdot}3H_2O$	65	19.14	18.202	0.25	52∼144	16
α-Fe_2O_3	37	45.43	44.73	0.65	22∼44	16

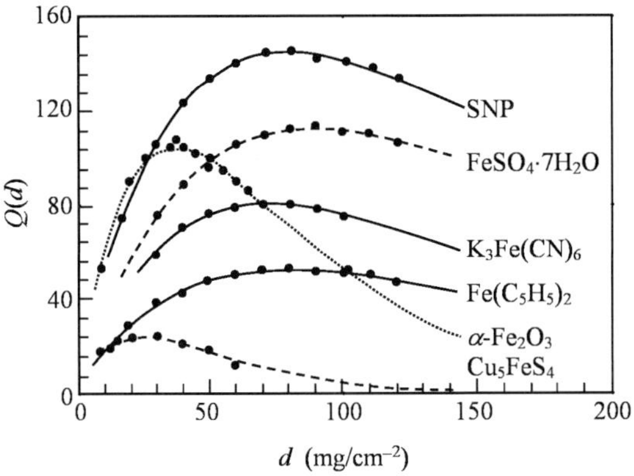

Fig. 3.5 Experimental $Q(d)$ data for the six materials indicated. The lines are fitted curves by Eq. (3.7).

First of all, the excellent fit between the experimental data and the calculated curve confirms that Eq. (3.7) describes Q as a function d very well. At room temperature, sodium nitroprusside (SNP, $Na_2Fe(CN)_5NO{\cdot}2H_2O$) has $f = 0.37$, which agrees exactly with the literature value [14]. For ferrocene ($Fe(C_5H_5)_2$), Table 3.2 gives $f = 0.08$, which is half of the value $f = 0.169$ reported in 1960 [15]. Later in Section 8.3, $f(295\ \text{K}) = 0.09$ is derived from known lattice dynamics parameters, in good agreement with this experiment. This will confirm that the above method for determining f is reliable, and the accuracy can reach 0.01 [16].

In Table 3.2, α-Fe_2O_3 is the only magnetic sample. As mentioned above, we only need to accurately measure the height of line 3 or line 4. Therefore, these experiments were done for velocities between -2.5 and $+2.5$ mm s^{-1}, and the result was $f = 0.65 \pm 0.03$.

The values for μ_a are available from Appendix H, but they can also be obtained from these experiments based on the relationship $I(\infty, d) = I(0)e^{-\mu_a d}$. Table 3.2 lists both the calculated and experimental values of μ_a for each sample, and they essentially agree with each other. If an experimental value turns out to be different from the calculated value beyond experimental uncertainty, there must have been some problems either with the sample composition or with the measurement process. These problems must be corrected before proceeding to using μ_a for estimating the optimal thickness.

When $\mu_r d$ is small, we may take the first two terms of the summation in Eq. (3.7), and take $e^{-\mu_r d/2} \approx 1 - \mu_r d/2$. At the maximum of the curve, $dQ/dx = 0$, we can derive the following approximate result for d_{opt}:

$$d_{opt} \approx \frac{2}{\mu_a}. \tag{3.9}$$

However, this is the upper limit of the sample thickness, and the optimal thickness should be thinner than $2/\mu_a$. An analysis has shown [11] that the optimal thickness is between $1/\mu_a$ and $2/\mu_a$. It is easy to see from data in Table 3.2 that the results evaluated using the above experimental method are indeed within this range.

Comparing the results in Table 3.1 with those in Table 3.2, one can find that neglecting the hyperfine splittings may cause the deduced d_{opt} to be underestimated when $b > 1$. Owing to this approximation the maximum of $Q(b, x)$ curve is generally shifted towards the coordinate origin. When $b < 1$, as we see below, such a shift does not affect d_{opt}, so the approximation just mentioned is admissible. For $b < 1$, those curves in Fig. 3.4 change very slowly in the vicinity of the maximum value Q_{max}. The smaller the b-value, the flatter is the curve. In this case, the thickness of the sample can be chosen in a relatively wide range between $Q_{max}(1 - \varepsilon)$ and $Q_{max}(1 + \varepsilon)$, with little difference in the quality of the spectrum. For example, if $\varepsilon = 0.01$, the corresponding sample thickness for ferrocene may range from 66 to 91 mg cm^{-2}. This provides certain flexibility in the sample preparation.

If the chemical composition of the sample is unknown, we may experimentally determine the lower limit of the sample thickness. Let $d = 1/\mu_a$, then by definition

$$\frac{I(\infty, d)}{I(0)} = \frac{1}{e}, \tag{3.10}$$

which means that when the transmitted intensity is $1/e$ of the incident intensity, the sample thickness is at the lower limit. However, the optimal thickness may be far above the lower limit (see Table 3.2).

3.3.2
Sample Preparation

Metallic or alloy samples are usually wrought or roll-milled into foils of the appropriate thickness. If only small pieces of sample foils are available, they may be arranged to cover the entire sample area with fewest gaps and overlaps possible.

Samples in chemistry and biology research are usually in powder form. Most solid-state materials are also prepared as powders. Samples may be immersed in petroleum ether during grinding to avoid oxidation. A particle size of about 200 mesh is appropriate, and the powder is pressed to the desired thickness between two pieces of thin plastic sheet. For low-temperature measurements, the powder sample is usually mixed with a compressible solid chemical of light atoms such as a sugar, and pressed into a "free-standing" sample. For high-temperature measurements, the powder sample is usually placed between two pieces of boron nitride sheet.

Sometimes, samples exhibit certain additional preferred orientation (e.g., texture), which causes the quadrupole split doublet to have different intensities. To reduce such an effect in a powder sample, one may prepare a sample by grinding it together with a little quartz. A small amount of chemically nonreactive additive such as vacuum grease or silicone may also be used.

Liquid samples are usually sealed in a sample holder and refrigerated until frozen. The sample holder's window for γ-rays must be made from a material of light atomic number elements.

^{57}Fe has a relatively low natural abundance. If the ^{57}Fe content in the sample is not sufficient to give a satisfactory spectrum, its enrichment in the sample may be necessary.

3.4
Detection and Recording Systems

If the 14.4 keV recoilless γ-rays were the only radiation emitted by a source containing ^{57}Co, as simply shown in Fig. 1.7, we would merely need to record the number of transmitted γ-photons at each source velocity with no need for the detector's energy resolution; thus, a Geiger–Muller counter with relatively high efficiency would do the job. But in reality, this is not the case. Take the ^{57}Co source, for example. It emits γ-rays of 136, 122, and 14.4 keV and x-rays of 6.3 keV (Fig. 3.6), with an approximate intensity ratio of 1:10:1:13. Therefore, the 14.4 keV Mössbauer radiation is only a small part of the total radiation, and what is worse is that the flux of 14.4 keV γ-rays is attenuated considerably after going through a typical sample, but the flux of the 122 keV γ-rays will be decreased very little.

Consequently, the detector must be highly efficient for the 14.4 keV γ-rays, but be as insensitive as possible to the 122 keV γ-rays. As to the γ-rays with energies below 14.4 keV, they will be discriminated against by the SCA if they have been

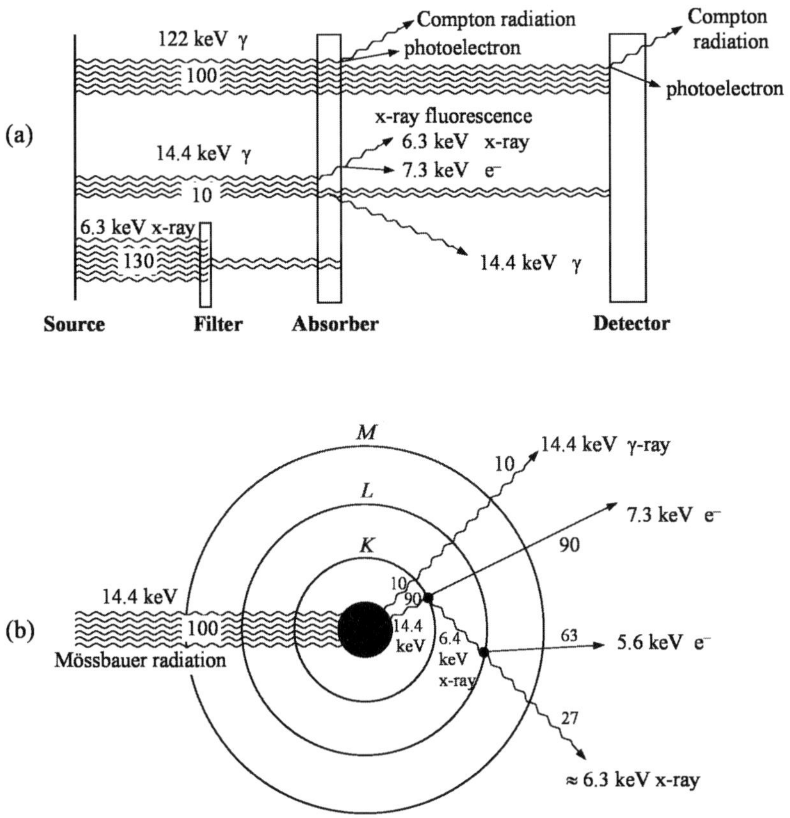

Fig. 3.6 (a) Schematic diagram showing various processes of secondary radiation as γ-rays from a ^{57}Co source travel through the absorber towards the detector. (b) The resonance absorption and internal conversion of the 14.4 keV radiation in a ^{57}Fe atom [17].

detected. The most widely employed detectors are proportional counters and NaI(Tl) scintillation counters, followed by semiconductor detectors. Their main characteristics are listed in Table 3.3. The choice of a detector also depends on the Mössbauer isotope in use.

Table 3.3 Characteristics of detectors in general use for Mössbauer spectroscopy.

Detector	Energy resolution (%)	Efficiency (%)	Maximum count rate (s^{-1})	Resolution time (s)
Gas proportional counters	10 ($E_\gamma = 14$ keV)	80	7×10^4	10^{-6}
NaI (Tl) counters	20 ($E_\gamma = 50$ keV)	100	2×10^5	10^{-6}–10^{-8}
Semiconductor counters	2 ($E_\gamma = 50$ keV)	100	$\sim 10^4$	

3.4.1
Gas Proportional Counters

Typically, a gas proportional counter has a cylindrical metal tube (cathode) and a metal wire on the axis (anode). It is filled with a gas, about 90% of it being a noble gas such as xenon, krypton, argon, or neon and about 10% being a quench gas such as methane or butane. A high voltage of 1500 to 3000 V is applied to the anode. A krypton counter filled to a pressure of 100 kPa has a good efficiency for the 14.4 keV γ-rays. This is because the krypton x-ray absorption edge is at 14.32 keV, which would largely absorb the 14.4 keV photons and mostly reject photons with higher energies. Its energy resolution is about 10%, sufficient enough to resolve the 14.4 keV γ-rays from the 6.3 keV x-rays. In addition, the gas proportional counter has a high signal-to-noise ratio and an upper count rate; its cross-section for Compton scattering is lower than that of an NaI(Tl) counter and is only 30% of that of a semiconductor detector. Therefore, the proportional counter is the popular detector for γ-rays of $E_\gamma < 20$ keV.

3.4.2
NaI(Tl) Scintillation Counters

These are also widely used counters with detection efficiency as high as 100%. When used in Mössbauer spectroscopy, the thickness of the NaI(Tl) crystal must be reduced to between 0.1 and 0.2 mm, which will ensure a high efficiency for the 14.4 keV γ-rays and a low efficiency for any higher energy γ-rays. The 14.4 keV γ-rays have an energy resolution of about 35%, and can be just resolved from the 6.3 keV x-rays. Since the NaI(Tl) scintillation counter is highly efficient and simple to use, it is often chosen for γ-rays of $E_\gamma > 15$ keV (e.g., ^{119}Sn).

3.4.3
Semiconductor Detectors

Each semiconductor detector is essentially a p–n junction, with a reverse-biased high voltage, creating a region that is sensitive to γ-rays. Selecting the correct thickness allows a great reduction of its sensitivity to high-energy γ-rays. Semiconductor detectors offer the best resolution (about a few percent) over the entire range of energies of interest in Mössbauer experiments, and their efficiency compares favorably with proportional or scintillation counters. At the present time, high-purity Ge detectors are available at lower cost than the lithium-drifted Ge(Li) or Si(Li) detectors. For detecting the 35.5 keV γ-rays or the 27.5 and 31.0 keV x-rays from ^{125}Te, proportional or scintillation counters would not be suitable, and a high-resolution semiconductor detector must be used.

The emission spectra from a ^{57}Co/Rh source recorded by the above three types of detectors are shown in Fig. 3.7.

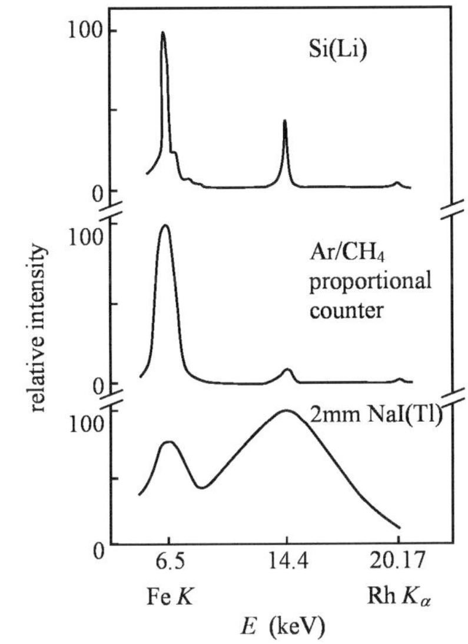

Fig. 3.7 ^{57}Co/Rh source emission spectra recorded by a semiconductor Si(Li) detector, an Ar/CH_4 proportional counter, and a NaI(Tl) scintillation counter.

3.4.4
Reduction and Correction of Background Counts

Background counts are what the detector would record if all "sources" causing the effect under investigation were removed. But separating the background from resonance absorption counts is not a simple matter, because both come from the same radiation source used in the experiment. Moreover, for different absorbers, the detected background counts are also different. Although there are methods of measuring background counts [18], it is advantageous to reduce the background as much as possible. One step is to reduce the detector efficiency for γ-rays of higher energies. Another step for reducing background is to set up a window (for 14.4 keV in the case of ^{57}Fe) in the SCA after the amplifier.

The background counts in the transmission spectrum come from the following events:

(1) Compton scattering of high-energy γ-rays in the sample or other components of the spectrometer produces secondary γ-rays, some of which fall into the 14.4 keV window. Also, high-energy γ-rays may enter the detector directly and produce a broad and nearly flat distribution of Compton electron energies from zero up to about 40 keV (for the 120 keV γ-rays), and those electrons within the window contribute to the main part of the background.

Therefore, we would reduce the Compton scattering cross-section for high-energy γ-rays in both the detector and other parts around the detector. Removing the shields farther away from the detector can help, but completely eliminating Compton secondary radiation is very difficult. This makes up the major portion of the background counts. It is especially difficult when the radiation to be discriminated against is only slightly higher in energy than the Mössbauer radiation energy. For instance, the ^{119}Sn Mössbauer radiation energy is 23.87 keV, but the source also emits Sn K_α x-rays of 25.2 keV, which a NaI(Tl) detector would not be able to resolve. Fortunately, the K_α absorption edge of Pd is at 24.35 keV, situated between the above two energy values. When a Pd filter of thickness of 0.05 or 0.10 mm is used, the x-rays can be largely absorbed with little attenuation of the γ-rays.

(2) The source also emits radiation with a continuous energy spectrum due to a recoiled 14.4 keV emission, and some of this radiation may fall into the window. It may be reduced by lowering the temperature of the source.

(3) Portions of the radiation with energies lower than 14.4 keV (e.g., the 6.3 keV x-rays) are detected because of an imperfect detector resolution. For reducing this contribution to the background a piece of aluminum of 0.1 mm thickness (or Plexiglass of 4 mm thickness) would be able to attenuate the 6.3 keV x-rays to about 1/50, with the 14.4 keV γ-rays being reduced only by 3%.

3.4.5 Geometric Conditions

The two types of geometric arrangements are transmission and scattering. In transmission geometry, the source, the sample, and the detector should be collinear and any deviation from that will have a substantial influence on the outcome of the measurement. A good-quality collimator, as well as the shield used, should be made of minimal x-ray fluorescent materials, such as Plexiglass or aluminum on the surfaces with lead in the interior. The distances separating the source, the sample, and the detector should not be too short for two main reasons. As the source vibrates, the solid angle it spans with respect to the detector will change and the detector would record different γ-ray counts for different positions of the source. A simple calculation for ^{57}Fe indicates that if the minimum distance passed over by the source does not exceed 0.2 mm, and if the source-to-detector distance is $L \geq 10$ cm and the detector window radius is $R = 1$ cm, the solid angle change can be limited within 0.5% and this effect may be neglected. But for some Mössbauer nuclei, such as ^{169}Tm, larger Doppler velocities are required and the solid angle change would not be negligible. The second main reason for a proper distance from the source to the detector is the cosine effect [19]. If the γ-rays are emitted in a direction not exactly parallel with the source–absorber relative velocity, but make an angle θ, then the Doppler shift is not $(v/c)E_0$, but $(v/c)E_0 \cos\theta$. This will cause line broadening, even a shift [20, 21]. Calculations have indicated that when $R/L < 0.1$, the spectral shift and broadening would be less than 0.4%.

3.4.6 Recording Systems

In a typical spectrometer, the detected signal goes through a preamplifier, the main amplifier, the SCA with a 14.4 keV window, and finally the multiscaler consisting of several hundred registers (channels). The most usual way of data acquisition has been to utilize a multichannel analyzer (MCA) in the multiscaler operating mode. But more recent spectrometers use a NIM unit with a microprocessor or a personal computer with special interfaces to perform SCA's and multiscaler's tasks, etc.

3.5 Velocity Drive System

A velocity drive system is not only the most important component, but also a feature unique to Mössbauer spectroscopy. Although there are several different types of drive systems, the best and prevalent is an electromagnetic drive system composed of a waveform generator, a drive circuit, a feedback circuit, and a velocity transducer. Especially after the advent of digital technology, it exhibits excellent stability, linearity, and reliability. Although its performance has improved tremendously since the 1960s, it still operates on the same original principle. We now describe the components of this system.

3.5.1 Velocity Transducer

The electromagnetic velocity transducer works in the same way as a loudspeaker [22, 23]. The transducer converts an applied current into the velocity of the source through a drive coil and provides a signal proportional to the actual velocity through a pickup coil. It is equivalent to two back-to-back speakers but without magnetic coupling between them. A shaft goes through the common center of the coils. The radiation source is attached to one end of the shaft, and a prism (or a mirror) for measuring velocity is installed on the other end. The drive coil uses thick wires and a small number of turns, to allow enough current for driving the shaft. The pickup coil uses thin wires and many turns for increasing its sensitivity. The shaft is supported by two thin flat springs and can move along the axis within certain amplitude. To ensure effective control of this motion, the fundamental frequency of the reference waveform should be close to the shaft's natural frequency, typically ranging between 10 and 40 Hz.

Figure 3.8 shows a cross-sectional diagram of a high-performance transducer developed recently [24]. Its drive component is not different from the usual design, but the pickup and feedback loop have been drastically modified. The pickup coil of as many as 50 000 turns of copper wire wound on a spool is fixed

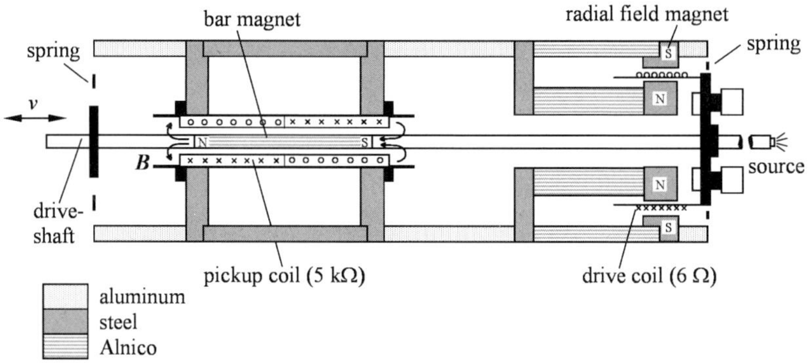

Fig. 3.8 Cross-section of a new type of velocity transducer [24].

inside the steel box and centered around the shaft, and its resistance is 5 kΩ, much higher than that of the 1.5 kΩ pickup coil manufactured by Wissel. Incorporated into the shaft inside the pickup coil is a cylindrical bar magnet with maximum field strength of about 0.5 T. Furthermore, the pickup coil is divided into two halves wound in opposite directions and connected in series to generate a feedback signal. This new design completely removes any magnetic coupling between the drive coil and the pickup coil. Another modification in the feedback loop is the introduction of a position feedback circuit for further improvement. This drive system has therefore high stability as well as linearity and immune to drifts in velocity and position, regardless of the duration of data acquisition. For instance, its nonlinearity is only 0.1% (Wissel's transducer nonlinearity is 0.3%), and the velocity at any point in the scan period has a maximum error ranging from $\pm 1\ \mu m\ s^{-1}$ (at 3.5 mm s^{-1}) to $\pm 3\ \mu m\ s^{-1}$ (at 21 mm s^{-1}).

3.5.2
Waveform Generator

The function of the waveform generator is to provide the drive system with a reference signal which determines the waveform of the source motion. Sinusoidal and triangular waves are the two most often used waveforms.

1. *Sinusoidal waveform.* The shaft can be easily driven by a sinusoidal waveform. As a result, the source motion could be controlled most accurately and this is the main advantage of this waveform. Because there is no abrupt change in acceleration, the effect of the system's inertia is reduced to its minimum. However, the sinusoidal waveform has some disadvantages. Owing to a nonlinear relationship between velocity and channel numbers, it is not easy to visualize what the spectrum would look like on the linear velocity scale. At present, this is not a serious problem because a simple computer program can easily manipulate the data and give a spectrum graphed against the linear velocity. If the shaft is relatively long, the mechanical load is heavy, or the velocity limits are high, the sinus-

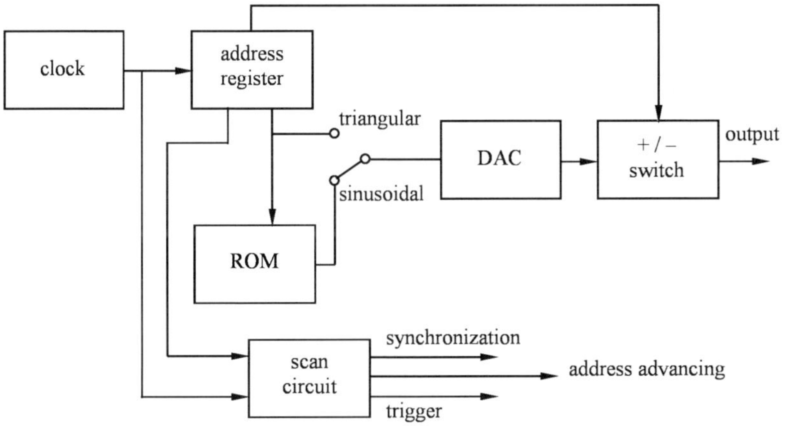

Fig. 3.9 Block diagram of a digital waveform generator.

oidal waveform should be used. By and large, sinusoidal waveform provides a good operation mode.

Figure 3.9 shows a block diagram of a digital waveform generator [25], whose principle of operation is completely different from the traditional one. Its key components include a read-only memory (ROM), an address register, and a digital-to-analog converter (DAC). For any address n ($0 \leq n \leq 1023$), the respective value of $\sin[(\pi/2)(n/1023)]$ is programmed into the ROM in the form of a 10-bit binary number. Each clock pulse triggers one up counting of the address register, and the address register advances from 0 to 1023. The values of $\sin x$ are read out in turn from the ROM, and converted to analog signals by the DAC, forming the first quarter of a sine wave. Next, each clock pulse triggers one down counting, and the circuit generates the second quarter of the sine wave. Execution of this "up" and "down" counting sequentially again, provided that the polarity is now inverted by the switching circuit, gives the third and fourth quarters, completing the entire period of a sine wave. In the meantime, the clock pulses are also divided to give the synchronization signals whenever the velocity is at $+v_{\max}$ or $-v_{\max}$.

2. *Triangular waveform.* This is another commonly used waveform. In this mode, the motion of the shaft in the transducer has a constant acceleration. The velocity starts at $-v_{\max}$, goes through zero, and increases linearly to $+v_{\max}$. It then uniformly decreases back to $-v_{\max}$. In the circuit of Fig. 3.9, the DAC successively reads in the address codes (from 0 to 1023) instead of the $\sin x$ value from the ROM, and outputs a linearly increasing voltage in the first half and decreasing voltage in the second half, completing a triangular waveform. A digital triangular waveform generator is better than an analog generator, not only because of its better linearity and stability, but also due to its versatility in providing the "region of interest" waveforms and constant-velocity waveforms. In the latter case, the counting of the address register is stopped for an exactly defined time precisely

at the maximum of the output waveform. Using the triangular form, the system records the spectrum twice as mirror images of each other, but a computer program can easily fold the two parts together. In order to reduce the abrupt change in acceleration, the triangular wave must be smoothed near $\pm v_{max}$. Such a waveform reduces impulsive forces and is easier for the driving system to follow.

There is also a sawtooth waveform, but it has no advantage over the triangular waveform, and therefore recent spectrometers have eliminated the sawtooth option all together.

3.5.3 Drive Circuit and Feedback Circuit

The purpose of the drive circuit is to produce a current signal in the drive coil to drive the shaft with the required velocity. In general, the circuit consists of three op-amps (A_1, A_2, and A_3) as shown in Fig. 3.10. A_1 is for amplifying the difference between the pickup signal and the reference signal, A_2 is an integrator, and A_3 is a power amplifier.

The basic principle of the drive circuit operation is as follows. The pickup coil is a negative feedback loop, electromagnetically decoupled from the drive coil, and the amplitude and shape of the feedback signal depend only on the shaft's motion. The difference between the pickup signal and the reference signal, known as an error signal, is proportional to the deviation of the actual velocity from its reference value. As we know, introducing a negative feedback into an amplifier can greatly improve its linearity and stability, and this is extremely important in velocity control [26]. Only when there is a suitable feedback would the shaft move precisely according to the reference signal, because if there is a slight deviation from the reference signal, the error signal instantly corrects the difference.

Let us now look at the function of the integrator. Suppose the waveform is triangular, and a Lorentz force experienced by the drive coil balances the springs' restoring force, $F = ks = ci(t)$, where c and k are constants, $i(t)$ is the current through the drive coil, and s is the displacement of the shaft with respect to its

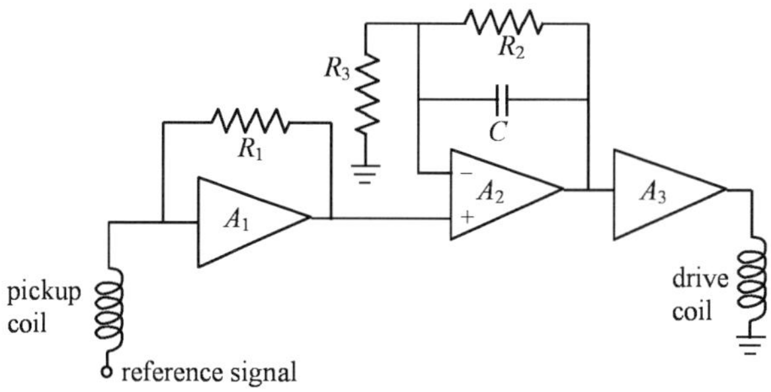

Fig. 3.10 Schematic diagram of the drive and feedback circuit system.

equilibrium position. If we require the shaft to execute a constant acceleration motion, then $s = at^2/2$, and

$$v = at = \frac{c}{k}\frac{\mathrm{d}i(t)}{\mathrm{d}t}.$$

The frequency of the drive voltage $e_0(t)$ is very low and, as a result, it is almost in phase with the current $i(t)$. Therefore

$$at = \frac{c_1}{k}\frac{\mathrm{d}e_0(t)}{\mathrm{d}t}$$

where c_1 is another constant. It is clear now that the output voltage of A_3 should be the integration of a signal that is linearly proportional to time t. This is why an integrator is normally used in the triangular waveform mode.

Careful attention must be paid to the stability of the negative feedback loop. Analyses have shown that when the frequency of the reference signal is low and near the natural resonance frequency of the shaft, its motion can be controlled most effectively. When the frequency is high, the pickup signal will lag behind the input reference signal. If such phase shift approaches 180°, the feedback is no longer negative, but positive, resulting in self-sustained oscillations at a relatively high frequency. This has been documented extensively in the literature [22, 27]. In a practical instrument, there are controls for adjusting the feedback gain and the frequency response so that the feedback circuit works under stable conditions with a large gain, small error signals, and no undesirable oscillations.

3.5.4 Velocity Calibration

The horizontal axis of a Mössbauer spectrum is initially labeled by the address codes of the data registers or the channel numbers of the MCA. To express the position parameters such as δ, ΔE_Q, B_{hf}, etc., in units of mm s^{-1}, the channel numbers must be converted to the velocities of the source, not only the values, but also the direction of the motion. This is velocity calibration, and is usually done by either a secondary standard method or an absolute velocity method.

3.5.4.1 Secondary Standard Calibration

In this method, the source velocity is calibrated using certain standard samples whose hyperfine splittings have been most accurately measured [28]. For velocities in the range of ± 10 mm s^{-1}, α-Fe or sodium nitroprusside is often used. For velocities higher than 100 mm s^{-1}, metallic Dy may be used as the standard sample.

Metallic iron has many advantages, including relatively large separations between the spectral lines in the sextet, a very large f-value at room temperature, and a cubic crystal structure (so $V_{zz} = 0$). However, there are slight differences among the α-Fe spectral results from various authors, mainly due to sample impurities [29]. Analyses have shown that in order to obtain a calibration accuracy of 0.1%, the α-Fe sample should have a purity of at least 9.99% [26], which is commercially available. This method is very simple and its accuracy is high enough for most research work. Many authors have made effort to measure the α-Fe hyperfine splittings as accurately and precisely as possible [30, 31], and the accepted values are 10.627 mm s^{-1} for the separation between line 1 and line 6.

3.5.4.2 Absolute Velocity Calibration

According to their underlying principles, the absolute calibration methods belong to two main categories, ultrasonic modulation and optical methods, which would directly give the velocity for each channel.

1. *Ultrasonic frequency modulation [32].* Suppose we have a single-line source (^{57}Co/Pd) and a single-line absorber, and mount either the source or the absorber on a piezoelectric crystal which vibrates with frequency ω (e.g., 16 MHz). The observed Mössbauer spectrum will have, in addition to the original single line (absorption intensity x_0 and frequency ω_0), a series of sidebands with frequencies $\omega_0 \pm n\omega$, where $n = 1, 2, 3$, etc. The corresponding absorption intensities can be predicted by the squares of Bessel function values $J_n^2(x_0/\lambdabar)$. Because the frequency of the piezoelectric crystal can be measured extremely precisely, this calibration method can be very accurate, often reaching 0.1%.

2. *Optical methods.* Laser interference fringes and moiré patterns are often used to measure the absolute values of the source velocity.

A simple Michelson interferometer is shown in Fig. 3.11, where M_2 is the stationary prism and M_1 the moving prism attached to the drive shaft. Let the wavelength of the laser light be λ. Every time the shaft moves a distance of $\lambda/2$, a bright fringe appears and the frequency of the emerging fringes is proportional to the velocity of the shaft. These fringes are transformed by a photodiode into

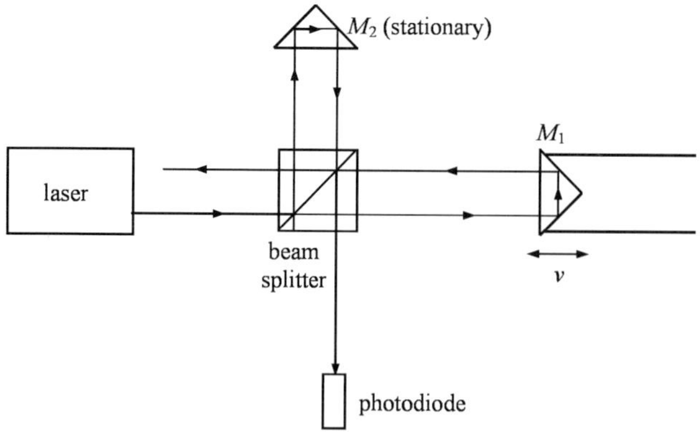

Fig. 3.11 Laser Michelson interferometer.

pulses, which are counted by the multiscaler. The velocity value can be expressed by

$$v_i = \frac{N_i}{nt_0}\frac{\lambda}{2} \tag{3.11}$$

where N_i is the fringe count (usually more than 10^5 accumulated) in the ith channel, n is the number of scans performed, and t_0 is the dwell time for each channel.

The interference method can only measure the magnitude of the instantaneous velocity of the source, so when the velocity uniformly changes from $-V_m$ and $+V_m$, the multiscaler records a V-shaped spectrum. The data points in first half of the "V" are flipped with respect to the horizontal axis. The two halves are then combined and numerically fitted by the function $ax^2 + bx + c$, from which the nonlinearity of the velocity scan (indicated by the value of a) and the precise position of the zero velocity are determined (see Fig. 3.12). The signs of velocities ($+v$ or $-v$) measured from the α-Fe absorption sextet spectrum are also given in Fig. 3.12.

This is a method of choice for many Mössbauer spectroscopists, and modern spectrometers often have an interferometer attachment. The problem with this method is that the fringe counts at low speeds are not as accurate, especially

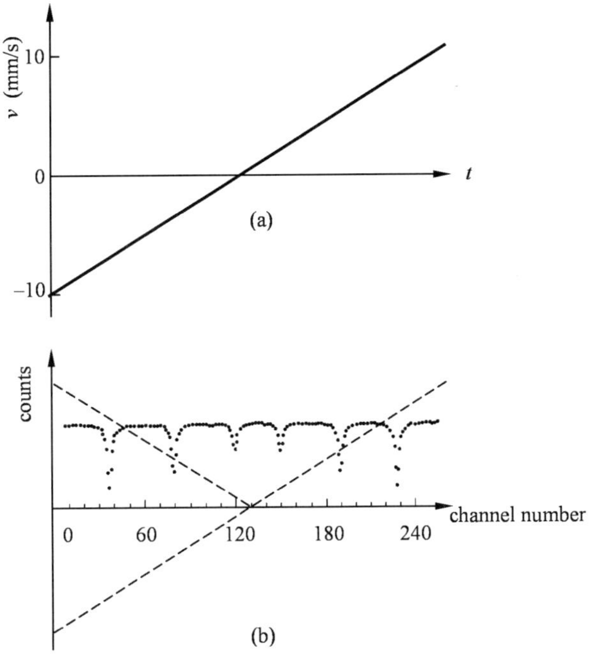

Fig. 3.12 (a) Triangular drive voltage. (b) A V-shaped interference fringe spectrum, superimposed on a sextet spectrum.

near zero velocity, because the error in fringe counts is ± 1. Therefore, the position of the zero velocity cannot be directly calculated using Eq. (3.11). Recently, there has been an improved version of the interferometer system capable of measuring velocity within a wide range of 0.5 to 1000 mm s^{-1} with a precision of 0.1% or better [33].

When the velocity values are vary high (around 700 mm s^{-1}), a method using gratings and observing changes in moiré patterns can be utilized [34]. In this method, we also record the number of bright and dark fringes to obtain the speed values, but the accuracy is not as good as in the Michelson interferometer method.

3.6 Data Analysis

In order to obtain reliable microscopic information from Mössbauer spectroscopy, it is imperative to analyze the measured spectra quantitatively. The experimental spectrum is often quite complicated, because it is usually a superposition of many sets of subspectra. Assignment of all subspectra before fitting the spectrum and interpretation of the results after fitting must be carried out based on a certain physical model for the sample. The process of spectral fitting can be tedious. There are many fitting methods, but they belong generally to two categories, either fitting individual single lines or fitting the entire spectrum. The former is based on the notion that the Mössbauer spectrum is a superposition of Lorentzian lines, while the latter is done by calculating the nuclear energy splittings and transition probabilities based on the model for the sample. We will briefly describe several methods, among which the single-line fitting using the Gauss–Newton method or a modified version is probably the most widely used.

In recent years, there have emerged a few novel methods for data analysis, such as the genetic algorithm [35] and the artificial neural network (ANN) [36]. ANNs are composed of elements that perform in a manner that is analogous to the elementary functions of the biological neurons. The elements are organized in a way that may or may not be related to the cerebral anatomy. ANNs learn from experience, generalize from previous examples to new ones, and extract essential characteristics from inputs containing relevant data. ANNs are used for the analysis of experimental data over a wide range of scientific disciplines and are now applied to fitting Mössbauer spectra. It may develop into a very convenient and highly accurate method, and may become quite attractive to Mössbauer effect researchers.

3.6.1 Fitting Individual Lorentzian Lines

Based on the thin absorber approximation, the Mössbauer spectrum is a superposition of Lorentzian lines. There are some differences in the fitting procedures for crystalline and amorphous samples.

3.6.1.1 Spectra from Crystalline Samples

If the experimental spectrum is a superposition of n Lorentzian lines (for α-Fe, $n = 6$), then

$$y(x_i, \mathbf{c}) = (E + Fx_i + Gx_i^2) + \sum_{k=1}^{n} \frac{A_k}{1 + \left(\dfrac{x_i - x_k(0)}{\Gamma_k/2}\right)^2} \qquad i = 1, 2, 3, \ldots, N \tag{3.12}$$

where x_i is the Doppler velocity, the quadratic term $(E + Fx_i + Gx_i{}^2)$ describes the baseline, $\mathbf{c}$ represents a total of $(3n + 3)$ parameters (A_k, $x_k(0)$, Γ_k, and E, F, G) to be determined, and N is the number of data points in the spectrum. Here A_k, $x_k(0)$, and Γ_k are the height, position (in mm s^{-1}), and linewidth (in mm s^{-1}) of the kth Lorentzian line, respectively. This expression gives a theoretical model and is used to fit the experimental spectrum. The usual method is the least squares fitting, minimizing the following function:

$$\chi^2 = \sum_{i=1}^{N} w_i [y(x_i, \mathbf{c}) - y_i]^2 = \min \tag{3.13}$$

where y_i is the γ-ray counts in the ith channel and $w_i = 1/y_i$ is its weight factor. A necessary condition for a minimum of the above quantity is

$$\frac{\partial \chi^2}{\partial c_j} = 2\sum_{i=1}^{N} w_i [y(x_i, \mathbf{c}) - y_i] \frac{\partial y(x_i, \mathbf{c})}{\partial c_j} = 0, \quad j = 1, 2, 3, \ldots, 3n + 3. \tag{3.14}$$

This actually represents a set of nonlinear simultaneous equations, and the parameters c_j cannot be solved analytically. One approach to this problem is to estimate the initial values of the parameters (zeroth approximation)

$$\mathbf{c}^{(0)} = (c_1{}^{(0)}, c_2{}^{(0)}, c_3{}^{(0)}, \ldots, c_m{}^{(0)}), \quad (m = 3n + 3)$$

and to approximate the function $y(x_i, \mathbf{c})$ by its Taylor expansion about $\mathbf{c}^{(0)}$. We then keep all the linear terms, and neglect all higher ones. The function $y(x, \mathbf{c})$ is linear with respect to the unknown parameters $c_1, c_2, c_3, \ldots, c_m$:

$$y(x_i, \mathbf{c}) \approx y(x_i, \mathbf{c}^{(0)}) + \left[\frac{\partial y}{\partial c_1}\right]_{c_1^{(0)}} \delta_1^{(1)} + \cdots + \left[\frac{\partial y}{\partial c_m}\right]_{c_m^{(0)}} \delta_m^{(1)} \tag{3.15}$$

where $\delta_j{}^{(1)} = c_j - c_j{}^{(0)}$. Substituting (3.15) into (3.14), the simultaneous equations become

$$\sum_{j=1}^{m} \delta_j^{(1)} \sum_{i=1}^{N} w_i \left[\frac{\partial y(x_i, \mathbf{c})}{\partial c_j}\right]_{c_j^{(0)}} \left[\frac{\partial y(x_i, \mathbf{c})}{\partial c_k}\right]_{c_k^{(0)}}$$

$$= \sum_{i=1}^{N} w_i \left[\frac{\partial y(x_i, \mathbf{c})}{\partial c_k}\right]_{c_k^{(0)}} [y_i - y(x_i, \mathbf{c}^{(0)})] \quad k = 1, 2, 3, \ldots, m. \tag{3.16}$$

This may be written in the matrix form

$$(\boldsymbol{F}^{\mathrm{T}} \mathbf{W}_y \boldsymbol{F}) \boldsymbol{\delta}^{(1)} = \boldsymbol{F}^{\mathrm{T}} \mathbf{W}_y (\mathbf{Y} - \mathbf{Y}_0),$$

and the solution is

$$\boldsymbol{\delta}^{(1)} = (\boldsymbol{F}^{\mathrm{T}} \mathbf{W}_y \boldsymbol{F})^{-1} \boldsymbol{F}^{\mathrm{T}} \mathbf{W}_y (\mathbf{Y} - \mathbf{Y}_0) \tag{3.17}$$

where

$$\mathbf{Y} - \mathbf{Y}_0 = \begin{bmatrix} y_1 - y(x_1, \mathbf{c}^{(0)}) \\ y_2 - y(x_2, \mathbf{c}^{(0)}) \\ \vdots \\ y_N - y(x_N, \mathbf{c}^{(0)}) \end{bmatrix}$$

$$\boldsymbol{F} = \begin{bmatrix} f_{11} & f_{12} & \cdots & f_{1m} \\ f_{21} & f_{22} & \cdots & f_{2m} \\ \vdots & \vdots & & \vdots \\ f_{N1} & f_{N2} & \cdots & f_{Nm} \end{bmatrix}$$

$$\mathbf{W}_y = \begin{bmatrix} w_1 & & & \\ & w_2 & & \mathbf{0} \\ & & \ddots & \\ & \mathbf{0} & & \\ & & & w_N \end{bmatrix}$$

and

$$f_{ij} = \left[\frac{\partial y(x_i, \mathbf{c})}{\partial c_j}\right]_{c_j^{(0)}},$$

with $i = 1, 2, \ldots, N$ and $j = 1, 2, \ldots, m$. Once the values in the matrix $\boldsymbol{\delta}^{(1)}$ are calculated, we obtain the first approximation of the parameters $\mathbf{c}$:

$$\mathbf{c}^{(1)} = \mathbf{c}^{(0)} + \boldsymbol{\delta}^{(1)}.$$

Now we use $\mathbf{c}^{(1)}$ as our initial values for the next round of calculations, which will give us $\boldsymbol{\delta}^{(2)}$ and $\mathbf{c}^{(2)}$. The iteration continues until a satisfactory convergence is reached. The convergence condition could be (1) when the relative change in c values from the kth iteration to the $(k+1)$th iteration is smaller than a given fraction (e.g., 0.01), or (2) when the function χ^2 shows almost no change and satisfies $0.8 < \frac{\chi^2}{v} < 1.2$, where v denotes the degrees of freedom of the χ^2 distribution. The above description is only the basic principle of the Gauss–Newton method. Because it has some disadvantages, several modified versions are available.

Mathematically, good fitting results can be easily achieved with this method. But without an appropriate physical model, it would often be laborious to find a satisfactory interpretation of the fitted spectrum.

3.6.1.2 **Spectra from Amorphous Samples**

In an amorphous material, the atoms occupy random sites, and consequently the isomer shift δ, quadrupole splitting ΔE_Q, and magnetic hyperfine field B will have continuous distributions. If all three interactions are involved, the spectrum would be extremely difficult to analyze. In many cases, fortunately, only one is dominant. For example, from $Fe_{40}Ni_{40}P_{14}B_6$ under certain conditions, only the continuous distribution of quadrupole splitting was observed [37]. In some amorphous iron alloys, if the spectrum is a symmetric and broad sextet, it may be taken as caused by a continuous distribution $P(B)$ of the magnetic hyperfine fields with $\Delta E_Q = 0$ and an identical isomer shift for all Fe nuclei. In addition, it is assumed that the area ratio within each sextet is the same, regardless of the B-value, and that the recoilless fraction is also the same thus the total area of a sextet is proportional to $P(B)$. Let $S(v)$ represent the experimental spectrum from the amorphous material. Under the thin absorber approximation, it can be expressed as

$$S(v) = \int_0^{\infty} P(B)\mathscr{L}_6(B, v)\,\mathrm{d}B \tag{3.18}$$

where v is the Doppler velocity, $\mathscr{L}_6(B, v)$ is the Lorentzian sextet with the magnetic hyperfine field B, and $P(B)$ is the normalized distribution function:

$$\int_0^{\infty} P(B)\,\mathrm{d}B = 1.$$

For calculating $P(B)$ from the experimental spectrum $S(v)$, one may use one of the following two methods developed by Hesse and Window, respectively.

(1) The Hesse method [38]. The range of the magnetic hyperfine fields in question is equally divided by n, each interval being ΔB. The maximum field value is therefore $B_{\max} = n\Delta B$ and the hth field value is $B_h = h\Delta B$. The spectrum can be now approximated by a finite series:

$$S(v_j) = \sum_{h=0}^{n} P(B_h)\mathscr{L}_6(B_h, v_j) + \varepsilon_j \quad j = 1, 2, \ldots, N \tag{3.19}$$

where ε_j is a term representing the influence of the statistical error at v_j. The expressions in (3.19) may be collectively written in the matrix form

$$\boldsymbol{S} = \boldsymbol{LP} + \varepsilon. \tag{3.20}$$

In order to avoid unrealistic fluctuations in the calculated hyperfine field distribution due to the statistical errors in the experimental spectrum, the data should be smoothed. To do this, we introduce a smoothing factor γ and require

$$\frac{\partial}{\partial P_i}\left[\sum_{k=2}^{n-1}\gamma(P_{k-1} - 2P_k + P_{k+1})^2 + \sum_{j=1}^{N} w_j\varepsilon_j^2\right] = 0.$$

This is a set of simultaneous equations with the probability values P_k as the unknowns, and may also be written in the matrix form

$$\gamma\boldsymbol{DP} + \boldsymbol{L}^{\mathrm{T}}\boldsymbol{W}\varepsilon = 0 \tag{3.21}$$

where

$$D = \begin{bmatrix} 1 & -2 & 1 & 0 & 0 & 0 & \cdots & 0 & 0 & 0 & 0 & 0 \\ -2 & 5 & -4 & 1 & 0 & 0 & \cdots & 0 & 0 & 0 & 0 & 0 \\ 1 & -4 & 6 & -4 & 1 & 0 & \cdots & 0 & 0 & 0 & 0 & 0 \\ 0 & 1 & -4 & 6 & 4 & 1 & \cdots & 0 & 0 & 0 & 0 & 0 \\ \vdots & \vdots & \vdots & \vdots & \vdots & \vdots & & \vdots & \vdots & \vdots & \vdots & \vdots \\ 0 & 0 & 0 & 0 & 0 & 0 & \cdots & 1 & -4 & 6 & -4 & 1 \\ 0 & 0 & 0 & 0 & 0 & 0 & \cdots & 0 & 1 & -4 & 5 & -2 \\ 0 & 0 & 0 & 0 & 0 & 0 & \cdots & 0 & 0 & 1 & -2 & 1 \end{bmatrix}.$$

We now substitute ε from (3.20) into Eq. (3.21), and solve for $\boldsymbol{P}$:

$$\boldsymbol{P} = (\boldsymbol{L}^{\mathrm{T}}\boldsymbol{WL} - \gamma\boldsymbol{D})^{-1}(\boldsymbol{L}^{-1}\boldsymbol{WS}), \tag{3.22}$$

which gives a set of values for the magnetic hyperfine field distribution.

(2) The Window method [39]. The probability $P(B)$ is expressed as a Fourier series in the range from 0 to $B_{\max}$:

$$P(B) = \frac{a_0}{2} + \sum_{n=1}^{N} a_n \cos\frac{n\pi B}{B_{\max}}.$$

Imposing a boundary condition $P(B_{\max}) = 0$, the above series is reduced to

$$P(B) = \sum_{n=1}^{N} a_n f_n(B) \tag{3.23}$$

where

$$f_n(B) = \cos\frac{n\pi B}{B_{\max}} - (-1)^n. \tag{3.24}$$

When Eq. (3.23) is substituted into (3.18)

$$S(v) = \sum_{n=1}^{N} a_n \int_0^{B_{\max}} f_n(B)\mathscr{L}_6(B, v)\,\mathrm{d}B.$$

The integral in this expression can be calculated because each f_n is a known function, and the coefficients a_n can be determined by minimizing the following expression:

$$\sum_v [S(v) - S_{\exp}(v)]^2 = \min. \tag{3.25}$$

Once values of a_n are calculated, Eq. (3.23) gives the magnetic hyperfine field distribution $P(B)$.

These two methods usually give similar results, both having the advantage of not requiring *a priori* knowledge or constraints on the shape of the distribution. But the calculated $P(B)$ often exhibits oscillations at the low field end, and sometimes even shows unrealistic negative values. Several improved versions of these methods are also available [40–42].

3.6.2
Full Hamiltonian Site Fitting

In this method, we assume that the entire spectrum is a superposition of several subspectra of Lorentzian lines. Each subspectrum corresponds to one particular crystal site (or one particular environment). The hyperfine interactions that would result in a subspectrum have been discussed in Section 2.4. Using this method, we are no longer mathematically fitting the Mössbauer spectral lines to obtain certain parameters, but are studying physical problems such as the splitting of the nuclear energy levels and the corresponding transitions in randomly oriented magnetic fields or in a low-symmetry EFG. This means that, before carrying out any calculations for fitting the spectrum, a Hamiltonian for the hyperfine interactions in the sample should be established. In a general case, it is the sum of all three types hyperfine interactions. Solving the secular equations would give the eigenvalues and eigenvectors of the ground state and the excited state. The energies involved in the allowed transitions between these states will determine the relative positions of the absorption lines. The transition probabilities will provide information on the relative intensities of the absorption lines. Such a spectrum is then characterized by a set parameters including isomer shift δ, magnetic hyperfine field B, quadrupole splitting ΔE_Q, the asymmetry parameter η, linewidth, area of the absorption line, external magnetic field orientation angles θ and ϕ, and γ-ray orientation angles α and β. During iterations using the Gauss–Newton method, these parameters can only be adjusted under the constraints of quantum mechanics to ensure that the fitting conforms to the proper physical model. Therefore, fitting the entire spectrum is theoretically a much better method than the earlier methods that fit only individual lines.

3.6.3 Fitting Thick Absorber Spectra

In reality, the thin absorber approximation may not be always practical. For instance, the thickness of a metallic Fe foil corresponding to $t_a = 1$ is only 2.4 mg cm^{-2}, and when $t_a < 1$ is required, the physical sample would be too thin. Therefore, it has been concluded that the above approximation is not valid for most ^{57}Fe absorbers [43]. As the thickness increases beyond the thin absorber limit, the spectral lines will be broadened. The finite absorber thickness is not likely to produce shifts in a Lorentzian distribution, i.e., it does not affect the line positions but will bring appreciable changes in the line area or the line height. Therefore, in this case it is necessary to evaluate accurately the transmission integral $T(v)$, defined in Eq. (1.19). Usually, fitting the data is either by numerical method directly [44] or by analytic representations [45–47]. Fourier transform is another way, which is relatively simple and easy to follow [48]. Especially after the recent development of fast Fourier transform algorithms, this method has attracted a great deal of attention and has been further studied [49–53].

If we assume that the emission spectral line is normalized and the background has been corrected, Eq. (1.24) may be rewritten as

$$\frac{-\varepsilon(v)}{f_s} = \frac{-I(\infty, d) + I(v, d)}{I(\infty, d) f_s} = \int_{-\infty}^{\infty} \mathscr{L}\left(E - \frac{v}{c} E_0\right) [A(E) - 1]\, dE. \qquad (3.26)$$

The right-hand side of this equation is a convolution of two functions. $\mathscr{L}$ is the Lorentzian distribution function and $A(E) = \exp[-\sigma(E) t_a]$ modifies the Lorentzian shape, thus distorts the Mössbauer line, due to a finite thickness of the sample. However, $\sigma(E)$ in the exponent is the absorber Lorentzian line shape. We are aiming at calculating $\sigma(E)$ from the experimental spectrum $\varepsilon(v)$ using Eq. (3.26), and in this process, the non-Lorentzian spectrum is converted to a Lorentzian spectrum, eliminating the effect of finite thickness.

In order to calculate $\sigma(E)$, we Fourier transform both sides of Eq. (3.26), and use a theorem for convolution integrals:

$$\mathscr{F}\left\{-\frac{\varepsilon(v)}{f_s}\right\} = \mathscr{F}\{\mathscr{L}\} \bullet \mathscr{F}\{\exp[-\sigma(E) t_a] - 1\}.$$

Let

$$\mathscr{F}\{\exp[-\sigma(E)t_a] - 1\} = \mathscr{F}\left\{-\frac{\varepsilon(v)}{f_s}\right\}\Big/\mathscr{F}\{\mathscr{L}\} \equiv n. \tag{3.27}$$

Applying inverse Fourier transform $\mathscr{F}^{-1}$ to Eq. (3.27):

$$\exp[-\sigma(E)t_a] = \mathscr{F}^{-1}\{n\} + 1,$$

and solving for $\sigma(E)$,

$$-\sigma(E)t_a = \ln[\mathscr{F}^{-1}\{n\} + 1]. \tag{3.28}$$

Based on this method, the experimental spectrum is first fitted to give $\varepsilon(v)$, which is then treated according to Eq. (3.28) to obtain $\sigma(E)$.

References

1 R.L. Cohen and G.K. Wertheim. Experimental methods in Mössbauer spectroscopy. In *Methods of Experimental Physics*, vol. 11, R.V. Coleman (Ed.), pp. 307–369 (Academic Press, New York, 1974).

2 G. Longworth. Instrumentation for Mössabuer spectroscopy. In *Advances in Mössbauer Spectroscopy: Applications to Physics, Chemistry and Biology*, B.V. Thosar and P.K. Iyengar (Eds.), pp. 122–158 (Elsevier, Amsterdam, 1983).

3 D.A. Shirley, M. Kaplan, and P. Axel. Recoil-free resonant absorption in Au^{197}. *Phys. Rev.* 123, 816–830 (1961).

4 S. Margulies and J.R. Ehrman. Transmission and line broadening of resonance radiation incident on a resonance absorber. *Nucl. Instrum. Methods* 12, 131–137 (1961).

5 S.L. Ruby and J.M. Hicks. Line shape in Mössbauer spectroscopy. *Rev. Sci. Instrum.* 33, 27–30 (1962).

6 U. Shimony. Condition for maximum single-line Mössbauer absorption. *Nucl. Instrum. Methods* 37, 348–350 (1965).

7 P.J. Blamey. The area of a single line Mössbauer absorption spectrum. *Nucl. Instrum. Methods* 142, 553–557 (1977).

8 G.M. Bancroft. *Mössbauer Spectroscopy: An Introduction for Inorganic Chemists and Geochemists* (Wiley, New York, 1973).

9 I.D. Weisman, L.J. Swartzendruber, and L.H. Bennett. Nuclear resonances in metals: nuclear magnetic resonance and Mössbauer effect. In *Measurement of Physical Properties*, Techniques of Metal Research, Vol. 6, Part 2, E. Passaglia (Ed.), pp. 165–504 (Wiley, New York, 1973).

10 P.R. Sarma, V. Prakash, and K.C. Tripathi. Optimization of the absorber thickness for improving the quality of a Mössbauer spectrum. *Nucl. Instrum. Methods* 178, 167–171 (1980).

11 G.J. Long, T.E. Cranshaw, and G. Longworth. The ideal Mössbauer effect absorber thickness. *Mössbauer Effect Ref. Data J.* 6, 42–49 (1983).

12 A. Kumar, M.R. Singh, P.R. Sarma, and K.C. Tripathi. Optimised thickness of diffusive Mössbauer absorbers. *J. Phys. D* 22, 465–466 (1989).

13 F.L. Zhang, F. Yi, Y.L. Chen and B.F. Xu. Determination of the optimum thickness of an absorber in Mössbauer spectroscopy. *J. Wuhan University (Natural Sci. Edition)* 43, 348–352 (1997) [in Chinese].

14 R.W. Grant, R.M. Housley, and U. Gonser. Nuclear electric field gradient and mean square displacement of the

iron sites in sodium nitroprusside. *Phys. Rev.* 178, 523–530 (1969).
15 R.H. Herber and G.K. Wertheim. Mössbauer effect in ferrocene and related compounds. In *The Mössbauer Effect*. D.M.J. Compton and A.H. Schoen (Eds.), pp. 105–111 (Wiley, New York, 1962).
16 Y.L. Chen, F. Yi, and F.L. Zhang. Optimum thickness of Mössbauer absorber. *Nucl. Sci. Techn.* 11, 91–95 (2000). Y.L. Chen and F. Yi. Unpublished data (2001).
17 J.J. Spijkerman. Conversion Electron Mössbauer Spectroscopy. In *Mössbauer Effect Methodology*, vol. 7, I.J. Gruverman (Ed.), p. 89 (Plenum Press, New York, 1971).
18 R.M. Housley, N.E. Erickson, and J.G. Dash. Measurement of recoil-free fractions in studies of the Mössbauer effect. *Nucl. Instrum. Methods* 27, 29–37 (1964).
19 J.J. Spijkerman, F.C. Ruegg, and J.R. DeVoe. Standardization of the differential chemical shift for Fe^{57}. In *Mössbauer Effect Methodology*, vol. 1, I.J. Gruverman (Ed.), pp. 115–120 (Plenum Press, New York, 1965).
20 N. Hershkowitz. Velocity shifts in Mössbauer spectroscopy. *Nucl. Instrum. Methods* 53, 172 (1967).
21 F. Aramu and V. Maxia. Shift and broadening of Mössbauer peaks by lack of collimation. *Nucl. Instrum. Methods* 80, 35–39 (1970).
22 E. Kankeleit. Feedback in electromechanical drive systems. In *Mössbauer Effect Methodology*, vol. 1, I.J. Gruverman (Ed.), pp. 47–66 (Plenum Press, New York, 1965).
23 H.P. Wit, G. Hoeksema, L. Niesen, and H. de Waard. A multi-purpose velocity transducer for Mössbauer spectrometers. *Nucl. Instrum. Methods* 141, 515–518 (1977).
24 W.F. Filter, R.H. Sands, and W.R. Dunham. An ultra-high-stability Mössbauer spectrometer drive using a type-2 feedback system. *Nucl. Instrum. Methods B* 119, 565–582 (1996).
25 N. Halder and G.M. Kalvius. A digital sine generator for Mössbauer spectrometers. *Nucl. Instrum. Methods* 108, 161–165 (1973).
26 H. Shechter, M. Ron, and S. Niedzwiedz. Mössbauer spectrometer calibration using ^{57}Fe enriched metallic iron. *Nucl. Instrum. Methods* 44, 268–272 (1966).
27 D. St. P. Bunbury. The design of apparatus for the measurement of Mössbauer spectra. *J. Sci. Instrum.* 43, 783–790 (1966).
28 J.G. Stevens, V.E. Stevens, and W.L. Gettys (Eds.). *Mössbauer Effect Ref. Data J.* 3, 99 (1980).
29 G.K. Wertheim. Impurity shift in the NMR of iron. *Can. J. Phys.* 48, 2751–2752 (1970).
30 T.E. Cranshaw. Three problems associated with the interferometric calibration of Mössbauer spectrometers. *J. Phys. E* 6, 1053–1057 (1973).
31 M.A. Player and F.W.D. Woodhams. An improved interferometric calibrator for Mössbauer spectrometer drive systems. *J. Phys. E* 9, 1148–1152 (1976).
32 T.E. Cranshaw and P. Reivari. A Mössbauer study of the hyperfine spectrum of ^{57}Fe, using ultrasonic calibration. *Proc. Phys. Soc. (London)* 90, 1059–1064 (1967).
33 M. Kwater. Laser calibration of the Mössbauer spectrometer velocity: improvements in design. *Hyperfine Interactions* 116, 53–66 (1998).
34 H.P. Wit. Simple moiré calibrator for velocity transducers used in Mössbauer effect measurements. *Rev. Sci. Instrum.* 46, 927–928 (1975).
35 H. Ahonen, P.A. de Souza Jr., and V.K. Garg. A genetic algorithm for fitting Lorentzian line shapes in Mössbauer spectra. *Nucl. Instrum. Methods B* 124, 633–638 (1997).
36 H. Paulsen, R. Linder, F. Wagner, H. Winkler, S.J. Pöppl and A.X. Trautwein. Interpretation of Mössbauer spectra in the energy and time domain with neural networks. *Hyperfine Interactions* 126, 421–424 (2000).
37 M. Kopcewicz, H.-G. Wagner, and U. Gonser. Mössbauer investigations of

ferromagnetic amorphous metals in radio frequency fields. *J. Magn. Magn. Mater.* 40, 139–146 (1983).
38 J. Hesse and A. Rübartsch. Model independent evaluation of overlapped Mössbauer spectra. *J. Phys. E* 7, 526–532 (1974).
39 B. Window. Hyperfine field distributions from Mössbauer spectra. *J. Phys. E* 4, 401–402 (1971).
40 G. Le Caer and J.M. Dubois. Evaluation of hyperfine parameter distributions from overlapped Mössbauer spectra of amorphous alloys. *J. Phys. E* 12, 1083–1090 (1979).
41 C. Wivel and S. Morup. Improved computational procedure for evaluation of overlapping hyperfine parameter distributions in Mössbauer spectra. *J. Phys. E* 14, 605–610 (1981).
42 H. Keller. Evaluation of hyperfine field distributions from Mössbauer spectra using Window's Fourier method. *J. Appl. Phys.* 52, 5268–5273 (1981).
43 D.G. Rancourt. Accurate site populations from Mössbauer spectroscopy. *Nucl. Instrum. Methods B* 44, 199–210 (1989).
44 T.E. Cranshaw. The deduction of the best values of the parameters from Mössbauer spectra. *J. Phys. E* 7, 122–124 (1974).
45 B.T. Cleveland. An analytic method for the least-squares fitting of single-line thick-absorber Mössbauer spectra. *Nucl. Instrum. Methods* 107, 253–257 (1973).
46 M. Capaccioli, L. Cianchi, F. Del Giallo, P. Moretti, F. Pieralli, and G. Spina. A method for measurement of the Debye–Waller factor *f*. *Nucl. Instrum. Methods B* 101, 280–286 (1995).
47 H. Flores-Llamas. Chebyshev approximations for the transmission integral for one single line in Mössbauer spectroscopy. *Nucl. Instrum. Methods B* 94, 485–492 (1994).
48 M.C.D. Ure and P.A. Flinn. A technique for the removal of the "blackness" distortion of Mössbauer spectra. In *Mössbauer Effect Methodology*, vol. 7, I.J. Gruverman (Ed.), pp. 245–262 (Plenum Press, New York, 1971).
49 W. Stieler, M. Hillberg, F.J. Litterst, Ch. Böttger, and J. Hesse. Numerical evaluation of Mössbauer spectra from thick absorbers. *Nucl. Instrum. Methods B* 95, 235–242 (1995).
50 S. Nikolov and K. Kantchev. Deconvolution of Lorentzian broadened spectra: I. Direct deconvolution. *Nucl. Instrum. Methods A* 256, 161–167 (1987).
51 K. Kantchev and S. Nikolov. Deconvolution of Lorentzian broadened spectra: II. Stepped deconvolution and smoothing filtration. *Nucl. Instrum. Methods A* 256, 168–173 (1987).
52 J.G. Mullen, A. Djedid, G. Schupp, D. Cowan, Y. Cao, M.L. Crow, and W.B. Yelon. Fourier-transform method for accurate analysis of Mössbauer spectra. *Phys. Rev. B* 37, 3226–3245 (1988).
53 E.V. Voronina, A.L. Ageyev, and E.P. Yelsukov. Using an improved procedure of fast discrete Fourier transform to analyse Mössbauer spectra hyperfine parameters. *Nucl. Instrum. Methods B* 73, 90–94 (1993).

4
The Basics of Lattice Dynamics

In a solid, atoms and ions are tightly bound to lattice points so that, when the temperature is not extremely high, they can only execute small vibrations (thermal motion) around their equilibrium positions. Lattice dynamics is a branch of solid-state physics that studies such lattice vibrations. It was first initiated in the 1930s by the Born–von Karman theory, the details of which have be described in the text by Max Born and Kun Huang [1]. Lattice dynamics is important in understanding various phenomena and properties of solids, such as thermodynamic properties, phase transitions, soft modes, etc. However, further research has revealed some of the difficulties with the Born–von Karman theory, especially concerning the periodic boundary condition and the basic equation of motion. Neglecting surface effects, the periodic boundary condition is only good for large and perfectly ordered crystals. In many cases, the crystals are imperfect and surface effects must be considered. In the basic equation of motion, the classical lattice dynamics neglects the role of electrons. Nevertheless, the Born–von Karman theory laid the foundations for lattice dynamics and, therefore, we first introduce this theory, followed by a brief description of first-principles lattice dynamics, which is a computational method developed in recent years. The distinctive feature of this method is its inclusion of the explicit effect of electrons on lattice dynamics.

4.1
Harmonic Vibrations

4.1.1
Adiabatic Approximation

The adiabatic approximation, also known as the Born–Oppenheimer theorem [1], allows one to decouple the motion of the atom (more precisely, ion core) from the motion of the valence electrons. The essential idea of this approximation is that the nucleus, being at least 10^3 times heavier, moves much more slowly than the electrons. At any moment the electrons "see" the nuclei fixed in some (generally displaced) configuration. During the atomic motion the electrons move as though

Mössbauer Effect in Lattice Dynamics. Yi-Long Chen and De-Ping Yang

ISBN: 978-3-527-40712-5

the nuclei were fixed in their instantaneous positions. We say that the electrons follow the atomic motion adiabatically. In an adiabatic approximation, an electron does not make abrupt transition from one state to others but will be in its ground state for that particular instantaneous atom configuration. In the following, we give an outline of some expressions in this approximation. The properties of a crystal consisting ion cores and valence electrons are derived from the solution of the Schrödinger equation

$$\mathscr{H}\Psi(\boldsymbol{r}, \boldsymbol{R}) = \varepsilon\Psi(\boldsymbol{r}, \boldsymbol{R}) \tag{4.1}$$

with the total Hamiltonian

$$\begin{aligned}\mathscr{H} &= T_{\mathrm{n}} + T_{\mathrm{e}} + V_{\mathrm{nn}}(\boldsymbol{R}) + V_{\mathrm{ee}}(\boldsymbol{r}) + V_{\mathrm{en}}(\boldsymbol{r}, \boldsymbol{R}) \\ &= \sum_{l} \frac{-\hbar^2}{2M_l} \nabla^2_{\boldsymbol{R}_l} + \sum_{i} \frac{-\hbar^2}{2m_i} \nabla^2_{\boldsymbol{r}_i} + \frac{e^2}{2} \sum_{l \neq l'} \frac{Z_l Z_{l'}}{|\boldsymbol{R}_l - \boldsymbol{R}_{l'}|} \\ &\quad + \frac{e^2}{2} \sum_{i \neq j} \frac{1}{|\boldsymbol{r}_i - \boldsymbol{r}_j|} - \sum_{i \neq j} \frac{Z_l e^2}{|\boldsymbol{r}_i - \boldsymbol{R}_l|}\end{aligned} \tag{4.2}$$

where $\boldsymbol{r}$ and $\boldsymbol{R}$ represent the coordinates of valence electrons (e) and nuclei (n), respectively.

To obtain an exact solution of the many-body equation defined by (4.1) is hopeless, but we can approximately break it down into two subsystems of the valence electrons and the core due to the large difference in their masses, and the problem can be substantially simplified. From (4.2) we abstract an electron Hamiltonian

$$\mathscr{H}_{\mathrm{e}} = T_{\mathrm{e}} + V_{\mathrm{ee}}(\boldsymbol{r}) + V_{\mathrm{en}}(\boldsymbol{r}, \boldsymbol{R}) \tag{4.3}$$

and diagonalize it for a given atomic configuration (i.e., $\boldsymbol{R}$ is not considered as a variable but a parameter). The equation is

$$\mathscr{H}_{\mathrm{e}}\psi(\boldsymbol{r}, \boldsymbol{R}) = E_{\mathrm{e}}\psi(\boldsymbol{r}, \boldsymbol{R}) \tag{4.4}$$

where $\psi(\boldsymbol{r}, \boldsymbol{R})$ is the wave function for the entire system of electrons. Assume that the total wave function can be written as a product

$$\Psi(\boldsymbol{r}, \boldsymbol{R}) = \psi(\boldsymbol{r}, \boldsymbol{R})\chi(\boldsymbol{R}) \tag{4.5}$$

where $\chi(\boldsymbol{R})$ is the wave function for the entire system of nuclei.

Substituting Eq. (4.5) in (4.1) using (4.4), one obtains the following Schrödinger equation for $\chi(\boldsymbol{R})$, determining the lattice-dynamical properties of a solid:

$$\left[\sum_{l} \frac{-\hbar^2}{2M_l} \nabla^2_{\boldsymbol{R}_l} + V(\boldsymbol{R})\right]\chi(\boldsymbol{R}) = \varepsilon\chi(\boldsymbol{R}) \tag{4.6}$$

with

$$V(\boldsymbol{R}) = V_{\mathrm{nn}}(\boldsymbol{R}) + E_{\mathrm{e}}(\boldsymbol{R}), \tag{4.7}$$

provided that the mixed terms

$$\frac{\hbar^2}{2M}\langle\psi|\nabla_{\boldsymbol{R}}^2|\psi\rangle \quad \text{and} \quad \frac{\hbar^2}{2M}\langle\psi|\nabla_{\boldsymbol{R}}|\psi\rangle\cdot\nabla_{\boldsymbol{R}}$$

can be neglected [2]. Thus we obtain the adiabatic lattice equation (Eq. (4.6)) with $V(\boldsymbol{R})$ as an effective potential consisting of interatomic potential and electron eigenvalue $E_{\mathrm{e}}(\boldsymbol{R})$, the latter being the mean contribution from the electrons. However, the evaluation of $E_{\mathrm{e}}(\boldsymbol{R})$ is very difficult. A more practical approach is to assume certain phenomenological potential for $V(\boldsymbol{R})$ which involves only a few parameters (see Section 4.2.6).

The adiabatic approximation may break down in some cases, but we will assume the validity of this approximation in our discussions here.

4.1.2 Harmonic Approximation

When an atom is at its equilibrium position (a lattice point), the attractive and repulsive forces on it are exactly balanced. Because of thermal motion, the atom moves away from the equilibrium position, the forces are no longer canceled, and the net force tends to bring the atom back to its equilibrium position. The farther from the equilibrium the atom is, the larger the restoring force. The displacement of one atom also changes the potential energies of the surrounding atoms and consequently causes them to move. A wave motion is produced because of this type of vibrational motion propagating through the entire solid. In a perfect solid, such a wave is known as the lattice wave. Since we are interested in atomic vibrational displacements much smaller than the interatomic distances, the method of small oscillations in classical mechanics can be applied to atomic vibrations in a crystal.

Unless stated otherwise, in the following sections a crystal with a Bravais lattice (one atom per cell) is used. Suppose the equilibrium position vector of each lattice point (or each atom) is

$$\boldsymbol{l} = l_1\boldsymbol{a}_1 + l_2\boldsymbol{a}_2 + l_3\boldsymbol{a}_3 \tag{4.8}$$

where l_1, l_2, and l_3 are any three integers, while $\boldsymbol{a}_1$, $\boldsymbol{a}_2$, and $\boldsymbol{a}_3$ are three noncoplanar basis vectors (Fig. 4.1). Suppose that the lth atom deviates from its equilibrium position by $\boldsymbol{u}(l)$. Its actual position is

$$\boldsymbol{R}_l = \boldsymbol{l} + \boldsymbol{u}(l). \tag{4.9}$$

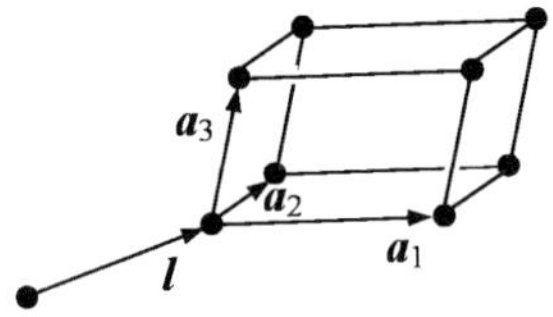

Fig. 4.1 Primitive cell in a crystal.

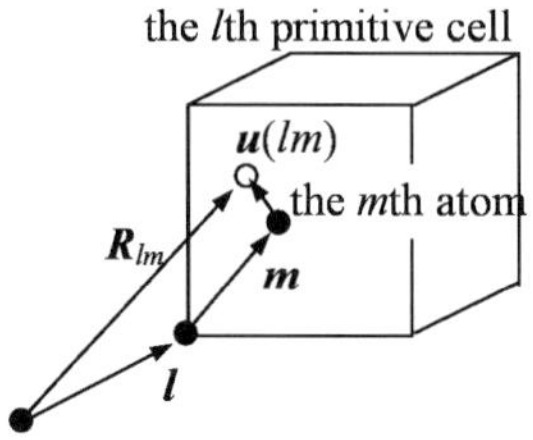

Fig. 4.2 Displacement of an atom in a non-Bravais crystal.

For the mth atom in the lth unit cell within a non-Bravais lattice (Fig. 4.2), it would be

$$\boldsymbol{R}_{lm} = \boldsymbol{l} + \mathbf{m} + \mathbf{u}(lm). \tag{4.10}$$

The total kinetic energy of the crystal is

$$T = \frac{1}{2}\sum_{\alpha,l} M\dot{u}_\alpha^2(l) \tag{4.11}$$

where M is the mass of the atom, and $\alpha = x, y, z$. The potential energy $V(\boldsymbol{R})$ is a function of the instantaneous positions of all the atoms. The displacements $u_\alpha(l)$, as mentioned above, are small and V can be expanded into a Taylor series:

$$V = V_0 + \sum_{\alpha,l} \Phi_\alpha(l)u_\alpha(l) + \frac{1}{2}\sum_{\alpha,l}\sum_{\beta,l'} \Phi_{\alpha\beta}(l,l')u_\alpha(l)u_\beta(l') + \cdots \tag{4.12}$$

where

$$\Phi_\alpha(l) = \left.\frac{\partial V}{\partial u_\alpha(l)}\right|_0, \tag{4.13}$$

$$\Phi_{\alpha\beta}(l,l') = \left.\frac{\partial^2 V}{\partial u_\alpha(l)\partial u_\beta(l')}\right|_0. \tag{4.14}$$

When each atom is at its equilibrium position, the potential energy value is chosen to be zero, i.e., $V_0 = 0$. Under equilibrium, the net forces are zero ($\Phi_\alpha(l) = 0$), thus the second term in (4.12) is also zero. Keeping only the quadratic terms, we are taking the harmonic approximation, which is the basis for treating small oscillations [2, 3]. Terms higher than the second order are known as anharmonic, and they need to be included when studying problems such as thermal expansion.

Under the harmonic approximation, the vibrational Hamiltonian and the net forces on the atoms are

$$\mathscr{H} = \frac{1}{2}\sum_{\alpha,l} M\dot{u}_\alpha^2(l) + \frac{1}{2}\sum_{\alpha,l}\sum_{\beta,l'} \Phi_{\alpha\beta}(l,l')u_\alpha(l)u_\beta(l'), \tag{4.15}$$

$$F_\alpha(l) = -\frac{\partial V}{\partial u_\alpha(l)} = -\sum_{\beta,l'} \Phi_{\alpha\beta}(l,l')u_\beta(l'), \tag{4.16}$$

and the equation of motion for the lth atom in the direction $\alpha(x, y, z)$ takes the form

$$M\ddot{u}_\alpha(l) = -\sum_{\beta,l'} \Phi_{\alpha\beta}(l,l')u_\beta(l'). \tag{4.17}$$

4.1.3
Force Constants and Their Properties

According to Eq. (4.16), the net force on atom l is a linear function of the displacements $u_\beta(l')$, and the coefficients $\Phi_{\alpha\beta}(l,l')$ are called atomic force constants. Their physical meanings are very simple. Suppose only one atom l' has a displacement (Fig. 4.3) while all other atoms are still at their equilibrium positions, then Eq. (4.16) becomes

$$-\Phi_{\alpha\beta}(l,l') = \frac{F_\alpha(l)}{u_\beta(l')}. \tag{4.18}$$

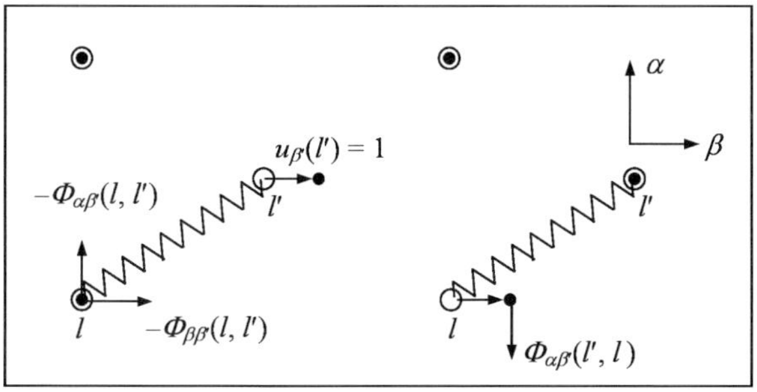

Fig. 4.3 Schematic illustration of the meaning of force constants.

It is clear that $-\Phi_{\alpha\beta}(l,l')$ is the α direction force acting on atom l when atom l' has moved a distance of unit length in the β direction. The force constants have the following properties.

1. $$\Phi_{\alpha\beta}(l,l') = \Phi_{\beta\alpha}(l',l) \tag{4.19}$$

   which is due to the fact that the partial derivatives are independent of the order in which the derivatives are taken.

2. $$\sum_{l'} \Phi_{\alpha\beta}(l,l') = 0. \tag{4.20}$$

   This is true because when each $u_\beta(l')$ in (4.16) is replaced by an arbitrary constant c_β, corresponding to a motion of the crystal as a whole, there would be no changes in the relative positions of the atoms, total potential energy, and its derivatives. Therefore, $F_\alpha(l) = 0$, which means

   $$\sum_{\beta} c_\beta \sum_{l'} \Phi_{\alpha\beta}(l,l') = 0$$

   and because c_β can be arbitrarily chosen, we have Eq. (4.20).

3. $$\Phi_{\alpha\beta}(l,l) = -\sum_{l'\neq l} \Phi_{\alpha\beta}(l,l') \tag{4.21}$$

   which comes directly from Eq. (4.20).

4. $$\Phi_{\alpha\beta}(l,l') = \Phi_{\alpha\beta}(0,l'-l) = \Phi_{\alpha\beta}(l-l',0). \tag{4.22}$$

   This indicates that a force constant depends only on the difference between l and l', i.e., only on the relative position between the atom pair.

5. When the force constants between atoms l and l' are represented by a 3×3 matrix, they exhibit certain symmetry properties. When applied to a specific crystal, these symmetry properties can greatly simplify the force constant matrix. Some matrix elements may be equal to one another, and others may be zero. Symmetry considerations are very important in lattice dynamics.

As an example [4], suppose two lattice points l and l' in a simple cubic (sc) crystal are at the origin O and at $a(100)$, respectively. Let us obtain the force constant matrix $\mathbf{\Phi}(l,l')$ between these two atoms. Because the potential energy is invariant under translations and rotations, $\mathbf{\Phi}(l, l')$ should remain the same when operated on by a symmetric orthogonal rotational operator (matrix) $\boldsymbol{S}$:

$$\boldsymbol{S}^{-1}\mathbf{\Phi}(l, l')\boldsymbol{S} = \mathbf{\Phi}(l, l'). \tag{4.23}$$

First, we assume

$$\boldsymbol{S} = \begin{bmatrix} 1 & 0 & 0 \\ 0 & -1 & 0 \\ 0 & 0 & 1 \end{bmatrix}$$

which describes an inversion with respect to the xz plane, leaving the bond between the two atoms unchanged. Substituting $\boldsymbol{S}$ into (4.23), carrying out the operations, and comparing the resultant matrix with $\Phi(l, l')$ on the right-hand side of (4.23), we find that $-\Phi_{12} = \Phi_{12}$, which means $\Phi_{12} = 0$. Also, we get $\Phi_{21} = \Phi_{23} = \Phi_{32} = 0$. Similarly, we may let $\boldsymbol{S}$ be

$$\begin{bmatrix} 1 & 0 & 0 \\ 0 & 1 & 0 \\ 0 & 0 & -1 \end{bmatrix} \quad \text{and} \quad \begin{bmatrix} 1 & 0 & 0 \\ 0 & 0 & -1 \\ 0 & 1 & 0 \end{bmatrix},$$

corresponding to an inversion with respect to the xy plane and a 90° rotation about the fourfold symmetry axis, respectively. These operations lead to $\Phi_{13} = \Phi_{31} = 0$ as well as $\Phi_{22} = \Phi_{33}$, and the force constant matrix for the neighboring atoms along the [100] direction must take the general form

$$\Phi = -\begin{bmatrix} \alpha & 0 & 0 \\ 0 & \beta & 0 \\ 0 & 0 & \beta \end{bmatrix} \tag{4.24}$$

where α describes the longitudinal (l) force and β the transverse (t) force. Because of the axial symmetry, the two transverse forces have equal magnitudes (Fig. 4.4(a)).

The force constants between two atoms form a second-order tensor, and there exists a principal coordinate system in which its matrix is diagonalized. But this coordinate system may not be the principal system for the force constant tensor between another pair of atoms. For example, if the coordinate system used in Fig. 4.4(a) is applied to a face-centered cubic (fcc) lattice, the force constant matrix between the atom at origin and the atom at $(a/2, a/2, 0)$ is

$$-\begin{bmatrix} \alpha & \gamma & 0 \\ \gamma & \alpha & 0 \\ 0 & 0 & \beta \end{bmatrix}$$

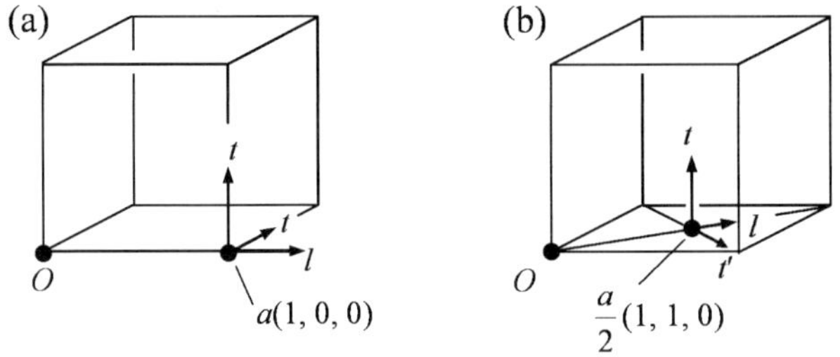

$$-\begin{bmatrix} f_l & 0 & 0 \\ 0 & f_t & 0 \\ 0 & 0 & f_t \end{bmatrix} \qquad -\begin{bmatrix} \alpha & \gamma & 0 \\ \gamma & \alpha & 0 \\ 0 & 0 & \beta \end{bmatrix}$$

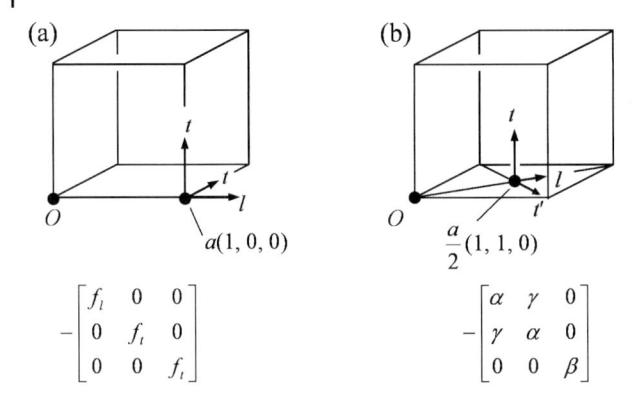

Fig. 4.4 Force constant matrices between the atom at origin and its nearest neighbor for crystals with simple cubic (sc) and face-centered cubic (fcc) lattices. For an fcc lattice, $f_l = \alpha + \gamma$, $f_{t'} = \alpha - \gamma$, and $f_t = \beta$.

which is not diagonal. The principal system for this matrix can be obtained by rotating the original system by 45° with respect to the vertical axis, and the diagonalized matrix is

$$-\begin{bmatrix} \alpha+\gamma & 0 & 0 \\ 0 & \alpha-\gamma & 0 \\ 0 & 0 & \beta \end{bmatrix}.$$

It is clear that the forces are no longer axially symmetric. If the forces between atoms are central forces, then $\beta = 0$ and $\alpha - \gamma = 0$.

The force constant matrices between various atom pairs in crystals with fcc and body-centered cubic (bcc) structures can be found in Appendix C.

4.1.4 Normal Coordinates

The kinetic energy in Eq. (4.15) is simply the sum of the quadratic terms, each of which involves only one atom in the crystal. But the potential energy is more complicated because of the cross products of atomic displacements, resulting in coupled equations of motion and making it difficult to find their solutions. However, a linear transformation will allow us to find a new coordinate system in which both kinetic and potential energies have only square terms and no cross terms. As a result, the equations of motion become uncoupled. These new coordinates are called the normal coordinates.

For a harmonic lattice, we try to solve equation (4.17) for an atom l. One possible oscillatory solution is

$$u_\alpha(l,t) = u_\alpha^0(l)e^{-i\omega t}. \tag{4.25}$$

Substituting (4.25) into (4.17) leads to the eigenvalue equation

$$\sqrt{M_l}\omega^2 u_\alpha^0(l) = \sum_{l',\beta} D'_{\alpha\beta}(l\,l')\sqrt{M_{l'}}u_\beta^0(l') \tag{4.26}$$

where

$$D'_{\alpha\beta}(l\,l') = \frac{\Phi_{\alpha\beta}(l\,l')}{\sqrt{M_l M_{l'}}} \tag{4.27}$$

are the elements of the matrix $\boldsymbol{D}'$.

For a total of N atoms in a solid, Eq. (4.26) has therefore $3N$ solutions labeled by an index s which runs from 1 to $3N$:

$$\sqrt{M_l}\omega_s^2 u_\alpha^0(l,s) = \sum_{l',\beta} D'_{\alpha\beta}(l\,l')\sqrt{M_{l'}}u_\beta^0(l',s). \tag{4.28}$$

Since $\boldsymbol{D}'$ is a real symmetric matrix, its eigenvalues ${\omega_s}^2$ must be real and, to keep the solid stable, ω_s must not be negative.

There is a unitary matrix $\boldsymbol{B}$ with elements $B_\alpha(l,s)$ which diagonalizes $\boldsymbol{D}'$ [3, 5]:

$$\sum_{\alpha,\beta,l,l'} B_\alpha^*(l,s)D'_{\alpha\beta}(l\,l')B_\alpha(l',s') = \omega_s^2\delta_{ss'}, \tag{4.29}$$

and has the following properties:

$$\begin{aligned} &\sum_{\alpha,l} B_\alpha^*(l,s)B_\alpha(l',s') = \delta_{ss'}, \\ &\sum_{s} B_\alpha^*(l,s)B_\beta(l',s') = \delta_{\alpha\beta}\delta_{ll'}. \end{aligned} \tag{4.30}$$

Since M_l is diagonal, after diagonalization of the matrix $\boldsymbol{D}'$ the expression in (4.28) becomes $3N$ separate equations, each of which describes a harmonic vibration.

It is this very matrix $\boldsymbol{B}$ that transforms the Cartesian coordinates to a set of normal coordinates q_s through the formula

$$u_\alpha(l) = \frac{1}{\sqrt{M_l}}\sum_s B_\alpha(l,s)q_s. \tag{4.31}$$

In terms of the normal coordinates, both the kinetic energy and the potential energy are without any cross terms:

$$T = \frac{1}{2}\sum_{s=1}^{3N} \dot{q}_s^2 = \frac{1}{2}\sum_{s=1}^{3N} p_s^2, \tag{4.32}$$

$$V = \frac{1}{2}\sum_{s=1}^{3N} \omega_s^2 q_s^2, \tag{4.33}$$

and the Hamiltonian (4.15) changes into

$$\mathcal{H} = \frac{1}{2}\sum_{s=1}^{3N} (p_s^2 + \omega_s^2 q_s^2). \tag{4.34}$$

With this Hamiltonian, one gets $3N$ uncoupled equations

$$\ddot{q}_s + \omega_s^2 q_s = 0, \quad s = 1, 2, 3, \ldots, 3N \tag{4.35}$$

where each equation is simply solved in the form of

$$q_s = A_s \mathrm{e}^{-i\omega_s t}. \tag{4.36}$$

The vibration with a particular ω_s is called a normal mode. If only one normal mode ω_s has amplitude A_s and all the others have zero amplitude, then (4.31) becomes

$$u_\alpha(l) = \frac{B_\alpha(l, s)}{\sqrt{M_l}} A_s \mathrm{e}^{-i\omega_s t}. \tag{4.37}$$

Here we see that different atoms in a crystal vibrate with the same frequency ω_s. In other words, a normal mode vibration is not the vibration of one single atom, but a collective vibration of all atoms in the crystal, forming a so-called lattice wave.

For a perfect crystalline solid, a normal mode s can be represented by the branch index j and wave vector $\boldsymbol{k}$ so that ω_s becomes $\omega_j(\boldsymbol{k})$, and the coefficient in (4.31) can be expressed as [6]

$$B_\alpha(l, \boldsymbol{k}j) = \frac{1}{\sqrt{N}} e_\alpha(\boldsymbol{k}j) \mathrm{e}^{i\boldsymbol{k}\cdot\boldsymbol{l}} \tag{4.38}$$

where $e_\alpha(\boldsymbol{k}j)$ is the α-component of the polarization vector. When Eq. (4.38) is substituted into (4.37), we can clearly see the meaning of the polarization vectors which describe the directions of the atomic vibrations in a normal mode.

4.2 Lattice Vibrations

4.2.1 Dynamical Matrix

Inserting (4.36) and (4.38) into (4.31), we get a general expression for the displacement of an atom in a Bravais lattice:

$$u_\alpha(l) = \frac{1}{\sqrt{NM}} \sum_{\boldsymbol{k},j} A(\boldsymbol{k}j) e_\alpha(\boldsymbol{k}j) \exp\{i[\boldsymbol{k} \cdot \boldsymbol{l} - \omega_j(\boldsymbol{k})t]\}. \tag{4.39}$$

A lattice wave traveling throughout the entire crystal results in displacements of neighboring atoms differing by a phase factor $\boldsymbol{k} \cdot \boldsymbol{l}$, while every atom will vibrate with the same frequency ω_j. Substituting (4.39) into (4.17) with $M = M_l$, we obtain, for each mode $\boldsymbol{k}j$

$$\omega_j^2(\boldsymbol{k}) e_\alpha(\boldsymbol{k}j) = \sum_\beta D_{\alpha\beta}(\boldsymbol{k}) e_\beta(\boldsymbol{k}j), \tag{4.40}$$

which is the eigenvalue equation determining the relation between the frequency ω_j and the wave vector $\boldsymbol{k}$. The eigenvectors $e_\alpha(\boldsymbol{k}j)$ satisfy the following orthonormality and closure conditions:

$$\begin{aligned} &\sum_\alpha e_\alpha^*(\boldsymbol{k}j) e_\alpha(\boldsymbol{k}j') = \delta_{jj'}, \\ &\sum_j e_\alpha^*(\boldsymbol{k}j) e_\beta(\boldsymbol{k}j) = \delta_{\alpha\beta}. \end{aligned} \tag{4.41}$$

The matrix element $D_{\alpha\beta}$ in (4.40) is

$$\begin{aligned} D_{\alpha\beta}(\boldsymbol{k}) &= \frac{1}{M} \sum_{l'-l} \Phi_{\alpha\beta}(0, l'-l) \exp[i\boldsymbol{k} \cdot (\boldsymbol{l}' - \boldsymbol{l})] \\ &= \frac{1}{M} \sum_L \Phi_{\alpha\beta}(0, L) \exp[i\boldsymbol{k} \cdot \boldsymbol{L}] \end{aligned} \tag{4.42}$$

where $\boldsymbol{L} = \boldsymbol{l}' - \boldsymbol{l}$ is the position vector from atom l to atom l'. This 3×3 matrix $\boldsymbol{D}$ is known as the dynamical matrix, which contains all the information about the particular normal mode. One of the tasks of lattice dynamics is to find an explicit expression of this matrix for a given crystal. Using the properties of $\Phi_{\alpha\beta}$, we can prove that $\boldsymbol{D}$ is Hermitian:

$$D_{\alpha\beta}(\boldsymbol{k}) = D^*_{\beta\alpha}(\boldsymbol{k}). \tag{4.43}$$

Based on (4.42), it can also be shown that

$$D_{\alpha\beta}(-\boldsymbol{k}) = D^*_{\alpha\beta}(\boldsymbol{k}). \tag{4.44}$$

The condition for the simultaneous Eqs. (4.40) to have nontrivial solutions is

$$\det[\omega^2\delta_{\alpha\beta} - D_{\alpha\beta}(\boldsymbol{k})] = 0. \tag{4.45}$$

For each $\boldsymbol{k}$ value, there are three eigenvalues $\omega_j{}^2(\boldsymbol{k})$ $(j = 1, 2, 3)$, which are guaranteed to be real because $\boldsymbol{D}$ is Hermitian. Also, $\omega_j(\boldsymbol{k})$ should be either positive or zero. For a particular j, the relation between the angular frequency ω and the wave vector $\boldsymbol{k}$ is known as the dispersion relation:

$$\omega = \omega_j(\boldsymbol{k}) \quad (j = 1, 2, 3). \tag{4.46}$$

Each j represents one branch of the vibration spectrum. In general, different branches have different dispersion relations. Within one branch, ω is a continuous function of the wave vector $\boldsymbol{k}$.

The vibration amplitude $A(\boldsymbol{k}j)$ depending on the average energy of mode $\boldsymbol{k}j$ can be expressed as [4]

$$|A(\boldsymbol{k}j)|^2 = \frac{E(\boldsymbol{k}j)}{\omega_j^2(\boldsymbol{k})} \tag{4.47}$$

where

$$E(\boldsymbol{k}j) = \hbar\omega_j(\boldsymbol{k})\left[\frac{1}{\exp(\hbar\omega_j(\boldsymbol{k})\beta) - 1} + \frac{1}{2}\right]. \tag{4.48}$$

Each atom's displacement from its equilibrium position can finally be written as

$$u_\alpha(l) = \sqrt{\frac{\hbar}{2NM}}\sum_{\boldsymbol{k},j}\left[\frac{\coth(\hbar\omega_j(\boldsymbol{k})\beta/2)}{\omega_j(\boldsymbol{k})}\right]^{1/2} e_\alpha(\boldsymbol{k}j)\exp\{i[\boldsymbol{k}\cdot\boldsymbol{l} - \omega_j(\boldsymbol{k})t]\}. \tag{4.49}$$

For a non-Bravais crystal, let each unit cell have r atoms. The above derivation is completely valid, provided that we use $u_\alpha(lm)$ to replace $u_\alpha(l)$, and also use $\Phi_{\alpha\beta}(lm, l'm')$, $e_\alpha(m|\boldsymbol{k}j)$, and $D_{\alpha\beta}(mm'|\boldsymbol{k}j)$ to replace the corresponding quantities, where $m = 1, 2, \ldots r$. The dynamical matrix is no longer 3×3, but $3r \times 3r$. For each wave vector $\boldsymbol{k}$, there will be $3r$ eigenvalues $\omega_j{}^2(\boldsymbol{k})$, $j = 1, 2, \ldots, 3r$.

4.2.2 Reciprocal Lattice and the Brillouin Zones

4.2.2.1 Reciprocal Lattice

A perfect crystal is composed of periodically arranged primitive cells; each is described by three basis vectors $\boldsymbol{a}_1$, $\boldsymbol{a}_2$, and $\boldsymbol{a}_3$. We now define the following three new vectors from the basis vectors:

$$\boldsymbol{b}_1 = 2\pi\frac{\boldsymbol{a}_2 \times \boldsymbol{a}_3}{\boldsymbol{a}_1 \cdot (\boldsymbol{a}_2 \times \boldsymbol{a}_3)}, \quad \boldsymbol{b}_2 = 2\pi\frac{\boldsymbol{a}_3 \times \boldsymbol{a}_1}{\boldsymbol{a}_1 \cdot (\boldsymbol{a}_2 \times \boldsymbol{a}_3)}, \quad \boldsymbol{b}_3 = 2\pi\frac{\boldsymbol{a}_1 \times \boldsymbol{a}_2}{\boldsymbol{a}_1 \cdot (\boldsymbol{a}_2 \times \boldsymbol{a}_3)}. \tag{4.50}$$

These new vectors $\boldsymbol{b}_i$ can be used to construct a new lattice, known as the reciprocal lattice. The original lattice is called the direct lattice. The position of each reciprocal lattice point is

$$\boldsymbol{\tau} = h_1\boldsymbol{b}_1 + h_2\boldsymbol{b}_2 + h_3\boldsymbol{b}_3 \tag{4.51}$$

where h_1, h_2, and h_3 are integers and $\boldsymbol{\tau}$ is referred to as the reciprocal lattice vector. The reciprocal lattice has the following properties:

1. The reciprocal basis vectors $\boldsymbol{b}_i$ satisfy

$$\boldsymbol{a}_i \cdot \boldsymbol{b}_j = 2\pi\delta_{ij} \quad (i, j = 1, 2, 3). \tag{4.52}$$

2. The dimensions of the direct lattice and reciprocal lattice are L and L^{-1}, respectively. Since the dimension of any wave vector is also L^{-1}, it may be represented in terms of the basis vectors of the reciprocal lattice, and therefore the reciprocal space is also known as the *k*-space. If the volume of the direct primitive cell is V_a, the reciprocal primitive cell volume is $(2\pi)^3/V_\mathrm{a}$.
3. The scalar product of any reciprocal lattice vector with any direct lattice vector yields an integer multiple of 2π.

$$\begin{aligned}\boldsymbol{\tau} \cdot \boldsymbol{L} &= (h_1\boldsymbol{b}_1 + h_2\boldsymbol{b}_2 + h_3\boldsymbol{b}_3) \cdot (L_1\boldsymbol{a}_1 + L_2\boldsymbol{a}_2 + L_3\boldsymbol{a}_3) \\ &= 2\pi(h_1L_1 + h_2L_2 + h_3L_3) = 2\pi m \end{aligned} \tag{4.53}$$

where h_i, L_i, and m are all integers.

4. The vector $\boldsymbol{\tau}$ defined in Eq. (4.51) is perpendicular to a family of lattice planes with Miller indices h_1, h_2, h_3 in the direct lattice. Each of the 14 different Bravais lattices has its specific reciprocal lattice type. For example, a bcc direct lattice has an fcc reciprocal lattice.

It should be noted that the number of unit cells in the direct crystal is very large and, therefore, the allowed $\boldsymbol{k}$ values form densely and uniformly distributed points in the k-space. For example, in a one-dimensional atom chain of 1 cm in length, the size of its Brillouin zone is about 10^8 cm^{-1}. There are 10^8 allowed $\boldsymbol{k}$-values and they can be regarded as quasi-continuous.

4.2.2.2 Brillouin Zones

In the dynamical matrix, the wave vector $\boldsymbol{k}$ appears only in the exponent of $e^{i\boldsymbol{k}\cdot\boldsymbol{L}}$. If $\boldsymbol{k}$ is increased by any reciprocal lattice vector, we have

$$e^{i(\boldsymbol{k}+\boldsymbol{\tau})\cdot\boldsymbol{L}} = e^{i(\boldsymbol{k}\cdot\boldsymbol{L}+2\pi m)} = e^{i\boldsymbol{k}\cdot\boldsymbol{L}} \tag{4.54}$$

and therefore

$$\boldsymbol{D}(\boldsymbol{k}+\boldsymbol{\tau}) = \boldsymbol{D}(\boldsymbol{k}). \tag{4.55}$$

This shows that $\boldsymbol{k}$ and $(\boldsymbol{k}+\boldsymbol{\tau})$ are equivalent in both the dynamical matrix and the dispersion relation. Therefore, it would be sufficient to confine $\boldsymbol{k}$ within the following ranges:

$$\begin{aligned} -\frac{b_1}{2} &< k_1 \le \frac{b_1}{2}, \\ -\frac{b_2}{2} &< k_2 \le \frac{b_2}{2}, \\ -\frac{b_3}{2} &< k_3 \le \frac{b_3}{2}. \end{aligned} \tag{4.56}$$

This region is called the first Brillouin zone, a reciprocal primitive cell.

Equation (4.56) also tells us how to construct the first Brillouin zone. We start by selecting a particular reciprocal lattice point as the origin ($\boldsymbol{k} = 0$), and connect this point to the nearest neighbor and next nearest neighbor reciprocal lattice points with straight lines. Now we construct perpendicular bisector planes of the lines. These planes form the smallest polyhedron that encloses the origin, and the space inside the polyhedron is called the first Brillouin zone. Similarly, second and third Brillouin zones can be constructed. Each zone has the same volume, and none of the bisecting planes cuts through it. Figure 4.5 shows three Brillouin zones for a two-dimensional square reciprocal lattice. If a vector $\boldsymbol{\tau} = -\boldsymbol{b}_1$ is added to every point in area A_2, it will be superimposed exactly onto A_1. Similarly, B_2, C_2, and D_2 can be made to coincide with B_1, C_1, and D_1, respectively. In summary, any higher Brillouin zone is a repeat of the first Brillouin zone.

The fcc lattice primitive cell and its reciprocal lattice first Brillouin zone are shown in Fig. 4.6. The basis vectors of the primitive cell are

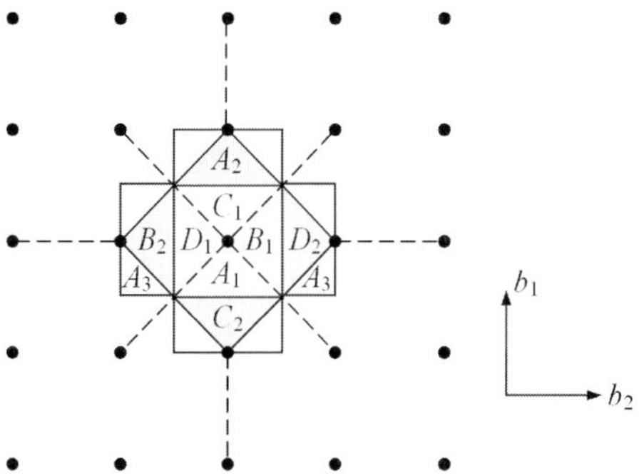

Fig. 4.5 First, second, and third Brillouin zones in a two-dimensional square reciprocal lattice.

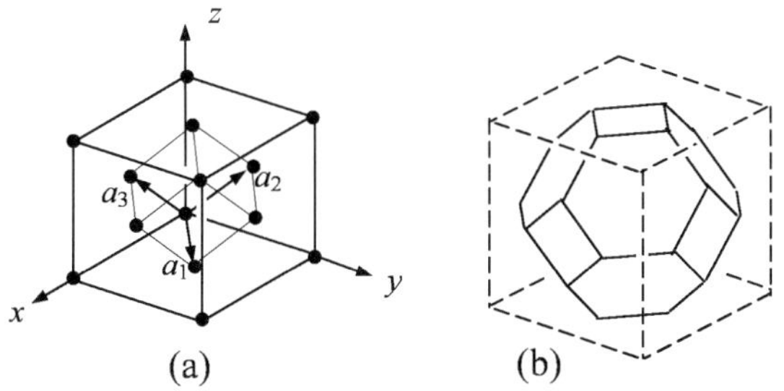

Fig. 4.6 (a) Fcc lattice and (b) the corresponding reciprocal primitive cell and the first Brillouin zone.

$$\boldsymbol{a}_1 = \frac{a}{2}(1, 1, 0), \quad \boldsymbol{a}_2 = \frac{a}{2}(0, 1, 1), \quad \boldsymbol{a}_3 = \frac{a}{2}(1, 0, 1) \tag{4.57}$$

with a volume $V_a = a^3/4$. According to (4.52), the corresponding reciprocal lattice basis vectors are

$$\boldsymbol{b}_1 = \frac{2\pi}{a}(1, 1, -1), \quad \boldsymbol{b}_2 = \frac{2\pi}{a}(-1, 1, 1), \quad \boldsymbol{b}_3 = \frac{2\pi}{a}(1, -1, 1). \tag{4.58}$$

Similarly, the basis vectors of the bcc primitive cell are

$$\boldsymbol{a}_1 = \frac{a}{2}(1, 1, -1), \quad \boldsymbol{a}_2 = \frac{a}{2}(-1, 1, 1), \quad \boldsymbol{a}_3 = \frac{a}{2}(1, -1, 1) \tag{4.59}$$

with a volume $V_a = a^3/2$. The corresponding reciprocal lattice basis vectors are

$$\boldsymbol{b}_1 = \frac{2\pi}{a}(1, 1, 0), \quad \boldsymbol{b}_2 = \frac{2\pi}{a}(0, 1, 1), \quad \boldsymbol{b}_3 = \frac{2\pi}{a}(1, 0, 1). \tag{4.60}$$

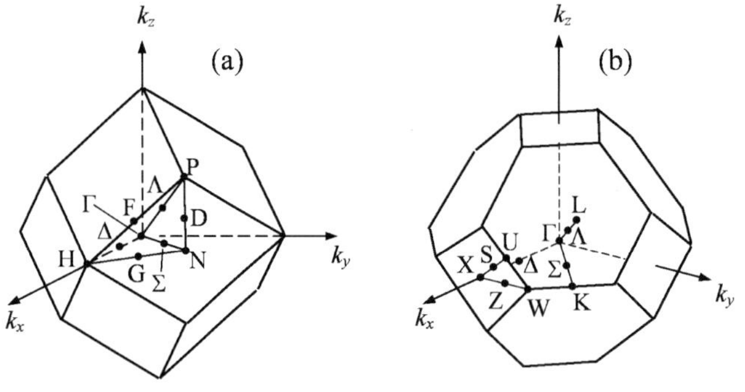

Fig. 4.7 First Brillouin zones of (a) a bcc lattice and (b) an fcc lattice, with some special reciprocal lattice points as indicated.

Comparing the four sets of vectors in Eqs. (4.57)–(4.60), we can see that the reciprocal lattice of fcc is bcc and the reciprocal lattice of bcc is fcc. The first Brillouin zones of these two types of lattices are shown in Fig. 4.7.

4.2.3 The Born–von Karman Boundary Condition

The general method for solving crystal vibrations discussed above can actually only be applied to an infinitely large crystal. A real crystal has a finite size and those atoms on the surface vibrate differently from the atoms in the interior. The forces between atoms are short ranged; thus there is only a small minority of surface atoms whose vibrations would be governed by equations of motion different from those discussed above. What complicates the matter is that the equations of motion for vibrations of atoms are all coupled together. In order to overcome the mathematical difficulty in solving the simultaneous equations, Born and von Karman proposed a periodic boundary condition [7]. For a one-dimensional chain of N atoms, it is modeled by a ring of N atoms as shown in Fig. 4.8. If N is very large, the curvature is small, and motion along the circumference is approximately the same as a one-dimensional motion along a straight

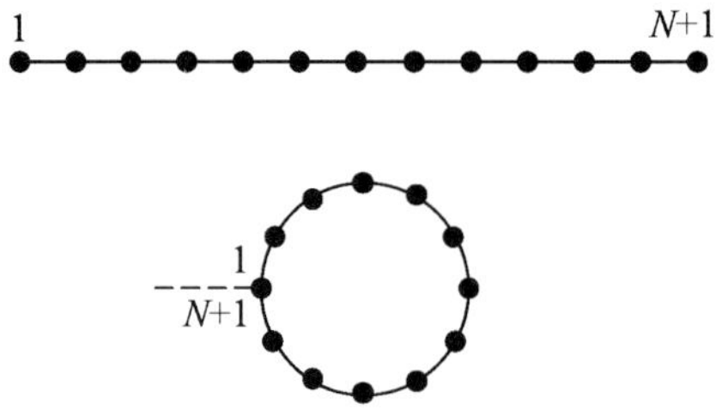

Fig. 4.8 Born–von Karman boundary condition for a one-dimensional chain.

line. The Born–von Karman boundary condition for a monatomic linear chain requires that the vibration of the first atom is the same as that of the $(N+1)$th atom:

$$u(1) = u(N+1).$$

According to Eq. (4.39), we have

$$e^{iNak} = 1 \tag{4.61}$$

which means

$$k = \frac{2\pi}{Na} h \tag{4.62}$$

where h is an integer and a is the nearest neighbor distance in the chain. Since k is restricted in the first Brillouin zone $(-\pi/a < k \leq \pi/a)$, there are only N possible integer values for h $(-N/2 < h \leq N/2)$. These N different k values correspond to N different lattice waves, which coincides with the total number of unit cells in the chain, indicating that we have obtained all the possible normal modes.

For two- or three-dimensional crystals, similar boundary conditions are imposed. For example, the basis vectors for a three-dimensional primitive cell are $\boldsymbol{a}_1$, $\boldsymbol{a}_2$, and $\boldsymbol{a}_3$, and there are a total of $N = N_1 N_2 N_3$ primitive cells where each N_i is a large number. The boundary condition requires that

$$e^{iN_i \boldsymbol{a}_i \cdot \boldsymbol{k}} = 1 \quad (i = 1, 2, 3).$$

The only difference is that a diagram depicting the three-dimensional boundary condition would be much more complicated than Fig. 4.8 [5].

The addition of the periodic boundary condition does not change the solutions to the equations of motion, nor the dispersion relations. The only difference is that k-values are no longer continuous, but discrete. The limited number of discrete values that k can take is equal to the total number of primitive cells in the crystal. The continuous lattice vibrational spectrum becomes a discrete one.

4.2.4
Acoustic and Optical Branches

A slightly complicated one-dimensional lattice is the diatomic chain, each unit cell containing two atoms. Solving the vibrations of this system is mathematically simple, but it is important because it already contains most of the essential concepts of the dynamics of atoms in a crystal [8].

Let m and M be the mass values of the two atoms, a be the interatomic distance at equilibrium, and u and v be their displacements along the chain (Fig. 4.9). Suppose each atom only interacts with its two nearest neighbors with a force constant α. The net forces on the atoms m and M in the lth cell are respectively

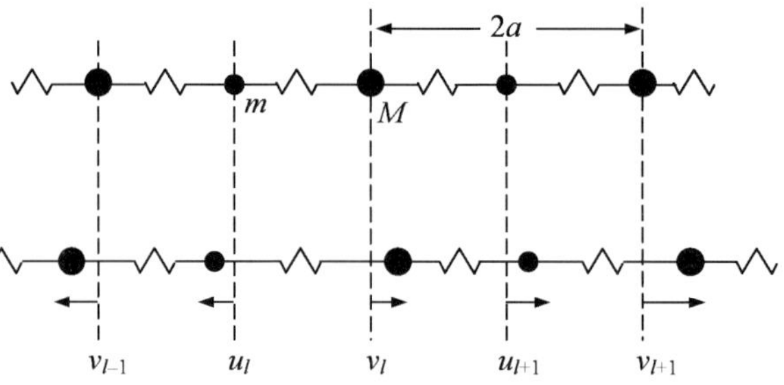

Fig. 4.9 Equilibrium positions and displacements of a one-dimensional diatomic chain.

$\alpha(v_l - u_l) - \alpha(u_l - v_{l-1})$ and $\alpha(u_{l+1} - v_l) - \alpha(v_l - u_l)$. The corresponding equations of motion for the two types of atoms are

$$\begin{aligned} m\ddot{u}_l &= \alpha(v_l + v_{l-1} - 2u_l) \\ M\ddot{v}_l &= \alpha(u_{l+1} + u_l - 2v_l). \end{aligned} \tag{4.63}$$

For a chain with N unit cells, there are a total of $2N$ simultaneous equations. We will try to solve Eq. (4.63) using traveling waves of the following forms:

$$\begin{aligned} u_l &= A\mathrm{e}^{\mathrm{i}[2lka-\omega t]} \\ v_l &= B\mathrm{e}^{\mathrm{i}[(2l+1)ka-\omega t]}. \end{aligned} \tag{4.64}$$

Substituting these into (4.63), we obtain

$$\begin{aligned} -m\omega^2 A &= \alpha[B(\mathrm{e}^{ika} + \mathrm{e}^{-ika}) - 2A] \\ -M\omega^2 B &= \alpha[A(e^{ika} + e^{-ika}) - 2B]. \end{aligned} \tag{4.65}$$

These two equations do not depend on l, indicating that all pairs of equations in (4.63) are reduced to the same equations in (4.65) as long as the solutions are in the form of a lattice wave. What we have in (4.65) are two homogeneous linear equations for A and B, and the condition for nontrivial solutions is

$$\begin{vmatrix} m\omega^2 - 2\alpha & 2\alpha\cos ka \\ 2\alpha\cos ka & M\omega^2 - 2\alpha \end{vmatrix} = 0, \tag{4.66}$$

from which we obtain two solutions for ω^2:

$$\omega_{\mathrm{LA}}^2 = \alpha\left(\frac{1}{m} + \frac{1}{M}\right) - \alpha\left[\left(\frac{1}{m} + \frac{1}{M}\right)^2 - \frac{4}{mM}\sin^2 ka\right]^{1/2}, \tag{4.67}$$

$$\omega_{\mathrm{LO}}^2 = \alpha\left(\frac{1}{m} + \frac{1}{M}\right) + \alpha\left[\left(\frac{1}{m} + \frac{1}{M}\right)^2 - \frac{4}{mM}\sin^2 ka\right]^{1/2}. \tag{4.68}$$

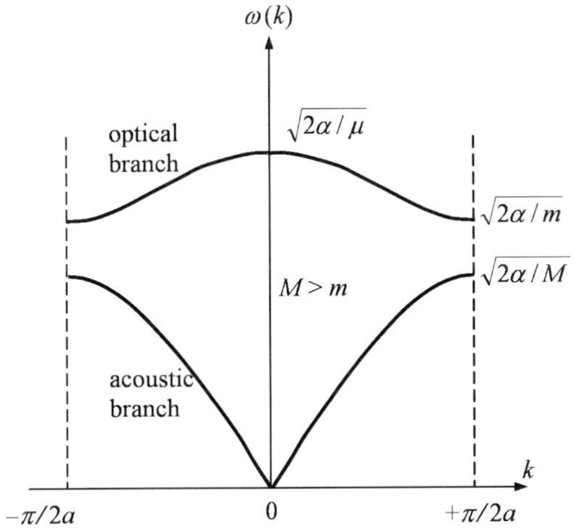

Fig. 4.10 Dispersion curves of a one-dimensional diatomic chain.

The first relation is for a longitudinal acoustic branch $\omega_{\rm LA}(k)$ and the second for a longitudinal optical branch $\omega_{\rm LO}(k)$. Their dispersion relations are shown in Fig. 4.10.

If $m = M$, the acoustic branch dispersion relation becomes

$$\omega_{\rm LA}(k) = 2\sqrt{\frac{\alpha}{M}}\left|\sin\frac{ka}{2}\right| \tag{4.69}$$

which is just the dispersion relation for a one-dimensional monatomic chain.

We now impose the same periodic boundary condition on the diatomic chain with N unit cells. There are N possible integer k-values in the first Brillouin zone:

$$-\frac{\pi}{2a} < k \leq \frac{\pi}{2a}.$$

For every k-value, (4.67) and (4.68) give two frequencies. Therefore, we obtain a total of $2N$ lattice waves, a complete set of normal modes.

Let us analyze some characteristics of lattice waves in the two branches $\omega_{\rm LA}$ and $\omega_{\rm LO}$.

1. In the limit of $k \to 0$. For the acoustic branch dispersion relation, $\sin^2 ka \approx (ka)^2$, and expanding Eq. (4.67) for small k gives

$$\omega_{\rm LA} \approx a\sqrt{\frac{2\alpha}{m+M}}k. \tag{4.70}$$

This shows two major characteristics of the acoustic branch: $\omega_{\rm LA}$ is proportional to k and when $k = 0$, $\omega_{\rm LA} = 0$. Furthermore, the ratio of the vibration amplitudes of the two atoms is

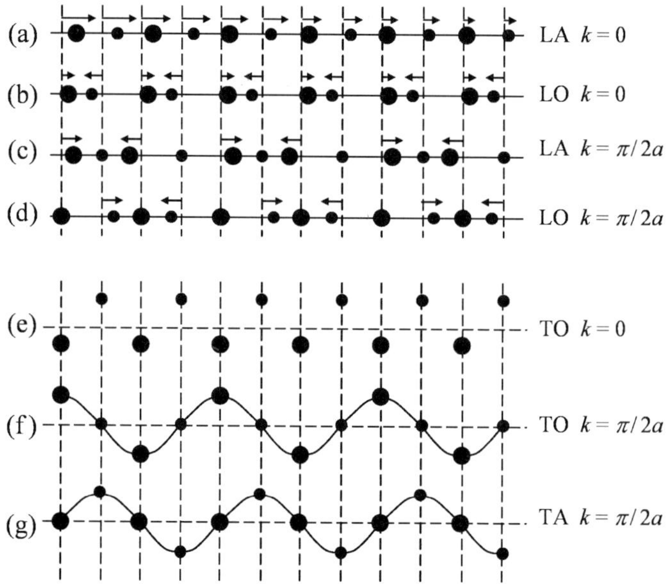

Fig. 4.11 Schematic illustration of displacements of a one-dimensional diatomic chain [9].

$$(A/B)_{\mathrm{LA}} \approx 1. \tag{4.71}$$

This indicates that when $k \to 0$ (long wavelength), the two types of atoms vibrate with nearly the same amplitude and the same phase. The wavelength is much longer than the dimension of the unit cell. If all the atoms are moving towards the right within one half wavelength, then the atoms in the next half wavelength are all moving towards the left, with the linear atomic density varying like a wave, as shown in Fig. 4.11(a). In this case, the lattice may be treated as a continuum, and a long-wavelength lattice wave can be regarded as an elastic wave. It can be shown that the phase velocities of the long waves and the continuum elastic waves (sound waves) are the same, and hence the name "acoustic branch."

For the optical branch, when $k \to 0$ (long wavelength), we obtain the following dispersion relation:

$$\omega_{\mathrm{LO}} \approx \sqrt{2\alpha \Big/ \left(\frac{mM}{m+M}\right)} \tag{4.72}$$

and the ratio of amplitudes

$$(A/B)_{\mathrm{LO}} = -M/m. \tag{4.73}$$

The dispersion relation shows that the frequency is independent of wave vector k. The amplitude ratio indicates that the two types of atoms vibrate in opposite directions and the center of mass of each unit cell remains stationary as shown in Fig. 4.11(b).

Long-wavelength optical modes in ionic crystals can absorb infrared waves, and light can be used to excite vibrations in the optical branch, hence the name "optical branch."

2. When $k = \pm\pi/2a$. In this case

$$\begin{aligned} \omega_{\mathrm{LA}} &= \sqrt{2\alpha/M}, \quad \frac{A}{B} = 0, \text{ or } A = 0, \\ \omega_{\mathrm{LO}} &= \sqrt{2\alpha/m}, \quad \frac{A}{B} = \infty, \text{ or } B = 0. \end{aligned} \tag{4.74}$$

For the acoustic branch, atom m is stationary while for the optical branch, M remains stationary (Figs. 4.11(c) and (d)). Compared to a monatomic chain, the most important feature of a diatomic chain is the addition of an optical branch. Furthermore, there is a gap between the dispersion curves ($\omega_{\mathrm{LO}} - \omega_{\mathrm{LA}}$), and the Brillouin zone is only half as large.

4.2.5
Longitudinal and Transverse Waves

The lattice waves described in the last section are longitudinal waves (denoted by the subscript L), propagating along the direction of atomic vibrations. If a wave is propagating in a direction perpendicular to atomic vibrations, it is a transverse wave (denoted by T). A perfect one-dimensional chain of atoms does not produce transverse waves, but a real crystal can have both longitudinal and transverse waves [9].

So far, a one-dimensional lattice vibration has been solved by applying Newton's laws. For a three-dimensional crystal, the method given in Section 4.2.1 should be used. In this section, we use an fcc crystal as an example to show how a transverse wave can exist and, more importantly, to demonstrate the process of solving a general problem of lattice dynamics.

In order to solve Eq. (4.40) we must first find the dynamical matrix. In an fcc crystal, each atom (or ion) has 12 nearest neighbors (Fig. 4.12), and the corresponding 12 force constant matrices are given in Appendix C (actually only 6 distinctive matrices). The latter fact indicates that the dynamical matrix can be expressed as follows:

$$\boldsymbol{D} = \sum_{i=1}^{6} \boldsymbol{D}_i + \frac{1}{M}\boldsymbol{\Phi}(0,0) \tag{4.75}$$

where $\boldsymbol{\Phi}(0,0)$ is the self-force constant matrix of the atom at the origin, and it can be calculated by Eq. (4.21):

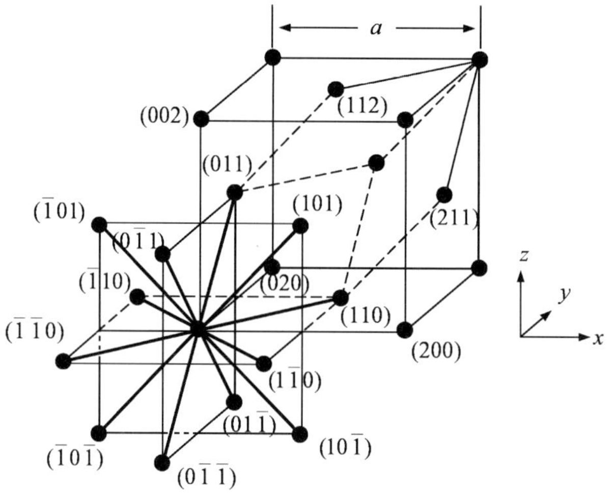

Fig. 4.12 Coordinates (x, y, z) of the 12 nearest neighbors of the atom at the origin in an fcc crystal.

$$\boldsymbol{\Phi}(0,0) = 2\begin{bmatrix} 4\alpha+2\beta & 0 & 0 \\ 0 & 4\alpha+2\beta & 0 \\ 0 & 0 & 4\alpha+2\beta \end{bmatrix}.$$

Suppose $\boldsymbol{D}_1$ is the contribution to $\boldsymbol{D}$ from two nearest neighbors at $\pm\frac{a}{2}(1,1,0)$, then [4]

$$\begin{aligned} \boldsymbol{D}_1 &= -\frac{1}{M}\begin{bmatrix} \alpha & \gamma & 0 \\ \gamma & \alpha & 0 \\ 0 & 0 & \beta \end{bmatrix}\left\{\exp\left[i\boldsymbol{k}\cdot\frac{a}{2}(1,1,0)\right] + \exp\left[-i\boldsymbol{k}\cdot\frac{a}{2}(1,1,0)\right]\right\} \\ &= -\frac{2}{M}\cos\left(\frac{ak_1}{2}+\frac{ak_2}{2}\right)\begin{bmatrix} \alpha & \gamma & 0 \\ \gamma & \alpha & 0 \\ 0 & 0 & \beta \end{bmatrix} \end{aligned} \tag{4.76}$$

where k_1, k_2, and k_3 are the components of $\boldsymbol{k}$ along the cubic crystal axes. The other five matrices $\boldsymbol{D}_2$ to $\boldsymbol{D}_6$ can also be written down in a similar manner. Making a sum of these contributions, we obtain $\boldsymbol{D}$ with the following elements:

$$\begin{aligned} D_{11} &= \frac{4\alpha}{M}\left[2 - \cos\frac{ak_1}{2}\left(\cos\frac{ak_2}{2}+\cos\frac{ak_3}{2}\right)\right] \\ &\quad + \frac{4\beta}{M}\left[1-\cos\frac{ak_2}{2}\cos\frac{ak_3}{2}\right], \end{aligned} \tag{4.77}$$

and D_{22} and D_{33} have similar expressions, the off-diagonal components being

$$D_{12} = D_{21} = \frac{4\gamma}{M} \sin\frac{ak_1}{2} \sin\frac{ak_2}{2},$$

$$D_{13} = D_{31} = \frac{4\gamma}{M} \sin\frac{ak_1}{2} \sin\frac{ak_3}{2}, \tag{4.78}$$

$$D_{23} = D_{32} = \frac{4\gamma}{M} \sin\frac{ak_2}{2} \sin\frac{ak_3}{2}.$$

The first Brillouin zone is shown in Fig. 4.7. Let us calculate the vibrational modes along the [100] direction, where $\boldsymbol{k} = \frac{2\pi}{a}(\zeta, 0, 0)$. Here ζ is the reduced wave number along the [100] direction, and it changes from 0 (at the origin Γ) to 1 (at the zone boundary X). Using this notation

$$\frac{ak_1}{2} = \pi\zeta, \quad \frac{ak_2}{2} = \frac{ak_3}{2} = 0,$$

the corresponding dynamic matrix can then be simplified to

$$\boldsymbol{D} = \frac{4}{M}(1 - \cos\pi\zeta)\begin{bmatrix} 2\alpha & 0 & 0 \\ 0 & \alpha+\beta & 0 \\ 0 & 0 & \alpha+\beta \end{bmatrix}. \tag{4.79}$$

In this case, $\boldsymbol{D}$ is already diagonal, and we can immediately write down the eigenvalues:

$$\omega_1^2 = \frac{8\alpha}{M}(1 - \cos\pi\zeta),$$

$$\omega_2^2 = \omega_3^2 = \frac{4(\alpha+\beta)}{M}(1 - \cos\pi\zeta).$$

Substituting each eigenvalue back into the original equation (Eq. (4.40)) to find the corresponding eigenvector:

$$\mathbf{e}(1) = \begin{bmatrix} 1 \\ 0 \\ 0 \end{bmatrix} \quad \text{(L mode)},$$

$$\mathbf{e}(2) = \begin{bmatrix} 0 \\ 1 \\ 0 \end{bmatrix} \quad (\mathrm{T_1\ mode}), \quad \mathbf{e}(3) = \begin{bmatrix} 0 \\ 0 \\ 1 \end{bmatrix} \quad (\mathrm{T_2\ mode}). \tag{4.80}$$

These are three unit vectors, none depending on $\boldsymbol{k}$, in the x-, y-, and z-directions, respectively. As mentioned above, α is the longitudinal force constant, so ω_1 is the

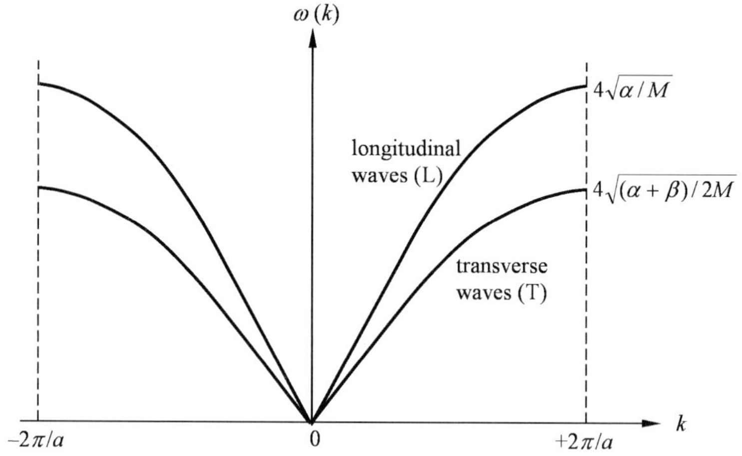

Fig. 4.13 Dispersion curves along the [100] direction in an fcc crystal.

longitudinal wave frequency. Inspecting Fig. 4.4 and the matrix in (4.79), we can see that ω_2 and ω_3 represent vibrations perpendicular to the [100] direction, i.e., transverse waves. Therefore, the dispersion relations are

$$\omega_{\mathrm{LA}} = 4\sqrt{\frac{\alpha}{M}}\sin\frac{ak}{4}, \tag{4.81}$$

$$\omega_{\mathrm{TA}} = 4\sqrt{\frac{\alpha+\beta}{2M}}\sin\frac{ak}{4}, \tag{4.82}$$

and the dispersion curves are shown in Fig. 4.13.

It is interesting to look at the patterns of atomic displacements for a mode at the zone center ($\zeta = 0$) or the zone boundary ($\zeta = 1$). Using the basis vectors $\boldsymbol{a}_i$ for the fcc lattice and Eq. (4.8), we can calculate the vector $\boldsymbol{l}$ for any atom in the crystal, $l = \frac{a}{2}(l_1 + l_3, l_1 + l_2, l_2 + l_3)$, where l_1, l_2, and l_3 are integers. In the case of $\zeta = 1$, the atomic displacement is given by Eq. (4.39) whose exponent is

$$\boldsymbol{k}\cdot\boldsymbol{l} = \frac{2\pi}{a}(1,0,0)\cdot\frac{a}{2}(l_1 + l_3, l_1 + l_2, l_2 + l_3) = (l_1 + l_3)\pi = \pm s\pi. \tag{4.83}$$

Here $s = l_1 + l_3$ is also an integer labeling successive lattice planes all parallel to the yz-plane and perpendicular to the wave propagation direction. When this is substituted into (4.39), we obtain the displacement for an atom in the sth plane:

$$\boldsymbol{u}(s) = \boldsymbol{u}(0)\mathrm{e}^{is\pi}\mathrm{e}^{-i\omega t} \tag{4.84}$$

which happens to be a standing wave, not a traveling wave. All the atoms in the same plane have the same phase; for even s-values, $\mathrm{e}^{is\pi} = 1$, and for odd s-values,

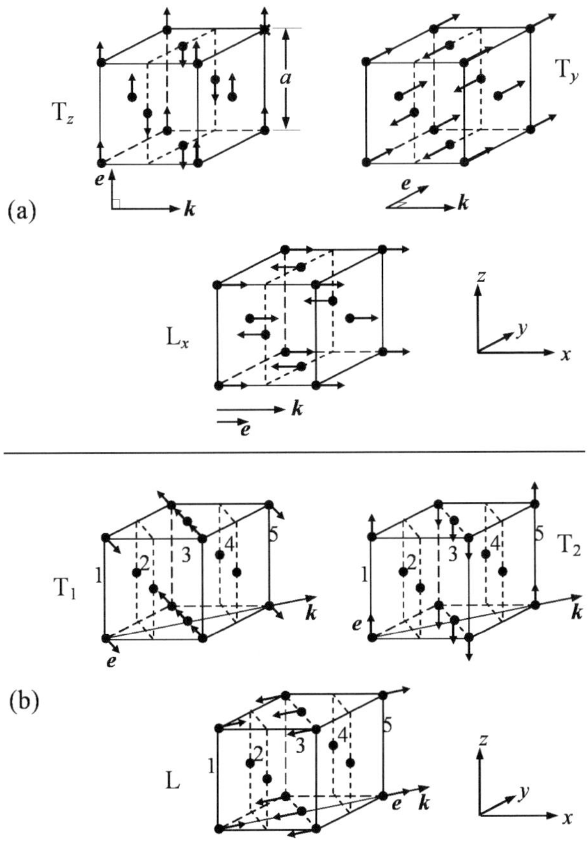

Fig. 4.14 Schematic representations of atomic displacement vectors in an fcc crystal [8]: (a) at the Brillouin zone boundary $\boldsymbol{k} = \frac{2\pi}{a}(1, 0, 0)$; (b) for $\boldsymbol{k} = \frac{2\pi}{a}\left(\frac{1}{2}, \frac{1}{2}, 0\right)$, alternate odd planes move in opposite directions while the even planes are stationary, $\lambda = \sqrt{2}a$.

$e^{is\pi} = -1$; therefore atoms in alternate planes have opposite phases (Fig. 4.14(a)). The wavelength of the standing wave is $\lambda = 2\pi/k = a$, which is consistent with Bragg's condition, $\lambda = 2d \sin\theta$. When $\theta = \pi/2$, $d = a/2$ is exactly the interplanar separation. We conclude that a vibration mode with a wave vector $\boldsymbol{k}$ at the Brillouin zone boundary does not propagate in the crystal, but is repeatedly reflected like a standing wave.

It should be noted that during the construction of $\boldsymbol{D}(\boldsymbol{k})$ only the first nearest neighbor atoms are involved, but in practice one has to include interactions at least out to the fifth nearest neighbors.

Now we turn our attention to the modes that propagate in the [110] direction. The wave vector is $\boldsymbol{k} = (2\pi/a)(\zeta, \zeta, 0)$ with ζ running from 0 (at the origin) to 3/4 (at point K). Following the above procedure, we obtain the corresponding dispersion relations:

$$\begin{aligned}
\omega_1^2 &= \omega_{\mathrm{LA}}^2 = \frac{8}{M}(\alpha+\beta)\sin^2\left(\frac{\pi\zeta}{2}\right) + \frac{4}{M}(\alpha+\gamma)\sin^2\pi\zeta, \\
\omega_2^2 &= \omega_{\mathrm{TA}_1}^2 = \frac{8}{M}(\alpha+\beta)\sin^2\left(\frac{\pi\zeta}{2}\right) + \frac{4}{M}(\alpha-\gamma)\sin^2\pi\zeta, \\
\omega_3^2 &= \omega_{\mathrm{TA}_2}^2 = \frac{16\alpha}{M}\sin^2\left(\frac{\pi\zeta}{2}\right) + \frac{4\beta}{M}\sin^2\pi\zeta,
\end{aligned} \tag{4.85}$$

and the eigenvectors

$$\begin{aligned}
&\mathbf{e}(1) = \frac{1}{\sqrt{2}}\begin{bmatrix}1\\1\\0\end{bmatrix} \quad (\text{LA mode}), \\
&\mathbf{e}(2) = \frac{1}{\sqrt{2}}\begin{bmatrix}-1\\1\\0\end{bmatrix} \quad (\text{TA}_1\ \text{mode}), \quad \mathbf{e}(3) = \begin{bmatrix}0\\0\\1\end{bmatrix} \quad (\text{TA}_2\ \text{mode}).
\end{aligned} \tag{4.86}$$

The patterns of atomic displacements for $\boldsymbol{k} = \frac{2\pi}{a}\left(\frac{1}{2},\frac{1}{2},0\right)$ are shown in Fig. 4.14(b).

The modes along the [111] direction can also be calculated, and the results are similar to those for [100], with sinusoidal functions for ω and the degenerate transverse waves [4].

The solutions in the above examples are relative simple, where the eigenvectors are all independent of the vectors $\boldsymbol{k}$, which are along three highly symmetric directions. Hence the direction of atomic displacement is determined by the crystal symmetry rather than the force constant, and both L-mode and T-mode are strictly pure.

In a general case, the eigenvectors will depend on both the magnitude and the direction of $\boldsymbol{k}$; that is, their orientation relative to $\boldsymbol{k}$ depends on the force constants and therefore the modes will not be purely longitudinal or purely transverse.

One final point is that if each primitive cell has r atoms, the number of mode branches is $3r$. Among these, there are three acoustic branches, one longitudinal (LA) and two transverse (TA) modes. The rest $(3r-3)$ branches are optical branches, also characterized by longitudinal (LO) and transverse (TO) modes.

4.2.6
Models of Interatomic Forces in Solids

Although this is one of the important issues in lattice dynamics, we will only briefly discuss the methodologies here. In the general principles of the Born–

von Karman theory described above, the force constants (α, β, γ, etc.) in the dispersion relations $\omega_j(\boldsymbol{k})$ are still unknown. There are generally two methods for calculating the vibration frequencies $\omega_j(\boldsymbol{k})$. One is a phenomenological approach and the other is a microscopic theory of lattice vibrations in which a general expression for the force constants is derived from first principles based on the electronic structure of the solid. Here we will confine ourselves to the phenomenological models of interatomic forces, the details of which can be found in Ref. [8]. In this approach the force constants (α, β, γ, etc.) are considered as parameters adjusted to fit experimentally observed vibration frequencies $\omega_j(\boldsymbol{k})$ or derived from an empirical potential model for the crystal under investigation to calculate $\omega_j(\boldsymbol{k})$. A good empirical potential model should contain a few feasible and physically meaningful parameters.

When inert gases Ne, Ar, Kr, and Xe solidify at low temperatures, they form fcc crystals. The interatomic interaction is characterized by a two-body central potential, the most successful being the Lennard-Jones potential expressed as

$$V(r) = 4\varepsilon\left[\left(\frac{\sigma}{r}\right)^{12} - \left(\frac{\sigma}{r}\right)^{6}\right] \tag{4.87}$$

where r is the distance between the atoms, the first term represents the van der Waals attraction, the second term represents the repulsive interaction, ε is the minimum value of the potential energy, and σ is the minimum distance between two atoms when $V(r) = 0$. The two parameters σ and ε can be determined from the measured lattice constant a_0 and the heat of sublimation $-L_0$, respectively.

From this empirical potential one can obtain the force constants and then phonon dispersion. For example, the observed $\omega_j(\boldsymbol{k})$ curves of ^{36}Ar at 10 K are satisfactorily consistent with the theoretical calculated dispersion relations [10]. This model has also worked quite well with a number of simple metals (such as aluminum and alkali metals).

For ionic crystals in which the ions are not polarizable, the Born–Mayer potential is generally applied. It consists of a short-range repulsive term and a Coulomb interaction term. But when the polarizabilities of ions and electrons need to be included, the shell model must be used. In the latter, an atom is represented by an unpolarizable ion core and a shell of valence electrons, and an electric dipole is generated by the relative displacement of the shell with respect to the core [11].

Covalent crystals are clearly distinct from other crystals, because the covalent bonds are highly anisotropic. Therefore, alternative models have been developed which use angular (the Keating model [12]) or bond charge (the bond-charge model [13]) to simulate the effects of the highly anisotropic distribution of electrons in these crystals. Recently, a multibody empirical potential has been proposed in the form of Morse pair potentials, which has provided more accurate descriptions of the interatomic forces in covalent systems [14, 15].

4.3 Quantization of Vibrations: The Phonons

In the previous sections we treated a normal mode in a crystal as equivalent to an oscillator. Quantum mechanically, the eigenvectors of the crystal vibration can be represented as a product of wave functions of a one-dimensional harmonic oscillator. The energy of a particular normal mode with ω is quantized and given by

$$E_n = \left(n + \frac{1}{2}\right)\hbar\omega, \quad n = 0, 1, 2, \ldots \tag{4.88}$$

where n is a positive integer (including zero), and $\frac{1}{2}\hbar\omega$ is the oscillator energy at absolution zero (zero point energy). At nonzero temperatures, a series of higher energy levels ($n = 1, 2, 3, \ldots$) are excited, the energy difference between adjacent levels being $\hbar\omega$ (Fig. 4.15). These quantum states for a particular normal mode can be described by a set of integers $n = 0, 1, 2, 3, \ldots$, in units of $\hbar\omega$. Analogous to the photon for the electromagnetic field, the energy quantum $\hbar\omega$ for the lattice waves is called a phonon. We see that one lattice wave or one type of vibration produces one type of phonons, and $n = 1, 2$, or $3, \ldots$ is the number of phonons of frequency ω. Two equivalent ways for describing a normal mode are shown in Fig. 4.15. The minimum energy exchanged between a γ-photon (or an electron) and the lattice is one phonon. The normal mode methodology is possible only because of the harmonic approximation. Therefore, the introduction of the phonon concept is a direct consequence of this approximation.

Phonons are nonlocalized quasi-particles. Atoms or nuclei are the real particles participating in the vibrations. Phonons are the energy quanta of their collective motion. A phonon as a quasi-particle does not have mass, and it is impossible to get a "phonon beam" out of a crystal. The wavelength of a phonon is usually quite long; therefore it is a nonlocalized state.

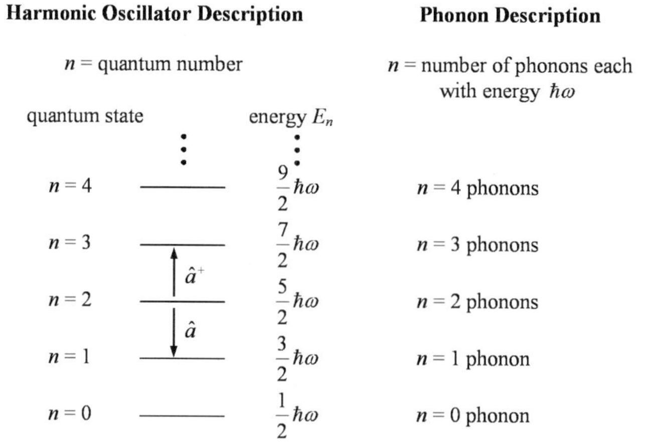

Fig. 4.15 Two different descriptions of a normal mode (ω).

A phonon in a crystal does possess momentum. When it interacts with a particle such as a photon or a neutron, it behaves as if it has a momentum of $\hbar\boldsymbol{k}$. Therefore, $\hbar\boldsymbol{k}$ is called the quasi-momentum of a phonon or the crystal momentum.

Phonons are bosons. Each lattice wave corresponds to one type of phonons, and they are all identical particles with zero spin (bosons). The higher the temperature is, the larger the amplitudes of lattice wave, and consequently the higher the average energy and the higher the average number of phonons. Therefore, the total number of phonons is not conserved. In a thermal equilibrium, the average phonon number $\langle n(\boldsymbol{k}j)\rangle$ is given by the Bose–Einstein statistics

$$\langle n(\boldsymbol{k}j)\rangle = \frac{1}{\exp[\hbar\omega_j(\boldsymbol{k})\beta] - 1} \tag{4.89}$$

where $\beta = 1/k_B T$. The average phonon energy for mode $\boldsymbol{k}j$ is

$$\langle E(\omega_j(\boldsymbol{k}), T)\rangle = \left[\langle n(\boldsymbol{k}j)\rangle + \frac{1}{2}\right]\hbar\omega_j(\boldsymbol{k}). \tag{4.90}$$

The total energy of the lattice is the sum of the above over all the vibration modes

$$\langle E(T)\rangle = \sum_{\boldsymbol{k},j}\langle E(\omega_j(\boldsymbol{k}), T)\rangle = \sum_{\boldsymbol{k},j}\frac{\hbar\omega_j(\boldsymbol{k})}{\exp[\hbar\omega_j(\boldsymbol{k})\beta] - 1} + \frac{1}{2}\sum_{\boldsymbol{k},j}\hbar\omega_j(\boldsymbol{k}) \tag{4.91}$$

where the second term is temperature independent.

The introduction of the phonon concept transforms the study of lattice vibrations to a problem similar to that of an ideal gas – a phonon gas system. When γ-photons or neutrons are scattered by a solid, phonons can be created or annihilated, and thus the dispersion relation $\omega_j(\boldsymbol{k})$ can be measured by such experiments. A process in which no phonons are created or annihilated is precisely the recoilless process in Mössbauer experiments.

Note that the phonon concept can be extended to disordered solids, where a phonon represents a quantum of atomic vibration energy but without the crystal momentum $\hbar\boldsymbol{k}$.

4.4 Frequency Distribution and Thermodynamic Properties

4.4.1 The Lattice Heat Capacity

The heat capacity c_V of a solid at constant volume is conventionally defined as

$$c_V = \left(\frac{\partial \bar{E}}{\partial T}\right)_V$$

where $\bar{E}$ is the average internal energy of the solid.

It has two contributions, one from lattice vibrations (lattice heat capacity) and the other from the thermal motion of electrons (electronic heat capacity). When the temperature is not too low, electronic heat capacity can be neglected because it is much smaller than lattice heat capacity. In this section, we discuss lattice heat capacity only.

According to the classical equipartition of energy, the average energy of each harmonic motion is $k_B T$. If there are N atoms in the solid, the total number of harmonic vibrations is $3N$, and the average total energy is $3Nk_B T$. From this, we obtain $c_v = 3Nk_B$, which indicates that the lattice heat capacity is independent of the material's properties and temperature. This is the well-known law of Dulong and Petit. For high temperatures, experiments agree with this law, but at low temperatures, c_v is no longer a constant and decreases as temperature drops. When $T \to 0$, c_v goes to zero for all solids. Quantum mechanics is required to explain the low-temperature behavior of lattice heat capacity.

From the previous section, we know that Eq. (4.90) is the energy of a lattice vibration mode. The temperature-dependent part of total energy in Eq. (4.91) may be replaced by an integral over the frequency distribution, if we treat values of $\omega_j(\boldsymbol{k})$ as almost continuous:

$$\langle E(T)\rangle = \sum_{\boldsymbol{k},j} \frac{\hbar\omega_j(\boldsymbol{k})}{\exp[\hbar\omega_j(\boldsymbol{k})\beta] - 1} = 3N \int_0^{\omega_m} \frac{\hbar\omega}{\exp[\hbar\omega\beta] - 1} g(\omega)\,d\omega \tag{4.92}$$

where $g(\omega)$ is the normalized frequency distribution function

$$\int_0^{\omega_m} g(\omega)\,d\omega = 1 \tag{4.93}$$

and ω_m is the maximum frequency. $g(\omega)$ is termed the phonon spectrum or the density of states (DOS). It describes the probability of lattice waves having frequencies between ω and $\omega + d\omega$, and hereafter it will be called DOS. The derivative of (4.92) with respect to temperature gives c_v:

$$\begin{aligned} c_v &= 3N \int_0^{\omega_m} k_B(\hbar\omega\beta)^2 \frac{\exp(\hbar\omega\beta)}{[\exp(\hbar\omega\beta) - 1]^2} g(\omega)\,d\omega \\ &= 3Nk_B \int_0^{\omega_m} \left[\frac{\hbar\omega\beta/2}{\sinh(\hbar\omega\beta/2)}\right]^2 g(\omega)\,d\omega. \end{aligned} \tag{4.94}$$

From this result, we see that when $T \to 0$, c_v indeed approaches zero. At high temperatures, $k_B T \gg \hbar\omega$, c_v approaches $3Nk_B$. Both of these limits agree with experiments. Now we need to find an appropriate frequency distribution function so that the integral predicts c_v for intermediate temperatures as well.

4.4.2
The Density of States

Generally speaking, as long as we know the dispersion relations $\omega_j(\boldsymbol{k})$, for all $\boldsymbol{k}$ in the first Brillouin zone, the density of states can be calculated according to

$$g(\omega) = \frac{1}{3N}\sum_{j}^{3}\sum_{\boldsymbol{k}}^{N}\delta(\omega - \omega_j(\boldsymbol{k})) \tag{4.95}$$

where $g(\omega)$ satisfies the normalization condition (4.93). Figure 4.16 shows the $\omega_j(\boldsymbol{k})$ curves and the corresponding calculated $g(\omega)$ curve for NaF.

Sometimes, it is more convenient to use another distribution function, defined as

$$g_1(\omega^2) = \frac{1}{3N}\sum_{j}^{3}\sum_{\mathrm{k}}^{N}\delta(\omega^2 - \omega_j^2(\boldsymbol{k})) \tag{4.96}$$

where

$$\int_0^{\omega_{\mathrm{m}}^2} g_1(\omega^2)\,\mathrm{d}\omega^2 = 1, \quad g_1(\omega^2) = g(\omega)/2\omega.$$

Recently, the first-principles quantum mechanical method, a very powerful tool, has been used to calculate $g(\omega)$, which appears to be somewhat complex. The two simplified models, namely the Einstein model and the Debye model, have been widely used for a long time. In many cases they can give results consistent with the experimental data.

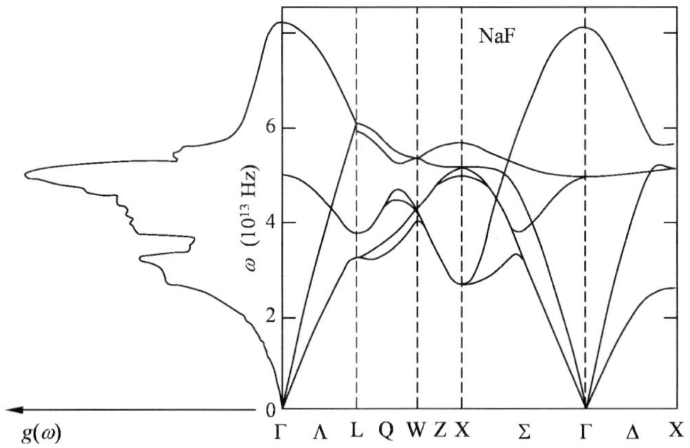

Fig. 4.16 Dispersion curves $\omega_j(\boldsymbol{k})$ and phonon frequency distribution $g(\omega)$ for NaF [16].

4.4.2.1 The Einstein Model

Einstein postulated that all atoms are vibrating independently with the same frequency ω_E. For a three-dimensional lattice at temperature T, the total vibration energy is

$$\bar{E} = \frac{3N\hbar\omega_E}{\exp(\hbar\omega_E\beta) - 1} + \frac{3N}{2}\hbar\omega_E, \tag{4.97}$$

from which the lattice heat capacity can be calculated as

$$c_V = \left(\frac{\partial \bar{E}}{\partial T}\right) = 3Nk_B\left[\frac{\hbar\omega_E\beta/2}{\sinh(\hbar\omega_E\beta/2)}\right]^2 = 3Nk_B\left[\frac{\theta_E/2T}{\sinh(\theta_E/2T)}\right]^2 \tag{4.98}$$

where

$$\theta_E = \hbar\omega_E/k_B \tag{4.99}$$

is called the Einstein temperature. At high temperatures, $k_BT \gg \hbar\omega$, $c_V \approx 3Nk_B$, the classical value. As temperature decreases, c_V decreases, consistent with the trend in the experimental results. But in the low temperature region, the predicted c_V values decrease too fast and do not exactly match the experimental curve (Fig. 4.17). The Einstein model is a good approximation for an optical branch where ω is a weak function of $\mathbf{k}$.

Although the Einstein model played an important role in the development of the quantum theory, it is over simplified. In a real crystal, the interactions between atoms are strong enough so that it is impossible for an atom to oscillate without affecting its neighbors. We now turn to the following more realistic model.

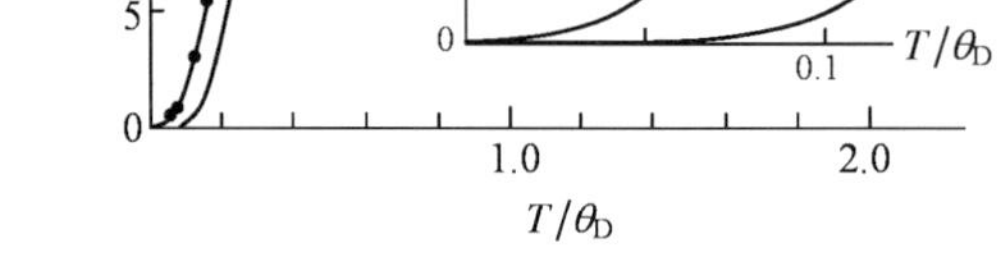

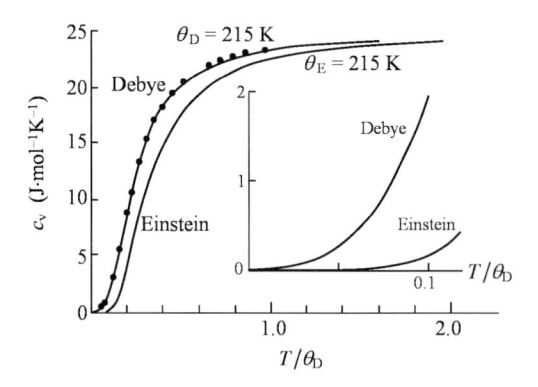

Fig. 4.17 Heat capacity of Ag as a function of temperature T: a comparison between the Einstein model and the Debye model [16].

4.4.2.2 The Debye Model

At low temperatures, the optical branch phonons have energies higher than $k_B T$, and therefore almost none of the optical branch waves is excited. Only acoustic waves (especially long-wavelength ones) contribute to the heat capacity. For an acoustic branch, $\omega \to 0$ as $k \to 0$, and the Einstein model obviously fails to include this feature. The main assumption of the Debye model is that the Bravais lattice is regarded as an isotropic continuum, and therefore the lattice waves are elastic waves (one longitudinal branch and two independent transverse branches). The frequency is not a constant but has a specific distribution with a cutoff frequency ω_D, above which no shorter wave phonons are excited. In the Debye model, $g(\omega)$ takes the following form:

$$g(\omega) = \begin{cases} 3\omega^2/\omega_D^3, & \text{when } \omega < \omega_D, \\ 0, & \text{when } \omega > \omega_D. \end{cases} \tag{4.100}$$

We may also define

$$\theta_D = \hbar\omega_D/k_B, \tag{4.101}$$

which is known as the Debye temperature, an important quantity in solid-state physics. One should note, however, that ω_D is merely a parameter, not the actual maximum phonon frequency in the solid.

Substituting (4.100) into (4.94), we obtain

$$c_V(T) = 9Nk_B \frac{T^3}{\theta_D^3} \int_0^{x_D} \frac{x^4 e^x \, dx}{(e^x - 1)^2} = 9Nk_B f_D(\theta_D/T) \tag{4.102}$$

where $x = \hbar\omega\beta$, $x_D = \hbar\omega_D\beta = \theta_D/T$, and $f_D(\theta_D/T)$ is called the Debye heat capacity function. The Debye model has been very successful in calculating the heat capacities for many solids, which agree well with the experimental results.

When $T \gg \theta_D$, c_V approaches the classical value of $3Nk_B$. In the low-temperature region, Debye's heat capacity is remarkable, because when $T \ll \theta_D$, Eq. (4.102) becomes

$$c_V = \frac{12\pi^4}{5} Nk_B \left(\frac{T}{\theta_D}\right)^3. \tag{4.103}$$

Here c_V is proportional to T^3, known as the Debye T^3 law. The lower the temperature, the better the Debye approximation, because almost all excited phonons belong to the long-wavelength waves in the acoustic branches and the crystal indeed behaves like a continuum. However, the T^3 law is applicable only for $T < \theta_D/5$.

The Debye temperature θ_D as defined in (4.101) is a temperature-independent parameter. As more sophisticated low-temperature techniques are now available, different θ_D-values have been observed at different temperatures. For many materials, θ_D is a constant for $T \geq \theta_D/2$, it decreases as T decreases, reaches a

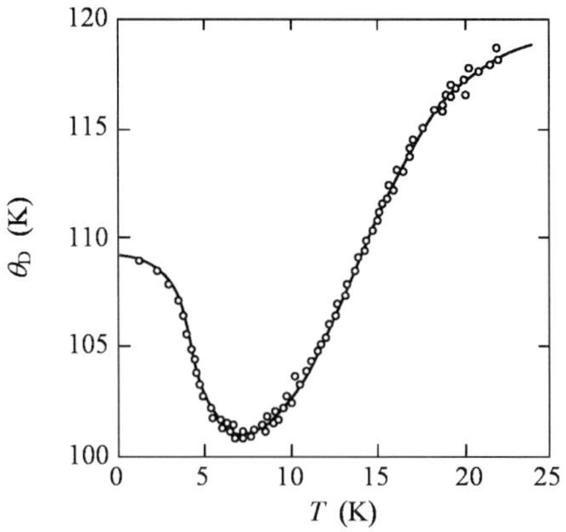

Fig. 4.18 Debye temperature θ_D of indium as a function of temperature T [17].

minimum around $\theta_D/10$, and rebounds at lower temperatures (Fig. 4.18). This indicates that the Debye model has its own limitations, and does not completely agree with experimental results.

In fact, the crystal cannot be completely treated as an elastic medium, but should be modeled based on its atomic structures as in the Born–von Karman theory. Neutron scattering experiments and theoretical calculations have shown that each crystal has its own frequency distribution $g(\omega)$, and c_V cannot be accurately calculated unless $g(\omega)$ is available.

4.4.3
Moments of Frequency Distribution

Today, because of the advances in computer science and neutron scattering, phonon spectra $g(\omega)$ have been obtained theoretically or experimentally for a number of perfect crystals, but for most of solids $g(\omega)$ is still unknown. There exists, however, an approximate method for obtaining thermodynamic quantities similar to c_V without knowing $g(\omega)$ itself. This is an average method using the moments of frequency distribution, and several authors contributed to this method in the early 20th century [18, 19].

The nth moment of the frequency distribution function is defined as

$$\mu(n) = \int_0^{\omega_m} \omega^n g(\omega)\, d\omega \tag{4.104}$$

where ω_m is the maximum allowed frequency. When (4.95) is substituted into (4.104), it becomes

$$\mu(n) = \frac{1}{3N} \sum_{\mathbf{k},j} \omega_j^n(\mathbf{k}). \tag{4.105}$$

The even moments $\mu(2n)$ can be expressed as [20]

$$\mu(2n) = \frac{1}{3N}\sum_{\boldsymbol{k},j}\omega_j^{2n}(\boldsymbol{k}) = \frac{1}{3N}\sum_{\boldsymbol{k}} \operatorname{tr} \boldsymbol{D}^n(\boldsymbol{k}) \tag{4.106}$$

where the trace of $\boldsymbol{D}^n(\boldsymbol{k})$ remains invariant under its diagonalization through a unitary transformation, and the diagonal elements are just $\omega_j^{2n}(\boldsymbol{k})$. The fact that the even moments $\mu(2n)$ can be evaluated from the dynamical matrix makes them vary useful in describing the characteristics of the unknown $g(\omega)$.

Now we derive c_V in terms of $\mu(2n)$, the moments of frequency distribution [19]. One uses the series expansion

$$\frac{x}{e^x - 1} = 1 - \frac{x}{2} - \sum_{n=1}^{\infty}(-1)^n B_{2n}\frac{x^{2n}}{(2n)!} \quad (|x| < 2\pi) \tag{4.107}$$

where B are the Bernoulli numbers:

$$B_2 = \frac{1}{6}, \quad B_4 = \frac{1}{30}, \quad B_6 = \frac{1}{42}, \quad B_8 = \frac{1}{30},$$

$$B_{10} = \frac{5}{66}, \quad B_{12} = \frac{691}{2730}, \quad B_{14} = \frac{7}{6}, \quad \text{etc.}$$

Expanding the lattice energy (4.92) according to (4.107) and taking a derivative with respect to T, we obtain an expression for c_V:

$$\frac{c_V}{3Nk_B} = 1 + \sum_{n=1}^{\infty}(-1)^n B_{2n}\frac{2n-1}{(2n)!}\left(\frac{\hbar}{k_B T}\right)^{2n}\mu(2n). \tag{4.108}$$

This series converges for $T > 50$ K [21]. In fact, only several low moments (e.g., up to $n = 3$) are usually required to obtain a relatively good accuracy. This is the most prominent feature of this method [22].

The moment method cannot give the low-temperature characteristics of c_V. Although Montroll [19] pointed out long ago that $g(\omega)$ could be calculated if all moments were known, high moments are usually not available because their calculations are very complicated. If only a limited number of moments are used, the resultant low-temperature phonon spectrum is only a poor approximation. In the 1970s, modified moments were introduced, and this method was further developed [23–25]. It has become a useful theoretical method in lattice dynamics [26–28].

Just like the lattice heat capacity, there are some other quantities that need to be evaluated as statistical averages using $g(\omega)$. Such quantities include $\langle u^2\rangle$ and $\langle v^2\rangle$, which are discussed in the next chapter.

Suppose that a Debye spectrum with a cutoff frequency $\omega_D(n)$ has its nth frequency moment $\mu_D(n)$ exactly equal to the nth moment of the actual phonon spectrum $\mu(n)$, we would have

$$\int_0^{\omega_D(n)} \omega^n \frac{3\omega^2}{\omega_D^3(n)} d\omega = \frac{3}{n+3}\omega_D^n(n) = \mu(n)$$

or

$$\omega_D(n) = \left[\frac{n+3}{2}\mu(n)\right]^{1/n}, \quad n > -3,\ n \neq 0. \tag{4.109}$$

The temperature corresponding to this cutoff frequency is called the weighted Debye temperature $\theta_D(n)$, written this way to distinguish it from the usual Debye temperature θ_D:

$$\theta_D(n) = \frac{\hbar\omega_D(n)}{k_B} = \frac{\hbar}{k_B}\left[\frac{n+3}{2}\mu(n)\right]^{1/n}. \tag{4.110}$$

This is how $\mu(n)$ can be calculated [29] using the parameter $\theta_D(n)$, and the importance of $\theta_D(n)$ is analyzed as follows.

Each dynamical or thermal property of the solid depends on a different way in which the phonon frequency spectrum is weighted. For instance, $\langle u^2 \rangle$ and recoilless fraction f are mainly determined by the low-frequency phonons whereas $\langle v^2 \rangle$ is more sensitive to the high-frequency phonons. Furthermore, the Debye temperatures θ_D obtained by measuring the entropy, the thermal energy, and the heat capacity are not in general the same. A typical example is the study of KBr crystals [30]. It is therefore more appropriate to use $\theta_D(n)$ of different n-values for describing dynamical and thermal quantities than to use just θ_D. We now point out the relations between several specific quantities and $\theta_D(n)$ with different n-values [31–35].

At high temperatures, heat capacity, entropy, and the mean square atomic displacement $\langle u^2 \rangle$ depend on $\theta_D(2)$, $\theta_D(0)$, and $\theta_D(-2)$, respectively. Under the limit $T \to 0$, the heat capacity is related to $\theta_D(-3)$ while $\langle u^2 \rangle$ is related to $\theta_D(-1)$. The θ_D derived from the elastic constant measurements should be equal to $\theta_D(-3)$ [36].

For an ideal Debye solid, $\theta_D(n) = \theta_D$ for all n-values. But for a real solid, $\theta_D(n)$ depends on n, which gives a measure of the difference between the actual phonon spectrum and the Debye spectrum.

In Mössbauer spectroscopy, $\theta_D(-1)$ and $\theta_D(-2)$ can be obtained by measuring the recoilless fraction, while $\theta_D(1)$ and $\theta_D(2)$ can be deduced from the second-order Doppler shift (see Chapter 5).

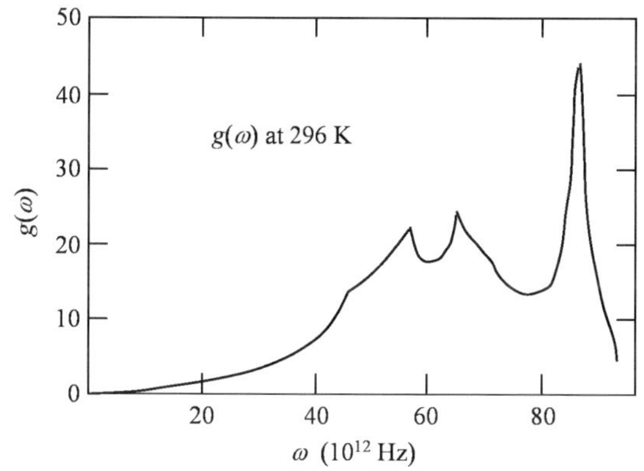

Fig. 4.19 Phonon DOS $g(\omega)$ for α-Fe [37].

4.4.4 The Debye Temperature θ_D

We now focus our attention on θ_D to obtain a better understanding of its physical meaning. It will facilitate the analysis of Debye temperature θ_D (or θ_M) from Mössbauer experiments and the comparison of θ_D with results from other methods.

In the Debye model, a maximum cutoff frequency ω_D was introduced so that the total number of vibration modes in a Bravais lattice is exactly $3N$, and consequently the Debye temperature was defined $\theta_D = \hbar\omega_D/k_B$. The Debye model is most successful in describing the vibration frequency distribution $g(\omega)$ in crystals such as Fe, Cu, K, and Na (Fig. 4.19). There is clearly a sharp cutoff point, but the cutoff frequency is not ω_D, and the distribution in the high-frequency portion has large deviations from the Debye model. For most crystals, $g(\omega)$ differs from the Debye model significantly; however, the parameter θ_D can still be obtained. It makes us wonder as to the exact meaning of θ_D. However, large amounts of experimental data indicate that the Debye model is essentially correct and θ_D is an important parameter of the solid. As already mentioned above, because many thermodynamic quantities are expressed as averages over the frequency distribution $g(\omega)$, they are not sensitive to its details. This may be one of the reasons for the Debye model's success. Therefore, the Debye temperature should be understood as a parameter that may not correspond to the actual cutoff frequency ω_D on DOS curve. Since θ_D is related to many other physical quantities through a variety of expressions, we can investigate it using several theoretical and experimental methods.

4.4.4.1 The Physical Meaning of θ_D

Only after a material's θ_D is determined would the terms "high temperature" and "low temperature" be meaningful. High temperature means $T > \theta_D$ with all the vibration modes excited, whereas low temperature means $T < \theta_D$ with some of the vibration modes beginning to be suppressed. For many solids, θ_D is obtained through measuring c_v, but the relation between θ_D and c_v is very complicated. The following two points would help us to find out which quantities are connected with θ_D.

1. When $k \to 0$, acoustic branch waves resemble elastic waves. The classical theory of elasticity may be applied to study the elastic properties of a crystal. Experimentally obtained elastic constants and elastic wave speeds can be used, eventually leading to θ_D according to Eq. (4.101). As an example, consider an fcc crystal with short-range central forces between atoms. The elastic constants $c_{44} = c_{11}/2$, and $v_l = v_t = c_{44}/\rho$, where ρ is the density of the solid. In the long-wavelength limit, θ_D can be calculated [4, 5, 16, 38] as

$$\theta_D = cv_l \approx c\left(\frac{c_{44}}{\rho}\right)^{1/2} = c'\left(\frac{\alpha}{M}\right)^{1/2} \tag{4.111}$$

where c and c' are constants, α is the force constant between adjacent atoms, and M is the mass of each atom. It is clear that θ_D is proportional to the square root of the force constant α and inversely proportional to the square root of the mass of each atom. For example, diamond is light and hard, because the mass of the carbon atom is low and the interatomic covalent bond is extremely strong, and consequently its Debye temperature θ_D is very high (2200 K). On the contrary, lead is heavy and soft, and its θ_D is very low (102 K). However, the force constant may vary over several orders of magnitude and therefore plays a larger role in determining θ_D. Solid neon, for instance, has a θ_D of only 63 K because the van der Waals force between the Ne atoms is very weak. By the way, we could also derive (4.111) directly from (4.81).

2. When the temperature of a solid rises, the amplitudes of atomic vibrations increase, the forces between atoms fail to hold the atoms in the solid form, and melting begins to take place. Since θ_D is proportional to the square root of the force constant, it is not surprising that θ_D is related to the melting temperature T_m of the solid and given by

$$\theta_D = c\left(\frac{T_m}{MN_A V^{2/3}}\right)^{1/2} \tag{4.112}$$

where c is a constant, having values of 137 and 200 for metals and nonmetals, respectively. When the temperature approaches T_m, the atomic motion can no longer be treated as small oscillations, and thus the anharmonic effect becomes significant. As a result, the θ_D value obtained from (4.112) usually has a poor agreement with the θ_D values from other methods.

4.4.4.2 Comparison of Results from Various Experimental Methods

The methods for determining θ_D include elastic constant measurements, heat capacity measurements, x-ray and neutron scattering, and the Mössbauer effect.

Table 4.1 Debye temperature θ_D values (in K) of several solids obtained from elastic constant, heat capacity, and melting temperature measurements [39, 40].

Solid	Elastic constant method	Heat capacity method	Melting temperature method
Al	438	428	~400
Cu	365	345	~300
Ni	446	450	
Pb	135	105	
Zn	307	327	
C (diamond)		2230	~2000
Na	164	158	~160

The first two are macroscopic methods. Although they are based on different principles, they give similar low-temperature results for θ_D (Table 4.1). As mentioned in the last section, different quantities such as c_V, entropy and $\langle u^2 \rangle$ are related to different $\theta_D(n)$ in different temperature regions, not simply to a single parameter θ_D. Therefore, θ_D values determined using different methods would never be exactly the same. The last three are microscopic methods, in which the Debye temperature θ_D is obtained through the measurements of either the Debye–Waller factor or the recoilless fraction. Both these factors are exponential functions of the mean-square displacement $\langle u^2 \rangle$, and replacing $g(\omega)$ in the expression of $\langle u^2 \rangle$ with the Debye model distribution would give θ_D. From this argument, it seems that the θ_D values obtained using neutron scattering and the Mössbauer effect should be in good agreement, whereas the x-ray scattering results for θ_D in most cases are only slightly higher than the Mössbauer results due to the possible deviation of the adiabatic approximation. When the Mössbauer atom is only one of the constituent elements in a sample, the x-ray result may be noticeably larger than that from the Mössbauer effect (see Table 4.2).

Table 4.2 Debye temperature θ_D values (in K) of three compounds obtained from the Mössbauer effect and x-ray diffraction [41].

Compound	θ_D (Mössbauer effect)	θ_D (x-ray diffraction)
$Fe[Co(CN)_6]$	146(30)	245(25)
$Fe[Rh(CN)_6]$	153(9)	274(30)
$Fe[Ir(CN)_6]$	177(13)	287(30)

4.5 Localized Vibrations

So far, we have been dealing with crystals of perfect periodic atomic arrangements, and their vibrations form lattice waves. In reality, such ideally perfect crystals are rare, and most crystals have impurities or other defects, which greatly influence their vibration properties. As the impurity concentration increases, the vibrations become very complicated and lattice waves can no longer exist. In this section, we will only discuss situations of extremely low concentrations of substitutional atoms, namely, isolated impurity atoms. Due to these impurity atoms, the physical picture for lattice vibrations is not easy to visualize; the mathematical treatment also becomes very difficult and one usually resorts to the Green's function method [42] or the molecular vibration method.

The Mössbauer effect is a suitable method for studying the dynamics of impurity atoms, because it has absolute isotope selectivity and the Mössbauer nucleus is often the impurity atom in a host crystal. The discovery of the Mössbauer effect has greatly advanced the research of impurity dynamics. The development of imperfect crystal dynamics has been divided into two stages, and the advent of the Mössbauer effect has been recognized as the beginning of the second stage [43]. While more details can be found in the next few chapters, here we will only describe the vibrations of isolated impurity atoms and their effects, in order to have a basic understanding of the phenomenon of local vibrations.

As an example, suppose we have a diatomic chain of 48 atoms, $M_0 = 31$ u and $m = 70$ u (compound GaP) [44]. First, simultaneous equations (4.65) are solved and 48 modes are obtained. Modes 1 through 24 form the acoustic branch, and the remaining modes form the optical branch, with the maximum frequency of $\omega_\mathrm{m} = 370\ \mathrm{cm}^{-1}$. (Wavenumber frequency units are used here, $1\ \mathrm{cm}^{-1} = 3 \times 10^{10}\ \mathrm{Hz} = 18.8 \times 10^{10}\ \mathrm{rad\ s}^{-1}$.) Now a lighter atom ($M$) replaces an atom ($M_0$) in the chain, and we assume that this substitution does not change the force constant. A computer program can numerically solve a set of equations similar to (4.65) and obtain also 48 modes, some of which are shown in Fig. 4.20. For clarity, the atomic displacements along the chain are all drawn perpendicular to the chain.

The most obvious change after substitution of the atom M is the emergence of a new mode with $\omega_\mathrm{L} = 416.4\ \mathrm{cm}^{-1}$, which is higher than the maximum frequency ω_m of the lattice waves. The atomic displacement $\boldsymbol{u}$ in this mode is no longer a sinusoidal function in space, but has a maximum at the position of M and decays quickly to zero at a distance only a few atoms from M. This is known as localized vibrations or a local mode, because it is restricted in a region near the impurity atom.

For the above situation where there is only one impurity atom, the solutions to the equations of motion can be categorized as either symmetric ($u_n = u_{-n}$) or antisymmetric ($u_n = -u_{-n}$). The impurity atom only influences the symmetric modes, because the impurity atom is at rest in antisymmetric modes and thus has no effect on them. As can be seen in Fig. 4.20, modes 4 and 24 are antisym-

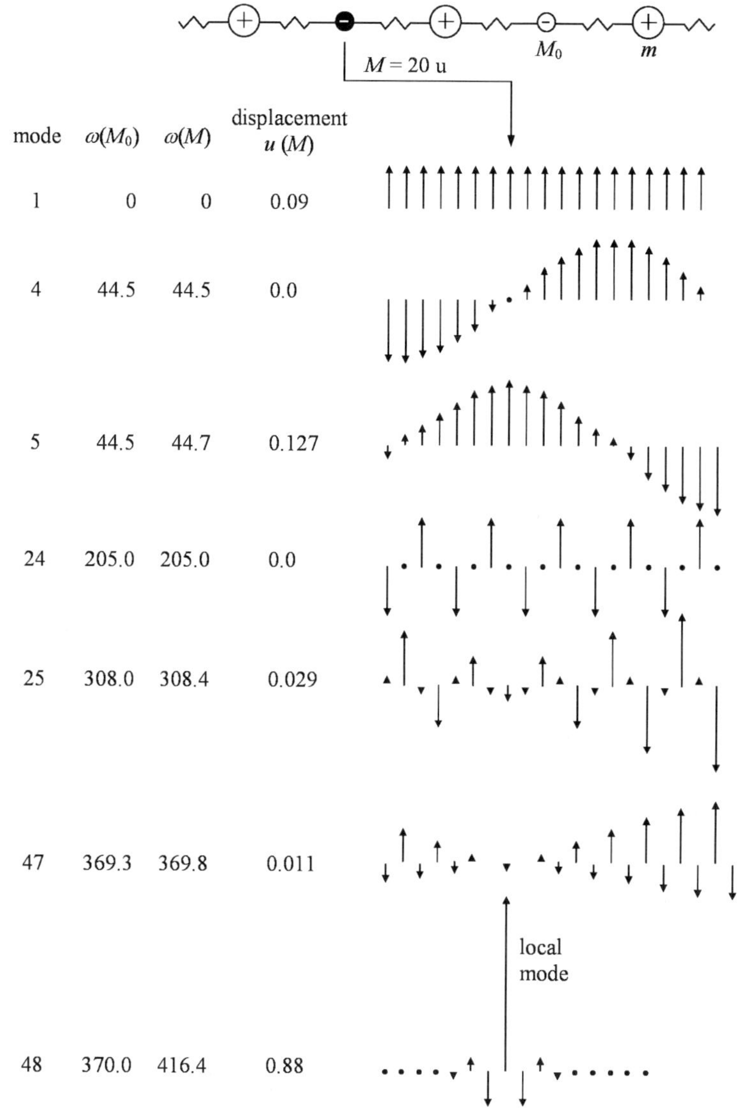

Fig. 4.20 Selected vibration modes of a 48-atom linear diatomic chain with one impurity atom [44].

metric and their frequencies do not change, and modes 5, 25, and 47 are symmetric and their frequencies all become higher. As shown in Fig. 4.21, if M decreases, the localized vibration displacement $\boldsymbol{u}$ at M site increases, whereas the spatial spread decreases accordingly. A measure of localization is usually defined as

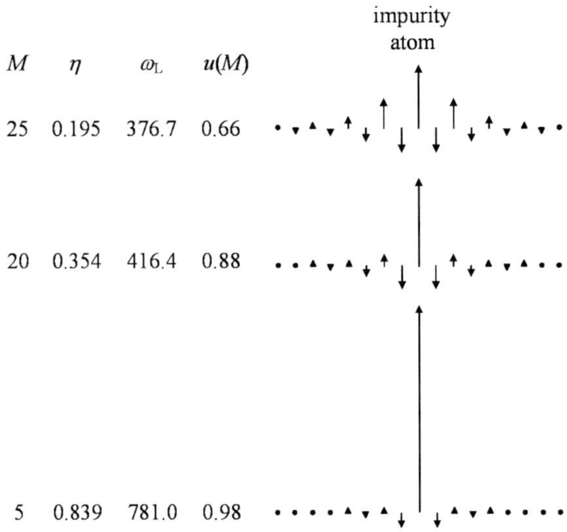

Fig. 4.21 Displacement vectors of local modes when $M = 25, 20$, and 5 u.

$$\eta = 1 - M/M_0. \tag{4.113}$$

Figure 4.21 indicates that besides the atom with $M = 5$ u, the others hardly move. The frequency of the localized mode is approximately

$$\omega_L \approx \sqrt{2\alpha/M}. \tag{4.114}$$

The decrease of M has another important effect, namely it increases the intensity of the entire acoustic branch (see Fig. 12 in Ref. [44]), which will be very useful.

When $M > M_0$, the localized mode is situated between the optical and acoustic branches, and thus known as a gap mode ω_{gap}.

As one of the heavier atoms m is replaced by m' and $m' > m$, the localized mode falls within the acoustic branch, causing increases in vibration amplitudes of nearby original atoms. This mode is therefore called a resonant mode, and it is a quasi-localized mode.

The above discussion involves only a change in the atomic mass. If the force constant also changes, a high-frequency localized mode can also be produced even with a heavy defect ($M > M_0$) as long as its interactions with the surrounding atoms are much stronger than those in the original perfect crystal.

In a real solid, the localized vibrations are much more complicated than the above model. The frequencies of localized modes are in the infrared region, and thus the existence of high-frequency modes or gap modes can be verified using infrared absorption experiments. Figure 4.22 shows infrared absorption spectra of AgBr with its Ag replaced by natural lithium (92.6% ^{7}Li and 7.4% ^{6}Li) and by lithium with enriched ^{6}Li [45]. Figure 4.22(a) shows two absorption peaks at

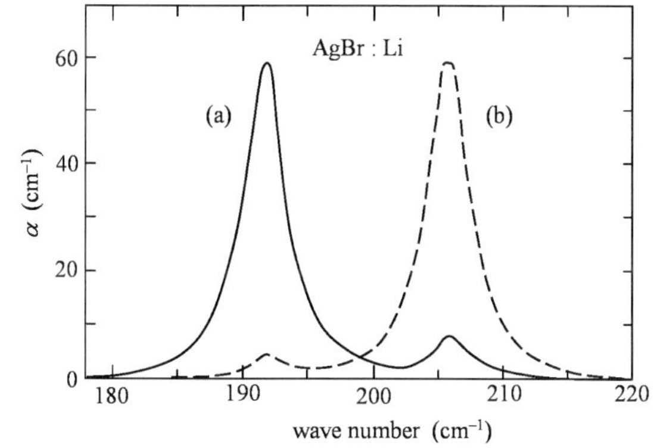

Fig. 4.22 Infrared absorption spectra of AgBr doped (a) with natural Li and (b) with Li enriched by ^{6}Li.

191.8 and 205.9 cm^{-1}, with an intensity ratio consistent with the natural abundance ratio of ^{7}Li to ^{6}Li. When ^{6}Li enrichment was used in the impurity atoms, the spectrum becomes that of Fig. 4.22(b), also showing two peaks at the same frequencies except a change in the relative intensities, as expected. Since the lighter defect atom should result in a localized mode with a higher frequency, this isomer effect verifies the existence of high-frequency modes.

4.6 Experimental Methods for Studying Lattice Dynamics

The experimental methods for studying lattice dynamics are all based on the interactions of electromagnetic radiation or particles with the solid. According to the subjects of investigations, the methods can be divided into three groups. The first group involves the measurements of phonon dispersion curves $\omega_j(\boldsymbol{k})$ or DOS curve $g(\omega)$. Currently, the best method is inelastic coherent neutron scattering, which can give complete $\omega_j(\boldsymbol{k})$ curves in the entire Brillouin zone. Before neutron scattering was available, dispersion curves for most crystals were unknown. Inelastic x-ray scattering and inelastic nuclear resonant scattering as means of obtaining $\omega_j(\boldsymbol{k})$ or $g(\omega)$ have been developed only after synchrotron radiation became accessible. The second group involves the studying of vibration modes near the center of the Brillouin zone ($k \to 0$). The experimental methods include infrared spectroscopy and Raman scattering to study optical modes and Brillouin scattering or the method of elastic coefficients to study acoustic modes. The third group is based on principles of statistical mechanics, obtaining the Debye–Waller factor, and hence the atomic mean-square displacement $\langle u^2\rangle$and the mean-square velocity $\langle v^2\rangle$, as well as $\theta_{\rm D}$. These quantities are weighted statistical averages over the phonon spectrum $g(\omega)$. The method of specific heat and later trans-

Table 4.3 Energies, wavelengths, and wave vectors of several radiations and particles.

	Energy (eV)	Approximate wavelength (Å)	Approximate wave vector ($Å^{-1}$)
Phonons	0.013	3	2
Thermal neutrons	0.025	2	3
Electrons	16	3	2
X-rays	4100	3	2
γ-rays	14 400	0.86	1.6
Visible light	3	4000	10^{-3}

mission Mössbauer spectroscopy, etc., all belong to this group. These are indirect methods compared with those in the first group.

As can be seen from Table 4.3, only thermal neutrons so far are found to be suitable for inelastic scattering to create or annihilate phonons in a solid with considerable momentum transfer. They have energies comparable to typical phonon energies and wavelengths comparable to atomic distances in crystals. As to x-rays, for an appropriate wavelength, say 3 Å, the corresponding energy $(E_x = hc/\lambda_x)$ is as high as 4136 eV. There is a large mismatch between the incident energy of x-rays and the phonon energy. Thus, to get the dispersion curve it requires an energy resolution of 10^{-7} or better in the x-rays. In short, lattice dynamical studies by conventional x-rays are limited.

The energy of visible light is much less than that of x-rays, and the wave vector is proportionally reduced to about 10^{-3} Å. So visible light scattering is only used for measuring those vibration modes of extremely long wavelengths.

In this section, we discuss neutron scattering, followed by x-ray scattering. In Chapter 7, new methods using synchrotron radiation in combination with the Mössbauer effect are described.

4.6.1 Neutron Scattering

Neutrons can be scattered by an atom through two mechanisms, nuclear scattering and magnetic scattering. Here we focus on the first mechanism. Through nuclear scattering, a neutron is scattered by the nucleus within the range of strong nuclear force (10^{-12} cm). Neutron scattering may be elastic or inelastic, coherent or incoherent. Coherent elastic scattering is usually called neutron diffraction, which is mainly used for determining the crystal structure and the Debye–Waller factor. To obtain the dispersion curve $\omega_j(\boldsymbol{k})$, we need to detect inelastic scatterings which involve the creation or absorption of phonons.

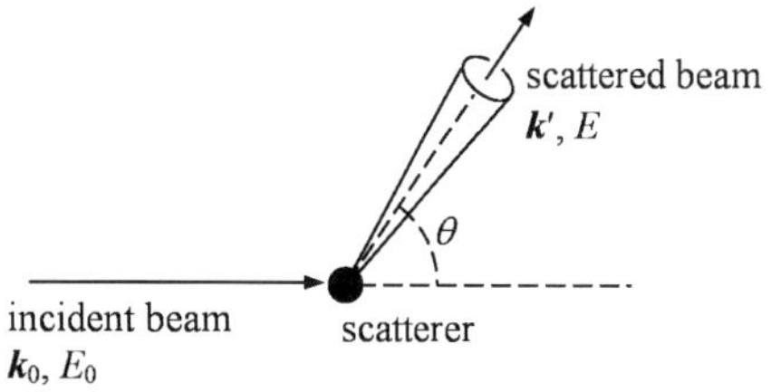

Fig. 4.23 Neutron scattering process.

4.6.1.1 **Theory**

1. *Basic formulation.* For the sake of simplicity, we will neglect neutron spin for the time being. Before scattering, the neutron has energy E_0 and wave vector $\boldsymbol{k}_0$. The scatterer has a wave function $|\lambda\rangle$ and energy E_λ. After scattering, these quantities become respectively E, $\boldsymbol{k}'$, $|\lambda'\rangle$, and $E_{\lambda'}$ (Fig. 4.23). The changes in the neutron energy and wave vector are, respectively

$$\hbar\omega = E_0 - E = \frac{\hbar^2}{2m}(k_0^2 - k'^2), \quad \boldsymbol{Q} = \boldsymbol{k}' - \boldsymbol{k}_0 \tag{4.115}$$

where m is the mass of neutron. Within a solid angle $\Delta\Omega$ in the θ direction, the neutron flux $\mathrm{d}\Phi$ in an energy range from E to $E + \Delta E$ recorded by the detector is

$$\mathrm{d}\Phi \approx \Phi\Delta\Omega\Delta E\eta \tag{4.116}$$

where Φ is the incident neutron flux and η is the detector efficiency.

The proportionality constant in Eq. (4.116) is called the double differential cross-section, denoted by $\mathrm{d}^2\sigma/(\mathrm{d}\Omega\,\mathrm{d}E)$. According to the Born approximation [46–48],

$$\frac{\mathrm{d}^2\sigma}{\mathrm{d}\Omega\,\mathrm{d}E} = \frac{k'}{k_0}\left(\frac{m}{2\pi\hbar^2}\right)^2 \sum_\lambda p_\lambda \sum_{\lambda'} |\langle \boldsymbol{k}'\lambda'|V(\boldsymbol{r})|\boldsymbol{k}_0\lambda\rangle|^2 \delta(\hbar\omega + E_\lambda - E_{\lambda'}) \tag{4.117}$$

where the δ-function is due to energy conservation. Here a summation is first carried out over λ', and it is then averaged over the initial states $|\lambda\rangle$. The weight

$$p_\lambda = \frac{\exp(-E_\lambda/k_B T)}{\sum_\lambda \exp(-E_\lambda/k_B T)}$$

is the probability of having the initial state $|\lambda\rangle$ in the scatterer at temperature T. The neutron–nucleus interaction V has the form of a Fermi pseudopotential:

$$\frac{2\pi\hbar^2}{m} b\delta(\boldsymbol{r} - \boldsymbol{R})$$

where b is the scattering length, whose square is the cross-section of the low-energy neutron–nucleus scattering

$$\frac{d\sigma}{d\Omega} = b^2.$$

The interaction potential between the neutron and the entire scatterer is

$$V(\mathbf{r}) = \frac{2\pi\hbar^2}{m}\sum_l b_l\delta(\mathbf{r} - \mathbf{R}_l)$$

where $\boldsymbol{R}_l$ and b_l are the position vector and the scattering length of the lth nucleus, respectively. Substitution of this potential into Eq. (1.117) yields

$$\frac{d^2\sigma}{d\Omega\, dE} = \frac{k'}{k_0}\sum_\lambda p_\lambda \sum_{\lambda'}\sum_{l,l'} b_l^* b_{l'}\langle\lambda|e^{-i\boldsymbol{Q}\cdot\boldsymbol{R}_{l'}}|\lambda'\rangle \times \langle\lambda'|e^{i\boldsymbol{Q}\cdot\boldsymbol{R}_l}|\lambda\rangle\delta(\hbar\omega + E_\lambda - E_{\lambda'}). \tag{4.118}$$

In order to carry out the summation over λ' and the average over λ, the δ-function needs to be expressed as a time integral:

$$\delta(\hbar\omega + E_\lambda - E_{\lambda'}) = \frac{1}{2\pi\hbar}\int_{-\infty}^{\infty} \exp[-i(E_\lambda - E_{\lambda'})t/\hbar]e^{-i\omega t}\, dt. \tag{4.119}$$

After substituting this into (4.118) and summing over λ', we have [47]

$$\begin{aligned}\frac{d^2\sigma}{d\Omega\, dE} &= \frac{k'}{k_0}\frac{1}{2\pi\hbar}\sum_{l,l'} b_l^* b_{l'}\sum_\lambda p_\lambda\int_{-\infty}^{\infty} \langle\lambda|e^{-i\boldsymbol{Q}\cdot\boldsymbol{R}_l(0)}e^{i\boldsymbol{Q}\cdot\boldsymbol{R}_{l'}(t)}|\lambda\rangle e^{-i\omega t}\, dt\\ &= \frac{k'}{k_0}\frac{1}{2\pi\hbar}\sum_{l,l'} b_l^* b_{l'}\int_{-\infty}^{\infty} \langle e^{-i\boldsymbol{Q}\cdot\boldsymbol{R}_l(0)}e^{i\boldsymbol{Q}\cdot\boldsymbol{R}_{l'}(t)}\rangle_T e^{-i\omega t}\, dt\\ &= \frac{N}{\hbar}\frac{k'}{k_0}\sum_{l,l'} b_l^* b_{l'} S_{ll'}(\boldsymbol{Q}\omega)\end{aligned} \tag{4.120}$$

where

$$S_{ll'}(\boldsymbol{Q}\omega) = \frac{1}{2\pi N}\int_{-\infty}^{\infty} \langle e^{-i\boldsymbol{Q}\cdot\boldsymbol{R}_l(0)}e^{i\boldsymbol{Q}\cdot\boldsymbol{R}_{l'}(t)}\rangle_T e^{-i\omega t}\, dt \tag{4.121}$$

is known as the scattering function which describes the dynamic properties of the target system, N is the number of nuclei in the Bravais scatterer, and $\langle\ldots\rangle_T$ represents a thermal average over the initial states $|\lambda\rangle$. The scattering length b is not a dynamic quantity, and therefore is taken out of the thermal average. Equation (4.120) is then a simplified expression for the double differential cross-section.

2. *Coherent and incoherent scattering.* The summation in Eq. (4.120) is to be done over pairs of nuclei (l, l'). The scattering function S contains the complete information about the physics of the scatterer. Nuclear scattering lengths b are different for different elements, and even for the same element in the scatterer, b also depends on the nuclear spin orientations and the amount of different isotopes, which should be randomly distributed in general. Therefore, we should use an average of $b_l^* b_{l'}$ over all possible nuclear states. If the b-values for different nuclei are independent of one another, then

$$\begin{aligned} \overline{b_l^* b_{l'}} &= |\bar{b}|^2, \quad l \neq l' \\ \overline{b_l^* b_{l'}} &= \overline{|b|^2}, \quad l = l'. \end{aligned} \tag{4.122}$$

Using these in Eq. (4.120), the double differential cross-section can be written as the sum of two terms:

$$\frac{d^2\sigma}{d\Omega\, dE} = \frac{N\sigma_{\text{coh}}}{4\pi\hbar} \frac{k'}{k_0} \sum_{l,l'} S_{ll'}(\boldsymbol{Q}\omega) + \frac{N\sigma_{\text{inc}}}{4\pi\hbar} \frac{k'}{k_0} \sum_{l} S_{ll}(\boldsymbol{Q}\omega) \tag{4.123}$$

where

$$\sigma_{\text{coh}} = 4\pi|\bar{b}|^2, \quad \sigma_{\text{inc}} = 4\pi(\overline{|b|^2} - |\bar{b}|^2).$$

The first term in Eq. (4.123) is the coherent scattering cross-section, which depends on the correlation between positions of different atoms at different times and therefore has an interference effect. The second term is the incoherent scattering cross-section, which depends on the correlation between positions of the same atom at different times, and has no interference effect.

4.6.1.2 **Neutron Scattering by a Crystal**

Suppose the scatterer is a Bravais crystal, in which atoms are vibrating around their equilibrium positions. We see from Eq. (4.120) that neutron scattering by a solid is essentially described by the S function. Substituting $\boldsymbol{R}_l$ from Eq. (4.9) into the S function in (4.123), we have

$$S_{\text{coh}} = \sum_{l,l'} S_{ll'} = \frac{1}{2\pi N} \sum_{l,l'} \int_{-\infty}^{\infty} \langle e^{-i\boldsymbol{Q}\cdot\boldsymbol{u}(l,0)} e^{i\boldsymbol{Q}\cdot\boldsymbol{u}(l',t)} \rangle_T e^{i\boldsymbol{Q}\cdot(\boldsymbol{l}'-\boldsymbol{l})} e^{-i\omega t}\, dt, \tag{4.124}$$

$$S_{\text{inc}} = \sum_{l} S_{ll} = \frac{1}{2\pi N} \sum_{l} \int_{-\infty}^{\infty} \langle e^{-i\boldsymbol{Q}\cdot\boldsymbol{u}(l,0)} e^{i\boldsymbol{Q}\cdot\boldsymbol{u}(l,t)} \rangle_T e^{-i\omega t}\, dt. \tag{4.125}$$

If we carry out the thermal averages according to the quantum theory of harmonic oscillators, we would obtain [46]

$$S_{\text{coh}} = \frac{1}{2\pi} e^{-2W} \sum_{l} e^{iQ\cdot l} \int_{-\infty}^{\infty} e^{\langle Q\cdot u(0,0) Q\cdot u(l,t)\rangle_T} e^{-i\omega t}\, dt, \tag{4.126}$$

$$S_{\text{inc}} = \frac{1}{2\pi} e^{-2W} \int_{-\infty}^{\infty} e^{\langle Q\cdot u(0,0) Q\cdot u(0,t)\rangle_T} e^{-i\omega t}\, dt. \tag{4.127}$$

Suppose that the atomic displacements $\boldsymbol{u}$ are very small (harmonic approximation), then the first exponential factor can be expanded into polynomials:

$$\begin{aligned} e^{\langle\cdots\rangle_T} &= 1 + \langle \boldsymbol{Q}\cdot\boldsymbol{u}(0,0)\boldsymbol{Q}\cdot\boldsymbol{u}(0,t)\rangle_T \\ &\quad + \frac{1}{2}\langle \boldsymbol{Q}\cdot\boldsymbol{u}(0,0)\boldsymbol{Q}\cdot\boldsymbol{u}(0,t)\rangle_T^2 + \cdots \end{aligned} \tag{4.128}$$

When this is substituted into (4.126) or (4.127), it will have three terms, describing zero-phonon (elastic scattering), one-phonon, and two-phonon processes, respectively.

1. *The Debye–Waller factor.* In the above expressions for scattering cross-sections, there is a common factor e^{-2W}, known as the Debye–Waller factor. For a cubic crystal, its exponent can be written as

$$2W = \langle [\boldsymbol{Q}\cdot\boldsymbol{u}(0,0)]^2\rangle_T = Q^2\langle u^2(0,0)\rangle_T$$

or

$$2W = \frac{\hbar}{2MN}\sum_s \frac{(\boldsymbol{Q}\cdot\boldsymbol{e}_s)^2}{\omega_s}\langle 2n_s + 1\rangle \tag{4.129}$$

where $s = \boldsymbol{k}j$ and M is the atomic mass. For a cubic Bravais crystal

$$(\boldsymbol{Q}\cdot\boldsymbol{e}_s)^2 = \frac{1}{3}Q^2, \tag{4.130}$$

and the above expression becomes

$$\begin{aligned} 2W &= \frac{\hbar^2 Q^2}{2M}\frac{1}{3N}\sum_s \frac{1}{\hbar\omega_s}\coth\left(\frac{1}{2}\hbar\omega_s\beta\right) \\ &= \frac{\hbar^2 Q^2}{2M}\int \frac{1}{\hbar\omega}\coth\left(\frac{1}{2}\hbar\omega\beta\right) g(\omega)\, d\omega. \end{aligned} \tag{4.131}$$

To understand this factor e^{-2W}, let us look at the process of coherent elastic scattering. Here we have $|\boldsymbol{k}_0| = |\boldsymbol{k}'|$, the neutron energy does not change after scattering, namely there is no creation or annihilation of a phonon in the crystal (the zero-phonon process). The corresponding cross-section is

$$\left(\frac{d\sigma}{d\Omega}\right)_{\text{coh, el}} = \frac{\sigma_{\text{coh}}}{4\pi} N \frac{(2\pi)^3}{V_a} e^{-2W} \sum_{\tau} \delta(\boldsymbol{Q} - \boldsymbol{\tau}) \tag{4.132}$$

where V_a is the unit cell volume of the crystal and $\boldsymbol{\tau}$ is a reciprocal lattice vector. Since $e^{-2W} \leq 1$, the intensity of the diffraction peak is reduced. The reason for this is that as temperature rises, the atomic mean-square displacement $\langle u^2 \rangle$ increases, causing e^{-2W} to decrease drastically. When $T = 0$, the atoms are at rest, $\langle u^2 \rangle = 0$, and in this case $e^{-2W} = 1$. Therefore, the Debye–Waller factor is used to describe how the diffraction intensity changes with temperature.

Inspecting the quantity $\hbar^2 Q^2 / 2M$ in Eq. (4.131), we recognize that it equals the recoil energy of a free nucleus when a neutron is scattered by it [46, 49]. Comparing (4.131) with (1.82), we also see that, except for $\boldsymbol{Q}$ instead of the wave vector $\boldsymbol{k}$ of the Mössbauer radiation, the Debye–Waller factor and the recoilless fraction have exactly the same form, both having the same temperature dependence.

2. *Coherent inelastic scattering.* Let us now substitute the expansion (4.128) into (4.126), and discuss the second term

$$S_{\text{coh}} = \frac{1}{2\pi} e^{-2W} \sum_{l} e^{i\boldsymbol{Q}\cdot\boldsymbol{l}} \int_{-\infty}^{\infty} \langle \boldsymbol{Q} \cdot \boldsymbol{u}(0,0) \boldsymbol{Q} \cdot \boldsymbol{u}(l,t) \rangle_T e^{-i\omega t}\, dt \tag{4.133}$$

which describes the process of the creation or annihilation of a phonon. After some manipulations, we obtain

$$\left(\frac{d^2\sigma}{d\Omega\, dE}\right)^{\pm}_{\text{coh}} = \frac{\sigma_{\text{coh}}}{4\pi} \frac{k'}{k_0} \frac{(2\pi)^3}{V_a} \frac{1}{2M} e^{-2W} \sum_{s} \sum_{\tau} \frac{(\boldsymbol{Q} \cdot \boldsymbol{e}_s)^2}{\omega_s} \times \left\langle n_s + \frac{1}{2} \pm \frac{1}{2} \right\rangle \delta(\omega \mp \omega_s) \delta(\boldsymbol{Q} \mp \boldsymbol{k} - \boldsymbol{\tau}) \tag{4.134}$$

where the $\pm$ sign represents either the creation or the annihilation of one phonon of the sth mode during the scattering of a neutron. The two δ-functions indicate the requirement of simultaneously satisfying the conservation of energy and momentum:

$$\pm \frac{\hbar^2}{2m} (k_0^2 - k'^2) = \hbar\omega, \tag{4.135}$$

$$\boldsymbol{Q} = \boldsymbol{k}' - \boldsymbol{k}_0 = \boldsymbol{\tau} \pm \boldsymbol{k}. \tag{4.136}$$

Later in this section, we will use these relations to measure the dispersion relations $\omega_j(\boldsymbol{k})$.

3. *Incoherent inelastic scattering.* Similarly, if we focus on the second term in the expansion (4.127), we obtain the cross-section of one-phonon incoherent neutron scattering:

$$\left(\frac{d^2\sigma}{d\Omega\, dE}\right)^{\pm}_{\text{inc}} = \frac{\sigma_{\text{inc}}}{4\pi}\frac{k'}{k_0}\frac{1}{2M}e^{-2W}\sum_s \frac{(\mathbf{Q}\cdot\mathbf{e}_s)^2}{\omega_s} \times\left\langle n_s + \frac{1}{2} \pm \frac{1}{2}\right\rangle \delta(\omega \mp \omega_s). \tag{4.137}$$

It contains only one δ-function, $\delta(\omega \mp \omega_s)$. Thus only the energy conservation needs to be satisfied, i.e.,

$$\pm\omega = \omega_s, \quad \text{or} \quad \pm\frac{\hbar^2}{2m}(k_0^2 - k'^2) = \hbar\omega_s. \tag{4.138}$$

Using Eqs. (4.89), (4.95), and (4.128) we can evaluate (4.137) as

$$\left(\frac{d^2\sigma}{d\Omega\, dE}\right)^{\pm}_{\text{inc}} = \frac{\sigma_{\text{inc}}}{4\pi}\frac{k'}{k_0}\frac{N}{4M}Q^2 e^{-2W}\frac{g(\omega)}{\omega}\left[\coth\left(\frac{1}{2}\hbar\omega\beta\right) \pm 1\right]. \tag{4.139}$$

This indicates that the phonon DOS $g(\omega)$ can be measured experimentally by one-phonon incoherent neutron scattering.

4. *Measurement of dispersion curves.* The experimental apparatus for this purpose is the triple-axis neutron spectrometer, invented in 1955 by Bertram Brockhouse [49] who won the 1994 Nobel Prize in Physics for developing this apparatus and the constant-$\boldsymbol{Q}$ method. The triple-axis spectrometer is illustrated in Fig. 4.24. A beam of neutrons, obtained by a Bragg reflection (through an angle $2\theta_M$) from a monochromator crystal, is scattered by the sample (through an angle θ) and the energy of this scattered beam is determined by a second Bragg reflection (through an angle $2\theta_A$) from an analyzer. The orientation of the sample is defined

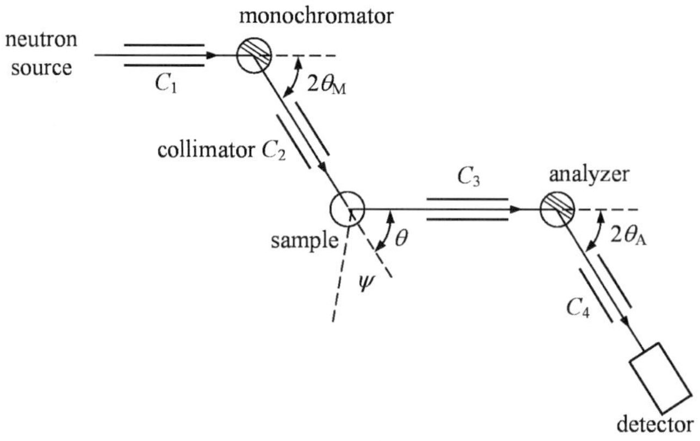

Fig. 4.24 Schematic diagram of the triple-axis neutron spectrometer. C_1, C_2, C_3, and C_4 are collimators.

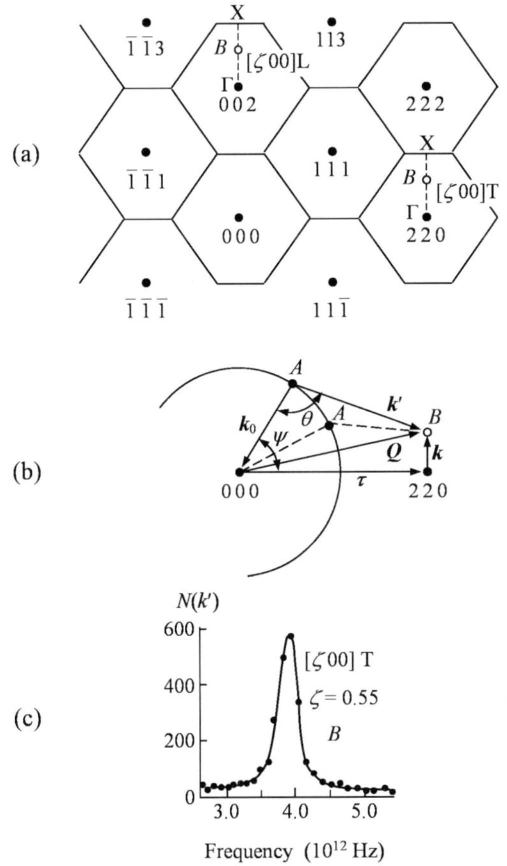

Fig. 4.25 (a) Brillouin zones within the $(1\bar{1}0)$ plane of a Cu crystal; (b) a vector diagram of the constant-$\boldsymbol{Q}$ method; (c) scattered neutron number N as a function of k'.

by the angle ψ between a reciprocal lattice vector $\boldsymbol{\tau}$ and the incident neutron beam.

Suppose we take a single-crystal Cu sample (fcc) and choose its reciprocal lattice plane $(1\bar{1}0)$ as the scattering plane, where the neutron beam, sample, and analyzer are placed. All the phonons to be measured have their wave vectors lying in this plane. The various Brillouin zones that this plane goes through are shown in Fig. 4.25(a). As discussed in Section 4.2, the modes along the [100], [110], and [111] directions are all pure longitudinal or pure transverse. The transverse modes along the [100] and [111] directions are degenerate, while the two transverse modes T_1 and T_2 along the [110] direction are non-degenerate. Therefore, in the $(1\bar{1}0)$ plane, we will be able to measure the three longitudinal modes L[100], L[110], and L[111], as well as the transverse modes T[100], T_2[110], and T[111]. The transverse branch T_1[110] will have to be measured in the scattering plane (100). Here three polarization vectors are mutually perpendicular $\boldsymbol{e}(1) \perp \boldsymbol{e}(2) \perp \boldsymbol{e}(3)$. Since $\boldsymbol{Q} \cdot \boldsymbol{e}(3) = 0$, the corresponding transverse mode will not show up, according to (4.118). This simplifies the analysis of the results.

Now we briefly describe the constant-$\boldsymbol{Q}$ method for measuring the $\omega_j(\boldsymbol{k})$ curve along the [100] direction in a Cu crystal. The energy of the incident neutron E_0 is kept constant, i.e., $|\boldsymbol{k}_0|$ is fixed. A phonon wave vector $\boldsymbol{k}$ at point B in the direction $\Gamma \to$ X (Fig. 4.25(a)) is selected with $|\boldsymbol{k}| = 0.55(2\pi/a)$. The vector diagram of Eq. (4.136) is illustrated in Fig. 4.25(b), and a circle of radius $|\boldsymbol{k}_0|$ is centered at (000).

So far, the sole unknown $\boldsymbol{k}'$ must be measured. Angles ψ and θ are changed such that point A as the common origin of $\boldsymbol{k}$ and $\boldsymbol{k}'$ moves but is confined on the circle. At the same time, the number of scattered neutrons $N(\boldsymbol{k}')$ is collected as a function of $\boldsymbol{k}'$. When all conservation laws are satisfied, the curve $N(\boldsymbol{k}')$ will show one peak, which yields a particular $\boldsymbol{k}'$ value. Therefore, we can calculate ω for a fixed $\boldsymbol{k}$ and complete one experimental point on the dispersion curve $\omega(\boldsymbol{k})$. During the measurement, $\boldsymbol{Q}$ is kept constant, hence the name constant-$\boldsymbol{Q}$ method. Now we choose other $\boldsymbol{k}'$ values in the same direction, and eventually, point by point, a complete dispersion curve of the T branch is measured [50]. Rotating the sample about the crystalline axis $[1\bar{1}0]$ will allow us to measure the L branch. Figure 4.26 shows the Cu dispersion curves in four major symmetry directions [51].

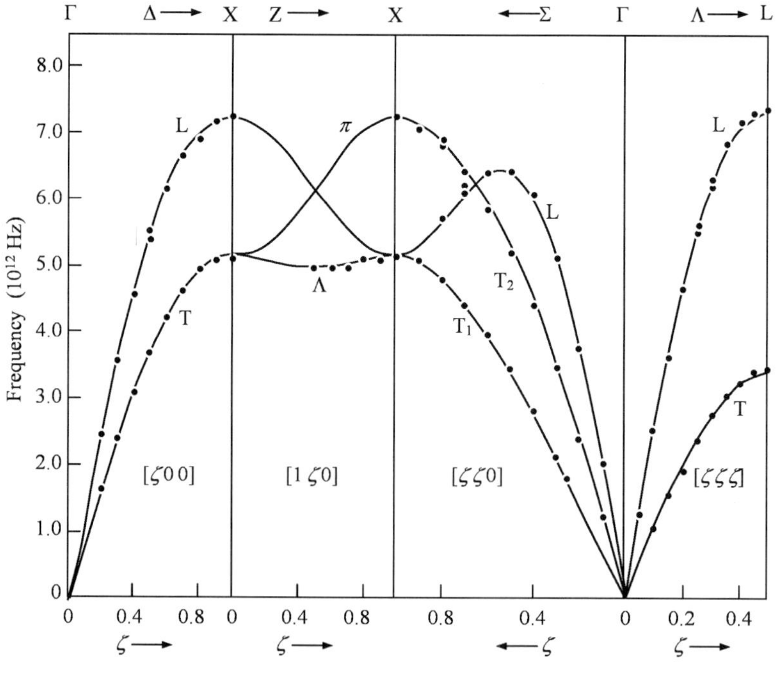

Fig. 4.26 Dispersion curves for fcc copper [51].

5. *DOS curves.* As Eq. (4.139) indicates, a $g(\omega)$ curve can be obtained by incoherent inelastic scattering provided σ_{inc} is large enough. Unfortunately, the number of materials having sufficiently large σ_{inc} is limited. For the metals Cu and Fe, the ratio $\sigma_{coh}/\sigma_{inc}$ is as high as 15.6 and 29.9, respectively. In these cases, one has to measure the dispersion curves first, then calculate $g(\omega)$. Calculations based on (4.95) cannot be carried out because the modes easily observed are only in certain high-symmetry directions and only consist of a few of all the normal modes. At present, there are several approaches to the calculation of $g(\omega)$ from experimental dispersion curves. A fast and more accurate one is the so-called "extrapolation method" [52, 53], which is outlined here.

The first step is fitting the experimental dispersion curves by the Born–von Karman theory to get the force constants, assuming the obtained constants can be used to calculate the DOS curve throughout all Brillouin zones (BZs). Then, the dynamical matrix $\boldsymbol{D}(\boldsymbol{k})$ is diagonalized for a relatively small number of $\boldsymbol{k}$ evenly spaced in the irreducible section of the first BZ. The frequencies between two successive wave vectors are obtained by linear extrapolation. The frequency gradients required for this extrapolation are given by standard perturbation theory. For bcc Fe, the calculated DOS curve is shown in Fig. 4.19.

4.6.2
X-ray Scattering

The theoretical description of inelastic x-ray scattering (IXS) is essentially the same as for neutron scattering, so all formulas derived in neutron coherent scattering are valid for x-ray scattering if the scattering length b is replaced by the atomic form factor $f(Q)$. This reflects the adiabatic approximation in which the electron density follows the nuclear motion instantaneously. In the early years, the lattice dynamics study by inelastic x-ray scattering was limited because of two main problems: insufficient radiation intensity and poor energy resolution. As mentioned above, x-rays with wavelengths comparable to interatomic distances have relatively high energies and such hard x-rays can only be found in a continuous spectrum where the intensity is very low even from high-power rotating anodes.

High intensities of x-rays emitted by synchrotrons provide the possibility of IXS with an energy resolution of meV. Phonon dispersion curves measured by coherent IXS were first reported in 1987 [54, 55]. In the last 10 years IXS has become a powerful spectroscopic tool, complementing the well-established coherent neutron scattering [56, 57].

4.7
First-Principles Lattice Dynamics

Recently, a first-principles quantum mechanical method (*ab initio*) based on the density-functional theory has become one of the most promising tools for studying structural and dynamical properties of real materials. This advanced theoretical method is impressive because it only requires input information on the material's composition, such as the atomic number, atomic mass of the constituent elements, and the lattice structure, with no need of any experimental results. The agreement between the calculated phonon frequencies and the experimentally observed dispersion curves is incredibly good. Therefore, it is now possible to map accurate phonon frequencies onto a fine grid of wave vectors within BZs for larger and larger lattice systems. Here, we describe the main features of this first-principles method applied to lattice dynamics. Readers are referred to many excellent reviews [58–60].

4.7.1 Linear Response and Lattice Dynamics

The first-principles theory takes into account the effect of electrons on lattice dynamics within the validity of adiabatic approximation (see Section 4.1.1), where the electron system is assumed to be in the ground state with respect to the instantaneous nuclear positions. The aim is to determine the interatomic force constants through minimizing the total energy of a crystal with "frozen" nuclear coordinates at any particular instance during the lattice vibration. The total energy E_{tot} is an eigenvalue of the equation

$$\mathscr{H}_{\text{tot}}\Psi(\boldsymbol{r}, \boldsymbol{R}) = E_{\text{tot}}\Psi(\boldsymbol{r}, \boldsymbol{R}) \tag{4.140}$$

where

$$\mathscr{H}_{\text{tot}} = T_{\text{e}} + V_{\text{ee}}(\boldsymbol{r}) + V_{\text{en}}(\boldsymbol{r}, \boldsymbol{R}) + V_{\text{nn}}(\boldsymbol{R}). \tag{4.141}$$

Here the meaning of each term is given in Section 4.1.1, but the Hamiltonian $\mathscr{H}_{\text{tot}}$ in (4.141) differs slightly from that in (4.2). Under equilibrium, the net force acting on each individual nucleus vanishes, so

$$\boldsymbol{F}(l) = -\nabla_{\boldsymbol{R}_l}[E_{\text{tot}}(\boldsymbol{r}, \boldsymbol{R})] = 0. \tag{4.142}$$

Consequently, one can calculate the vibrational frequencies of nuclei by the Born–von Karman theory within the harmonic approximation. This involves evaluating the force constant matrix (also known as the Hessian) by taking the second derivatives of E_{tot}, constructing dynamical matrix $\boldsymbol{D}$ at a given point in the BZ, and solving a secular equation such as (4.45).

Systematic studies of the effect of electrons on lattice dynamics were carried out in the 1960s [61, 62], revealing a linear response of electron charge density $\rho(\boldsymbol{r})$ to perturbation caused by a change in the nuclear positions in the crystal. Information on the harmonic force constants of a crystal is then clearly imbedded in this linear response. It was also found that by employing the Hellmann–Feynman theorem [63, 64], one can simply obtain the forces on individual nuclei through the electron charge density and its linear response without having to calculate the total energy. This theorem is expressed as

$$\frac{\partial E_\lambda}{\partial \lambda} = \langle \Psi_\lambda | \partial \mathscr{H}_\lambda / \partial \lambda | \Psi_\lambda \rangle \tag{4.143}$$

where λ is a parameter (or a set of parameters) and Ψ_λ is the eigenvector of a Hamiltonian $\mathscr{H}_\lambda$ corresponding to the eigenvalue E_λ, i.e., $\mathscr{H}_\lambda \Psi_\lambda = E_\lambda \Psi_\lambda$. Since the nuclear coordinates are treated as frozen, the Hamiltonian in (4.141) is a parametric function of the electronic coordinates $\boldsymbol{r}$, with the nuclear coordinates $\boldsymbol{R}$ as parameters playing the role of λ in Eq. (4.143). Taking derivatives with respect to $\boldsymbol{R}$ is equivalent to taking derivatives with respect to $\boldsymbol{u}(l)$, because $\boldsymbol{R}_l = \boldsymbol{l} + \boldsymbol{u}(l)$, where $\boldsymbol{l}$ is the equilibrium position of the lth nucleus. Based on (4.143), the expression for the force will be

$$\boldsymbol{F}(l) = -\nabla_{\boldsymbol{u}} E_{\text{tot}} = -\langle \Psi_{\boldsymbol{u}(l)} | \nabla_{\boldsymbol{u}} \mathscr{H}_{\text{tot}} | \Psi_{\boldsymbol{u}(l)} \rangle . \tag{4.144}$$

In the Hamitonian (4.141), the first two terms are independent of $\boldsymbol{R}$, the third term depends on $\boldsymbol{R}$ through the electron–ion interaction that couples to the electronic degrees of freedom only via the electron charge density, and the fourth depends on $\boldsymbol{R}$ but not on $\boldsymbol{r}$. Thus the expectation value in Eq. (4.144) can be calculated:

$$\boldsymbol{F}(l) = -\int \rho(\boldsymbol{r}, \boldsymbol{R}) \nabla_{\boldsymbol{u}} V_{\text{en}}(\boldsymbol{r}, \boldsymbol{R}) \, \mathrm{d}\boldsymbol{r} - \nabla_{\boldsymbol{u}} V_{\text{nn}}(\boldsymbol{R}) \tag{4.145}$$

where $\rho(\boldsymbol{r}, \boldsymbol{R})$ is the ground-state electron charge density with respect to the nuclear positions $\boldsymbol{R}$. Therefore, it is easy to obtain the force constants by taking again the derivative of (4.145):

$$\begin{aligned}
\Phi_{\alpha\alpha'}(l, l') &= -\frac{\partial F_\alpha(l)}{\partial u_{\alpha'}(l')} = \frac{\partial^2 E_{\text{tot}}}{\partial u_\alpha(l) \partial u_{\alpha'}(l')} \\
&= \int \frac{\partial \rho(\boldsymbol{r}, \boldsymbol{R})}{\partial u_{\alpha'}(l')} \frac{\partial V_{\text{en}}(\boldsymbol{r}, \boldsymbol{R})}{\partial u_\alpha(l)} \mathrm{d}\boldsymbol{r} + \int \rho(\boldsymbol{r}, \boldsymbol{R}) \frac{\partial^2 V_{\text{en}}(\boldsymbol{r}, \boldsymbol{R})}{\partial u_\alpha(l) \partial u_{\alpha'}(l')} \mathrm{d}\boldsymbol{r} \\
&\quad + \frac{\partial^2 V_{\text{nn}}(\boldsymbol{r}, \boldsymbol{R})}{\partial u_\alpha(l) \partial u_{\alpha'}(l')} .
\end{aligned} \tag{4.146}$$

The first two integrals are the contributions from the valence electrons, and last term is from other nuclei in the crystal. The quantity $\partial \rho(\boldsymbol{r}, \boldsymbol{R}) / \partial u_{\alpha'}(l')$ indicates the linear response to a distortion of the nuclear geometry.

It should be noted that the force constants are obtained without imposing any analytical models for interatomic forces, which is necessary in traditional lattice dynamics. With the above force constants, the sum of the first two integrals in Eq. (4.146) and the dynamical matrix can be constructed. Its consequent diagonalization will give the phonon frequencies.

4.7.2
The Density-Functional Theory

The properties of a system of N interacting electrons can be obtained by solving a set of Schrödinger equations

$$\mathscr{H}\Psi = [T_e + V_{ee} + V]\Psi = E\Psi \tag{4.147}$$

where T_e, V_{ee}, and V are the kinetic energy, electron–electron interaction energy, and the external potential operators, respectively, and Ψ is a wave function of N electrons. This is a typical many-body problem. Using the density-functional theory (DFT), this set of equations can be simplified and solved.

As shown in (4.146), the effect of electrons on lattice dynamics in a solid is directly presented by its charge density and its linear response. It is a special case within a much more general theoretical framework, known as the density-functional theory [65], for which Walter Kohn won a Nobel prize in 1998. It is such a theory that provides a radically different approach, using the electron charge density as the central quantity describing electron interactions, thus avoiding dealing with N-electron wave functions. Note that each N-electron wave function is a complex function of all $3N$ electron coordinates, while the corresponding electron charge density is a simple function of three variables.

First, the DFT asserts that the external potential $v(\boldsymbol{r})$ corresponding to the operator V, and hence the total energy E_{tot}, is a unique functional of the electron density $\rho(\boldsymbol{r})$. This predicts the existence of a one-to-one correspondence between the external potential $v(\boldsymbol{r})$ and the ground-state electron density $\rho(\boldsymbol{r})$. In other words, the electron density uniquely determines the potential acting on the electrons, and vice versa. As V operator defines the Hamiltonian of the system, every observable quantity must also be determined by $v(\boldsymbol{r})$ or $\rho(\boldsymbol{r})$, i.e., it must also be a functional of $\rho(\boldsymbol{r})$.

The first two terms $(T_e + V_{ee})$ of the Hamiltonian (1.414) describe the electron interactions only, and do not depend on the specific system, whether it is an atom, a molecule, or a solid sample under consideration. Their combined expectation value can then be expressed as a universal functional of $\rho(\boldsymbol{r})$:

$$\langle\Psi_0|(T_e + V_{ee})|\Psi_0\rangle = F[\rho(\boldsymbol{r})] \tag{4.148}$$

where Ψ_0 is the electron wave function in the ground state. The last two terms V_{en} and V_{nn} represent the external potential operator V, which is not universal, but system specific. Based on (4.147) an expression of the ground-state energy can be obtained:

$$E[\rho(\boldsymbol{r})] = \langle\Psi_0|(T_e + V_{ee} + V)|\Psi_0\rangle = F[\rho(\boldsymbol{r})] + V[\rho(\boldsymbol{r})] \tag{4.149}$$

where

$$V[\rho(\boldsymbol{r})] = \langle \Psi_0 | V | \Psi_0 \rangle = \int \mathscr{V}(\boldsymbol{r})\rho(\boldsymbol{r})\,\mathrm{d}\boldsymbol{r} \tag{4.150}$$

is a functional of $\rho(\boldsymbol{r})$, specific to the system under investigation.

Secondly, the DFT states that the ground state energy can be obtained using the variational principle: the density that minimizes the total energy is the exact ground state density. Therefore, the functional (4.149) is minimized by the ground state electron density $\rho_0(\boldsymbol{r})$ corresponding to external potential $v(\boldsymbol{r})$ and the value of this minimum coincides with the true ground state energy $E(\rho_0(\boldsymbol{r}))$. This statement may be formally written as

$$E(\rho_0(\boldsymbol{r})) = \min_{\rho(\boldsymbol{r})} E[\rho(\boldsymbol{r})]. \tag{4.151}$$

DFT has now provided a variational methodology for obtaining $\rho_0(\boldsymbol{r})$ and $E(\rho_0(\boldsymbol{r}))$, but it does not specify those functionals $F[\rho(\boldsymbol{r})]$ and $V[\rho(\boldsymbol{r})]$ for carrying out the calculations. To do this, some approximation approaches are necessary. For $F[\rho(\boldsymbol{r})]$, a non-interacting electron system is adopted for $T_e[\rho(\boldsymbol{r})]$ and $V_{ee}[\rho(\boldsymbol{r})]$, with an added compensation term called the exchange-correlation energy. For $V[\rho(\boldsymbol{r})]$, the actual potential is often substituted with a pseudopotential that produces exactly the same behavior of the valence electrons as the original potential. These approaches have proven quite successful when applied to a variety of solid and molecular systems.

4.7.3
Exchange-Correlation Energy and Local-Density Approximation

The DFT states that all physical properties of a system of interacting electrons are uniquely determined by its ground state electron charge density. Such an assertion remains valid independently of the precise form of the electron–electron interaction. This fact was used by Kohn and Sham [66] to turn the problem of a system of interacting electrons into an equivalent non-interacting problem. As defined in Eq. (4.148), the functional $F[\rho(\boldsymbol{r})]$ has contributions from $T_e[\rho(\boldsymbol{r})]$ and $V_{ee}[\rho(\boldsymbol{r})]$. The first one is approximated by $T_0[\rho(\boldsymbol{r})]$, the kinetic energy corresponding to a non-interacting system of electrons. The second is usually approximated by the Hartree functional $E_H[\rho(\boldsymbol{r})]$, which expresses the Coulomb mean-field interaction among the electrons. To compensate the missing interactions in this approximation, we add the so-called exchange-correlation functional $E_{xc}[\rho(\boldsymbol{r})]$ [59]:

$$F[\rho(\boldsymbol{r})] = T_0[\rho(\boldsymbol{r})] + E_H[\rho(\boldsymbol{r})] + E_{xc}[\rho(\boldsymbol{r})]. \tag{4.152}$$

A practical calculation requires a specific, albeit approximate, expression for exchange-correlation energy $E_{xc}[\rho(\boldsymbol{r})]$. The most widely used is the local-density approximation, which states that $E_{xc}[\rho(\boldsymbol{r})]$ can be given by assuming, for each infinitesimal element of density $\rho(\mathbf{r})\,d\mathbf{r}$, the exchange-correlation energy is that of a uniform electron gas of density $\rho = \rho(\mathbf{r})$. Then

$$E_{xc}[\rho(\mathbf{r})] \cong \int \varepsilon_{xc}(\rho)\rho(\mathbf{r})\,d\mathbf{r} \tag{4.153}$$

where $\varepsilon_{xc}(\rho)$ is the exchange-correlation energy per electron in a uniform electron gas of density ρ.

Using the variational principle, the energy functional (4.149) is minimized with respect to all possible functions of $\rho(\mathbf{r})$, with the constraint that the total number of electrons should not change. Because of the non-interacting model for the electron kinetic energy, this variational procedure leads to a set of self-consistent equations:

$$\left[-\frac{\hbar^2}{2m}\nabla^2 + V_{SCF}(\mathbf{r})\right]\phi(\mathbf{r}) = \varepsilon\phi(\mathbf{r}), \tag{4.154}$$

but the wave funtion $\phi(\mathbf{r})$ and energy ε are only for one electron. Here $V_{SCF}(\mathbf{r})$ is an effective potential, known as the self-consistent field (SCF) potential, in which the electron seems to be immersed:

$$V_{SCF}(\mathbf{r}) = v(\mathbf{r}) + \frac{\delta E_H[\rho(\mathbf{r})]}{\delta\rho(\mathbf{r})} + \frac{\delta E_{xc}[\rho(\mathbf{r})]}{\delta\rho(\mathbf{r})}. \tag{4.155}$$

The Schrödinger-like equations in Eq. (4.155) are known as Kohn–Sham equations. A total of $N/2$ solutions ($\phi_n(\mathbf{r})$, $n = 1, 2, 3, \ldots N/2$) can be obtained, called the auxiliary Kohn–Sham orbitals. The ground state electron density and non-interacting kinetic energy functional can be then given in terms of these auxiliary orbitals, ϕ_n:

$$\rho(\mathbf{r}) = 2\sum_{n=1}^{N/2} |\phi_n(\mathbf{r})|^2, \tag{4.156}$$

$$T_0[\rho(\mathbf{r})] = -2\frac{\hbar^2}{2m}\sum_{n=1}^{N/2}\int \phi_n^*(\mathbf{r})[\nabla^2\phi_n(\mathbf{r})]\,d\mathbf{r}, \tag{4.157}$$

whereas the ground state energy is given in terms of the Kohn–Sham eigenvalues:

$$E[\rho(\mathbf{r})] = 2\sum_{n=1}^{N/2}\varepsilon_n - E_H[\rho(\mathbf{r})] + E_{xc}[\rho(\mathbf{r})] - \int \rho(\mathbf{r})\frac{\delta E_{xc}[\rho(\mathbf{r})]}{\delta\rho(\mathbf{r})}\,d\mathbf{r}. \tag{4.158}$$

4.7.4
Plane Waves and Pseudopotentials

Up to now, most implementations of $V(r)$ in the Kohn–Sham equations have been based on the pseudopotential method in conjunction with plane-wave expansion. The optimal choice of an orthonormal basis set to represent the electron wave functions is dependent on the physical properties of a concrete system. For solid-state calculations, the periodic character of wave functions naturally suggests a plane-wave basis set, which has many attractive features [59]. We are mainly interested in the valence electrons because they play a dominant role in chemical bonding. This implies that we may replace the real electron–ion interaction with a fictitious potential, acting on valence electrons only. Although this pseudopotential is not required to produce any energy states for the core electrons, it must satisfy the condition that it produces the same energy states of the valence electrons as if they were in the original potential. Under the pseudopotential approximation the core electrons, which are supposed to be frozen, exert an effective repulsion on the valence electrons due to mutual orthogonality of their wave functions. As a result, this repulsion reduces to a large extent the attraction from the atomic nuclei. In short, plane waves and pseudopotentials are a natural combination and have become a quite useful method.

4.7.5
Calculation of DOS in Solids

Two approaches to calculating the phonon frequencies in solids are currently in use: the linear response method and the direct method. In the first, the dynamical matrix is obtained from the modification of electron charge density resulting from the atomic displacements. The dynamical matrix can be determined at any wave vector in the Brillouin zone using computational procedures similar to that of a ground state optimization. However, this approach only allows studies of linear effects, such as harmonic phonons.

The direct approach is based on the solution of the Kohn–Sham equations and allows one to study both linear and nonlinear effects. For phonons in a periodic lattice, there exists a superlattice constructed from periodic arrangement of a three-dimensional supercell. The motions of corresponding atoms in different supercells are assumed to be identical. In this approach, therefore, a distorted crystal due to atomic displacements is treated as a crystal in a new structure with a lower symmetry than the undistorted one. We then treat the undistorted and the distorted crystals separately but using exactly the same method. All of the atoms (or ions) in both kinds of crystals are assumed to reside motionlessly at their equilibrium positions; consequently the phonons are "frozen." A comparison between the two crystals will provide the lattice dynamics information. For a selected normal mode, the force constants can be calculated in two ways: the second derivative of $E_{\text{tot}}(u)$ (per atom) with respect to the displacement u,

$$\Phi = \left(\frac{\partial^2 E_{\text{tot}}(u)}{\partial u^2}\right)_{u=0} \approx \frac{2\Delta E_{\text{tot}}(u)}{u^2}, \tag{4.159}$$

or using the Hellmann–Feynman force $F(u)$

$$\Phi = -\left(\frac{\partial F}{\partial u}\right)_{u=0} \approx -\frac{F(u)}{u} \tag{4.160}$$

where

$$\Delta E_{\mathrm{tot}}(u) = E_{\mathrm{tot}}(u) - E_{\mathrm{tot}}(0). \tag{4.161}$$

The first way involves the so-called frozen-phonon local density approximation (LDA) calculations. The phonon frequency is then

$$\omega = \sqrt{\frac{\Phi}{M}}. \tag{4.162}$$

The direct method is rather straightforward computationally and very accurate. But the supercell may only contain a small number of unit cells.

After the phonon frequencies at selected high-symmetry points of BZs are calculated, the dispersion curves and thus the phonon DOS are easily obtained. As an example, we illustrate the calculated results of crystal $CuInSe_2$ by the direct method [67]. The crystal structure of $CuInSe_2$ has a D_{2d}^{12} symmetry. A total of eight coordination shells were considered with 19 independent force constants and 136 independent potential parameters. A $1 \times 1 \times 1$ supercell (crystallographic unit cell) with 16 atoms was used in all calculations. The partial and total phonon DOS presented in Fig. 4.27 were obtained by sampling the dynamical matrix at 10 000 randomly selected wave vectors. The total DOS (lower right in Fig. 4.27) exhibits three well-separated bands: the acoustic region (0.0–2.5 THz), the low optical region (3.0–4.5 THz), and the high optical region (5.5–6.8 THz).

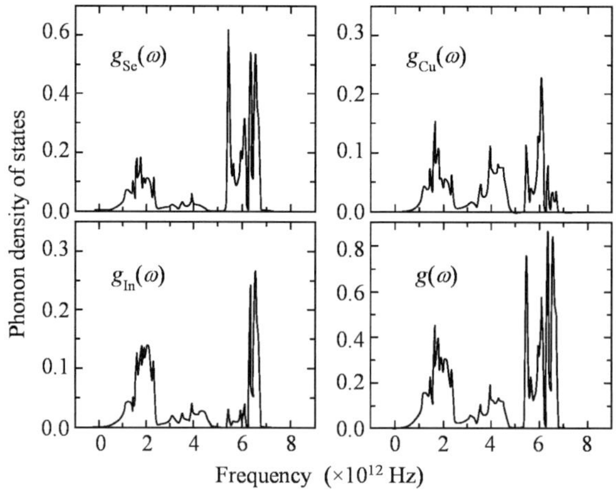

Fig. 4.27 Calculated atomic partial DOS and total phonon DOS for $CuInSe_2$ [67].

References

1 M. Born and K. Huang. *Dynamical Theory of Crystal Lattices* (Oxford University Press, London, 1954).

2 A.A. Maradudin. Elements of the theory of lattice dynamics. In *Dynamical Properties of Solids*, vol. 1, G.K. Horton and A.A. Maradudin (Eds.), pp. 1–82 (North-Holland, Amsterdam, 1974).

3 H. Böttger. *Principles of the Theory of Lattice Dynamics* (Physik Verlag, Weinheim, 1983).

4 B.T.M. Willis and A.W. Pryor. *Thermal Vibrations in Crystallography* (Cambridge University Press, London, 1975).

5 R.M. Housley and F. Hess. Analysis of Debye–Waller-factor and Mössbauer-thermal-shift measurements. *Phys. Rev.* 146, 517–526 (1966).

6 J.A. Reissland. *The Physics of Phonons* (Wiley, London, 1973).

7 M. Born and Th. von Kármán. Über Schwingungen in Raumgittern. *Physik. Zeitschr.* 13, 297–309 (1912). M. Born and Th. von Kármán. Über die Verteilung der Eigenschwingungen von Punktgittern. *Physik. Zeitschr.* 14, 65–71 (1913).

8 P. Brüesch. *Phonons: Theory and Experiments I. Lattice Dynamics and Models of Interatomic Forces* (Springer-Verlag, Berlin, 1982).

9 G. Burns. *Solid State Physics* (Academic Press, New York, 1985). N.W. Ashcroft and N.D. Mermin. *Solid State Physics* (Saunders College Publishing, Fort Worth, 1976).

10 Y. Fujii, N.A. Lurie, R. Pynn, and G. Shirane. Inelastic neutron scattering from ^{36}Ar. *Phys. Rev. B* 10, 3647–3659 (1974).

11 B.G. Dick, Jr., and A.W. Overhauser. Theory of the dielectric constants of alkali halide crystals. *Phys. Rev.* 112, 90–103 (1958).

12 P.N. Keating. Effect of invariance requirements on the elastic strain energy of crystals with application to the diamond structure. *Phys. Rev.* 145, 637–645 (1966).

13 W. Weber. Adiabatic bond charge model for the phonons in diamond, Si, Ge, and α-Sn. *Phys. Rev. B* 15, 4789–4803 (1977).

14 J. Tersoff. New empirical model for the structural properties of silicon. *Phys. Rev. Lett.* 56, 632–635 (1986).

15 K.E. Khor and S. Das Sarma. Proposed universal interatomic potential for elemental tetrahedrally bonded semiconductors. *Phys. Rev. B* 38, 3318–3322 (1988).

16 J.R. Hardy and A.M. Karo. *The Lattice Dynamics and Statics of Alkali Halide Crystals* (Plunum Press, New York, 1979). H.P. Myers. *Introductory Solid State Physics* (Taylor & Francis, London, 1990).

17 J.B. Clement and E.H. Quinnell. Atomic heat of indium below 20°K. *Phys. Rev.* 92, 258–267 (1953).

18 H. Thirring. Zur Theorie der Raumgitterschwingungen und der spezifischen Wärme fester Körper. *Physik. Zeitschr.* 14, 867–873 (1913). H. Thirring: Raumgitterschwingungen und spezifische Wärmen mehratomiger fester Körper. I. *Physik. Zeitschr.* 15, 127–133 (1914).

19 E.W. Montroll. Frequency spectrum of crystalline solids: II. General theory and applications to simple cubic lattices. *J. Chem. Phys.* 11, 481–495 (1943). E.W. Montroll and D.C Peaslee. Frequency spectrum of crystalline solids: III. Body-centered cubic lattices. *J. Chem. Phys.* 12, 98–106 (1944).

20 W. Jones and N.H. March. *Theoretical Solid State Physics*, vol. 1 (Wiley, London, 1973).

21 A.A. Maradudin, E.W. Montroll, and G.H. Weiss. *Theory of Lattice Dynamics in the Harmonic Approximation* (Solid State Physics Supplement 3) (Academic Press, New York, 1963).

22 G.K. Horton and H. Schiff. On the evaluation of equivalent Debye temperatures and related problems. *Proc. R. Soc. (London)* A250, 248–265 (1959).

23 J.C. Wheeler and C. Blumstein. Modified moments for harmonic solids. *Phys. Rev. B* 6, 4380–4382 (1972).

24 C. Blumstein and J.C. Wheeler. Modified-moments method: applications to harmonic solids. *Phys. Rev. B* 8, 1764–1776 (1973).

25 J.C. Wheeler, M.G. Prais, and C. Blumstein. Analysis of spectral densities using modified moments. *Phys. Rev. B* 10, 2429–2447 (1974).

26 C. Isenberg. Moment calculations in lattice dynamics: I. Fcc lattice with nearest-neighbor interactions. *Phys. Rev.* 132, 2427–2433 (1963).

27 C. Isenberg. Moment calculations in lattice dynamics II. *J. Phys. C* 4, 164–173 (1971).

28 C. Benoit. The moments method and damped systems. *J. Phys.: Condens. Matter* 6, 3137–3160 (1994).

29 G.J. Kemerink, N. Ravi, and H. de Waad. Debye–Waller factor of ^{129}I in CuI, SnTe, ZnTe and the alkali iodides LiI, NaI, KI, RbI and CsI determined by Mössbauer spectroscopy. *J. Phys. C* 19, 4897–4915 (1986).

30 M.P. Tosi and F.G. Fumi. Temperature dependence of the Debye temperatures for the thermodynamic functions of alkali halide crystals. *Phys. Rev.* 131, 1458–1465 (1963).

31 T.H.K. Barron, W.T. Berg, and J.A. Morrison. The thermal properties of alkali halide crystals: II. Analysis of experimental results. *Proc. R. Soc. (London)* A242, 478–492 (1957).

32 T.H.K. Barron, A.J. Leadbetter, J.A. Morrison. The thermal properties of alkali halide crystals: IV. Analysis of thermal expansion measurements. *Proc. R Soc. (London)* A279, 62–81 (1964).

33 T.H.K. Barron and J.A. Morrison. The thermal properties of alkali halide crystals: III. The inversion of the heat capacity. *Proc. R. Soc. (London)* A256, 427–439 (1960).

34 B. Yates. *Thermal Expansion* (Plenum Press, New York, 1972).

35 G.H. Wolf and R. Jeanloz. Vibrational properties of model monatomic crystals under pressure. *Phys. Rev. B* 32, 7798–7810 (1985).

36 M.V. Nevitt and G.S. Knapp. Phonon properties of vanadium-substituted lanthanum niobate derived from heat-capacity measurements. *J. Phys. Chem. Solids* 47, 501–505 (1986).

37 B.N. Brockhouse, H.E. Abou-Helal, and E.D. Hallman. Lattice vibrations in iron at 296 K. *Solid State Commun.* 5, 211–216 (1967).

38 M.A. Omar. *Elementary Solid State Physics: Principles and Applications* (Addison-Wesley, Reading, MA, 1975).

39 J.S. Blakemore. *Solid State Physics*, 2nd edn (Cambridge University Press, London, 1985).

40 E.S.R. Gopal. *Specific Heats at Low Temperatures* (Plenum Press, New York, 1966).

41 T. Nakazawa, H. Inoue, and T. Shirai. Comparison of characteristic temperatures determined by Mössbauer spectroscopy and x-ray diffraction. *Hyperfine Interactions* 55, 1145–1150 (1990).

42 A.A. Maradudin. Phonons and lattice imperfections. In *Phonons and Phonon Interactions*, T.A. Bak (Ed.), pp. 424–504 (Benjamin, New York, 1964).

43 A.A. Maradudin. Theoretical and experimental aspects of the effects of point defects and disorder on the vibrations of crystals. In *Solid State Physics*, vol. 18, F. Seitz and D. Turnbull (Eds.), pp. 273–420 (Academic Press, New York, 1966).

44 A.S. Barker, Jr., and A.J. Sievers. Optical studies of the vibrational properties of disordered solids. *Rev. Mod. Phys.* 47 (Suppl. 2), S1–S179 (1975).

45 T. Hattori, K. Ehara, A Mitsuishi, S. Sakuragi, and H. Kanzaki. Localized modes of Li and Na impurities in silver halides. *Solid State Commun.* 12, 545–548 (1973).

46 G.L. Squires. *Introduction to the Theory of Thermal Neutron Scattering*

(Cambridge University Press, London, 1978).
47 D.L. Price and K. Sköld. Introduction to neutron scattering. In *Neutron Scattering* (*Methods of Experimental Physics*, vol. 23, Part A), K. Sköld and D.L. Price (Eds.) (Academic Press, New York, 1986).
48 S.W. Lovesey. Introduction. In *Dynamics of Solids and Liquids by Neutron Scattering*, S.W. Lovesey and T. Springer (Eds.), pp. 1–75 (Springer-Verlag, Berlin, 1977).
49 B.N. Brockhouse. Energy distribution of neutrons scattered by paramagnetic substances. *Phys. Rev.* 99, 601–603 (1955). B.N. Brockhouse. Methods for neutron spectroscopy. In *Inelastic Scattering of Neutrons in Solids and Liquids* (Proceedings of the 1960 Symposium), pp. 113–157 (International Atomic Energy Agency, Vienna, 1961).
50 E.C. Svensson, B.N. Brockhouse, and J.M. Rowe. Crystal dynamics of copper. *Phys. Rev.* 155, 619–632 (1967).
51 R.M. Nicklow, G. Gilat, H.G. Smith, L.J. Raubenheimer, and M.K. Wilkinson. Phonon frequencies in copper at 49 and 298 °K. *Phys. Rev.* 164, 922–928 (1967).
52 G. Gilat and G. Dolling. A new sampling method for calculating the frequency distribution function of solids. *Phys. Lett.* 8, 304–306 (1964).
53 G. Gilat and L.J. Raubenheimer. Accurate numerical method for calculating frequency-distribution functions in solids. *Phys. Rev.* 144, 390–395 (1966).
54 E. Burkel, J. Peisl, and B. Dorner. Observation of inelastic x-ray scattering from phonons. *Europhys. Lett.* 3, 957–961 (1987).
55 B. Dorner, E. Burkel, Th. Illini, and J. Peisl. First measurement of a phonon dispersion curve by inelastic x-ray scattering. *Z. Phys. B* 69, 179–183 (1987).
56 S.K. Sinha. Theory of inelastic x-ray scattering from condensed matter. *J. Phys.: Condens. Matter* 13, 7511–7523 (2001).
57 E. Burkel. Determination of phonon dispersion curves by means of inelastic x-ray scattering. *J. Phys.: Condens. Matter* 13, 7627–7644 (2001).
58 P. Pavone. Old and new aspects in lattice-dynamical theory. *J. Phys.: Condens. Matter* 13, 7593–7610 (2001).
59 S. Baroni, S. de Gironcoli, A. Dal Corso, and P. Giannozzi. Phonons and related crystal properties from density-functional perturbation theory. *Rev. Mod. Phys.* 73, 515–562 (2001).
60 M.D. Segall, P.J.D. Lindan, M.J. Probert, C.J. Pickard, P.J. Hasnip, S.J. Clark, and M.C. Payne. First-principles simulation: ideas, illustrations and the CASTEP code. *J. Phys.: Condens. Matter* 14, 2717–2744 (2002).
61 P.D. DeCicco and F.A. Johnson. The quantum theory of lattice dynamics. *Proc. R. Soc. (London)* A310, 111–119 (1969).
62 R.M. Pick, M.H. Cohen, and R.M. Martin. Microscopic theory of force constants in the adiabatic approximation. *Phys. Rev. B* 1, 910–920 (1970).
63 H. Hellmann. *Einführung in die Quantenchemie* (Deuticke, Leipzig, 1937).
64 R.P. Feynman. Forces in molecules. *Phys. Rev.* 56, 340–343 (1939).
65 P. Hohenberg and W. Kohn. Inhomogeneous electron gas. *Phys. Rev.* 136, B864–B871 (1964). W. Kohn. Nobel lecture: Electronic structure of matter – wave functions and density functionals. *Rev. Mod. Phys.* 71, 1253–1266 (1999).
66 W. Kohn and L.J. Sham. Self-consistent equations including exchange and correlation effects. *Phys. Rev.* 140, A1133–A1138 (1965).
67 J. Lazewski, K. Parlinski, B. Hennion, and R. Fouret. First-principles calculations of the lattice dynamics of $CuInSe_2$. *J. Phys.: Condens. Matter* 11, 9665–9671 (1999).

5
Recoilless Fraction and Second-Order Doppler Effect

As is well known, obtaining information on lattice dynamics of solids using transmission Mössbauer spectroscopy is mainly through the measurements of the recoilless fraction f and the second-order Doppler shift δ_{SOD}, from which the atomic mean-square displacement $\langle u^2 \rangle$, the mean-square velocity $\langle v^2 \rangle$, the anharmonic effect, Einstein temperature θ_E or Debye temperature θ_D, and the effective vibrating mass M_{eff} are determined. All these quantities are discussed in detail in the following chapters. On the one hand, f and δ_{SOD} can be accurately measured experimentally using the Mössbauer effect. On the other hand, $\langle u^2 \rangle$, $\langle v^2 \rangle$, and θ_D can be calculated through several models and methods, and can be compared with the experimental results, allowing us to have a better understanding of the dynamical properties of solids. Therefore, the Mössbauer effect can play an important role in lattice dynamics research. Although $\langle u^2 \rangle$ may also be measured using elastic scattering of neutrons or x-rays, the Mössbauer method yields better accuracy. In cases where information on the $\langle u^2 \rangle$ of an impurity atom is needed, the Mössbauer effect is the only method, provided that this impurity atom is a Mössbauer isotope.

In this chapter, we focus on the common theoretical and experimental issues concerning the recoilless fraction f and the second-order Doppler shift δ_{SOD}, such as how f depends on temperature and pressure, its anisotropy, its anharmonic effect, and how to measure f using absolute and relative methods.

5.1
Mean-Square Displacement $\langle u^2 \rangle$ and Mean-Square Velocity $\langle v^2 \rangle$

Since we will frequently encounter these two quantities and they are also related to each other, let us discuss their general expressions.

During the lifetime (usually 10^{-7} to 10^{-10} s) of the excited state of a Mössbauer nucleus, an atom would have vibrated at least several hundred times around its equilibrium position, and therefore $\langle \mathbf{u} \rangle = 0$ and $\langle \mathbf{v} \rangle = 0$. However, $\langle u^2 \rangle$ and $\langle v^2 \rangle$ are nonzero, and may have large magnitudes.

Mössbauer Effect in Lattice Dynamics. Yi-Long Chen and De-Ping Yang

ISBN: 978-3-527-40712-5

Calculating $\langle u^2 \rangle$ as a thermal average [1, 2] according to (4.49), we obtain

$$\langle u^2 \rangle = \langle \mathbf{u} \cdot \mathbf{u}^* \rangle_{\mathrm{T}} = \frac{\hbar}{2M} \int \frac{1}{\omega} \coth\left(\frac{1}{2}\hbar\omega\beta\right) g(\omega)\, \mathrm{d}\omega, \tag{5.1}$$

and similarly

$$\langle v^2 \rangle = \frac{3\hbar}{M} \int \coth\left(\frac{1}{2}\hbar\omega\beta\right) g(\omega)\omega\, \mathrm{d}\omega. \tag{5.2}$$

In order to carry out the above integrals, concrete phonon DOS $g(\omega)$ must be specified. If the Debye distribution is used, the above expressions become, respectively

$$\langle u^2 \rangle = \frac{3\hbar^2}{4Mk_{\mathrm{B}}\theta_{\mathrm{D}}} \left[1 + 4\left(\frac{T}{\theta_{\mathrm{D}}}\right)^2 \int_0^{\theta_{\mathrm{D}}/T} \frac{x}{\mathrm{e}^x - 1} \mathrm{d}x\right], \tag{5.3}$$

$$\langle v^2 \rangle = \frac{9k_{\mathrm{B}}\theta_{\mathrm{D}}}{M} \left[\frac{1}{8} + \left(\frac{T}{\theta_{\mathrm{D}}}\right)^4 \int_0^{\theta_{\mathrm{D}}/T} \frac{x^3}{\mathrm{e}^x - 1} \mathrm{d}x\right]. \tag{5.4}$$

At high temperatures (i.e., $T > \theta_{\mathrm{D}}/2$), we have

$$\langle u^2 \rangle \approx \frac{3\hbar^2 T}{Mk_{\mathrm{B}}\theta_{\mathrm{D}}^2} \left[1 + \left(\frac{\theta_{\mathrm{D}}}{6T}\right)^2\right], \tag{5.5}$$

$$\langle v^2 \rangle \approx \frac{3k_{\mathrm{B}}T}{M} \left[1 + \frac{1}{20}\left(\frac{\theta_{\mathrm{D}}}{T}\right)^2\right]. \tag{5.6}$$

The expressions for low temperatures ($T \ll \theta_{\mathrm{D}}$) are

$$\langle u^2 \rangle \approx \frac{3\hbar^2}{4Mk_{\mathrm{B}}\theta_{\mathrm{D}}} \left[1 + \frac{2\pi^2}{3}\left(\frac{T}{\theta_{\mathrm{D}}}\right)^2\right], \tag{5.7}$$

$$\langle v^2 \rangle \approx \frac{9k_{\mathrm{B}}\theta_{\mathrm{D}}}{M} \left[\frac{1}{8} + \frac{\pi^4}{15}\left(\frac{T}{\theta_{\mathrm{D}}}\right)^4\right]. \tag{5.8}$$

Another way to express $\langle u^2 \rangle$ and $\langle v^2 \rangle$ is using the frequency moment method, the advantage of which is that the specific details of $g(\omega)$ are not required. At high temperatures ($T > \theta_{\mathrm{D}}/2$), $\langle u^2 \rangle$ and $\langle v^2 \rangle$ can be written as [3]

$$\begin{aligned} \langle u^2 \rangle &= \frac{k_{\mathrm{B}}T}{M} \left[\mu(-2) + \frac{1}{12}\left(\frac{\hbar}{k_{\mathrm{B}}T}\right)^2 - \frac{1}{720}\left(\frac{\hbar}{k_{\mathrm{B}}T}\right)^4 \mu(2) + \cdots\right] \\ &\approx \frac{3\hbar^2 T}{Mk_{\mathrm{B}}} \frac{1}{\theta_{\mathrm{D}}^2(-2)} \left[1 + \left(\frac{\theta_{\mathrm{D}}(-2)}{6T}\right)^2\right] \end{aligned} \tag{5.9}$$

and

$$\langle v^2 \rangle = \frac{3k_B T}{M}\left[1 + \frac{1}{12}\left(\frac{\hbar}{k_B T}\right)^2 \mu(2) - \frac{1}{720}\left(\frac{\hbar}{k_B T}\right)^4 \mu(4) + \cdots\right]$$

$$\approx \frac{3k_B T}{M}\left[1 + \frac{1}{20}\left(\frac{\theta_D(2)}{T}\right)^2\right]. \tag{5.10}$$

In the limiting case of $T \to 0$

$$\langle u^2 \rangle \approx \frac{\hbar}{2M}\mu(-1) = \frac{3\hbar^2}{4Mk_B}\frac{1}{\theta_D(-1)}, \tag{5.11}$$

$$\langle v^2 \rangle \approx \frac{3\hbar}{2M}\mu(1) = \frac{9k_B\theta_D(1)}{8M}. \tag{5.12}$$

Comparing (5.5) and (5.9), one notices that the only difference is θ_D and $\theta_D(-2)$, the Debye temperature and the weighted Debye temperature (see Section 4.4), respectively. Other expressions also have similar patterns. It seems that using the frequency moment method is closer to reality because it would give a different Debye temperature value when the measurement is done in a different temperature range.

If the Debye distribution is chosen for $g(\omega)$ and various moments $\mu(n)$ are calculated according to (4.104), it is easy to verify that when the $\mu(n)$ expressions are used in (5.9) and (5.10), they indeed reduce to (5.5) and (5.6), as expected.

To show the magnitude of $\langle u^2 \rangle$ in a solid, Table 5.1 lists the $\langle u^2 \rangle$ values of Eu atoms in the compound $Eu_{1.15}Ba_{1.85}Cu_3O_{7-\delta}$. It is easy to see that the amplitude of Eu atomic vibration is of the order of 0.1 Å, which is the typical u-value for most solids at room temperature.

Table 5.1 Mean-square displacement $\langle u^2 \rangle$ of Eu atoms in $Eu_{1.15}Ba_{1.85}Cu_3O_{7-\delta}$ [4].

Temperature, T (K)	f	$\langle u^2 \rangle$ (10^{-3} Å^2)
25	0.518(7)	5.5(1)
40	0.522(7)	5.4(1)
60	0.509(7)	5.6(1)
90	0.495(7)	5.9(1)
200	0.411(6)	7.4(1)
300	0.305(5)	9.9(1)

5.2 Temperature Dependence of the Recoilless Fraction f

Under the harmonic approximation, the recoilless fraction f in Eq. (1.71) can be simplified to

$$f = |\langle e^{i\boldsymbol{k}\cdot\boldsymbol{u}}\rangle|^2 \approx e^{-\langle(\boldsymbol{k}\cdot\boldsymbol{u})^2\rangle} = e^{-k^2\langle u^2\rangle}. \tag{5.13}$$

Therefore, the mean-square displacement $\langle u^2\rangle$ along the direction of γ-ray propagation can be readily obtained by measuring the recoilless fraction f. The above

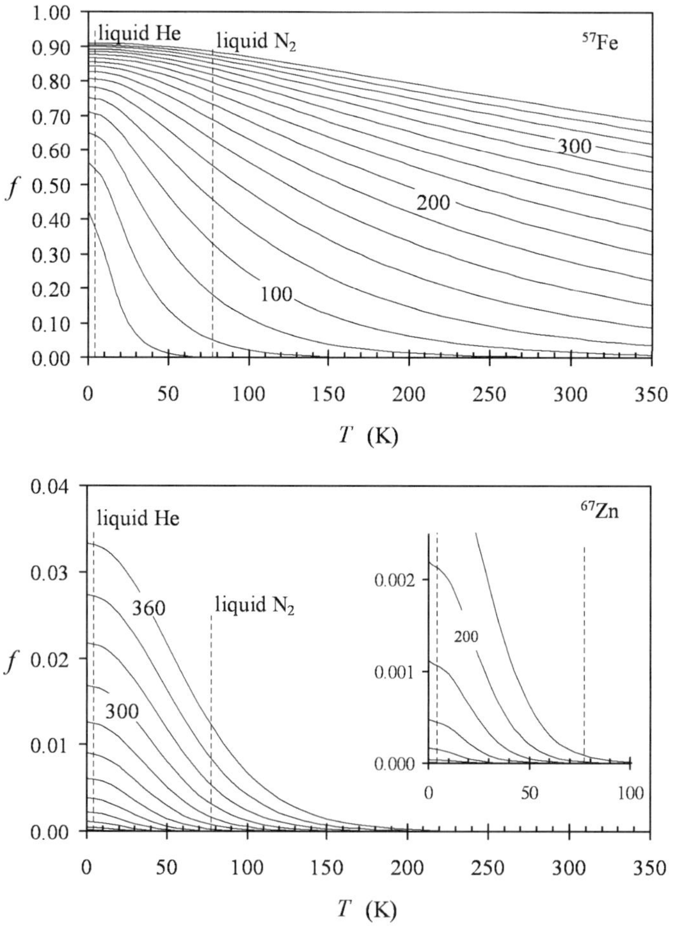

Fig. 5.1 Recoilless fraction f as a function of temperature T for the 14.4 keV transition in ^{57}Fe and the 93.3 keV transition in ^{67}Zn. θ_D is used as a parameter in calculating each curve, with the top curve corresponding to $\theta_D = 360$ K and the lower curves corresponding to decreasing θ_D-values with intervals of 20 K. Liquid helium and liquid nitrogen temperatures are represented by the vertical dashed lines.

expression is a general form, applicable to all Bravais crystals, and its strong dependence on temperature is embedded in $\langle u^2 \rangle$. When the explicit expression for $\langle u^2 \rangle$ in Eq. (5.3) is substituted into (5.13), the temperature dependence of f is now through the two parameters E_R and θ_D:

$$f = \exp\left\{ -\frac{3E_R}{2k_B\theta_D}\left[1 + 4\left(\frac{T}{\theta_D}\right)^2 \int_0^{\theta_D/T} \frac{x\,dx}{(e^x - 1)}\right]\right\}$$

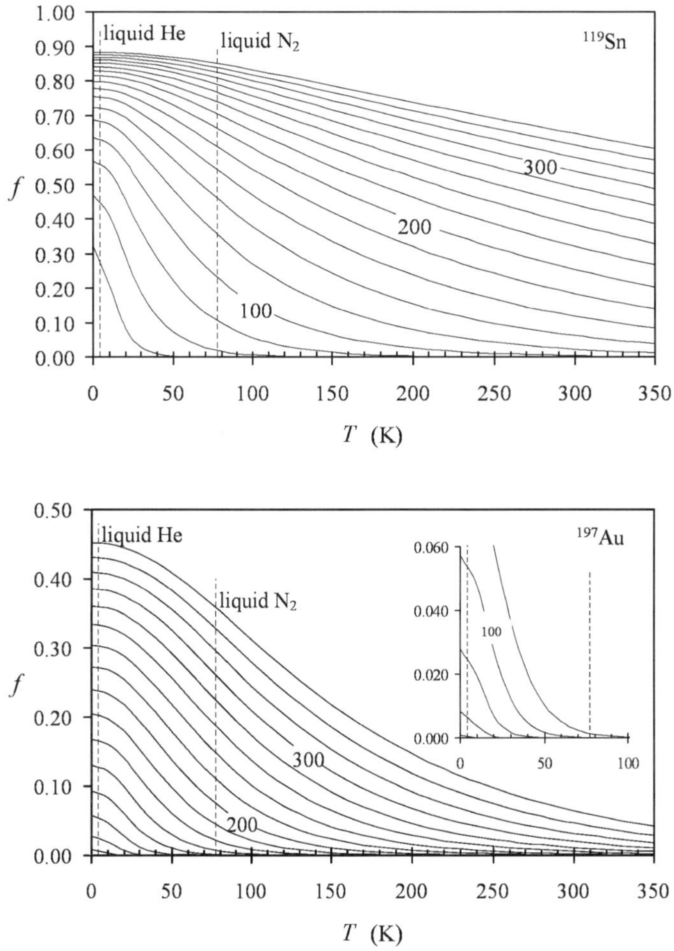

Fig. 5.2 Recoilless fraction f as a function of temperature T for the 23.9 keV transition in ^{119}Sn and the 77.3 keV transition in ^{197}Au. θ_D is used as a parameter in calculating each curve, with the top curve corresponding to $\theta_D = 360$ K and the lower curves corresponding to decreasing θ_D-values with intervals of 20 K. Liquid helium and liquid nitrogen temperatures are represented by the vertical dashed lines.

which is Eq. (1.84). In Ref. [5], the temperature dependence of the recoilless fraction f has been numerically calculated for 29 commonly used Mössbauer transitions, among which four are shown in Figs. 5.1 and 5.2. In each graph, the first curve from the top corresponds to $\theta_D = 360$ K, the lower curves are drawn for decreasing temperatures with intervals of 20 K, and the liquid nitrogen (77 K) and liquid helium (4.2 K) temperatures are indicated. Most Fe compounds and some other compounds have their θ_D above 300 K; therefore they have relatively large f-values even at room temperature. For ^{67}Zn, only at liquid nitrogen temperature is f large enough for observation of its Mössbauer effect.

For the convenience of future reference, we substitute various forms of $\langle u^2 \rangle$ into Eq. (5.13) to obtain explicit expressions for f in two difference temperature regions. When $T > \theta_D/2$, we have

$$-\ln f = \frac{6E_R T}{k_B \theta_D^2}\left[1 + \left(\frac{\theta_D}{6T}\right)^2\right] = \frac{6E_R T}{k_B \theta_D^2(-2)}\left[1 + \left(\frac{\theta_D(-2)}{6T}\right)^2\right], \tag{5.14}$$

and for the low-temperature limit ($T \to 0$)

$$\begin{aligned}-\ln f &= \frac{3E_R}{2k_B\theta_D}\left[1 + \frac{2\pi^2}{3}\left(\frac{T}{\theta_D}\right)^2\right] \\ &= \frac{3E_R}{2k_B\theta_D(-1)}\left[1 + \frac{2\pi^2}{3}\left(\frac{T}{\theta_D(-1)}\right)^2\right].\end{aligned} \tag{5.15}$$

5.3
The Anharmonic Effects

First, the Taylor series of the potential energy in Eq. (4.12) is represented by

$$V = V_0 + V_1 + V_2 + V_3 + V_4 + \cdots$$

The harmonic approximation ignores V_3 and higher order terms, and V_2 is just the nonzero term proportional to the square of the atomic displacement u^2 (i.e., the parabolic potential). This approximation fails to explain some phenomena such as thermal expansion in solids, and the anharmonic terms V_3 and V_4 need to be included. These terms couple one phonon to another, i.e., phonon–phonon interaction. For most solids, the potential energy curve has the parabolic shape only in a very small region near the atom's equilibrium position. As the temperature rises, the amplitude of atomic vibration increases, and the anharmonic effect becomes appreciable. For any potential energy deviating from the parabolic shape, the anharmonic effect should not be overlooked, even at very low temperatures.

The recoilless fraction f (or the second-order Doppler shift δ_{SOD}) is closely related to the harmonicity or anharmonicity of the solid. Because the expression for

f has the atomic mean-square displacement $\langle u^2 \rangle$ on the exponent, the anharmonicity of a solid can be sensitively detected by measuring the recoilless fraction f. Using the Mössbauer effect, several frequency moments of the phonon, including those moments directly related to the anharmonic effect, can be accurately measured.

5.3.1 The General Form of the Recoilless Fraction f

To study the size of the anharmonic effect, the usual procedure is to include V_3 and V_4 in the potential energy, calculate the corresponding $\langle u^2 \rangle$, and compare the results with experimental data. Over the years, many theoretical methods [3, 6–10] have been developed for calculating $\langle u^2 \rangle$. Here we describe a relatively simple theory based on the work in Refs. [11–14]. The goal is to derive an expression for the recoilless fraction f that is explicitly dependent on anharmonic parameters.

For an anharmonic crystal, the recoilless fraction f can be written as

$$-\ln f = \langle (\boldsymbol{k} \cdot \boldsymbol{u})^2 \rangle - \frac{1}{12} \langle (\boldsymbol{k} \cdot \boldsymbol{u})^4 \rangle + \frac{1}{4} \langle (\boldsymbol{k} \cdot \boldsymbol{u})^2 \rangle^2 + O(k^6). \tag{5.16}$$

For an fcc lattice, using the properties of the central forces between nearest neighbors, the bracketed part in the second term can be shown to be approximately

$$\langle (\boldsymbol{k} \cdot \boldsymbol{u})^4 \rangle \approx 3 \langle (\boldsymbol{k} \cdot \boldsymbol{u})^2 \rangle^2 \tag{5.17}$$

which happens to make the second and the third terms cancel each other. This means that for an anharmonic crystal, the recoilless fraction f can still be adequately described by Eq. (5.13), except that $\boldsymbol{u}$ should be the actual atomic displacement in the anharmonic vibration. We now focus on calculating $\boldsymbol{u}(l)$, based on Eq. (4.49) and using the creation and annihilation operators $\hat{a}^+_{-\boldsymbol{k}j}$ and $\hat{a}_{\boldsymbol{k}j}$:

$$\boldsymbol{u}(l) = \left(\frac{\hbar}{2MN} \right)^{1/2} \sum_{\boldsymbol{k}j} \frac{\boldsymbol{e}(\boldsymbol{k}j)}{\sqrt{\omega_j(\boldsymbol{k})}} (\hat{a}_{\boldsymbol{k}j} + \hat{a}^+_{-\boldsymbol{k}j}) e^{i\boldsymbol{k} \cdot \boldsymbol{R}_l}. \tag{5.18}$$

When (5.18) is substituted into the first term in (5.16),

$$\langle (\boldsymbol{k} \cdot \boldsymbol{u})^2 \rangle = \frac{\hbar}{2MN} \sum_{\substack{\boldsymbol{k},\boldsymbol{k}' \\ j,j'}} \frac{[\boldsymbol{k} \cdot \boldsymbol{e}(\boldsymbol{k}j)][\boldsymbol{k} \cdot \boldsymbol{e}(\boldsymbol{k}'j')]}{[\omega_j(\boldsymbol{k}) \omega_{j'}(\boldsymbol{k}')]^{1/2}} e^{i(\boldsymbol{k}+\boldsymbol{k}') \cdot \boldsymbol{l}} \langle A_{\boldsymbol{k}j} A^*_{\boldsymbol{k}'j'} \rangle \tag{5.19}$$

where

$$A_{\boldsymbol{k}j} = \hat{a}_{\boldsymbol{k}j} + \hat{a}^+_{-\boldsymbol{k}j}. \tag{5.20}$$

The self-correlation function $\langle A_{kj}A^*_{k'j'}\rangle$ can be expressed by the Green's functions (see Appendix F1). For a Bravais crystal, the anharmonic Hamiltonian is [11]

$$\mathscr{H} = \sum_{kj} \hbar\omega_j(\mathbf{k})\left(\hat{a}^+_{-kj}\hat{a}_{kj} + \frac{1}{2}\right)$$
$$+ \sum_{\substack{k_1k_2k_3\\ j_1j_2j_3}} \Phi^3(\mathbf{k}_1j_1, \mathbf{k}_2j_2, \mathbf{k}_3j_3)A_{\mathbf{k}_1j_1}A_{\mathbf{k}_2j_2}A_{\mathbf{k}_3j_3}$$
$$+ \sum \Phi^4(\mathbf{k}_1j_1, \mathbf{k}_2j_2, \mathbf{k}_3j_3, \mathbf{k}_4j_4)A_{\mathbf{k}_1j_1}A_{\mathbf{k}_2j_2}A_{\mathbf{k}_3j_3}A_{\mathbf{k}_4j_4} + \cdots \tag{5.21}$$

where Φ^n $(n \geq 3)$ represents the Fourier transform of the partial derivative of the potential function. The self-correlation function $\langle A_{kj}A^*_{k'j'}\rangle$ can be expressed by the retarded Green's function

$$G^{jj'}_{\mathbf{kk}'}(\omega) = -\frac{i}{\hbar}\theta(t)\langle[A_{\mathbf{k}j}(t), A_{\mathbf{k}'j'}]\rangle$$

which is similar to the definition (F.3). Using the two expressions

$$J^{jj'}_{\mathbf{kk}'}(\omega) = -\lim_{\varepsilon\to 0}\left[\frac{2}{e^{\beta\hbar\omega}-1}\operatorname{Im} G^{jj'}_{\mathbf{kk}'}(\omega+i\varepsilon)\right],$$

$$\langle A_{\mathbf{k}j}A^*_{\mathbf{k}'j'}\rangle = \int_{-\infty}^{\infty} J^{jj'}_{\mathbf{kk}'}(\omega)\,\mathrm{d}\omega = -\lim\int_{-\infty}^{\infty}\frac{2\,\mathrm{d}\omega}{e^{\beta\hbar\omega}-1}\operatorname{Im} G^{jj'}_{\mathbf{kk}'}(\omega+i\varepsilon),$$

which are also analogous to (F.21) and (F.27), respectively, the explicit form of the Green's function was achieved as follows [11, 14]:

$$G^{jj'}_{\mathbf{kk}'}(\omega) = \frac{\omega_j(\mathbf{k})}{\pi}\frac{\delta_{\mathbf{k},-\mathbf{k}'}\delta_{jj'}}{\omega^2-\omega_j^2(\mathbf{k})-2\omega_j(\mathbf{k})\sigma_{\mathbf{k}j}(\omega)}, \tag{5.22}$$

where $\sigma_{\mathbf{k}j}(\omega)$ is the energy shift in phonon $|\mathbf{k}j\rangle$ due to the anharmonic effect. It is composed of the real part and the imaginary part:

$$\sigma_{\mathbf{k}j}(\omega+i\varepsilon) = \Delta_{\mathbf{k}j}(\omega) - i\Gamma_{\mathbf{k}j}(\omega). \tag{5.23}$$

The real part $\Delta_{\mathbf{k}j}(\omega)$ is the actual shift in the phonon frequency $\omega_j(\mathbf{k})$, while $1/\Gamma_{\mathbf{k}j}$ is the average lifetime of the phonon.

Using the Green's function, Eq. (5.19) becomes [14]

$$-\ln f = \langle(\mathbf{k}\cdot\mathbf{u})^2\rangle = \frac{\hbar}{MN}\sum_{\mathbf{k}j}[\mathbf{k}\cdot\mathbf{e}(\mathbf{k}j)]^2\int_0^{\infty}\coth\left(\frac{\hbar\omega\beta}{2}\right)$$
$$\times\frac{2\omega_j(\mathbf{k})\Gamma_{\mathbf{k}j}(\omega)/\pi}{[\omega^2-\omega_j^2(\mathbf{k})-2\omega_j(\mathbf{k})\Delta_{\mathbf{k}j}]^2+[2\omega_j(\mathbf{k})\Gamma_{\mathbf{k}j}]^2}\mathrm{d}\omega. \tag{5.24}$$

Because this includes all anharmonic effects due to the third-order and higher terms in the Hamiltonian (5.21), it is a general expression for the recoilless fraction f, valid for all Bravais crystals.

However, the above integral can only be carried out after proper approximations have been taken for the phonon frequency shift. The usual methods of approximation include the quasiharmonic and pseudoharmonic methods. Under the harmonic approximation, the third and higher terms in the potential energy function are neglected; therefore, $\sigma_{\boldsymbol{k}j}(\omega) = 0$ and there is no thermal expansion. Under the quasiharmonic approximation, it is assumed that thermal expansion is the only cause for the temperature-dependences of phonon frequency and force constants. Based on this assumption, the frequency shift is directly proportional to the relative change in volume V:

$$\Delta_{\boldsymbol{k}j}^{\mathrm{qh}} = -\gamma_{\boldsymbol{k}j}\omega_j(\boldsymbol{k})\frac{\Delta V}{V} \tag{5.25}$$

where $\gamma_{\boldsymbol{k}j}$ is known as the Grüneisen constant. At room temperature, the quasiharmonic effect dominates and the resultant recoilless fraction f value agrees with experimental data quite well. When the temperature is higher, the coupling between phonons becomes significant and causes additional frequency shift $\delta\omega_{\boldsymbol{k}j}^{a}$, which can only be analyzed by the pseudoharmonic method. Therefore, the total frequency shift is the sum of the above two contributions:

$$\Delta_{\boldsymbol{k}j} = \delta\omega_{\boldsymbol{k}j}^{\mathrm{a}} - \gamma_{\boldsymbol{k}j}\omega_j(\boldsymbol{k})\frac{\Delta V}{V} = -\varepsilon_{\boldsymbol{k}j}\omega_j(\boldsymbol{k})T \tag{5.26}$$

where the superscript a stands for anharmonicity and $\varepsilon_{\boldsymbol{k}j}$ is the anharmonic constant, indicating the relative change in phonon frequency when temperature is increased by 1 K. We will now convert (5.24) into a more practical form using the pseudoharmonic approximation.

5.3.2 Calculating the Recoilless Fraction f Using the Pseudoharmonic Approximation

In the limit $\Gamma_{\boldsymbol{k}j} \to 0$, the new phonon frequency is

$$\omega_{\boldsymbol{k}j}^{\mathrm{a}} = \omega_j(\boldsymbol{k}) + \Delta_{\boldsymbol{k}j} = \omega_j(\boldsymbol{k})(1 - \varepsilon_{\boldsymbol{k}j}T). \tag{5.27}$$

The integrand in (5.24) is the Breit–Wigner type distribution, with a maximum at $\omega = \omega_{\boldsymbol{k}j}^{\mathrm{a}}$. This distribution can be replaced by the following δ-function:

$$\delta[\omega^2 - (\omega_{\boldsymbol{k}j}^{\mathrm{a}})^2] = \frac{1}{\pi}\lim_{\Gamma_{\boldsymbol{k}j}\to 0}\frac{2\omega_j(\boldsymbol{k})\Gamma_{\boldsymbol{k}j}}{[\omega^2 - (\omega_{\boldsymbol{k}j}^{\mathrm{a}})^2]^2 + [2\omega_j(\boldsymbol{k})\Gamma_{\boldsymbol{k}j}]^2} \tag{5.28}$$

where $(\omega_{\boldsymbol{k}j}^{\mathrm{a}})^2 \approx \omega_j^2(\boldsymbol{k}) + 2\omega_j(\boldsymbol{k})\Delta_{\boldsymbol{k}j}$. Substituting (5.28) into (5.24), we obtain the following expression of the recoilless fraction f for cubic lattices:

$$-\ln f = k^2\langle u^2\rangle = \frac{E_R}{3N}\sum_{kj}\frac{1}{\hbar\omega^{a}_{kj}}\coth\left(\frac{\hbar\omega^{a}_{kj}\beta}{2}\right). \tag{5.29}$$

This has the same form as the harmonic result (1.81), except for $\omega^{a}_{\boldsymbol{k}j}$ replacing $\omega_j(\boldsymbol{k})$ due to anharmonicity. As long as the temperature is not near the melting point and the potential function of atomic interaction is nearly a parabola, the recoilless fraction f values derived from the pseudoharmonic approximation are in good agreement with experimental results.

Equation (5.29) can be further simplified using the concept of frequency moments. First we define the anharmonic frequency moment, analogous to that in Eq. (4.105):

$$\mu^{a}(n,T) = \frac{1}{3N}\sum_{\boldsymbol{k}j}(\omega^{a}_{\boldsymbol{k}j})^n = \frac{1}{3N}\sum_{\boldsymbol{k}j}\omega_j^n(\boldsymbol{k})(1-\varepsilon_{\boldsymbol{k}j}T)^n \tag{5.30}$$

and the anharmonic characteristic temperature

$$\theta^{a}_{D}(n,T) = \frac{\hbar}{k_B}\left[\frac{n+3}{3}\mu^{a}(n,T)\right]^{1/n}. \tag{5.31}$$

Unlike the corresponding harmonic approximation parameters $\mu(n)$ and $\theta_D(n)$, these anharmonic parameters $\mu^{a}(n,T)$ and $\theta^{a}_{D}(n,T)$ are temperature-dependent and may be expanded into power series of T:

$$\mu^{a}(n,T) = \mu(n)\left[1 - n\varepsilon_1(n)T + \frac{n(n-1)}{2!}\varepsilon_2(n)T^2 + \cdots\right] \tag{5.32}$$

and

$$\theta^{a}_{D}(n,T) = \theta_D(n)\left\{1 - \varepsilon_1(n)T + \frac{n-1}{2!}[\varepsilon_2(n) - \varepsilon_1^2(n)]T^2 + \cdots\right\}, \tag{5.33}$$

where $\varepsilon_p(n)$ are the weighted anharmonic constants, defined as

$$\varepsilon_p(n) = \sum_{\boldsymbol{k}j}(\varepsilon_{\boldsymbol{k}j})^p\omega_j^n(\boldsymbol{k})\Big/\sum_{\boldsymbol{k}j}\omega_j^n(\boldsymbol{k}). \tag{5.34}$$

These coefficients diminish rapidly with decreasing temperature, and as long as the temperature is not too high, $\mu^{a}(n,T)$ is approximately a linear function of T. Because $\varepsilon_2(n) \approx \varepsilon_1^2(n)$, $\theta^{a}_{D}(n,T)$ also depends linearly on T. The following are two special cases.

1. When $T > \theta^{a}_{D}(-2)/(2\pi)$, the function $\coth(x)$ in (5.29) can be expanded for small x

$$\coth x = \frac{1}{x} + \frac{x}{3} + \cdots$$

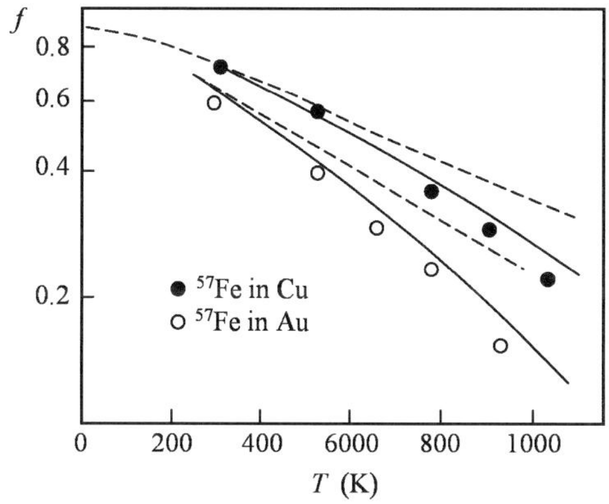

Fig. 5.3 Recoilless fraction f of ^{57}Fe in host materials Cu and Au as functions of temperature. The dashed and solid lines represent results using the harmonic and quasiharmonic approximations, respectively.

and (5.29) becomes

$$-\ln f = \frac{6E_R}{k_B}\frac{T}{[\theta_D(-2)]^2} \times \left\{1 + 2\varepsilon_1(-2)T + \left[\frac{\theta_D(-2)}{6T}\right]^2 + \cdots\right\}. \tag{5.35}$$

This is a very practical formula, because it can be used to fit the Mössbauer spectra for obtaining $\varepsilon_1(-2)$ and $\theta_D(-2)$, which characterize the size of the anharmonic effect. Figure 5.3 shows such a fitting example [12].

2. When $T \to 0$, the recoilless fraction f can be written as [13]

$$\langle(\boldsymbol{k}\cdot\boldsymbol{u})^2\rangle = \frac{3E_R}{2k_B}\frac{1}{\theta_D^a(-1)}. \tag{5.36}$$

Here the superscript a is the only difference between (5.36) and the first term in (5.15). This result shows that f is independent of T at low temperatures and the slope of the mean-square displacement versus T curve is zero near $T = 0$ [3]. Therefore, the corresponding f-value should not be exactly 1, and it is this deviation from 1 that provides a measure of the zero-point mean-square displacement. However, for intermediate temperatures, $5\text{ K} \le T \le \theta_D(-2)/(2\pi)$, the two formulas (5.35) and (5.36) are not adequate.

5.3.3
Low-Temperature Anharmonic Effect

For some solids, the potential energy curve is never parabolic. Even at $T = 0$, $\langle u^2 \rangle$ is still much larger than predicted by (5.36). This phenomenon is known as low-temperature anharmonicity, which was first observed [15] in light molecular solids such as Ne, D_2, H_2, ^{4}He, and ^{3}He.

For these light molecular solids, the atomic interactions are relatively weak. Since the cohesive energy is small, the zero-point energy becomes important, causing the interatomic distance to increase. In this case, there is a relatively large and force-free volume (a cavity), and the potential energy curve deviates significantly from a parabola (Fig. 5.4), causing low-temperature anharmonicity.

Generally speaking, whenever a crystal structure has cavities or atoms that are loosely bonded, low-temperature anharmonicity is likely to exist. Inclusion compounds, such as hydroquinone, $C_6H_4(OH)_2$, form regularly spaced cavities, capable of containing isolated foreign atoms or ions. Ionic crystals (or solid solutions) may also contain small impurity ions, such as Li^+ in $PtCl_2$ or PtB_2, resulting in several minima in the potential energy curve [15]. In these systems, the f-value is relatively small and depends only weakly on temperature.

Measuring the recoilless fraction f to investigate low-temperature anharmonicity is a straightforward method. In Fig. 5.5, curve (c) is a typical temperature dependence of f. In the high-temperature region where thermal expansion can be

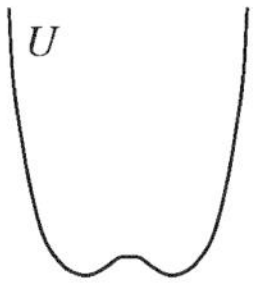

Fig. 5.4 Shape of potential energy between atoms is modified to wine-bottle-like when the interatomic distance is increased.

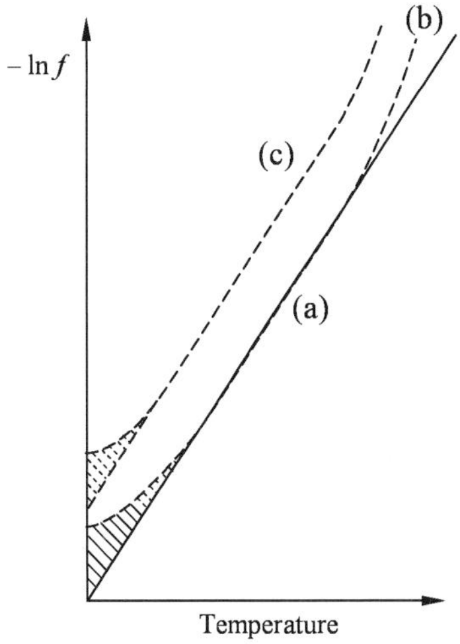

Fig. 5.5 Characteristic temperature dependences of f: (a) harmonic approximation, (b) high-temperature anharmonicity present, and (c) low-temperature anharmonicity present [17].

neglected, the curve is fitted with a straight line, whose intercept with the vertical axis is a measure of the size of the anharmonicity. The anharmonic effect causes f to decrease from the harmonic predictions. Therefore, measuring f allows us first to detect whether a solid is harmonic or anharmonic and then to study the properties of the force constants and potential energy between the atoms.

Using such a method, low-temperature anharmonicity was detected in many compounds, including $FeCl_2$ [16], superconducting Nb_3Sn [17] and $CuRh_{1.95}Sn_{0.05}Se_4$ [18], and the high T_c superconductor $EuBa_2Cu_3O_{7-\delta}$ [19, 20] discovered recently. Here we discuss some of the results from $FeCl_2$, which has a layered structure. The Cl^- ions form hexagonal layers, with Fe^{2+} hexagonal layers sandwiched between every two layers of chloride ions. The Fe^{2+} ions are located in the octohedral interstices of nearly perfect close-packed array of chloride ions. The radius of the octohedral interstices is larger than the radius of Fe^{2+} by about 0.05 Å, suggesting that Fe^{2+} may be loosely bound. Another result that supports this conclusion is that the stretching force constant in the molecule is 2.23×10^{-5} mN Å^{-1}, but the average force constant between Fe^{2+} and the six Cl^- ions measured by the Mössbauer effect is only 0.46×10^{-5} mN Å^{-1} [21]. Therefore, the size of the octahedron is determined not by the overlapping of Fe–Cl electron clouds, but by that of the Cl–Cl electron clouds (covalent bonds).

Results for $\langle u^2 \rangle$ from Mössbauer effect measurements [16] are shown in Fig. 5.6, where the high-temperature data are fitted by the harmonic approximation (solid curves). The slopes of the solid lines do not go through the origin, indicating the existence of low-temperature anharmonicity. When $T < 120$ K, the experimental data points for $FeCl_2$ are gradually higher than the solid line, demonstrating that the anharmonic effect becomes more significant at lower temperatures.

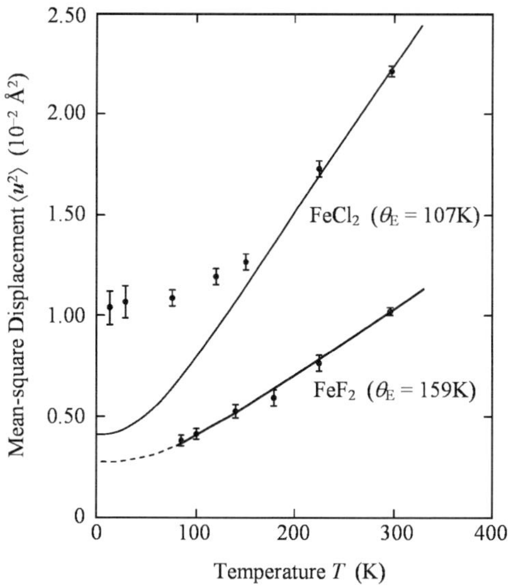

Fig. 5.6 Temperature dependences of mean-square displacement for $FeCl_2$ and FeF_2.

The radii of the Fe^{2+} and Cl^- ions are approximately 0.74 and 1.81 Å, respectively. This difference is the key factor for $FeCl_2$ to exhibit low-temperature anharmonicity. In the case of F^- instead of Cl^-, because the radius of F^- is 25% smaller than that of Cl^-, Fe^{2+} can no longer enter the octahedron space in FeF_2. This is why $FeCl_2$ and FeF_2 are not isostructural, the former having the layered $CaCl_2$-type structure and the latter having the rutile SnO_2-type structure. It is easy to see from the FeF_2 data in Fig. 5.6 that FeF_2 has very little low-temperature anharmonicity.

5.4
Pressure Dependence of the Recoilless Fraction f

Theoretical calculations [22] have predicted that the recoilless fraction f should be affected significantly by an external pressure. This effect can be studied by supposing that pressure causes a shift in each of the phonon frequencies from $\omega_j(\boldsymbol{k})$ to $\omega_j(\boldsymbol{k}) + \delta\omega_{\boldsymbol{k}j}$, with $\delta\omega_{\boldsymbol{k}j} > 0$ in most cases. However, there is a simpler way to treat the effect of pressure on lattice dynamics by a change in the volume of the solid, instead of a change in the phonon DOS. Using the Debye model, volume change will result in a change in the Debye temperature θ_D, and eventually f can be expressed as a function of pressure [23–26].

The volume of a solid V and its Debye temperature θ_D have the following simple relationship:

$$\frac{\partial \ln \theta_D}{\partial \ln V} = -\gamma \tag{5.37}$$

where γ is the Grüneisen constant, the average value of $\gamma_{\boldsymbol{k}j}$ for individual modes in Eq. (5.25). For different solids, the γ-values range from 1 to 3.

Assuming γ itself is independent of volume, then (5.37) can be written as

$$\frac{\partial \theta_D}{\partial V} = -\frac{\theta_D}{V}\gamma. \tag{5.38}$$

On the other hand, volume and pressure are related by the isothermal compressibility β:

$$\frac{1}{V}\left(\frac{\partial V}{\partial p}\right)_T = -\beta. \tag{5.39}$$

The β-values are very small for most solids, generally no larger than 10^{-11} Pa^{-1}. For example, metallic Au has $\beta \approx 5.5 \times 10^{-12}$ Pa^{-1} [27].

Now we substitute (5.38) and (5.39) into

$$\frac{\partial \theta_D}{\partial p} = \frac{\partial \theta_D}{\partial V}\frac{\partial V}{\partial p}$$

and integrate to obtain

$$\theta_D(p) = \theta_D(0)e^{\gamma\beta p} \tag{5.40}$$

where $\theta_D(p)$ and $\theta_D(0)$ are the Debye temperatures when pressure is at p and 0, respectively. Substituting this relation into Eq. (1.84), we arrive at an expression for the recoilless fraction f as a function of pressure p, $f(p)$, which can be used for analyzing high-pressure Mössbauer spectra. Since $f(p)$ is not a simple function, we will discuss the following limiting cases.

1. In the low-temperature case ($T \to 0$), we have

$$\ln f(p) = -\frac{3E_R}{2k_B}\frac{1}{\theta_D(p)} \approx -\frac{3E_R}{2k_B\theta_D(0)}(1-\gamma\beta p). \tag{5.41}$$

2. In the high-temperature case ($T \gg \theta_D(0)/2$), we have

$$\ln f(p) = -\frac{6E_R T}{k_B}\frac{1}{\theta_D^2(p)} \approx -\frac{6E_R}{k_B\theta_D^2(0)}(1-2\gamma\beta p). \tag{5.42}$$

In both of these limiting cases, $\ln f$ is approximately a linear function of pressure, except for different proportionality coefficients. However, the coefficients are positive in both cases, indicating that f increases as pressure increases, as shown by the examples in Fig. 5.7.

Recently, the phonon DOS of α-Fe and hcp-Fe have been observed by inelastic neutron scattering and inelastic nuclear resonance scattering of synchrotron radi-

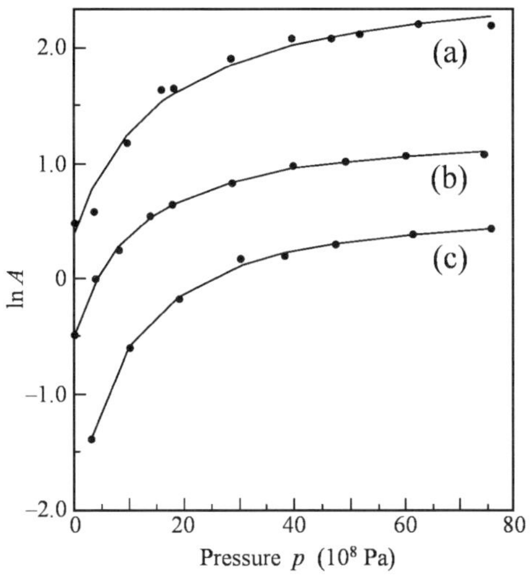

Fig. 5.7 Relations between absorption area A and pressure p for (a) $^{57}Fe[(ethyl)_2dtc]_3$, (b) $^{57}Fe[(methyl)_2dtc]_3$, and (c) $^{57}Fe[(benzyl)_2dtc]_3$ complexes. The vertical separation between data sets is arbitrary [26].

ation at high pressures up to 153 GPa [28, 29]. Both results show significant changes in the phonon DOS shifting towards the high-frequency region. The Debye temperature θ_D increases with pressure as described by Eq. (5.40), even though the DOS curves at ultrahigh pressures deviate appreciably from the Debye model. As for the recoilless fraction f, its dependence on pressure can be accurately determined using the DOS $g(\omega)$.

5.5
The Goldanskii–Karyagin Effect

When the bonding forces on a Mössbauer nucleus in a crystal do not possess cubic symmetry, the vibration amplitudes and thus $\langle u^2 \rangle$ values in different directions are not the same, resulting in an anisotropic behavior of the recoilless fraction f. This consequently leads to the relative absorption intensities in the subspectra split by hyperfine interactions having a different ratio, which was first observed by Goldanskii [30] in polycrystalline samples and was first explained theoretically by Karyagin [31]. Hence this phenomenon is known as the Goldanskii–Karyagin (G-K) effect.

Since Mössbauer spectroscopy is a unique method that can measure atomic mean-square displacements $\langle u^2 \rangle$ along different crystal axes, it is a method of choice for investigating anisotropic lattice vibrations. A good example is the measurement of the 81 keV ^{133}Cs Mössbauer recoilless fraction f as a function of θ, the angle between the γ-ray wave vector $\boldsymbol{k}$ and the c-axis of the cesium–graphite intercalation compound C_8Cs [32]. As shown in Fig. 5.8, $f(0°)$ was found to be astonishingly 20 times larger than $f(90°)$.

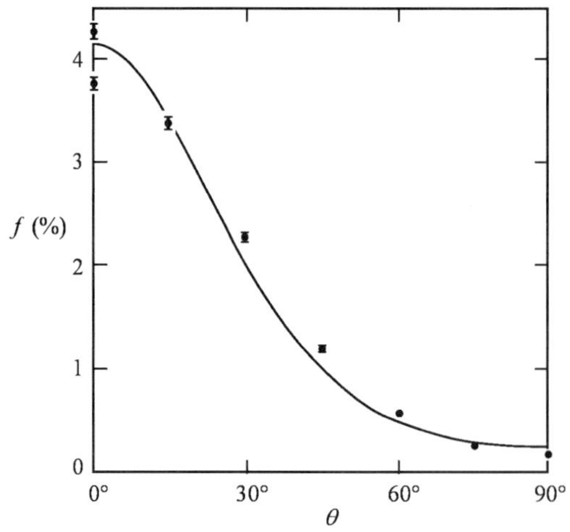

Fig. 5.8 ^{133}Cs recoilless fraction f as a function of the angle between γ-ray direction $\boldsymbol{k}$ and the c-axis of the intercalation compound C_8Cs.

Let us look at the influence of the G-K effect on the Mössbauer spectral intensities from single-crystal and polycrystalline samples.

5.5.1 Single Crystals

Anisotropic recoilless fraction f has been studied, for example, using the 103.18 keV Mössbauer transition ($3/2^+ \rightarrow 5/2^+$, $M1$ type) of ^{153}Eu in $Eu_2Ti_2O_7$ [33]. This crystal has a cubic structure of the pyrochlore type with a space group symmetry Fd3m. The Eu^{3+} ions occupy positions with a three-fold symmetry ($\bar{3}m$ point symmetry), where the electric field gradient is axially symmetric ($\eta = 0$) with its principal axis in the [111] direction. However, there are four equivalent [111] directions, and the cosine of the angle between them is 1/3. In the Mössbauer experiment, the single-crystal sample is oriented such that the incident γ-ray is parallel to a particular [111] direction ($\theta = 0$) and consequently there are two inequivalent Eu sites: one site with 1/4 of the population having $\theta = 0$ and another site with 3/4 of the population having $\cos\theta' = 1/3$.

According to Eq. (2.47), the angular distribution functions for a dipole radiation are

$$F_1^0(\theta) = |\chi_1^0|^2 = \sin^2\theta \quad \text{for } \Delta m = 0,$$

and

$$F_1^1(\theta) = |\chi_1^1|^2 = \frac{1}{2}(1 + \cos^2\theta) \quad \text{for } \Delta m = \pm 1.$$

For the first Eu site, $\theta = 0$, thus $F_1^0(0) = 0$ and $F_1^1(0) = 1$. For the other three sites with $\cos\theta' = 1/3$, $F_1^0(\theta') = 8/9$ and $F_1^1(\theta') = 5/9$. The intensities of $\Delta m = 0$ transitions, as determined by their respective Clebsch–Gordan (C-G) coefficients, would be multiplied by $a = 0f(0) + 3(8/9)f(\theta')$, and those of $\Delta m = \pm 1$ transitions by $b = 1f(0) + 3(5/9)f(\theta')$. Therefore, the relative intensities of the $\Delta m = 0$ transitions, as determined by the appropriate C-G coefficients, would be multiplied by

$$B = \frac{a}{b} = \frac{8}{5 + 3f(0)/f(\theta')}. \tag{5.43}$$

Suppose that the recoilless fraction f is axially symmetric and we use $f_\parallel$ and $f_\perp$ to represent $f(0)$ and $f(90°)$, respectively, then [34]

$$f(\theta) = \exp[-k^2\langle x^2\rangle - \varepsilon\cos^2\theta] = f_\perp \exp(-\varepsilon\cos^2\theta), \tag{5.44}$$

with

$$\varepsilon = k^2[\langle z^2\rangle - \langle x^2\rangle] = -\ln\left(\frac{f_\parallel}{f_\perp}\right) \tag{5.45}$$

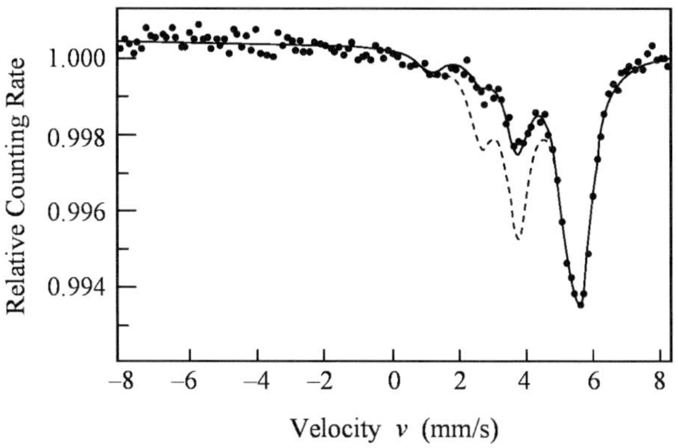

Fig. 5.9 ^{153}Eu Mössbauer spectrum of single-crystal $Eu_2Ti_2O_7$ at $T = 36$ K.

where $\langle z^2 \rangle$ and $\langle x^2 \rangle$ denote the mean-square displacements along directions parallel and perpendicular to the $\boldsymbol{k}$ direction, respectively. Applying this to the $Eu_2Ti_2O_7$ single crystal:

$$\frac{f(0)}{f(\theta')} = \frac{f_{\parallel}}{f_{\perp}\exp(-\varepsilon\cos^2\theta')} = \exp(-\varepsilon)\exp(\varepsilon\cos^2\theta') = \exp(-8\varepsilon/9), \qquad (5.46)$$

and (5.43) becomes

$$B = \frac{8}{5 + 3\exp(-8\varepsilon/9)}. \qquad (5.47)$$

The ^{153}Eu spectrum at $T = 36$ K is shown in Fig. 5.9. The dashed line is the calculated curve assuming an isotropic recoilless fraction f, and it obviously does not fit the experimental data. Now the intensities of all $\Delta m = 0$ transitions have been multiplied by an attenuating factor to obtain the best fit (the solid line). Because this factor should be B in Eq. (5.47), the parameter ε can be easily calculated. A nonzero ε-value indicates anisotropic lattice vibrations and the existence of the G-K effect. Table 5.2 lists values of B for the single-crystal $Eu_2Ti_2O_7$ and

Table 5.2 Parameters B and ε derived from the analysis of ^{153}Eu Mössbauer spectra of single-crystal and polycrystalline samples of $Eu_2Ti_2O_7$.

T (K)	B (single crystal)	ε (single crystal)	ε (polycrystal)
4.1	0.40(2)	−1.81(5)	−1.95(15)
36	0.27(5)	−2.4(1)	−2.6(2)

values of ε for the single-crystal and polycrystalline samples [35]. From the data at 4.2 K, it has been derived that $\langle z^2 \rangle - \langle x^2 \rangle = -0.00226\,\text{Å}^2$. From both the single-crystal and polycrystalline samples of $Eu_2Ti_2O_7$, the measured G-K effect amounts (ε-values) are consistent with each other.

5.5.2
Polycrystals

At a first glance, the G-K effect might not be apparent in polycrystalline samples. But in fact, measuring the intensities of the quadrupole splitting lines is quite straightforward to detect this effect. This is because when the recoilless fraction f is isotropic, the two spectral lines (e.g., $3/2 \to 1/2$ transition in ^{57}Fe) should have equal intensities. But if f is anisotropic, the two lines would have different intensities:

$$\frac{I_{3/2}}{I_{1/2}} = \frac{\displaystyle\int_0^{\pi} (1 + \cos^2\theta) f(\theta) \sin\theta \, d\theta}{\displaystyle\int_0^{\pi} \left(\frac{5}{3} - \cos^2\theta\right) f(\theta) \sin\theta \, d\theta} \neq 1 \qquad (5.48)$$

which are independent of the particular orientation of the sample. If the probability of the γ-transition was θ-independent, the G-K effect would not be observed in polycrystalline samples.

From Eq. (5.45), we know that ε is proportional to k^2, which is equal to $E_\gamma^2/(\hbar^2 c^2)$. Because E_γ of ^{133}Cs is about 6 times that of ^{57}Fe, the cesium recoilless fraction f is much more sensitive to the changes in its mean-square displacement than the iron recoilless fraction f. Therefore, high-energy Mössbauer transitions such as $2^+ \to 0$ (E2 transition) would be more advantageous for studying the anisotropy in lattice vibrations [36]. In this case, the quadrupole split spectrum has simply three absorption lines. Such Mössbauer isotopes include ^{152}Sm, ^{156}Gd, ^{160}Dy, ^{166}Er, ^{170}Yb, and ^{174}Yb, all of which have relatively large E_γ values and therefore any G-K effects can be sensitively detected. For example, the lattice vibrational anisotropy detected by the ^{170}Yb Mössbauer effect is as high as 30 times that by ^{57}Fe. In addition, E2-type radiation contains high-order harmonics, which are more sensitive to the vibrational anisotropy [37].

Figure 5.10 shows ^{156}Gd Mössbauer spectra from a polycrystalline $Gd_2Ti_2O_7$ sample, which exhibits a relatively large G-K effect [38]. Analogous to the previous example, for $\Delta m = 0$ and $\Delta m = \pm 2$ transitions, two attenuation factors B_0 and B_2 are used to fit the data. For each temperature, B_0 and B_2 lead to ε_0 and ε_2 values that are equal to each other, strongly supporting the G-K theory. The experimental results show that at $T = 4.2$ K, $\langle z^2 \rangle - \langle x^2 \rangle = -0.00076\,\text{Å}^2$.

There are two more points worth mentioning.

1. The G-K effect is sometimes very small and requires careful experimental planning for its observation. An important consideration is the saturation effect of the absorber [39]. To

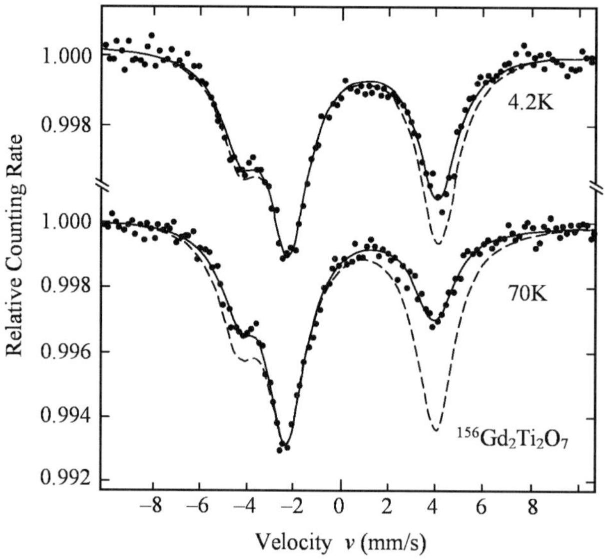

Fig. 5.10 ^{156}Gd Mössbauer spectra of polycrystalline $Gd_2Ti_2O_7$. The solid lines are fits to the experimental spectra, taking into account the anisotropic f. The dashed lines are theoretical spectra with an isotropic f.

reduce this effect, thin samples are preferred, but this results in smaller Mössbauer effect absorption and the accompanied large statistical errors. A better alternative is to use emission Mössbauer spectroscopy for studying the G-K effect [40–42].

2. Some polycrystalline samples may contain texture due to preferential orientation, which could affect the Mössbauer spectrum in a manner similar to the G-K effect. But the texture effect should be basically independent of temperature, while lattice vibration is strongly dependent on temperature and the G-K effect should be larger when the temperature is higher (see Fig. 5.10). Analysis of Mössbauer spectra from a sample at different temperatures would allow us to distinguish between these two effects.

5.6 Second-Order Doppler Shift

5.6.1 Transverse Doppler Effect

The second main methodology for obtaining information on lattice dynamics through Mössbauer spectroscopy is analysis of the shift of the entire spectrum due to the second-order Doppler effect.

Suppose we have two reference frames, one attached to the laboratory and the other to the vibrating Mössbauer nucleus. When this nucleus emits or absorbs a γ-photon ($E_\gamma = \hbar\omega_0$), according to the special theory of relativity, the photon's angular frequency ω as observed in the laboratory reference frame is

$$\omega = \omega_0 \frac{\sqrt{1 - v^2/c^2}}{1 - v\cos\theta/c} \tag{5.49}$$

where v is the speed of the nucleus and θ is the angle between the photon direction and the velocity of the nucleus. When $v \ll c$, the above equation can be expanded as

$$\omega \approx \omega_0 \left(1 + \frac{v}{c}\cos\theta - \frac{v^2}{2c^2} + \cdots\right)$$

or

$$\frac{\Delta E}{E_0} \approx \frac{v}{c}\cos\theta - \frac{v^2}{2c^2}. \tag{5.50}$$

The first term is the usual first-order Doppler effect, used for modulating the photon energy in Mössbauer experiments. The next term, which does not exist in the classical theory, is the second-order Doppler effect, as a consequence of the time dilation phenomenon in relativity theory [13, 43]. When the nucleus moves in a direction perpendicular to the photon direction, $\cos\theta = 0$, the first-order term vanishes and $\Delta E/E_0 = -v^2/2c^2$. Therefore, the second-order Doppler effect is also known as the transverse Doppler effect [44].

In a solid, the average atomic velocity is zero, $\langle v\rangle = 0$, so Eq. (5.50) becomes

$$\langle \Delta E\rangle = -E_0 \frac{\langle v^2\rangle}{2c^2}.$$

Since the typical value of atomic mean-square velocity for metallic iron at room temperature is $\langle v^2\rangle \approx 6 \times 10^{10}$ mm^2 s^{-2}, the energy shift due to the second-order Doppler effect is therefore $\langle\Delta E\rangle = -E_0\langle v^2\rangle/(2c^2) \approx -4.8 \times 10^{-9}$ eV. Before the discovery of the Mössbauer effect, this small change in energy could not be resolved by any other method.

Suppose that the source and the absorber are at different temperatures T_s and T. The second-order Doppler shift as observed in the Mössbauer spectrum (in units of mm s^{-1}) can be expressed as

$$\delta_{\mathrm{SOD}} = \frac{\langle v^2\rangle_{T_s} - \langle v^2\rangle_T}{2c}. \tag{5.51}$$

If T_s is fixed, the first term is a constant (c_1), and when the high- and low-temperature expressions (Eqs. (5.6) and (5.8)) for $\langle v^2\rangle$ are used in the second term, we have

$$\delta_{\text{SOD}} = c_1 - \frac{3k_B T}{2Mc}\left[1 + \frac{1}{20}\left(\frac{\theta_D}{T}\right)^2\right] \quad \text{for } T > \theta_D, \tag{5.52}$$

$$\delta_{\text{SOD}} = c_1 - \frac{9k_B \theta_D}{2Mc}\left[\frac{1}{8} + \frac{\pi^4}{15}\left(\frac{T}{\theta_D}\right)^4\right] \quad \text{for } T \ll \theta_D. \tag{5.53}$$

The observed shift δ of the entire Mössbauer spectrum is called the center shift, which is the sum of isomer shift δ_{IS} and the second-order Doppler shift δ_{SOD}:

$$\delta = \delta_{\text{IS}} + \delta_{\text{SOD}} = \delta_{\text{IS}} + \frac{\langle v^2 \rangle_{T_s}}{2c} - \frac{\langle v^2 \rangle_T}{2c}. \tag{5.54}$$

The second-order Doppler shift δ_{SOD} strongly depends on temperature while the isomer shift δ_{IS} is a measure of the s-electron density at the nucleus and thus is approximately independent of temperature. The specific details of $\langle v^2 \rangle$ are determined by the model chosen for the lattice vibration.

In 1960, Pound and Rebka [45] first proved the existence of the second-order Doppler effect using the Mössbauer effect. They measured the relation between second-order Doppler shift and temperature in ^{57}Fe γ-ray resonance absorption, and used the Debye model for the distribution function $g(\omega)$ in Eq. (5.2) with $\theta_D = 420$ K. Their theoretical curve has an excellent agreement with the experimental data (see Fig. 5.11). Since then, many studies on the temperature dependence of δ_{SOD} have been carried out to give lattice dynamics parameters such as θ_D or $\theta_D(n)$, $\langle v^2 \rangle$, and the effective vibrating mass M_{eff} [46–49].

In the meantime, Josephson [1] derived the second-order Doppler shift from the mass–energy relation, a different aspect of the special theory of relativity. This was based on the notion that the mass of the Mössbauer nucleus in the excited state is larger than that of the same nucleus in the ground state, and the energy of the emitted γ-photon corresponds to the difference in the mass values,

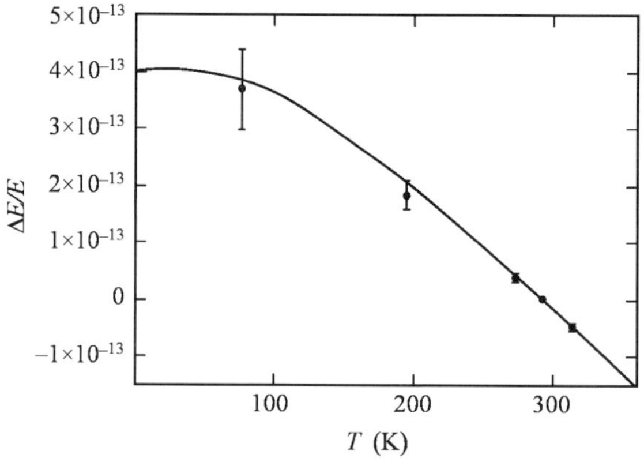

Fig. 5.11 Temperature dependence of second-order Doppler shift in the 14.4 keV γ-ray resonance absorption.

$E_0 = c^2\delta M$. Now if the nucleus has kinetic energy, its expectation value would be altered by the following amount:

$$\delta E = \frac{\langle p^2 \rangle}{2(M + \delta M)} - \frac{\langle p^2 \rangle}{2M} \approx -\frac{\langle p^2 \rangle}{2M^2}\delta M = -\frac{\langle v^2 \rangle}{2c^2}E_0 \tag{5.55}$$

causing the photon energy reduction, which is identical to the second-order Doppler shift. It should be noted that (1) these two apparently different origins of the second-order Doppler shift can be shown to be equivalent, and (2) this kinetic energy difference before and after the γ-emission is not to be confused with the recoil energy, which is an entirely separate quantity.

5.6.2 The Relation between f and δ_{SOD}

The recoilless fraction f and the second-order Doppler shift δ_{SOD} are related to $\langle u^2 \rangle$ and $\langle v^2 \rangle$, respectively. Once the phonon frequency distribution function $g(\omega)$ has been determined, both $\langle u^2 \rangle$ and $\langle v^2 \rangle$ (thus f and δ_{SOD}) can be accurately calculated. Experimentally, however, f and δ_{SOD} are measured differently, because f is related to the relative areas of the spectral lines and δ_{SOD} is related to the positions of the lines. For a spectrum without overlapping lines, the precision in measuring line positions is much higher than that in determining the areas. For example, the precision in line positions in a room temperature sodium nitroprusside spectrum is 0.2%, while that in the spectral areas is only 0.7%. This does not mean that the lattice dynamics parameters based on δ_{SOD} measurements are more reliable. In fact, it is difficult to separate δ_{SOD} from δ_{IS}. In most cases of ^{57}Fe work, the temperature variation of δ_{IS} is neglected, which brings certain amount of error to δ_{SOD}. Also, it is not uncommon to find discrepancies between the two θ_D-values from f and δ_{SOD} for the same solid, the reason being that the actual phonon distribution of most solids deviates significantly from the Debye distribution.

As we know from Eq. (1.84), the f-value can be determined from the Debye temperature and it is sensitive to any changes in this temperature [50]. Consequently, extraction of the f-value from δ_{SOD} data must be done with special care [51]. In general, analysis of f and δ_{SOD} data is not carried out by the Debye model only, even in the high-temperature range.

If we use a shorthand notation

$$\langle \omega^l \rangle = \int \coth\left(\frac{1}{2}\hbar\omega\beta\right)\omega^l g(\omega)\,d\omega,$$

we can rewrite the definitions in Eqs. (5.1) and (5.2) as follows:

$$\langle u^2 \rangle = \frac{\hbar}{2M}\langle \omega^{-1} \rangle,$$

$$\langle v^2 \rangle = \frac{3\hbar}{M}\langle \omega \rangle.$$

Also, we will define a new quantity S_T in order to facilitate the discussion of the relation between f and δ_{SOD}:

$$S_T = \frac{\ln f}{\delta_{SOD}} = \frac{-k^2\langle u^2\rangle}{-\langle v^2\rangle/2c} = \frac{ck^2}{3}\frac{1}{\langle\omega\rangle/\langle\omega^{-1}\rangle}, \tag{5.56}$$

where $\langle\omega\rangle/\langle\omega^{-1}\rangle$ is known as the McMillan ratio. In the high-temperature limit, experiments have shown that both $\langle u^2\rangle$ and $\langle v^2\rangle$ are nearly linear in temperature. Therefore, S_T or the McMillan ratio is approximately a constant [52]. However, S_T is not the same for different solids. Figure 5.12 shows the experimental

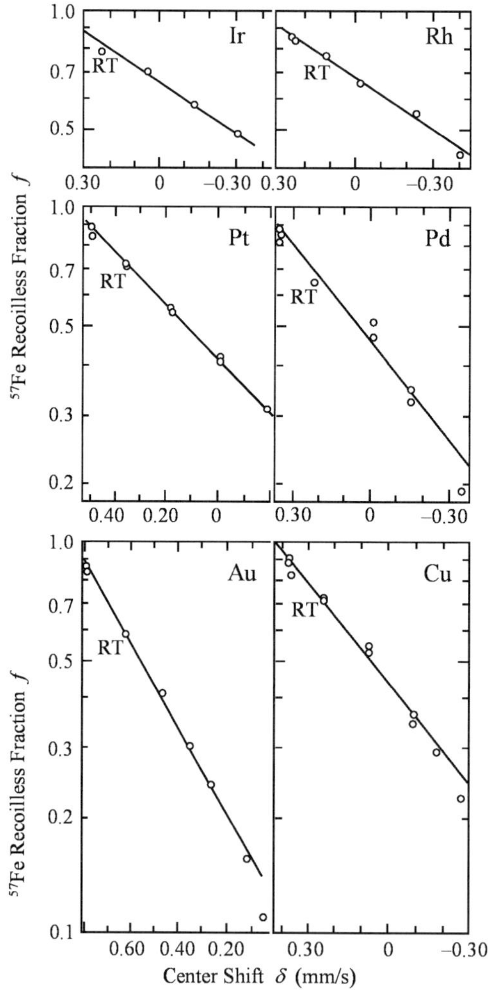

Fig. 5.12 Plots of ln f versus center shift δ for dilute ^{57}Fe in six different hosts. "RT" indicates the room temperature data point(s). δ is measured with respect to the center shift of room temperature α-Fe [52].

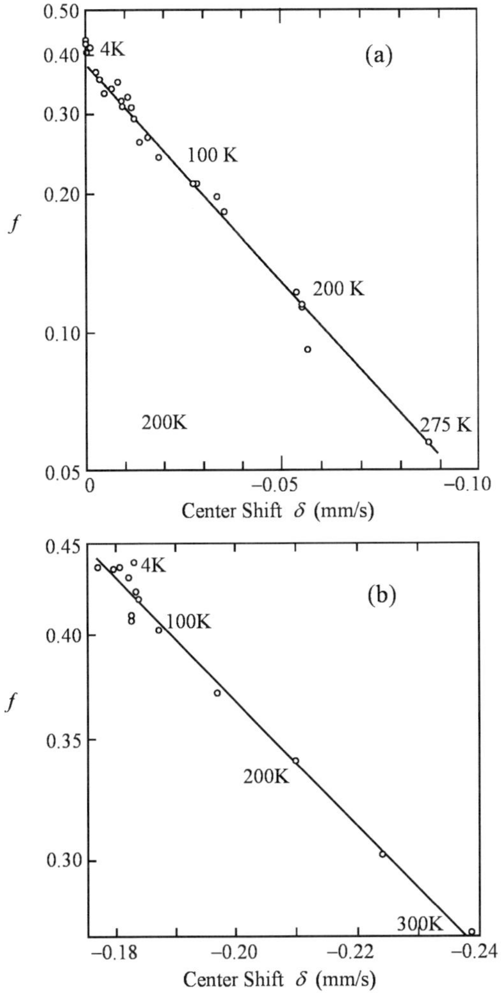

Fig. 5.13 Relations between the ^{119}Sn recoilless fraction f and center shift δ for (a) SnTe with a SnTe source at 19.4 K and (b) Nb_3Sn with a Pd_3Sn source at 19.4 K [52]. The temperature values next to the data points are absorber temperatures.

results of ^{57}Fe impurities in six different fcc crystals, and it is convincing that $\ln f$ and δ_{SOD} are linearly related in an extremely wide temperature range from 100 to 1020 K. Figure 5.13 shows the linear relations between $\ln f$ and δ_{SOD} for two ^{119}Sn compounds.

The above mentioned linearity can be predicted by the Debye model. When the absorber is at a high temperature, we can regard the source temperature $T_s \to 0$ and obtain from Eq. (5.52)

$$\delta_{SOD} = -\frac{\langle v^2 \rangle}{2c} = -\frac{3}{2}\frac{k_B}{Mc}T. \tag{5.57}$$

On the other hand, Eq. (5.14) gives

$$\ln f = -\frac{6E_R T}{k_B \theta_D^2}.$$

Therefore

$$S_T = \frac{2E_\gamma^2}{ck_B^2\theta_D^2}, \tag{5.58}$$

which indicates that S_T or the McMillan ratio depends on two constants E_γ and θ_D but not on temperature T. Because of this reason, studies of the second-order Doppler effect are usually carried out in a relatively high-temperature region.

If the $g(\omega)$ of a solid deviates substantially from the Debye model, the relation between them could become very complicated [51].

5.7
Methods for Measuring the Recoilless Fraction f

Special attention has always been paid to the precise measurement of the recoilless fraction f. The precision has reached 1% when radioactive sources are used and it can be better than 0.4% when synchrotron Mössbauer radiation is employed.

There are two main difficulties in the transmission method. One is the complicated background, which cannot be accurately calibrated as done in other radioactivity experiments. This severely limits the precision in determining the baseline counts $I(\infty)$, and has been regarded as the main source of error in measuring the recoilless fraction f [53, 54]. The second difficulty is that the sample always has a finite thickness, causing some amount of distortion in the spectral shape. Completely correcting the thickness effect is also very difficult. Fortunately, these limitations can all be overcome by a radically different experimental method in synchrotron Mössbauer spectroscopy (see Chapter 7).

There are many methods for measuring the recoilless fraction f, categorized mainly into to two groups: absolute methods and relative methods. All of them are based on the measurements of spectral intensities (areas or heights), shapes, and widths. Detailed descriptions of these methods can be found in Ref. [13].

5.7.1
Absolute Methods

In an absolute method, we obtain the recoilless fraction f by measuring $A(t_a)$, $\varepsilon(v_r)$, and $\Gamma_a{}^{exp}$ and utilizing their relations with t_a (note that t_a is proportional to f because $t_a = n_a f \sigma_0 d$).

The absorption area method is the most popular one, because the spectral area $A(t_a)$ is approximately independent of the line shape of source emission and the instrumental broadening [54]. When the sample thickness increases, $A(t_a)$ saturates more slowly than the spectral height. Therefore, $A(t_a)$ is more sensitive to the change in t_a. For a single line absorption, the normalized area $A(t_a)$ can be expressed as (see Appendix A):

$$A(t_a) = \int_{-\infty}^{\infty} \varepsilon(v)\,dv = \int_{-\infty}^{\infty} \frac{I(\infty) - I(v)}{I(\infty) - I_b}\,dv = \int_{-\infty}^{\infty} f_s[1 - T(v)]\,dv$$
$$= f_s\Gamma_a\pi\frac{t_a}{2}\exp\left(-\frac{t_a}{2}\right)\left[I_0\left(\frac{t_a}{2}\right) + I_1\left(\frac{t_a}{2}\right)\right] \quad (5.59)$$

where I_b represents the background counts. If f_s is known, we can calculate t_a from the measured absorption areas, thus obtaining the recoilless fraction f. When $t_a < 1$, Eq. (5.59) may be expanded into a polynomial series:

$$A(t_a) = \frac{\pi}{2} f_s\Gamma_a t_a(1 - 0.25t_a + 0.0625t_a^2 + \cdots). \quad (5.60)$$

In the first-order approximation, the spectral area $A(t_a)$ is directly proportional to t_a or f:

$$A(t_a) \approx \frac{\pi}{2} f_s\Gamma_a t_a = \left(\frac{\pi}{2} f_s\Gamma_a n_a\sigma_0 d\right) f. \quad (5.61)$$

The accuracy of the absorption area method is limited by statistical errors in counts $I(\infty)$ and I_b. A small error in the baseline counts $I(\infty)$ would cause a relative large uncertainty in the spectral area measurement. In addition, during the measurement or fitting of the spectrum, the chosen velocity range $\pm v_1$ not being large enough will also add more uncertainties in $A(t_a)$. As shown in Fig. 5.14, the shaded area is equal to $2\Gamma_a/(\pi v_1)$, and an $A(t_a)$ accuracy better than 1% would require $v_1 \approx 64\Gamma_a$ [13].

The "white source" method can be used for accurate measurements of $I(\infty)$ and $A(t_a)$ [55]. This method uses a separate counter recording the total transmitted γ-rays when the source executes a constant acceleration motion between $-v_1$

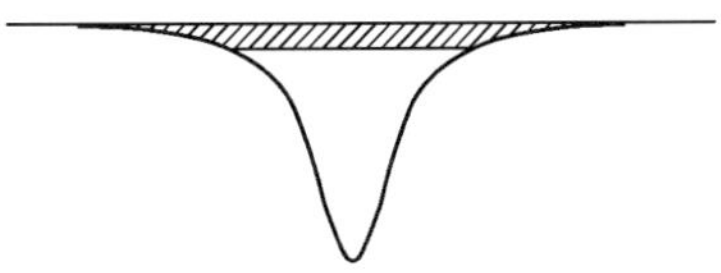

Fig. 5.14 Shaded area indicates the error introduced in the area of an absorption peak if the velocity range is not wide enough.

and $+v_1$. This is equivalent to the average of the counts for all velocities. According to Eq. (1.23),

$$\overline{I(v_1)} = \frac{1}{2v_1}\int_{-v_1}^{v_1} I(v)\,\mathrm{d}v = I(\infty) - I(\infty)\frac{1}{2v_1}\int_{-v_1}^{v_1} f_s[1 - T(v)]\,\mathrm{d}v. \qquad (5.62)$$

Obviously, when $v_1 \to \infty$, the last integral in the above equation is the area $A(t_a)$, and therefore

$$\overline{I(v_1)} = I(\infty) - I(\infty)\frac{A(t_a)}{2v_1}, \qquad (5.63)$$

which shows a linear relation between $\overline{I(v_1)}$ and $1/v_1$. Performing a linear regression on the experimental $\overline{I(v_1)}$ versus $1/v_1$ curve would then give $I(\infty)$ and $A(t_a)$, whose uncertainties can be as good as 1 and 0.7%, respectively, based on experiments using a $^{57}Co/Pd$ source and an Fe/Rh absorber.

Another technique for eliminating the influence of background is the selective modulation method [56]. Between the specimen absorber A and the source is inserted a so-called control absorber C (Fig. 5.15). For the sake of simplicity, suppose that the control absorber's isomer shift is the same as that of the source. The control absorber is driven to move along the γ-ray direction. When the control absorber is moving with a high speed, it will not resonantly absorb the incident γ-rays. Under such a condition, what registers in the detector is $I(\infty)$, which includes both recoilless and recoiled γ-rays as well as background. Next, when the control absorber is at rest, the recoiled γ-rays and the background in the detected intensity $I(0)$ should be identical to the previous case, but the recoilless part will be reduced due to resonance absorption. The difference $\Delta I = I(\infty) - I(0)$ represents the "pure" recoilless radiation, equivalent to a source that emits only Mössbauer radiation. Now the specimen absorber is driven with a constant acceleration mode, synchronous with the control absorber motion. During the increasing half of the triangular wave, the control absorber is moving with a high speed, while during the decreasing half of the triangular wave, the control absorber is at rest.

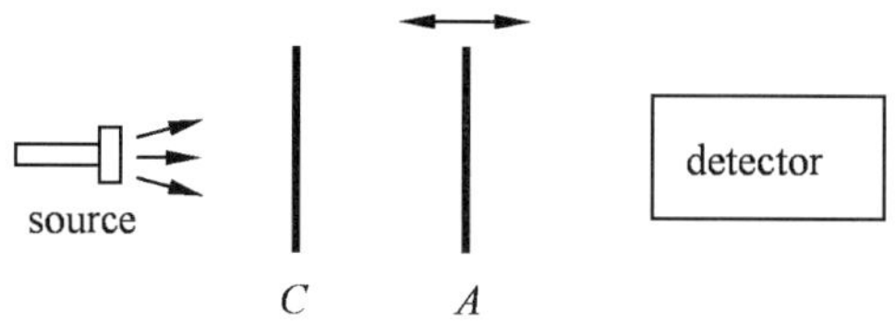

Fig. 5.15 Positions of the two absorbers in the selective modulation method.

Two different spectra are obtained in a single experiment, and the difference between them is a "pure" Mössbauer spectrum (with $f_s = 1$). The spectral area is

$$A(t_a) = \int_{-\infty}^{\infty} \frac{\Delta I(\infty) - \Delta I(v)}{\Delta I(\infty)} dv$$
$$= \Gamma_a \pi \frac{t_a}{2} \exp\left(-\frac{t_a}{2}\right)\left[I_0\left(\frac{t_a}{2}\right) + I_1\left(\frac{t_a}{2}\right)\right] \tag{5.64}$$

where ∞ and v represent the "infinite" and "finite" velocity values of specimen absorber. Drawbacks of this method include the requirement that the control absorber have the same isomer shift as the source and the inconvenience in high- or low-temperature experiments. Of course, a serious drawback is the low activity of such a "pure" recoilless γ-source. A couple of improved setups have also been developed [57, 58], and applied to the investigations of $BaSnO_3$ and $K_4Fe(CN)_6{\cdot}3H_2O$, with room temperature results of $f = 0.57 \pm 0.02$ and $f = 0.281 \pm 0.004$, respectively. Figure 5.16 shows a comparison between Mössbauer spectra obtained using different methods, and the results are also similar to those obtained by means of a "resonance" detector [59].

If f_s cannot be accurately known, a usual method to circumvent this difficulty is to use a series of specimens of the same material but with different thicknesses d. The spectra are fitted using f_s as one of the parameters. Reference [60] presents one such example, where five $EuBa_2Cu_3O_7$ samples were prepared with different thicknesses d and after fitting all the spectra, $f = 0.26$ was obtained. A compilation of f-values of various materials is given in Table 5.3.

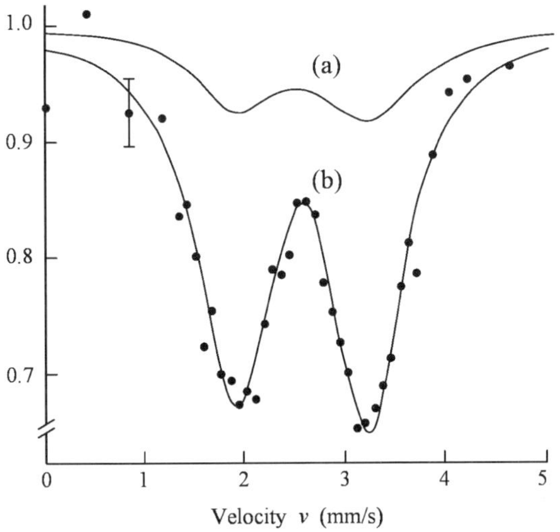

Fig. 5.16 ^{119}Sn Mössbauer spectra of SnO obtained by (a) using no control absorber and (b) using a control absorber of $BaSnO_3$ [56].

Table 5.3 List of recoilless fraction f values of various materials.

Mössbauer nucleus	Solid material	f	T (K)	Ref.
^{57}Fe	α-Fe	0.93(3)	4.2	61
		0.78	293	62
		0.67	300	63
		0.771(17)	298	64
		0.688	293	65
	α-Fe_2O_3	0.66	293	66
	$Na_2[Fe(CN)_5NO]\cdot 2H_2O$	0.468(7)		67
		0.359		68
		0.43(3)		65
		0.37		69
	$Fe(C_5H_5)_2$	0.169		70
		0.08	295	69
	$K_4Fe(CN)_6\cdot 3H_2O$	0.281(4)		58
	FeS_2	0.20(2)		71
^{119}Sn	SnO_2	0.28(3)		72
	$BaSnO_3$	0.57(2)		56
		0.52(2)		73
		0.65(1)		74
^{151}Eu	$EuBa_2Cu_3O_7$	0.26		60
^{159}Tb	$TbAl_2$	0.108(3)	115	75
	Tb_4O_7	0.237(15)	81	75
^{183}W	Metallic tungsten	0.299(1)	297	76
^{191}Ir	Metallic iridium	0.036(5)	80	76

5.7.2
Relative Methods

The absolute methods are suitable only under various restrictions on line shapes, widths, and experimental arrangements [54, 55]. When a relative method is employed, most restrictions can be removed and, especially, several sources of errors that occur in an absolute method may be avoided. Therefore, the accuracy in a relative method is usually higher than that in an absolute one. In fact, we are more interested in how f changes with temperature or pressure in lattice dynamics than its absolute value.

If the sample is very thin, Eq. (5.61) is valid, and the relation between spectral area and temperature T is essentially the same as the relation between f and T:

$$A(T) = \left[\frac{\pi}{2} f_s(T_s)\Gamma_a n_a \sigma_0 d\right] f(T) = cf(T) \tag{5.65}$$

where T_s is the temperature of the source, which is usually kept constant during experiments, and hence c is a constant. If the $A(T)$ values are divided by the area $A(T_0)$ deduced from a spectrum at a particular temperature T_0,

$$\frac{A(T)}{A(T_0)} = \frac{f(T)}{f(T_0)}, \tag{5.66}$$

the other factors cancel out and we obtain the relative change in the recoilless fraction f. Using this for fitting the spectral areas as a function of temperature will allow us to extract lattice dynamics parameters such as $\theta_D(-2)$, θ_D, and $\varepsilon(-2)$.

References

1 B.D. Josephson. Temperature-dependent shift of γ rays emitted by a solid. *Phys. Rev. Lett.* 4, 341–342 (1960).

2 H. Böttger. *Principles of the Theory of Lattice Dynamics* (Physik Verlag, Weinheim, 1983). B.T.M. Willis and A.W. Pryor. *Thermal Vibrations in Crystallography* (Cambridge University Press, London, 1975).

3 R.M. Housley and F. Hess. Analysis of Debye–Waller-factor and Mössbauer-thermal-shift measurements. *Phys. Rev.* 146, 517–526 (1966).

4 L. Cianchi, F. Del Giallo, F. Pieralli, M. Mancini, S. Sciortino, G. Spina, N. Ammannati, and R. Garré. Low temperature anharmonicity and phonon anomalies at ^{151}Eu sites in $Eu_{1+x}Ba_{2-x}Cu_3O_{7-\delta}$. *Solid State Commun.* 80, 705–708 (1991).

5 G.H.M. Calis and R.J. Baker. Debye model Mössbauer recoil-free fractions. In *Handbook of Spectroscopy*, vol. III, J.W. Robinson (Ed.), pp. 424–432 (CRC Press, Boca Raton, FL, 1981).

6 S.K. Roy and N. Kundu. Dynamical properties of ^{57}Fe impurities in different metallic solids from anharmonic recoilless fractions. *J. Phys. F: Met. Phys.* 17, 1051–1064 (1987).

7 R.C. Shukla and H. Hübschle. Anharmonic contributions to the Debye–Waller factor of aluminium. *Solid State Commun.* 72, 1135–1140 (1989).

8 R.C. Shukla and H. Hübschle. Atomic mean-square displacement of a solid: a Green's-function approach. *Phys. Rev. B* 40, 1555–1559 (1989).

9 R.C. Shukla and D.W. Taylor. Debye–Waller factor of sodium: a comparison of theory and experiment. *Phys. Rev. B* 45, 10765–10768 (1992).

10 R.C. Shukla and D.W. Taylor. Mean-square displacement from Mössbauer and x-ray data for solid krypton: a comparison of theory and experiment. *Phys. Rev. B* 49, 9966–9968 (1994).

11 A.A. Maradudin and P.A. Flinn. Anharmonic contributions to the Debye–Waller factor. *Phys. Rev.* 129, 2529–2547 (1963).

12 K.N. Pathak and B. Deo. Effect of lattice anharmonicity on the Debye–Waller factor. *Physica* 35, 167–176 (1967).

13 B.V. Thompson. Neutron scattering by an anharmonic crystal. *Phys. Rev.* 131, 1420–1427 (1963).

14 K.N. Pathak. Theory of anharmonic crystals. *Phys. Rev.* 139, A1569–A1580 (1965).

15 J.G. Dash, D.P. Johnson, and W.M. Visscher. Low-temperature anharmonicity and the Debye–Waller factor. *Phys. Rev.* 168, 1087–1094 (1968).

16 D.P. Johnson and J.G. Dash. Low-temperature anharmonicity in $FeCl_2$. *Phys. Rev.* 172, 983–990 (1968).

17 J.S. Shier and R.D. Taylor. Temperature-dependent isomer shift and anharmonic binding of Sn^{119} in Nb_3Sn. *Solid State Commun.* 5, 147–149 (1967).

18 P.P. Dawes, N.W. Grimes, and D.A. O'Connor. Direct experimental evidence for low temperature anharmonicity in superconducting spinels. *J. Phys. C* 7, L387–L389 (1974).

19 L. Cianchi, F. Del Giallo, F. Pieralli, M. Mancini, S. Sciortino, G. Spina, N. Ammannati, and R. Garré. Low temperature anharmonicity and phonon anomalies at ^{151}Eu sites in $Eu_{1+x}Ba_{2-x}Cu_3O_{7-\delta}$. *Solid State Commun.* 80, 705–708 (1991).

20 M. Capaccioli, L. Cianchi, F. Del. Giallo, F. Pieralli, and G. Spina. Low-temperature vibrational anharmonicity of ^{151}Eu in $EuBa_2Cu_3O_{7-\delta}$. *J. Phys.: Condens. Matter* 7, 2429–2438 (1995).

21 D.P. Johnson and J.G. Dash. Mössbauer effect of Fe^{57} in $FeCl_2$. *Bull. Am. Phys. Soc.* 12, 378 (1967).

22 D.N. Talwar and M. Vandevyver. Pressure-dependent phonon properties of III–V compound semiconductors. *Phys. Rev. B* 41, 12129–12139 (1990).

23 H. Frauenfelder. *The Mössbauer Effect, A Review – with a Collection of Reprints* (W.A. Benjamin, New York, 1963).

24 R.V. Hanks. Pressure dependence of the Mössbauer effect. *Phys. Rev.* 124, 1319–1320 (1961).

25 J.A. Moyzis Jr., G. DePasquali, and H.G. Drickamer. Effect of pressure on f number and isomer shift for Fe^{57} in Cu, V, and Ti. *Phys. Rev.* 172, 665–670 (1968).

26 J.M. Fiddy, I. Hall, F. Grandjean, U. Russo, and G.J. Long. Pressure dependence of the Mössbauer spectra of several iron(III) trisdithiocarbamate complexes. *J. Phys.: Condens. Matter* 2, 10109–10122 (1990).

27 L.D. Roberts, D.O. Patterson, J.O. Thomson, and R.P. Levey. Solid-state and nuclear results from a measurement of the pressure dependence of the energy of the resonance gamma ray of ^{197}Au. *Phys. Rev.* 179, 656–662 (1969).

28 S. Klotz and M. Braden. Phonon dispersion of bcc iron to 10 GPa. *Phys. Rev. Lett.* 85, 3209–3212 (2000).

29 H.K. Mao, J. Xu, V.V. Stuzhkin, J. Shu, R.J Hemley, W. Sturhahn, M.Y. Hu, E.E. Alp et al. Phonon density of states of iron up to 153 Gigapascals. *Science* 292, 914–916 (2001).

30 V.I. Goldanskii, G.M. Gorodinskii, S.V. Karyagin, L.A. Korytko, L.M. Krizhanskii, E.F. Makarov, I.P. Suzdalev, and V.V. Khrapov. The Mössbauer effect in tin compounds. *Proc. Acad. Sci. USSR (Phys. Chem. Sect.)* 147, 766–768 (1962) [Russian original: *Doklady Akad. Nauk* 147, 127–130 (1962)].

31 S.V. Karyagin. A possible cause for the doublet component asymmetry in the Mössbauer absorption spectrum of some powdered tin compounds. *Proc. Acad. Sci. USSR Phys. Chem. Sect.* 148, 110–112 (1963) [Russian original: *Doklady Akad. Nauk USSR* 148, 1102–1105 (1963)].

32 L.E. Campbell, G.L. Montet, and G.J. Perlow. Anisotropy of the Debye–Waller factor in cesium–graphite intercalation compounds by Mössbauer spectroscopy, and the quadrupole moment of the 81-keV state in ^{133}Cs. *Phys. Rev. B* 15, 3318–3324 (1977).

33 E.R. Bauminger, A. Diamant, I. Felner, I. Nowik, A. Mustachi, and S. Ofer. The Goldanskii–Karyagin effect in $Gd_2M_2O_7$ and $Eu_2M_2O_7$ compounds. *J. de Physique (Colloque)* 37, C6-49–C6-52 (1976).

34 V.I. Goldanskii and E.F. Makarov. Fundamentals of gamma-resonance spectroscopy. In *Chemical Applications of Mössbauer Spectroscopy*, V.I.

Goldanskii and R.H. Herber (Eds.), pp. 1–113 (Academic Press, New York, 1968).
35 H. Armon, E.R. Bauminger, A. Diamant, I. Nowik, and S. Ofer. Large Goldanskii effect in quadrupole Mössbauer spectra of the 103.2 keV gamma ray of ^{153}Eu. *Phys. Lett. A* 44, 279–280 (1973).
36 G.K. Shenoy. Rare-earth Mössbauer studies of chemical problems. In *Chemical Mössbauer Spectroscopy*, R.H. Herber (Ed.), pp. 343–354 (Plenum Press, New York, 1984).
37 G.K. Shenoy and J.M. Friedt. Dependence of Mössbauer resonance intensities on vibrational lattice anisotropy in case of an axial electric field gradient. *Nucl. Instrum. Methods* 136, 569–574 (1976).
38 H. Armon, E.R. Bauminger, A. Diamant, I. Nowik, and S. Ofer. Goldanskii effect in quadrupole Mössbauer spectra of the 89 keV gamma ray of ^{156}Gd. *Solid State Commun.* 15, 543–545 (1974).
39 R.W. Grant, R.M. Housley, and U. Gonser. Nuclear electric field gradient and mean square displacement of the iron sites in sodium nitroprusside. *Phys. Rev.* 178, 525–530 (1969).
40 R.M. Housley and R.H. Nussbaum. Mean-square nuclear displacement of Fe^{57} in Zn from the Mössbauer effect. *Phys. Rev.* 138, A753–A754 (1965).
41 L. Niesen and B. Stenekes. Anisotropic Debye–Waller factor of iodine impurities in p-type silicon. *J. Phys.: Condens. Matter* 3, 3617–3623 (1991).
42 M. Steiner, M. Köfferlein, W. Potzel, H. Karzel, W. Schiessl, G.M. Kalvius, D.W. Mitchell, N. Sahoo, H.H. Klauss, T.P. Das, R.S. Feigelson, and G. Schmidt. Investigation of electronic structure and anisotropy of the Lamb–Mössbauer factor in ZnF_2 single crystals. *Hyperfine Interactions* 93, 1453–1458 (1994).
43 C.W. Sherwin. Some recent experimental tests of the "clock paradox." *Phys. Rev.* 120, 17–21 (1960).
44 W. Kündig. Measurement of the transverse Doppler effect in an accelerated system. *Phys. Rev.* 129, 2371–2375 (1963).
45 R.V. Pound and G.A. Rebka Jr. Variation with temperature of the energy of recoil-free gamma rays from solids. *Phys. Rev. Lett.* 4, 274–275 (1960).
46 Y.L. Chen and Yan Xiaohua. Measurement of second order Doppler shifts. *J. Wuhan University (Natural Sci. Ed.)* 10, 36–38 (1988) [in Chinese].
47 Y.L. Chen, B.F. Xu, and J.G. Cheng. Mössbauer effect study of a natural chromite from Xinjiang deposit. *J. Wuhan University (Natural Sci. Ed.)* 4, 29–34 (1991) [in Chinese].
48 Y.L. Chen, B.F. Xu, J.G. Cheng, and Y.G. Ge. Study on iron distribution in natural chromite by Mössbauer technique. *Nucl. Sci. Techn.* 3, 135–138 (1992) (Chinese Nuclear Society).
49 Y.L. Chen, B.F. Xu, and J.G. Chen. Mössbauer evidence for Fe^{2+}–Fe^{3+} ordering in magnesioferrochromite. *Hyperfine Interactions* 70, 1029–1032 (1992).
50 J.W. Niemantsverdriet, C.F.J. Flipse, B. Selman, J.J. van Loef, and A.M. van der Kraan. Influence of particle motion on the Mössbauer effect in microcrystals α-FeOOH and α-Fe_2O_3. *Phys. Lett. A* 100, 445–447 (1984).
51 J.K. Dewhurst, H. Pollak, U. Karfunkel, and Z. Nkosibomvu. Recoil-free fractions from isomeric shift measurements. In *International Conference on the Applications of the Mössbauer Effect, ICAME-95*, (Italian Physical Society Conference Proceedings No. 50), I. Ortalli (Ed.), pp. 903–906 (Editrice Compositori, Bologna, 1996).
52 R.D. Taylor and P.P. Craig. Correlation between Mössbauer resonance strength and second-order Doppler shift: estimate of zero-point velocity. *Phys. Rev.* 175, 782–787 (1968).
53 D.A. O'Connor. The effect of line broadening of Mössbauer resonant sources and absorbers on the

resonance absorption. *Nucl. Instrum. Methods* 21, 318–322 (1963).
54 R.M. Housley, N.E. Erickson, and J.G. Dash: Measurement of recoil-free fractions in studies of the Mössbauer effect. *Nucl. Instrum. Methods* 27, 29–37 (1964).
55 D.A. O'Connor and G. Skyrme. The determination of the recoilless fraction of Mössbauer absorbers. *Nucl. Instrum. Methods* 106, 77–81 (1973).
56 P. Hannaford and R.G. Horn. Selective modulation of recoilless γ-rays. *J. Phys. C* 6, 2223–2233 (1973).
57 B. Manouchev, Ts. Bonchev, and D. Ivanov. Methods of determining the recoilless absorption probability of Mössbauer radiation. *Nucl. Instrum. Methods* 136, 261–265 (1976).
58 J. Ball and S.J. Lyle. A method for the determination of the Lamb–Mössbauer factor for a crystal lattice. *Nucl. Instrum. Methods* 163, 177–181 (1979).
59 K.P. Mitrofanov, N.V. Illarionova, and V.S. Shpinel. Counter with selective efficiency for the recording of gamma-rays which are emitted without recoil. *Instrum. Exp. Techn.* 3, 415–420 (1963) [Russian original: *Pribory i Tekhnika hksperimenta* 3, 49–54 (1963)].
60 M. Capaccioli, L. Cianchi, F. Del Giallo, P. Moretti, F. Pieralli, and G. Spina. A method for measurement of the Debye–Waller factor *f*. *Nucl. Instrum. Methods B* 101, 280–286 (1995).
61 H. Vogel, H. Spiering, W. Irler, U. Volland, and G. Ritter. The Lamb–Mössbauer factor of metal iron foils at 4.2 K. *J. de Physique (Colloque)* 40, C2-676 (1979).
62 S.L. Ruby and J.M. Hicks. Line shape in Mössbauer spectroscopy. *Rev. Sci. Instrum.* 33, 27–30 (1962).
63 J.M. Williams and J.S. Brooks. The thickness dependence of Mössbauer absorption line areas in unpolarized and polarized absorbers. *Nucl. Instrum. Methods* 128, 363–372 (1975).
64 U. Bergmann, S.D. Shastri, D.P. Siddons, B.W. Batterman, and J.B. Hastings. Temperature dependence of nuclear forward scattering of synchrotron radiation in α-^{57}Fe. *Phys. Rev. B* 50, 5957–5961 (1994).
65 Z.X. Quan, R.Q. Chang, H.J. Jin, G.Q. Xu, J.Y. Ping, and R.Z. Ma. A Method for the determination of the Mössbauer recoil-free factor. *Chinese Sci. Bull.* 32, 1392–1397 (1987).
66 F. Yi, F.L. Zhang, Y.L. Chen, and B.F. Xu. Determination of the recoilless fraction of magnetic splitting material. *J. Wuhan University (Natural Sci. Ed.)* 44, 609–611 (1998) [in Chinese].
67 V.N. Belogurov and V.A. Bilinkin. Determination of the absolute Mössbauer fraction in the absorption spectra of an arbitrary form. *Nucl. Instrum. Methods* 175, 495–501 (1980).
68 R.W. Grant, R.M. Housley, and U. Gonser. Nuclear electric field gradient and mean square displacement of the iron sites in sodium nitroprusside. *Phys. Rev.* 178, 523–530 (1969).
69 F.L. Zhang, F. Yi, Y.L. Chen, and B.F. Xu. Determination of the optimum thickness of an absorber in Mössbauer spectroscopy. *J. Wuhan University (Natural Sci. Ed.)* 43, 348–352 (1997) [in Chinese].
70 R.H. Herber and G.K. Wertheim. Mössbauer effect in ferrocene and related compounds. In *The Mössbauer Effect*, D.M.J. Compton and A.H. Schoen (Eds.), p. 105–111 (Wiley, New York, 1962).
71 E. Fritzsch and C. Pietzsch. Lamb Mössbauer fraction and mean square displacement of iron in pyrite (FeS2). *Phys. Stat. Sol. (a)* 79, K113–K115 (1983).
72 G. Hembree and D.C. Price. Determination of hyperfine interaction parameters from unresolved Mössbauer spectra. *Nucl. Instrum. Methods* 108, 99–106 (1973).
73 J. Sitek. Non-resonant background and recoil-free fraction of a $BaSnO_3$ ^{119m}Sn Mössbauer source. *Nucl. Instrum. Methods* 114, 163–164 (1974).

74 R.K. Puri. Recoil-free fraction and maximum resonance cross section of ^{119m}Sn in a Mössbauer $BaSnO_3$ source employing a Si(Li) detector. *Nucl. Instrum. Methods* 117, 381–383 (1974).

75 B.R. Bullard and J.G. Mullen. Mössbauer line-shape parameters for ^{159}Tb in $TbAl_2$ and Tb_4O_7. *Phys. Rev. B* 43, 7416–7421 (1991).

76 B.R. Bullard, J.G. Mullen, and G. Schupp. Mössbauer line-shape parameters for ^{183}W and ^{191}Ir in metallic tungsten and iridium. *Phys. Rev. B* 43, 7405–7415 (1991).

6
Mössbauer Scattering Methods

So far, we have been using Mössbauer spectroscopy in transmission geometry. The scattering of Mössbauer radiation is another method, but based on a different principle. Compared with neutron and x-ray scattering methods, the development of Mössbauer scattering has been slow and incomplete, mainly limited by the lack of strong Mössbauer radioactive sources in the early days. Since synchrotron Mössbauer sources became available in the mid-1980s, research on Mössbauer scattering has been substantially reinvigorated.

In this chapter, we describe the basic principles of scattering methods and some early experimental results using Mössbauer radioactive sources. It is shown that the coherence phenomenon can play a crucial role in the nuclear resonance scattering of Mössbauer radiation. Such scattering experiments can be implemented using synchrotron radiation, which is discussed in detail in Chapter 7.

6.1
The Characteristics and Types of Mössbauer γ-ray Scattering

6.1.1
The Main Characteristics

Generally speaking, compared with the transmission method, the scattering method is much more complex, both conceptually and experimentally, but it can provide more information. There are many excellent monographs and articles available [1–4]. Figure 6.1 shows several arrangements for scattering experiments [5, 6]. In Fig. 6.1(a), the scatterer does not contain the Mössbauer isotope, but in Figs 6.1(b)–(d), the Mössbauer isotope must be present. For Figs 6.1(a) and (c), a resonant absorber A in front of the detector serves as the energy analyzer by using the Mössbauer effect. In all of these arrangements, the source may be stationary or be driven to constant velocity or constant acceleration.

In order to reduce background counts in scattering experiments, shielding in different parts of the apparatus is extremely important (not shown in Fig. 6.1). For ^{57}Fe, shielding may attenuate background counts to as few as 0.05 per min-

Mössbauer Effect in Lattice Dynamics. Yi-Long Chen and De-Ping Yang

ISBN: 978-3-527-40712-5

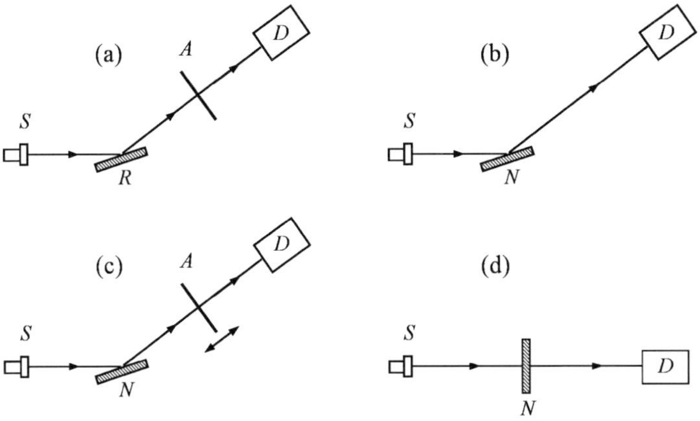

Fig. 6.1 Schematic diagrams of various Mössbauer scattering experiments: (a) Rayleigh scattering, (b) nuclear resonance scattering (NRS), (c) selective excitation double Mössbauer (SEDM) spectroscopy, where the source S undergoes constant-velocity motion and the absorber A undergoes constant-acceleration motion, and (d) nuclear forward scattering (NFS). R represents a Rayleigh scatterer, N represents a nuclear scatterer, and D represents the detector.

ute. In addition, good collimators are required in order to ensure low dispersion (within several arc seconds). In diffraction experiments, especially, the angular measurements must be precise, within $\pm 0.1''$ [7]. For ^{57}Fe, the source activity usually falls between 3.7×10^9 and 9.25×10^9 Bq. It may take at least 100 hours, and sometimes as many as 600 hours, to obtain a scattering spectrum.

In scattering experiments, several different types of scattered particles can be detected. The most usual are the γ-rays re-emitted by the Mössbauer nuclei after resonance absorption. But if the excited state of the Mössbauer nucleus has a large internal conversion coefficient, the instrument will also detect the conversion electrons (an incoherent process) and accompanying products (see Chapter 3, Fig. 3.6), e.g., K-fluorescence photons.

In addition, there may be some interference phenomena between different scattering processes, such as the interference between Mössbauer resonance scattering and Rayleigh scattering, and the interference between conversion electrons and photoelectrons. These constitute another characteristic of the Mössbauer scattering method.

It is because of the complexity associated with the scattering method that more information can be gathered by this method than by the emission or absorption method. Mössbauer scattering has distinctive features in comparison with neutron or x-ray diffraction. It can overcome the difficulties encountered by neutron or x-ray diffraction to determine the phase in structure factors. Mössbauer scattering is capable of clearly resolving hyperfine interactions for elucidating magnetic structures of crystals, which can provide important complementary information to neutron diffraction results (whereas x-ray diffraction detects no magnetic structure). Although most Mössbauer radiations share the same part of the electromagnetic radiation with x-rays (wavelengths between 0.1 and 1 Å), their scattering behaviors are very different, chiefly due to the fact that the Mössbauer radiation linewidth is about 8 orders of magnitude narrower than that of x-rays. As a consequence, conventional x-ray diffraction can only provide spatial information due to its lack of energy resolution.

In some Mössbauer scattering experiments (e.g., Rayleigh scattering), the scatterer does not necessarily contain the Mössbauer nucleus and is not limited to solids either, so long as there is an additional resonance absorber (A). Therefore, the scattering method can make use of the advantages of the Mössbauer effect, and in principle extend the applications of Mössbauer spectroscopy. Because of the ingenious method of Rayleigh scattering, which separates the elastic and inelastic γ-ray scattering components, it has become an important technique in studying dynamics in solids and liquids.

In nuclear resonance scattering (NRS) the γ-ray is first recoillessly absorbed, then re-emitted by a nucleus in its exited state with a half-life of the order of 10^{-7} s. This is much longer than ${\omega_m}^{-1} \approx 10^{-14}$–$10^{-13}$ s, where ω_m is the maximum vibration frequency of the nucleus about its equilibrium position. Thus, NRS is often regarded as a "slow" process. In contrast, Rayleigh scattering, a non-resonance scattering by bound electrons, takes place in a time interval of 10^{-16}–10^{-15} s $< {\omega_m}^{-1}$ and it is regarded as a "fast" process.

6.1.2
Types of Scattering Processes

In general, the scattering of Mössbauer γ-rays with energies $E_\gamma < 200$ keV may be mainly categorized into three groups of processes, each being coherent or incoherent, elastic or inelastic [6]:

Elastic nuclear and Rayleigh	Coherent:	recoilless (elastic)
		with recoil (inelastic or quasi-elastic)
	Incoherent:	recoilless (elastic)
		with recoil (inelastic or quasi-elastic)
Inelastic nuclear	Incoherent:	recoilless (energy shifted)
		with recoil (inelastic)
Compton scattering	Incoherent:	recoilless (inelastic)

We are familiar with the concepts of elastic and inelastic scattering. As to coherent and incoherent scattering, they are discussed in Section 4.6.1. Essentially, coherence is the result of periodic arrangement of those scattering centers (atoms or nuclei) that have the same scattering properties. Any crystal imperfections, such as random distribution of the Mössbauer isotope ^{57}Fe amongst natural iron, or spin effects, would result in an incoherent component contributing to a diffused background between the Bragg peaks.

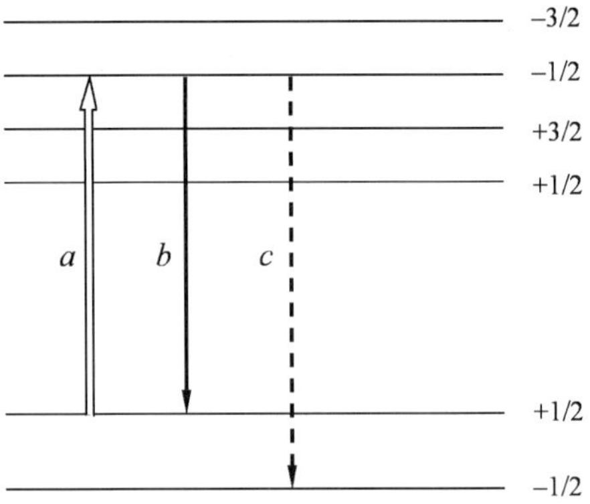

Fig. 6.2 When there is a magnetic hyperfine field in the scatterer, there are two possible de-excitation modes after a resonance absorption with a transition from $m_g = +1/2$ to $m_e = -1/2$. Mode a–b is an elastic coherent process, while mode a–c is an inelastic incoherent one.

Figure 6.2 illustrates how inelastic and incoherent nuclear resonance scattering can occur due to magnetic hyperfine splittings in ^{57}Fe scatterers. During the scattering process, a ^{57}Fe nucleus in the ground state ($m_g = +1/2$) makes a transition to the excited state ($m_e = -1/2$), followed by emission of a γ-photon and returning to the $m_g = +1/2$ state (transitions a and b). This is obviously an elastic process. If the process follows transitions a and c instead, the scattering is clearly inelastic and leads to a change in the nuclear spin state ($m_g = +1/2 \rightarrow m_g = -1/2$). Since Rayleigh scattering does not change nuclear spin, the fact that the scattering process with transitions a and c cannot have interference with Rayleigh scattering indicates that it is incoherent.

Mössbauer diffraction, as an alternative method, has been chosen to verify the above coherent and incoherent nuclear resonance scatterings [8]. In Fig. 6.3, a ^{57}Co/Cr source is attached to the first vibrator (vb_1) and the scatterer is an α-Fe_2O_3 single crystal with an 85% ^{57}Fe enrichment. A weak magnetic field is applied perpendicular to the scattering plane. When the scattering is chosen to be from the (8 8 8) plane, the Bragg angle is $\theta_B = 49°$. The absorber A is a stainless steel foil of thickness 10 μm, with ^{57}Fe enriched to more than 20%, and is used as an analyzer for energies of the diffracted rays. During the first part of the experiment, the analyzer A is removed, vb_1 works in the constant-acceleration mode, and the detector measures the diffraction intensity as a function of the source velocity. This gives the resonance scattering Mössbauer spectrum under the Bragg condition, as shown in Fig. 6.4(a). From this spectrum, the source velocity required to cause transition a can be precisely determined to be -0.60 mm s^{-1}. The slight asymmetry in the spectral lines is due to the interference between resonance scattering and Rayleigh scattering processes. Now the second part of the experiment is performed, where vb_1 works in the constant-velocity mode with a speed of -0.60 mm s^{-1}. The absorber A is installed on the second vibrator (vb_2)

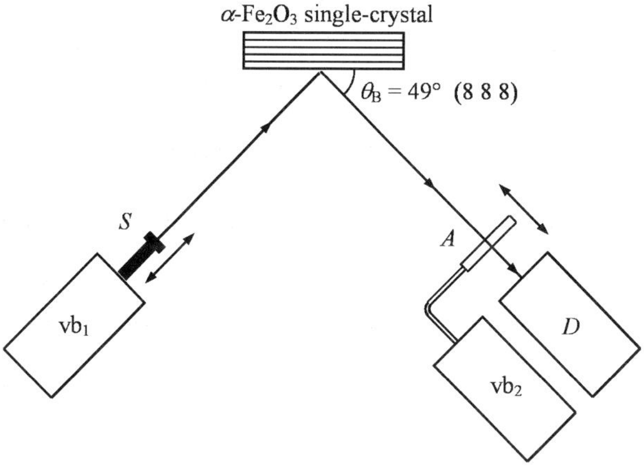

Fig. 6.3 Schematic diagram of a Mössbauer diffraction experiment.

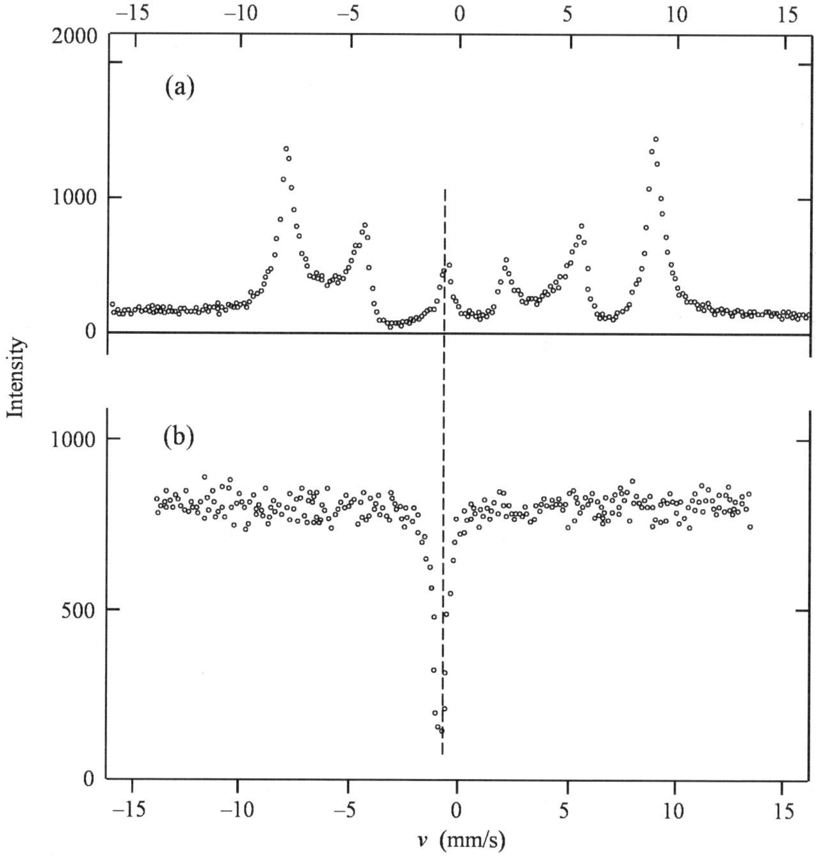

Fig. 6.4 Intensity of diffracted ^{57}Fe γ-rays: (a) as a function of source velocity and (b) as a function of analyzer velocity.

which works in the constant-acceleration mode and scans the velocity range of ± 13 mm s^{-1}. The resultant spectrum is a single line shown in Fig. 6.4(b), which convincingly verifies that only γ-rays associated with transition b are coherent while those associated with transition c are incoherent. The coherently diffracted γ-rays are diffracted at the Bragg angle. As for the incoherently diffracted γ-rays, the change in energy is extremely small, $\Delta E/E = \Delta\lambda/\lambda \approx 2 \times 10^{-12}$, and thus the Bragg condition is still satisfied. However, to a first approximation, these incoherent diffracted γ-rays have an isotropic distribution so that very little enters the detector.

It should be noted that an experiment of this type is impossible with neutrons or x-rays.

As the classical theory has pointed out [9], when the frequency of the incident radiation ω is much lower than the characteristic frequency ω_r, $\omega \ll \omega_r$, Rayleigh scattering is predominant. In the other extreme, i.e., $\omega \gg \omega_r$, the process is known as Thomson scattering. For most Mössbauer transition energies, the latter can be neglected and is not included in the above categorization. When the incident γ-rays are exactly at resonance for a single "unsplit" ^{57}Fe nucleus, the differential scattering cross-sections of nuclear resonant scattering, Rayleigh scattering, and Compton scattering are, respectively, about 10^{-20}, 10^{-24}, and 10^{-24} cm^2 [6]. The exact values depend on scattering angle, polarization, recoil effects, and the abundance of the Mössbauer isotope. In the case of Fe, for instance, with the natural abundance of ^{57}Fe taken into consideration, the resonance scattering cross-section will be effectively reduced to 1/45 of that if all Fe atoms were ^{57}Fe [1].

In Compton scattering, the photons are scattered by those electrons that are independent and loosely bound. Therefore, the scattering process is incoherent and inelastic. This contributes to the background counts in various scattering experiments. When $E_\gamma < 200$ keV, the cross-sections of both Compton scattering and Rayleigh scattering depend on the scattering angle θ, as shown in Fig. 6.5 [10]. For Pb, σ_C and σ_R are equal when $\theta \approx 11°$, and σ_R dominates at smaller scattering angles. For Cu, similar behavior has been observed. For ^{57}Fe, σ_C is smaller than σ_R by a factor of about 8 [1].

At small scattering angles, Rayleigh scattering is predominant and Compton scattering comes next. It has been pointed out [11] that more than three-quarters of Rayleigh scattering is concentrated between $\theta = 0$ and $\theta = \theta_0$, where

$$\theta_0 = 2 \sin^{-1}\left[0.026 Z^{1/3} \frac{m_e c^2}{E_\gamma}\right]. \tag{6.1}$$

For a Pb scatter with $E_\gamma = 410$ keV, we have $\theta_0 = 14°$, consistent with the results shown in Fig. 6.5. Since θ_0 can be calculated by Eq. (6.1) only if the quantity in the brackets is less than 1, this formula may not be used for the majority of Mössbauer radiation energies. In those cases, the largest portion of Rayleigh scattering is no longer limited to the small scattering angle region [1].

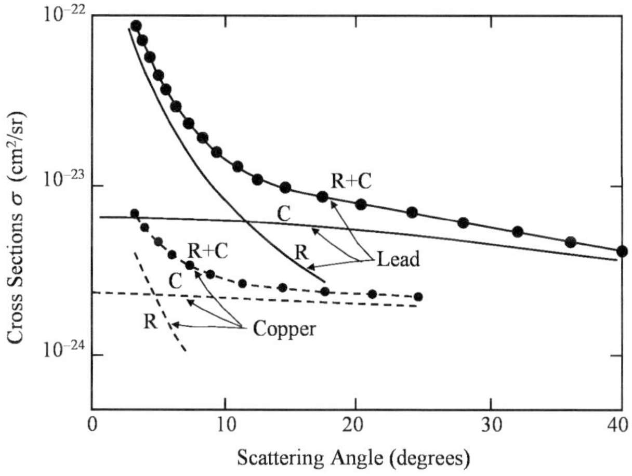

Fig. 6.5 Cross-sections per unit solid angle for Rayleigh scattering (R) and Compton scattering (C) of 410 keV γ-rays by lead and copper, as functions of the scattering angle. R+C represents the respective total cross-section.

6.2
Interference and Diffraction

Interference and diffraction are the main characteristics of waves. Classical theory can successfully describe the interference and diffraction phenomena of visible light waves and electromagnetic waves (x-rays). Two interfering light sources must satisfy coherent conditions, which require that they have not only the same wavelength, but also a fixed phase difference. After the verification of the wave nature of particles such as electrons and neutrons, quantum theory also has a similar definition for coherence. At a point $\boldsymbol{r}$ in space, if the probability of finding a photon from one source is $|a|^2$ and from a second source is $|b|^2$, then when both sources radiate, the probability of finding a photon at that point is not $|a|^2 + |b|^2$ but

$$|a|^2 + |b|^2 + 2\,\mathrm{Re}(a \cdot b^*) \tag{6.2}$$

where the last term represents interference, indicating that the two waves have some degree of coherence. This concept is not as simple as it looks, and confusion may arise in its applications. Unless one is very clear about the concept of coherence and its physical requirements, one could be mislead to erroneous conclusions [12, 13].

In 1960, Black and Moon [14] demonstrated the Bragg reflection of the 14.4 keV Mössbauer radiation from an enriched ^{57}Fe crystal, the main γ-ray scattering being nuclear resonance scattering. A maximum intensity recorded at the Bragg angle shows that the scattering is coherent. This is Mössbauer diffraction, which is also called nuclear Bragg scattering (NBS). As will be shown, nuclear forward scattering (NFS) is also elastic and coherent.

But such a diffraction is much more complex than optical diffraction, due to the fact that recoilless resonance scattering is a "slow" process. As a result, the spontaneous character of the nuclear decay becomes so marked that it is not easy to decide theoretically whether the scattered γ-rays due to transition b (Fig. 6.2) are coherent. But experimental facts have provided an affirmative answer. Thus, we will begin with a description of some experimental results.

6.2.1
Interference between Nuclear Resonance Scattering and Rayleigh Scattering

Let $f^{\mathrm{R}}(hh_0)$ and $f^{\mathrm{N}}(hh_0)$ be the coherent Rayleigh and nuclear resonance scattering amplitudes, respectively, for incident radiation of polarization h_0 and scattered radiation of polarization h (where both h_0 and h are $+1$ for right- and -1 for left-circular polarizations). Also, assume that the scatterer is a single crystal composed entirely of Mössbauer atoms. The scattering intensity for an unpolarized beam can be written as [15]

$$I = \frac{1}{2}\sum_{hh_0} |f^{\mathrm{N}}(hh_0) + f^{\mathrm{R}}(hh_0)|^2$$

$$= \frac{1}{2}\sum_{hh_0} [|f^{\mathrm{N}}(hh_0)|^2 + |f^{\mathrm{R}}(hh_0)|^2 + 2c\,\mathrm{Re}(f^{\mathrm{N}}(hh_0) \cdot f^{\mathrm{R}^*}(hh_0))] \tag{6.3}$$

where a factor c is used to represent the degree of coherence between the two scattered rays. The amplitude of Rayleigh scattering is [6]

$$f^{\mathrm{R}} = -r_{\mathrm{e}} F(\theta_i)\mathbf{e} \cdot \mathbf{e}_0 \tag{6.4}$$

where r_{e} is the classical radius of the electron, $F(\theta_i)$ is the atomic scattering factor, θ_i is the incident angle, and $\mathbf{e}_0$ and $\mathbf{e}$ are the polarization unit vectors of the incident photon and the scattered photon, respectively. In order to calculate $f^{\mathrm{R}}(hh_0)$, it is necessary to use circular polarizations. The above expression can be written as [15, 16]

$$f^{\mathrm{R}}(hh_0) = f_0^{\mathrm{R}} d^{(1)}_{hh_0}(\theta) \tag{6.5}$$

where $d^{(1)}_{hh_0}(\theta)$, the reduced rotation matrix elements, take the followings values:

h_0	h	$d^{(1)}_{hh_0}(\theta)$
1	-1	$\frac{1}{2}(1 - \cos\theta)$
1	1	$\frac{1}{2}(1 + \cos\theta)$
-1	1	$\frac{1}{2}(1 - \cos\theta)$
-1	-1	$\frac{1}{2}(1 + \cos\theta)$

(6.6)

As will be shown in Eq. (6.14), the amplitude for nuclear resonance scattering can also be similarly written as

$$f^{\mathrm{N}}(hh_0) = hh_0 f_0^{\mathrm{N}} d_{hh_0}^{(1)}(\theta). \tag{6.7}$$

Here f_0^{R} and f_0^{N} are independent of polarization. Substituting Eqs. (6.5) and (6.7) into Eq. (6.3), we obtain

$$I = \frac{1}{2}|f_0^{\mathrm{R}}|^2(1+\cos^2\theta) + \frac{1}{2}|f_0^{\mathrm{N}}|^2(1+\cos^2\theta) + 2c\,\mathrm{Re}(f_0^{\mathrm{N}} \cdot f_0^{\mathrm{R}^*})\cos\theta \tag{6.8}$$

where $\theta = (\mathbf{k}', \mathbf{k}_0)$ is the scattering angle. For $M1$ type transitions (e.g., ^{57}Fe), the angular dependences for both resonance scattering and Rayleigh scattering are in the form of $\cos\theta$, and thus the interference term vanishes when $\theta = 90°$.

The coherent nature of nuclear resonance scattering is also demonstrated by its ability to interfere with Rayleigh scattering. In the experiment by Black and Moon mentioned above [14], the range of source velocity was less than the separation between lines 3 and 4 in the sextet, and the observed diffraction maximum indeed presented an asymmetric profile, as shown in Fig. 6.6(b). Such an asymmetric resonance peak can only be explained by the interference between nuclear resonance scattering and Rayleigh scattering by the same ^{57}Fe atom, i.e., an intra-atomic interference. These two scattering processes are in phase above resonance and are antiphase below resonance, resulting in constructive and destructive interference, respectively. Since Rayleigh scattering is coherent, this experiment again provides evidence that nuclear resonance scattering is also coherent.

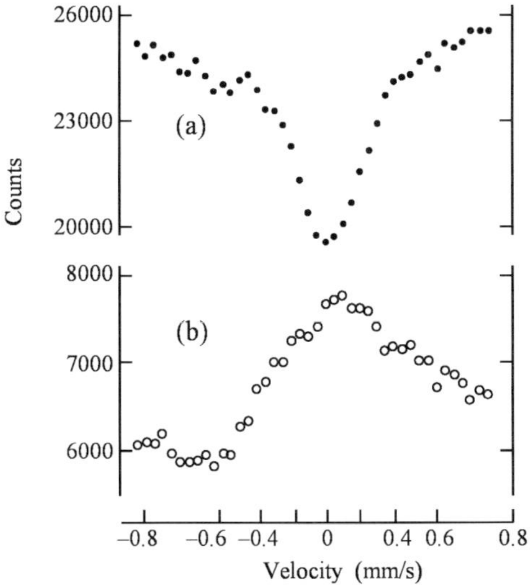

Fig. 6.6 Third Mössbauer absorption line (a), and the corresponding scattering spectrum (b). The source is ^{57}Co/Fe and the scatterer is an Fe foil with ^{57}Fe enriched to 56%.

Soon after, Bernstein and Campbell [17] obtained total reflection of the 14.4 keV γ-rays with a glancing angle of 2 mrad by an optically flat ^{57}Fe mirror and then proved the interference between nuclear resonance scattering and Rayleigh scattering. In this case, the interference took place between different atoms, indicating spatial coherence of the nuclear resonance scattering for an ensemble of nuclei.

Using various reflection planes of an iron single crystal of natural abundance in a diffraction experiment [18, 19], one can also clearly see the interference effect between nuclear resonance scattering and Rayleigh scattering processes (Fig. 6.7).

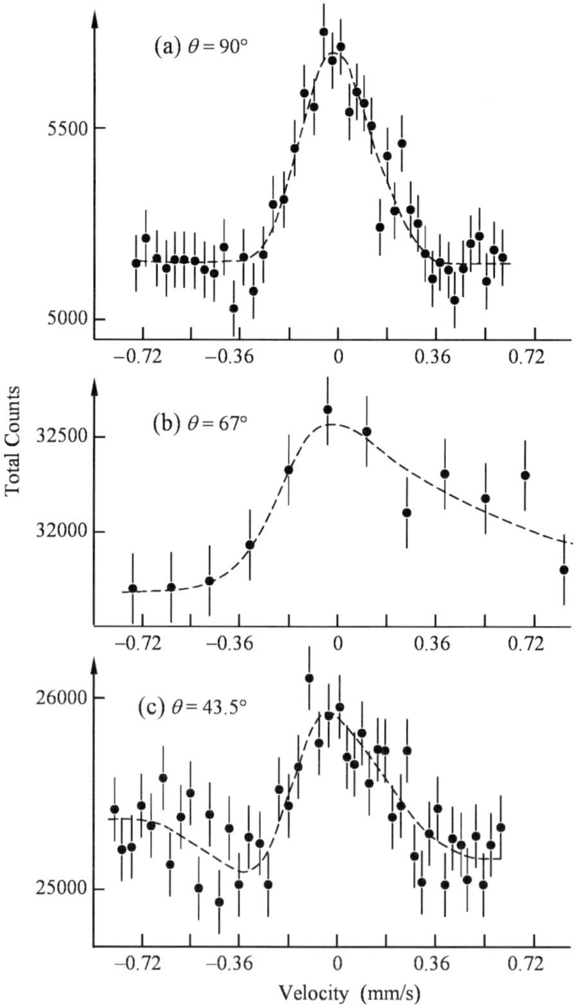

Fig. 6.7 ^{57}Fe Mössbauer scattering spectra (14.4 keV γ-rays) of metallic Fe at three scattering angles corresponding to Bragg reflections from (a) the (3 3 2) plane, (b) the (3 2 1) plane, and (c) the (2 1 1) plane. The θ values are scattering angles, $\theta = \theta_B$.

For the (3 3 2) plane, $\theta = 90^\circ$, the spectrum is symmetric, indicating absence of interference. At lower angles, as shown in Figs. 6.7(b) and (c), the spectral shapes are more asymmetric.

Using the 23.8 keV ^{119}Sn Mössbauer radiation, the γ-rays scattered from the (0 2 0) plane ($\theta_B = 5^\circ\ 7'$) of a single-crystal tin foil (88% ^{119}Sn) produce a typical interference pattern [20, 21]. The shape of the spectrum depends on the ratio $\xi = f^N/f^R$. One method to vary ξ is to lower the scatterer temperature; for example, at $T = 110$ K, the recoilless fraction f^N increases drastically from its room temperature value, causing a 7-fold increase in the ξ-value. Another method for varying ξ is to use second- or third-order reflections, so that the corresponding f^R-value is lower. In addition, the spectral shape also depends on the ratio $\zeta = \mu_r/\mu_a$. Figure 6.8 shows spectra obtained with several different ξ-values. As the f^N-value increases, the curves sharpen and the peak positions shift leftward gradually.

When there exists a magnetic hyperfine field, the spectral shape becomes very complex. This can be illustrated using an α-Fe_2O_3 single crystal [22], by applying

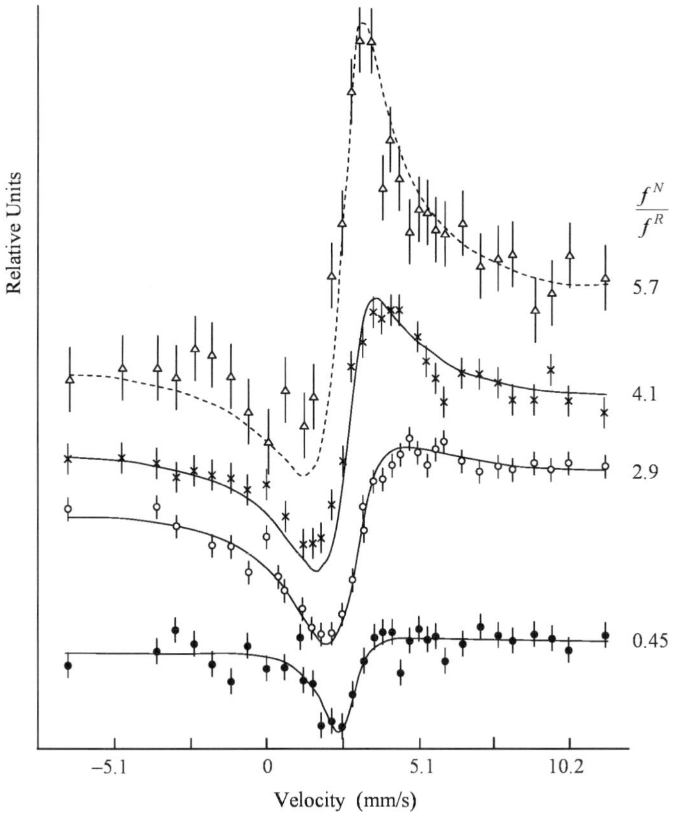

Fig. 6.8 Normalized Mössbauer diffraction spectra of 23.8 keV γ-rays by the (0 2 0) plane of a single-crystal ^{119}Sn foil (thickness 2 μm) [20].

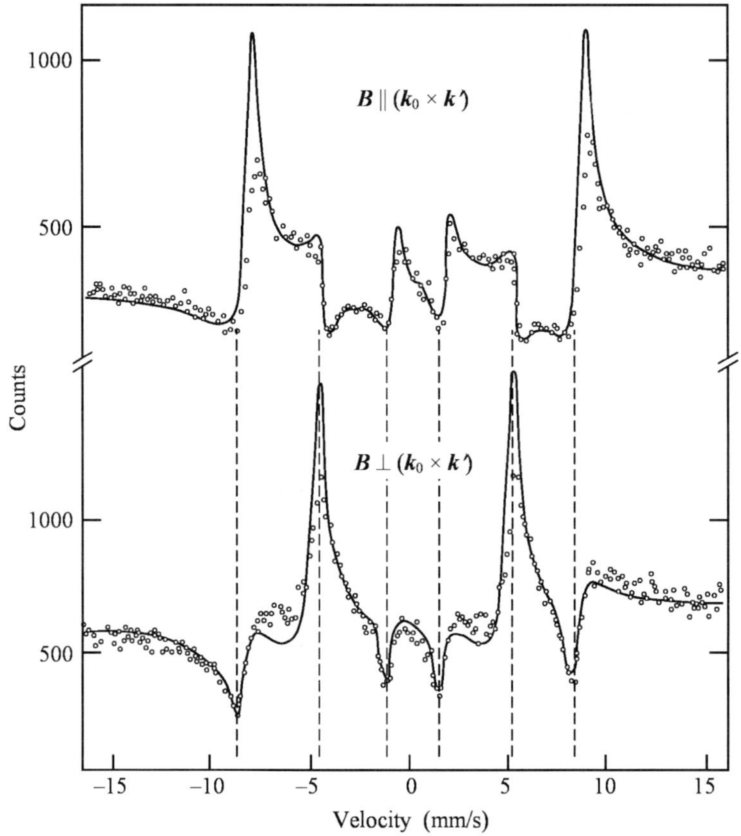

Fig. 6.9 Bragg diffraction spectra of 14.4 keV Mössbauer γ-rays scattered from the (6 6 6) plane of single-crystal α-Fe_2O_3.

a magnetic field $B = 0.1$ T along the (1 1 1) plane, either parallel or perpendicular to the scattering plane $(\boldsymbol{k}'\boldsymbol{k}_0)$. Interference between resonance scattering and Rayleigh scattering exists in even-order Bragg reflections $(2n\ 2n\ 2n)$. Figure 6.9 shows the curves of the dispersion type in two spectra from the (6 6 6) reflections, where the dashed lines indicate the positions of absorption lines in transmission geometry for the same sample and the solid curves are fitted results. The spectra from the (4 4 4) reflections also show asymmetric dips, and those from the (8 8 8) and (10 10 10) reflections show complex asymmetric peaks.

The main reason for such a complexity is that the scattering polarization factor is no longer simply $(1 + \cos^2 \theta)$ as in Eq. (6.8), but depends on the type of transition Δm and the direction of the hyperfine field as well as on the scattering angle θ.

It is quite clear that the coherence phenomenon can play a crucial role in the nuclear resonance scattering of Mössbauer or synchrotron radiation by an ensemble of Mössbauer nuclei. In addition to what has been described so far, other interesting phenomena also occur, such as interference effects in the presence of quadrupole splitting, interference between photoelectrons and internal conversion electrons, anomalies in the width of resonance peak in dynamic Mössbauer diffraction, and suppression of inelastic channels. The reader may find details in Refs. [23–30].

It is often necessary to separate nuclear resonance scattering from Rayleigh scattering, to avoid possible complications caused by the interference between them. In the following sections, we discuss each of the two scattering processes in detail.

6.2.2
Observation of Mössbauer Diffraction

The chief interest in Mössbauer diffraction comes from the usually large cross-section attainable at resonance. In order to observe pure Mössbauer γ-ray diffraction, the amplitude of Rayleigh scattering should be reduced to a negligibly small amount or completely eliminated. Mössbauer diffraction was first observed in $K_4Fe(CN)_6 \cdot 3H_2O$ (90% ^{57}Fe) [31]. Using the (0 6 0) reflection and an off-resonance source velocity $v \neq v_r$, the angular dependence of scattered intensity has been measured and is shown in Fig. 6.10(a). Because in this case there is no nuclear Bragg scattering, the relatively large diffraction peak at $\theta_B = 8° \ 50'$ is due to Rayleigh scattering. When the (0 8 0) reflection is chosen to repeat the above measurements, the detected counts are very low and independent of the scattering angle, as in Fig. 6.10(b). This indicates that Rayleigh scattering of the ^{57}Fe atoms is canceled by Rayleigh scattering of the other atoms in the unit cell [31]. When the source velocity is on-resonance $v = v_r$, the nuclear Bragg scattering, by

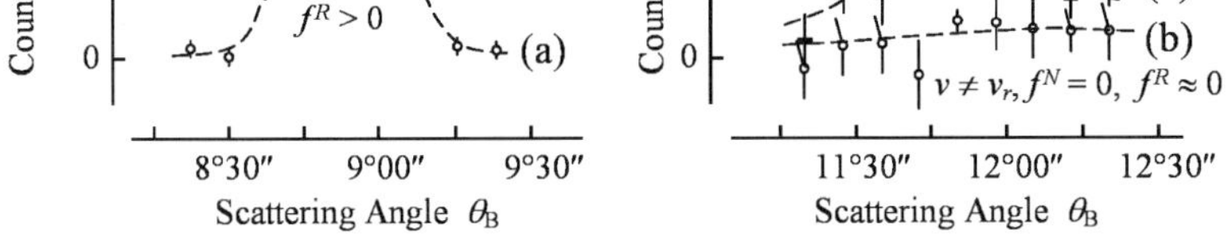

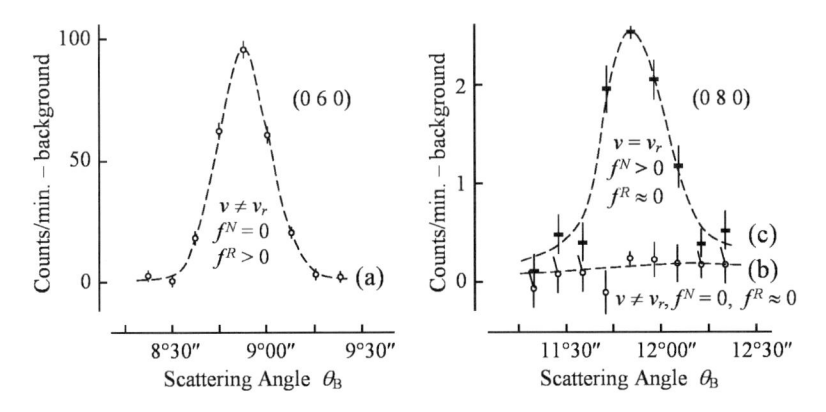

Fig. 6.10 Scattering intensity versus scattering angle for 14.4 keV ^{57}Fe γ-rays from the (0 6 0) and (0 8 0) planes of single-crystal $K_4Fe(CN)_6 \cdot 3H_2O$.

contrast, is not canceled and gives a diffraction peak as in Fig. 6.10(c), which is pure Mössbauer diffraction, centered at $\theta_B = 11^\circ\ 50'$.

The above example of separating Mössbauer diffraction from Rayleigh scattering is a rare, fortuitous case. A more systematic method is based on the relation between the resonance scattering amplitude and orientation of nuclear spins. For an antiferromagnetic α-Fe_2O_3 single crystal [32, 33], it is known that for an odd-order Bragg reflection from the (1 1 1) plane, there should be no Rayleigh scattering due to extinction. This is verified by the (1 1 1) reflection result for an off-resonance radiation ($v \neq v_r$) in Fig. 6.11(a), which is similar to that in Fig. 6.10(b). When the source velocity is $v = v_r = 8.6$ mm s^{-1}, the resonance absorption of $-1/2 \rightarrow -3/2$ takes place, and a pure nuclear diffraction peak appears at $\theta_B = 5^\circ\ 20'$, as shown in Fig. 6.11(b). The corresponding Mössbauer spectrum in Fig. 6.11(c) is obtained at the fixed Bragg angle of $\theta_B = 5^\circ\ 20'$. The fact that the

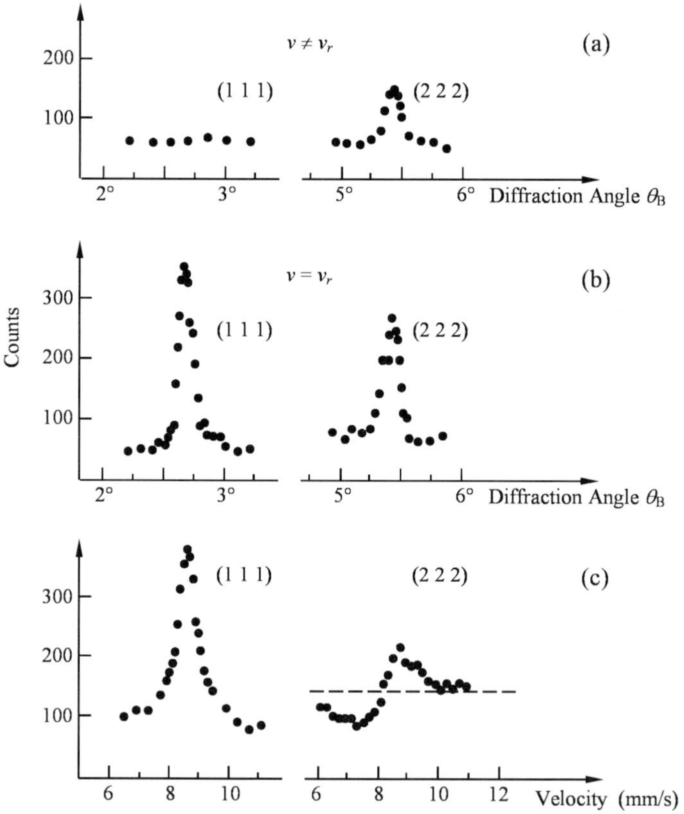

Fig. 6.11 14.4 keV γ-ray Bragg reflections from the (1 1 1) and (2 2 2) planes of single-crystal α-Fe_2O_3 (enriched to 85% ^{57}Fe): (a) $\nu \neq \nu_r$, (b) $\nu = \nu_r$, and (c) the corresponding Mössbauer spectra for fixed Bragg angles.

peak is symmetric indicates that this diffraction peak is solely due to the radiation from the $-3/2 \rightarrow -1/2$ transition, with no Rayleigh scattering. The right-hand side of Fig. 6.11 shows the results from the (2 2 2) reflections as a comparison. In this case, Rayleigh scattering occurs, the amplitude of which can be seen in Fig. 6.11(a) and whose interference with Mössbauer diffraction results in the asymmetric peak in Fig. 6.11(c).

Mössbauer diffraction has attracted a great deal of attention, and it may be used to obtain unique results not available from any other diffraction methods for studying crystalline and magnetic structures of solids. For example, x-ray diffraction cannot provide information related to magnetic hyperfine interactions. At the present time, this technique relies on the availability of synchrotron Mössbauer radiation, and there have been many reports of interesting applications, such as the measurements of the magnetic structures and the phase of structure factors.

The use of Mössbauer diffraction for determining the magnetic structures in Fe_3BO_6 is given here as an example. By the magnetic structure of a magnetically ordered crystal, we mean how the magnetic moments of magnetic atoms (ions) are periodically arranged. The intensity and shape of the Mössbauer diffraction spectrum are closely related to the orientations of the magnetic hyperfine fields at the nuclear position. The hyperfine fields are in turn related to the atomic magnetic moments; thus information on the magnetic structure can be extracted from a diffraction spectrum. Let us look at how this method is applied to Fe_3BO_6 [34], an antiferromagnet below its Néel temperature of $T_N = 508$ K. Fe_3BO_6 has an orthorhombic unit cell with $a = 10.5$ Å, $b = 8.55$ Å, $c = 4.47$ Å, and belongs to the space group $D_{2h}^{16}(P_{nma})$. The Fe atoms (ions) are located at two nonequivalent crystallographic sites $4c$ and $8d$. Two different magnetic structures (Fig. 6.12) were proposed after bulk magnetization measurements. Neutron diffraction by a polycrystalline sample indicated that structure I is likely to be the correct one [35]. In structure I, the moments are ferromagnetically coupled within each of the two planes, and they are antiferromagnetically coupled between the planes. If the moments of the $4c$ sites are inverted, structure I becomes structure II.

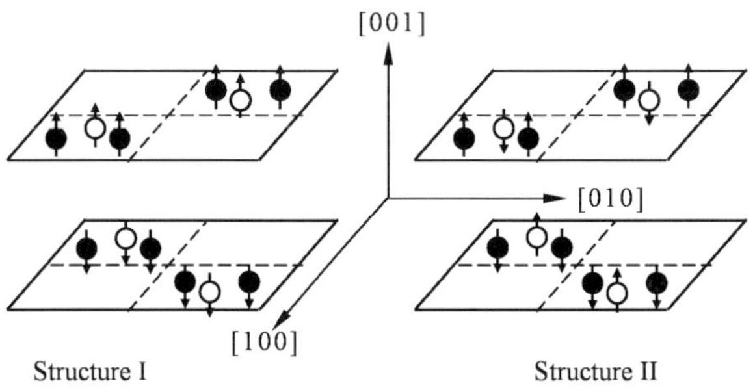

Fig. 6.12 Two possible magnetic structures allowed by the symmetry of the Fe_3BO_6 crystal. Black circles represent Fe ions in $8d$ sites and white circles represent Fe ions in $4c$ sites.

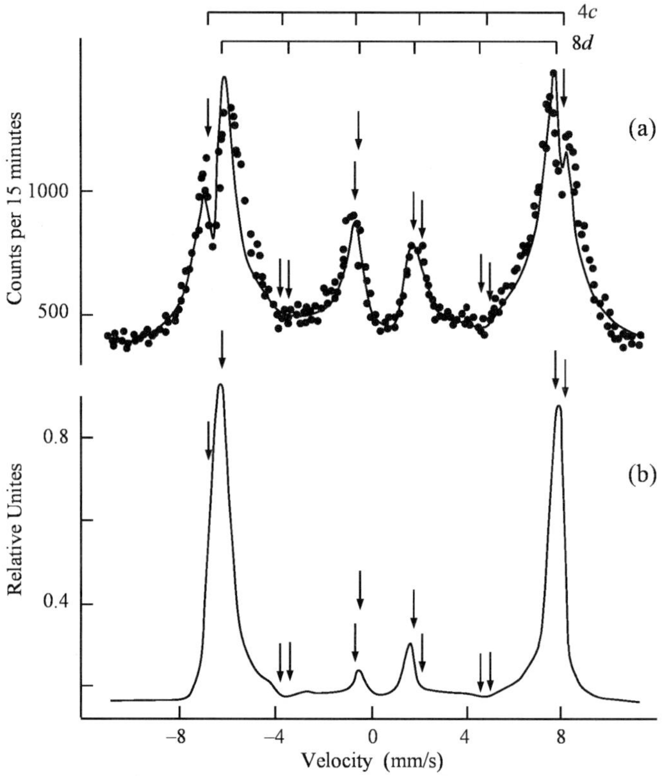

Fig. 6.13 Experimental and theoretical nuclear Bragg diffraction spectra of (7 0 0) reflection from a Fe_3BO_6 single crystal using 14.4 keV γ-rays. The solid curves in (a) and (b) are calculated based on magnetic structures I and II, respectively. Arrows show where the resonance absorptions are expected for ^{57}Fe nuclei in 4*c* and 8*d* positions.

In the diffraction experiment, the 14.4 keV γ-rays were allowed to diffract from the (7 0 0) plane of a $^{57}Fe_3BO_6$ single crystal, with the antiferromagnetic axis in the scattering plane $(\boldsymbol{k}'\boldsymbol{k}_0)$. The diffraction spectrum is shown in Fig. 6.13(a). The solid curves in Figs. 6.13(a) and (b) represent theoretically calculated results based on structures I and II, respectively. It is obvious that structure I agrees with the experimental result whereas structure II is in conflict. Therefore, nuclear resonance diffraction unequivocally verified the correctness of structure I, achieving what neutron diffraction was not able to do. Also, for neutron diffraction, it was necessary to use a ^{11}B-enriched sample of Fe_3BO_6 to reduce the absorption by ^{10}B.

This example illustrates that Mössbauer diffraction is extremely sensitive to the magnetic structure of the material, an important characteristic of this method. Between the two possible magnetic structures, the only difference is the magnetic moment reversal of the 4*c* Fe ions, which amount to only 1/3 of the total, but the difference causes substantial changes in the spectral shape.

6.3 Coherent Elastic Scattering by Bound Nuclei

In studying either nuclear Bragg scattering (NBS), or nuclear forward scattering (NFS) of Mössbauer γ-rays, it is necessary to consider carefully the scattering amplitude, scattering cross-section, Lamb–Mössbauer factor f_{LM} and the Debye–Waller factor f_D. We now discuss each of these quantities in detail. Incoherent inelastic nuclear resonance scattering, an equally important topic, is considered in Chapter 7.

6.3.1 Nuclear Resonance Scattering Amplitude

Suppose there is a magnetic hyperfine field (or an electric field) at a Mössbauer nucleus in the scatterer. The amplitude of the γ-ray scattered by this nucleus for a given incident radiation is [36]

$$
\begin{aligned}
f^{N}(\boldsymbol{k}'\boldsymbol{k}_0, \boldsymbol{e}\boldsymbol{e}_0) = {} & \frac{2\pi}{k_0} \frac{\langle \chi | e^{-i\boldsymbol{k}'\cdot \boldsymbol{R}_l} | \chi_0 \rangle \langle \chi_0 | e^{i\boldsymbol{k}_0\cdot \boldsymbol{R}_l} | \chi_0 \rangle}{E_\gamma - E_0(m_e, m_g) + i\Gamma/2} \\
& \times \sum_{L'\lambda'} \sum_{L\lambda} \langle I_g m_g' L' M' | I_e m_e \rangle \langle I_g m_g L M | I_e m_e \rangle \\
& \times \mathbf{e}^* Y_{L'M'}^{(\lambda')}(\boldsymbol{k}') \cdot Y_{LM}^{(\lambda)^*}(\boldsymbol{k}_0) \boldsymbol{e}_0 [\Gamma_\gamma(L'\lambda')\Gamma_\gamma(L\lambda)]^{1/2} \\
& \times \exp[i(\eta L'^{\lambda'} - \eta L^{\lambda})]
\end{aligned}
\tag{6.9}
$$

where $|\chi_0\rangle$ and $|\chi\rangle$ represent the wave functions of initial and final states of the crystal, $\lambda = 1$ and 0 in $(L\lambda)$ represent electric and magnetic multipole radiations, and $\Gamma_\gamma(L\lambda)$ is the γ-radiation linewidth for the 2^L-order multipole radiation (proportional to the emission probability of this radiation). The ratio of the linewidths is equal to the mixing parameter δ^2 of the multipole radiations, i.e., $\delta^2 = \Gamma_\gamma(E2)/\Gamma_\gamma(M1)$. If the invariance of time T is correct, $\eta L'^{\lambda'} - \eta L^{\lambda}$ should be equal to either 0 or π. When the problem depends on angular momentum, it would be more convenient to transform Eq. (6.9) using the circular polarization unit vectors ($\mathbf{e}_{+1}$ and $\mathbf{e}_{-1}$) as defined in Eq. (2.75) and the $\boldsymbol{D}$ function to replace $\mathbf{e}$, $\mathbf{e}_0$, and $Y_{LM}^{(\lambda)}$, respectively, and hence

$$
\begin{aligned}
& \mathbf{e}^* Y_{L'M'}^{(\lambda')}(\boldsymbol{k}') \cdot Y_{LM}^{(\lambda)^*}(\boldsymbol{k}_0) \mathbf{e}_0 \\
& \quad = \frac{1}{8\pi} h^{\lambda'+1} h_0^{\lambda+1} [(2L'+1)(2L+1)]^{1/2} D_{hM'}^{(L')^*}(\boldsymbol{k}'\boldsymbol{z}) D_{h_0 M}^{(L)}(\boldsymbol{k}_0\boldsymbol{z}).
\end{aligned}
\tag{6.10}
$$

The Mössbauer isotopes currently used for diffraction experiments, such as $^{57}Fe(M1)$, $^{119}Sn(E1)$, $^{125}Te(M1)$, $^{141}Pr(M1)$, and $^{183}Ta(M1)$, all have pure dipole transitions, thus $\delta^2 = 0$. For ^{57}Fe, $L' = L = 1$, and $\lambda = \lambda' = 0$. The elastic scattering amplitude per nucleus in the $|m_g\rangle$ state can be then calculated using Eqs. (6.9) and (6.10) [37, 38]:

$$f_{hh_0}^{\mathrm{N}} = hh_0 \sum_M \frac{3}{2} D_{hM}^{(1)^*}(\boldsymbol{k}'\boldsymbol{z}) D_{h_0M}^{(1)}(\boldsymbol{k}_0\boldsymbol{z}) |\langle I_\mathrm{g} m_\mathrm{g} 1M | I_\mathrm{e} m_\mathrm{e} \rangle|^2$$

$$\times \frac{\Gamma_\gamma}{2k_0} \frac{f_{\mathrm{LM}}}{E_\gamma - E_0(m_\mathrm{e}, m_\mathrm{g}) + i\Gamma/2} \tag{6.11}$$

where f_{LM} is discussed in Section 6.3.3, and h_0, $h = \pm 1$ denote right and left circular polarizations. The elements of the rotation matrix $\boldsymbol{D}$ are

$$D_{hM}^{(1)}(\phi, \theta, \psi) = \mathrm{e}^{-ih\phi} d_{hM}^{(1)}(\theta) \mathrm{e}^{-iM\psi} \tag{6.12}$$

where the values of $d_{hM}^{(1)}$ are given in Eq. (6.6). The summation in Eq. (6.11) over M means adding up contributions from all $m_\mathrm{e} \to m_\mathrm{g}$ transitions.

6.3.2
Coherent Elastic Nuclear Scattering

6.3.2.1 Scattering Amplitude

Suppose the solid scatterer is a perfect crystal composed entirely of Mössbauer atoms, the ground state nuclear spin of the Mössbauer isotope is zero, and the scatterer is very thin. In this case, the elastic scattering amplitude (6.11) is also the elastic coherent scattering amplitude.

When the above conditions are not satisfied, the process contains a certain degree of incoherence. As has been mentioned before, there are mainly two reasons for the incoherence. The first reason is due to the presence of different isotopes, even though the crystal is perfect. For example, the random distribution of ^{57}Fe in a perfect bcc Fe metal does not give a fixed phase for the scattered waves. The result in Eq. (6.11) should be multiplied by the Mössbauer isotope abundance a_m to take the isotope incoherence into account. The second reason is due to spin incoherence, because the elastic scattering amplitude depends on the m_g-component of the nuclear ground state spin I_g. If $I_\mathrm{g} \neq 0$, we need to average Eq. (6.11) over different m_g states to obtain the coherent elastic scattering amplitude [4]:

$$f_{\mathrm{coh}}^{\mathrm{N}} = a_m \sum_{m_\mathrm{g}} p_{m_\mathrm{g}} f^{\mathrm{N}}(hh_0) \tag{6.13}$$

where p_{m_g} is the relative occupation of the sublevel m_g. If the nuclear spins in the solid are randomly oriented, $p_{m_\mathrm{g}} = (2I_\mathrm{g} + 1)^{-1}$.

When the magnetic splittings (ΔE) of the energy level are either very large or negligibly small compared to Γ, Eq. (6.13) can be simplified. In the first case ($\Delta E \gg \Gamma$), the amplitude $f_{\mathrm{coh}}^{\mathrm{N}}$ has only one term due to a particular transition between the excited and the ground sublevels. In the second case ($\Delta E \ll \Gamma$), we can sum over m_g, replace all $E_0(m_\mathrm{e}, m_\mathrm{g})$ by E_0, and obtain [36]

$$f_{\rm coh}^{\rm N}(\boldsymbol{k}'\boldsymbol{k}_0, hh_0) = \frac{3}{2}\frac{(2I_{\rm e}+1)a_m}{3(2I_{\rm g}+1)} d_{hh_0}^{(1)}(\boldsymbol{k}'\boldsymbol{k}_0)\frac{\Gamma_\gamma}{2k_0}\frac{f_{\rm LM}}{E_\gamma - E_0 + i\Gamma/2} \tag{6.14}$$

where the values of $d_{hh_0}^{(1)}$ are given in Eq. (6.6).

6.3.2.2 Nuclear Bragg Scattering (NBS)

When $\boldsymbol{k}_0 - \boldsymbol{k}' = \boldsymbol{\tau}$, coherent elastic nuclear scattering becomes Bragg scattering. If the incident radiation is unpolarized, the differential cross-section of the scattered radiation by a unit cell is

$$\frac{{\rm d}\sigma}{{\rm d}\Omega} = \frac{(2\pi)^3}{V_{\rm a}}\overline{|F_r|^2}\delta(\boldsymbol{k}_0 - \boldsymbol{k}' - \boldsymbol{\tau}) \tag{6.15}$$

where $V_{\rm a}$ is the unit cell volume, $\boldsymbol{\tau}$ is a reciprocal lattice vector, and

$$F_r = \sum_l f_{\rm coh}^{\rm N}(\boldsymbol{k}'\boldsymbol{k}_0, hh_0)\exp[i(\boldsymbol{k}_0 - \boldsymbol{k}')\cdot \boldsymbol{R}_l]. \tag{6.16}$$

The summation in (6.16) is for all Mössbauer nuclei, and $\overline{|F_r|^2}$ represents the average over photon polarization. Based on Eq. (6.8), we obtain

$$\overline{|F_r|^2} = |F_r|^2\frac{1+\cos^2\theta}{2}. \tag{6.17}$$

6.3.2.3 Nuclear Forward Scattering (NFS)

In forward scattering $\boldsymbol{k}' = \boldsymbol{k}_0$, and, according to (6.6), $d_{hh_0}^{(1)}(0) = \delta_{hh_0}$. The four amplitudes for forward scattering are given by Eq. (6.14):

$$\begin{aligned} &f_{\rm coh}^{\rm N}(0,11) = f_{\rm coh}^{\rm N}(0,-1-1) \neq 0,\\ &f_{\rm coh}^{\rm N}(0,1-1) = f_{\rm coh}^{\rm N}(0,-11) = 0. \end{aligned} \tag{6.18}$$

These indicate that the circular polarization of incident radiation is not changed after forward scattering. This is true even when the condition $\Delta E \ll \Gamma$ is not satisfied [39]. The forward coherent scattering amplitude calculated according to Eq. (6.14) is

$$f_{\rm coh}^{\rm N}(0, hh_0) = \frac{k_0 a_m}{8\pi} d_{hh_0}^{(1)}(0)\sigma_0 f_{\rm LM}\frac{\Gamma}{E_\gamma - E_0 + i\Gamma/2}. \tag{6.19}$$

Forward scattering experiments using isotopic Mössbauer sources are very difficult, and they should be performed using synchrotron Mössbauer radiation.

6.3.2.4 Scattering Cross-Sections

If we integrate the square of Eq. (6.14) and sum over the h-values, we obtain the coherent elastic scattering cross-section for that nucleus:

$$\sigma_{\text{coh}} = \frac{2\pi a_m^2}{3}\left(\frac{2I_\text{e}+1}{2I_\text{g}+1}\right)^2 \left|\frac{\Gamma_\gamma f_{\text{LM}}}{2k_0(E_\gamma - E_0 + i\Gamma/2)}\right|^2. \tag{6.20}$$

Furthermore, according to an optical theorem, we can calculate the total cross-section $\sigma_\text{T} = (-4\pi/k_0)\,\text{Im}[f(0)]$ by using Eq. (6.19):

$$\sigma_\text{T} = 2\pi a_m \frac{2I_\text{e}+1}{2I_\text{g}+1}\frac{\Gamma_\gamma\Gamma}{4k_0^2}\frac{f_{\text{LM}}}{(E_\gamma - E_0)^2 + \Gamma^2/4}. \tag{6.21}$$

For comparing cross-section values, the numerical results for α-Fe at resonance are calculated according to Eqs. (6.20) and (6.21):

$$\sigma_{\text{coh}} \approx 8a_m{}^2 \times 10^3 b,$$

$$\sigma_\text{T} \approx 4a_m \times 10^6 b$$

where $\Gamma = \Gamma_\gamma(1+\alpha)$, $f_{\text{LM}}(T = 300\text{ K}) = 0.67$, and $\alpha = 8.20$ were used. It is obvious that the coherent scattering cross-section is smaller than the total cross-section by 2 or 3 orders of magnitude.

6.3.3
Lamb–Mössbauer Factor and Debye–Waller Factor

As we know, in a scattering process the Lamb–Mössbauer factor f_{LM} very closely resembles the Debye–Waller factor f_D. In fact, they originate from the same expression under the condition of either the "slow" or the "fast" scattering process. This expression is $\langle\chi|e^{-i\boldsymbol{k}'\cdot\boldsymbol{R}_l}|\chi_0\rangle\langle\chi_0|e^{i\boldsymbol{k}_0\cdot\boldsymbol{R}_l}|\chi_0\rangle$ as in Eq. (6.9). In order to illustrate the differences between the two factors f_{LM} and f_D, it is better to transform the resonance elastic scattering amplitude f^N into a time-dependent representation [40]:

$$f^\text{N} = \frac{\Gamma_\gamma}{2ik}\int_0^\infty \text{d}t e^{i(\omega-\omega_0)t} e^{-(\Gamma/2\hbar)t}\langle e^{-i\boldsymbol{k}'\cdot\boldsymbol{R}_l(t)} e^{i\boldsymbol{k}_0\cdot\boldsymbol{R}_l(0)}\rangle. \tag{6.22}$$

Due to the long lifetime of its excited state, Mössbauer resonance scattering is a "slow" process. Typical scattering times are $\hbar/\Gamma \approx 10^{-9}$–$10^{-6}$ s $\gg \omega_\text{m}{}^{-1}$, where ω_m is the maximum vibration frequency of the nucleus about its equilibrium position and is of the order of 10^{13} rad s^{-1}. Therefore, for all practical purposes $t \to \infty$ and the correlation positions $\boldsymbol{R}_l(\text{t} \to \infty)$ and $\boldsymbol{R}_l(\text{t} = 0)$ can be neglected. This permits one to carry out the thermal average of the γ-ray absorption and re-emission processes separately. As a result, the Lamb–Mössbauer factor can be reduced to

$$\begin{aligned} f_{\text{LM}} &= \langle\exp[-i\boldsymbol{k}'\cdot\boldsymbol{R}_l(t\to\infty)]\exp[i\boldsymbol{k}_0\cdot\boldsymbol{R}_l(t=0)]\rangle \\ &= \exp\left\langle -\frac{1}{2}(\boldsymbol{k}'\cdot\boldsymbol{R}_l)^2\right\rangle \exp\left\langle\frac{1}{2}(\boldsymbol{k}_0\cdot\boldsymbol{R}_l)^2\right\rangle = \sqrt{f(\boldsymbol{k}')f(\boldsymbol{k}_0)} \end{aligned} \tag{6.23}$$

where $f(\boldsymbol{k}_0)$ and $f(\boldsymbol{k}')$ are the recoilless fractions of γ-ray absorption and emission in the directions $\boldsymbol{k}_0$ and $\boldsymbol{k}'$, respectively. In nuclear resonance scattering the above expression is usually called the Lamb–Mössbauer factor. If the motion of scattering nuclei is isotropic, then

$$f_{\mathrm{LM}} = [f(\boldsymbol{k}')]^2 = [f(\boldsymbol{k}_0)]^2 = f. \tag{6.24}$$

On the other hand, for non-resonance Rayleigh scattering and x-ray scattering, the characteristic scattering times are $\hbar/\Gamma \approx 10^{-16}$–$10^{-15}$ s and hence these processes are fast compared to $\omega_{\mathrm{m}}{}^{-1}$. In this case, we have effectively $t \approx 0$, which gives the Debye–Waller factor:

$$\begin{aligned} f_{\mathrm{D}} &= \langle \exp[-i\boldsymbol{k}' \cdot \boldsymbol{R}_l(t \approx 0)] \exp[i\boldsymbol{k}_0 \cdot \boldsymbol{R}_l(t = 0)] \rangle \\ &\approx \langle \exp[-i(\boldsymbol{k}' - \boldsymbol{k}_0) \cdot \boldsymbol{R}_l] \rangle. \end{aligned} \tag{6.25}$$

Therefore, the two factors are not identical but provide the same information concerning the dynamics of a given crystal, because they contain the following common factor due to the same thermal averaging:

$$\int \frac{1}{\omega} \coth\left(\frac{1}{2}\hbar\omega\beta\right) g(\omega)\, \mathrm{d}\omega.$$

For a harmonic lattice, the Lamb–Mössbauer factor f_{LM} in forward scattering or in transmission geometry is simply related to the Debye–Waller factor f_{D} by

$$\ln f_{\mathrm{LM}} = \frac{k_0^2}{Q^2} \ln f_{\mathrm{D}} \tag{6.26}$$

where vectors $\boldsymbol{k}_0$ and $\boldsymbol{Q}$ are in the same direction.

6.4 Rayleigh Scattering of Mössbauer Radiation (RSMR)

6.4.1 Basic Properties of RSMR

Rayleigh scattering is an elastic scattering of electromagnetic waves, and it occurs widely in light, x-ray and γ-ray scatterings by electron shells of atoms in a variety of condensed matter materials. Its scattering cross-section, proportional $1/\lambda^4$, is considered as a characteristic feature. In RSMR experiments, the scattered γ-ray is analyzed by a resonant absorber inserted between the scatterer and the detector. In other words, RSMR can detect an energy change comparable to the typical width of a Mössbauer line ($\sim 10^{-9}$ eV for ^{57}Fe). So, RSMR may be regarded as x-ray spectroscopy with high energy resolution. Therefore, the elastic and inelastic scatterings can be separated with a high precision. As a result, this method has been used widely for studying liquids as well as solids, not containing the Mössbauer isotope. As discussed above, RSMR is not only elastic, but also coherent.

There have been published a special monograph [41] and several review articles [6, 42–45] on this new methodology. A disadvantage of this technique is its very low count rate when a radioactive Mössbauer source is used, but this has been recently overcome by using high-intensity synchrotron radiation sources.

The general theory of Rayleigh scattering is essentially the same as that of conventional x-ray scattering. The scattering differential cross-section is written in a form similar to Eq. (4.120):

$$I(\boldsymbol{Q},\omega) = \frac{\mathrm{d}^2\sigma}{\mathrm{d}\Omega\,\mathrm{d}E} = N|F(\boldsymbol{Q})|^2 S(\boldsymbol{Q},\omega) \tag{6.27}$$

where N is the number of atoms in the crystal, the factor $F(\boldsymbol{Q})$ is given by the Thomas–Rayleigh formula, and $S(\boldsymbol{Q},\omega)$ is the scattering function. We are mostly interested in $S(\boldsymbol{Q},\omega)$, which describes the dynamic properties of the scatterer, as has been introduced earlier in Eq. (4.121). Unlike neutron scattering, the energy transferred in x-ray (and γ-ray) scattering (i.e., the phonon energy) is much smaller than the incident energy. Therefore, $S(\boldsymbol{Q},\omega)$ is difficult to measure, and one usually obtains the total integrated function instead:

$$S(\boldsymbol{Q}) = \int_{-\infty}^{\infty} S(\boldsymbol{Q},\omega)\,\mathrm{d}\omega. \tag{6.28}$$

Accordingly, integrating Eq. (6.27) gives

$$\frac{\mathrm{d}\sigma}{\mathrm{d}\Omega} = N|F(\boldsymbol{Q})|^2 S(\boldsymbol{Q}). \tag{6.29}$$

To find an explicit expression for $S(\boldsymbol{Q})$, we substitute (4.124) into (6.28) and get

$$S(\boldsymbol{Q}) = \frac{1}{N}\sum_{ll'} \langle \mathrm{e}^{-i\boldsymbol{Q}\cdot\boldsymbol{u}(l,0)}\mathrm{e}^{i\boldsymbol{Q}\cdot\boldsymbol{u}(l',0)}\rangle \mathrm{e}^{i\boldsymbol{Q}\cdot(l'-l)}. \tag{6.30}$$

If we take the harmonic approximation again by expanding Eq. (6.30) according to Eq. (4.128):

$$S(\boldsymbol{Q}) = S_0(\boldsymbol{Q}) + S_1(\boldsymbol{Q}) + S_2(\boldsymbol{Q}) + \cdots \tag{6.31}$$

where the first term corresponds to elastic scattering, $S_{\mathrm{el}}(\boldsymbol{Q}) = S_0(\boldsymbol{Q})$, while the other terms involve one or more phonons and correspond to inelastic scattering processes, $S_{\mathrm{in}}(\boldsymbol{Q}) = S_1(\boldsymbol{Q}) + S_2(\boldsymbol{Q}) + \cdots$

The scattering function $S_{el}(\mathbf{Q})$ can be derived directly from (6.30) for the zero-phonon process:

$$S_{el}(\mathbf{Q}) = \frac{e^{-2W}}{N} \sum_{ll'} e^{i\mathbf{Q}\cdot(l'-l)}. \tag{6.32}$$

For a single crystal, $S_{el}(\mathbf{Q})$ is nonzero only when $\mathbf{Q} = \boldsymbol{\tau}$, which predicts a diffraction peak at the Bragg angle.

Inelastic scattering is basically the familiar thermal diffuse scattering (TDS) caused by lattice vibrations. Because TDS overlaps with the Bragg diffraction peak and is not easy to separate, it is difficult to extract the lattice dynamics information contained in it. Fortunately, RSMR provides a means to separate inelastically scattered radiation from elastically scattered radiation, and this has rekindled research interest in TDS.

It can be shown that the scattering function $S_{in}(\mathbf{Q})$ takes the following form:

$$S_{in}(\mathbf{Q}) = 1 - e^{-2W} + \frac{1}{N} e^{-2W} \sum_{l \neq l'} [e^{Q^2 \langle \mathbf{u}(l)\cdot\mathbf{u}(l')\rangle} - 1] e^{i\mathbf{Q}\cdot(l-l')}. \tag{6.33}$$

For a perfect crystal, we may use the general expression for $\mathbf{u}(l)$ to show how photons are scattered by phonons. If we consider only the single-phonon process:

$$S_1(\mathbf{Q}) = \frac{1}{N} e^{-2W} \sum_{l \neq l'} \langle \mathbf{Q}\cdot\mathbf{u}(l)\mathbf{Q}\cdot\mathbf{u}(l')\rangle e^{-i\mathbf{Q}\cdot(l'-l)} \tag{6.34}$$

and after substituting for $\mathbf{u}(l)$, it becomes [41]

$$S_1(\mathbf{Q}) = \frac{1}{2M} e^{-2W} \sum_j [\mathbf{Q}\cdot\mathbf{e}(\mathbf{k}j)]^2 \frac{\langle 2n_j(\mathbf{k}) + 1\rangle}{\omega_j(\mathbf{k})}. \tag{6.35}$$

Now we will be able to calculate the scattered γ-ray intensity in the single-phonon process. If we first assume $T = 0$, then according to Eqs. (6.27) and (6.35) we obtain

$$I_{in}^1 = N|F(\mathbf{Q})|^2 \frac{1}{2M} e^{-2W} \sum_j [\mathbf{Q}\cdot\mathbf{e}(\mathbf{k}j)]^2 \frac{1}{\omega_j(\mathbf{k})}. \tag{6.36}$$

When the temperature is raised such that $k_B T \gg \hbar\omega_j(\mathbf{k})$, which means $\langle n_j(\mathbf{k})\rangle + 1/2 \approx k_B T/\hbar\omega_j(\mathbf{k})$, then

$$I_{in}^1 = N|F(\mathbf{Q})|^2 e^{-2W} \frac{k_B T}{\hbar M} \sum_j [\mathbf{Q}\cdot\mathbf{e}(\mathbf{k}j)]^2 \frac{1}{\omega_j^2(\mathbf{k})}. \tag{6.37}$$

The above two intensity expressions describe TDS in the reciprocal space, and both intensity distributions are relatively flat. But for small k-values and when $\boldsymbol{Q}$ falls near a reciprocal lattice vector $\boldsymbol{\tau}$, TDS increases drastically. This indicates that the TDS contribution originates mainly from the phonons in the acoustic branch ($\omega \propto k$); thus the intensity is proportional to $1/k$. An important conclusion is that the maxima of TDS occur at the reciprocal lattice points, which correspond to the Bragg diffraction positions. In most cases, the relation between TDS and temperature can be expressed as

$$I_{\text{in}}^{1} \approx T\mathrm{e}^{-2W}. \tag{6.38}$$

Regarding deriving the Debye–Waller factor $f_{\mathrm{D}} = \mathrm{e}^{-2W}$ from TDS, it is exactly the same as that in neutron scattering. The γ-photon energy change is usually less than 10^{-3} eV in TDS. The energy resolution of ordinary x-ray spectroscopy is very poor, not enough to separate the small TDS from elastic scattering, both of which contribute to the same Bragg peak. Consequently, using Eq. (6.32) to analyze x-ray diffraction intensity will not give an accurate result for the Debye–Waller factor e^{-2W}, and for the subsequent determination of the mean-square displacement $\langle u^2 \rangle$ and the Debye temperature θ_{D}. Furthermore, because of the incomplete separation, the phonon information contained in TDS is also very difficult to extract. But these two scattering components can be separated by the RSMR method.

6.4.2 Separation of Elastic and Inelastic Scatterings

An experimental method for separating the Bragg peak and TDS was developed in 1963 [46], and its main principle is indicated in Fig. 6.1(a) where the analyzer A is at rest. The source velocity is adjusted to v_{r} so that after the incident γ-ray is scattered from the single crystal it may be resonantly absorbed by the analyzer A. If the γ-ray is inelastically scattered and loses or gains an amount of energy larger than 10^{-9} eV (for ^{57}Fe), the energy will not be resonantly absorbed, but transmitted through the analyzer A. Therefore, this method cleanly separates the elastic and inelastic scattering components. Also, the probability for a non-resonant incident γ-photon having the resonance energy after inelastic scattering is extremely small.

First, let $I_{\infty}(\theta)$ and $I_{\mathrm{r}}(\theta)$ be the counts of scattered photons at angle θ for off- and on-resonance between the source and analyzer, respectively. The scattering angle θ is related to the Bragg angle θ_{B} by $\theta = 2\theta_{\mathrm{B}}$. The following ratio is proportional to the fraction of recoilless γ-rays which are scattered elastically by the crystal:

$$\xi = \frac{I_{\infty}(\theta) - I_{\mathrm{r}}(\theta)}{I_{\infty}(\theta)}. \tag{6.39}$$

Next, the scatterer is removed, and the analyzer A together with the detector is rotated counterclockwise by an angle θ. The counts of transmitted photons are then recorded as $I(\infty)$ and $I(v_r)$, from which the parameter $\varepsilon(v_r)$ is evaluated according to Eq. (1.24):

$$\varepsilon(v_r) = \frac{I(\infty) - I(v_r)}{I(\infty)}.$$

It is easy to see that the elastically scattered fraction of incident radiation is $\xi/\varepsilon(v_r)$, and the inelastically scattered fraction is $1 - \xi/\varepsilon(v_r)$. Therefore, the respective scattered intensities are

$$I_{el} = I_\infty(\theta)\frac{\xi}{\varepsilon(v_r)} = \frac{I_\infty(\theta) - I_r(\theta)}{\varepsilon(v_r)} \tag{6.40}$$

$$I_{in} = I_\infty(\theta)\left[1 - \frac{\xi}{\varepsilon(v_r)}\right] = I_\infty(\theta) - I_{el}. \tag{6.41}$$

When using these two expressions, one must pay attention to the following two points. (1) The accurate measurement of $\varepsilon(v_r)$ may be complicated by background counts. A 122 keV photon from a ^{57}Co source after Compton scattering may cause background counts. Although background in the numerator of the $\varepsilon(v_r)$ expression is canceled, the one in the denominator remains. It has been pointed out [49] that $\varepsilon(v_r) = 0.39$ and 0.47 for without and with background correction, respectively. Therefore, background correction must be included or the resultant I_{el} value would be overestimated. (2) Even when the background in a scattering experiment is very small, it may still be significant when the inelastic portion is to be evaluated, because the fraction of inelastic Rayleigh scattering is also very small.

In order to circumvent the difficulties of measuring background, an approach using four measurements was developed [41, 47], and it is briefly described below.

When the source velocity is large so that no resonant absorption takes place in the analyzer A placed between the scatterer and the detector, the photon count is

$$I_\infty(\theta) = I(0)Pe^{-\mu_a d} + I_b \tag{6.42}$$

where μ_a is the atomic mass absorption coefficient, d is the thickness of analyzer A, I_b is the background count, $I(0)$ is the total number of photons from the source due to all Mössbauer transitions, and

$$P = \frac{I_\infty(\theta)}{I(\infty)} = \frac{I_{el} + I_{in}}{I(\infty)} = P_{el} + P_{in} \tag{6.43}$$

is the total scattering probability of the incident γ-rays being scattered in the θ direction.

The second measurement is carried out after the source velocity is adjusted to a resonant energy, and the photon count is

$$I_r(\theta) = I(0) f_s P_{el} e^{-(\mu_a+\mu_r)d} + I(0) f_s P_{in} e^{-\mu_a d} + I(0)(1 - f_s) P e^{-\mu_a d} + I_b \quad (6.44)$$

where the first and second terms correspond to the intensities of elastic and inelastic scatterings of the recoilless radiation, and the third term is the scattered intensity of the non-recoilless radiation. Comparing this with Eq. (1.17), we see that the energy distribution of the incident radiation has not been taken into consideration.

We then move the analyzer so that it is now between the source and the scatterer, and carry out the third and fourth measurements, obtaining

$$I'_\infty(\theta) = I(0) e^{-\mu_a d} P + I'_b \quad (6.45)$$

and

$$I'_r(\theta) = I(0) f_s e^{-(\mu_a+\mu_r)d} P + I(0)(1 - f_s) e^{-\mu_a d} P + I'_b. \quad (6.46)$$

To facilitate the understanding of each term, we have written the factors in each of the terms in Eqs. (6.42), (6.44), (6.45), and (6.46) in the same order as the sequence of events in each process. For example, $I(0)e^{-\mu_a d}P$ indicates that the radiation is first absorbed then scattered, while $I(0)Pe^{-\mu_a d}$ indicates the reverse order. Also, the detector is assumed to have a 100% efficiency.

According to the results of these four measurements

$$\Delta I = I_\infty - I_r = I(0) P_{el} f_s e^{-\mu_a d}(1 - e^{-\mu_r d}), \quad (6.47)$$

$$\Delta I' = I'_\infty - I'_r = I(0) P f_s e^{-\mu_a d}(1 - e^{-\mu_r d}). \quad (6.48)$$

Therefore, the final results are

$$\frac{I_{el}}{I_{el} + I_{in}} = \frac{P_{el}}{P} = \frac{\Delta I}{\Delta I'}, \quad (6.49)$$

$$\frac{I_{in}}{I_{el} + I_{in}} = 1 - \frac{\Delta I}{\Delta I'}. \quad (6.50)$$

Because measuring background is not required in this method and parameters such as μ_a, μ_r, f, and $I(0)$ do not appear in (6.49) and (6.50), the separation of elastic and inelastic scatterings is much more accurate. Strictly speaking, if the energy distribution of the incident radiation is considered, I_{el} should be proportional to the area under the Mössbauer spectrum. If the scattering spectrum does not change its shape, then the above separation method is still valid except that the intensity should be understood as the integrated intensity.

The experimental procedure of the four-measurement method is very tedious. A well-planned two-measurement experiment has been reported [48], and a separation accuracy was achieved of better than 1% with ^{57}Co/Rh as the source and ^{57}Fe/Rh as the analyzer. The source and the scatterer could be rotated in two mutually perpendicular planes, with accurate goniometers. Shielding was by multilayers of lead, brass, and aluminum, and the collimator was made of pure aluminum (99.999%). The background count rate was reduced to 0.02 per second. The output of a high energy-resolution Si(Li) detector was processed by two multichannel analyzers, one in multichannel scaling (MCS) mode and the other in pulse-height analysis (PHA) mode, the latter used for background determination. The distance between source and scatterer was 15.5 cm, and that between scatterer and detector was 9.5 cm. The divergence of the scattered beam was 0.65° in the horizontal direction and 1.4° in the vertical direction. Figure 6.14 shows the Rayleigh scattering results from the single-crystal Si(4 0 0) plane, and Table 6.1 lists the elastic fractions and their accuracies for four different scattering angles.

From the above example, we see that the Mössbauer effect can be used to isolate the elastic scattering, accurately measure the Bragg peak intensity, and therefore provide more reliable values of the Debye–Waller factor, $\langle u^2 \rangle$, and the Debye temperature θ_D, all very useful parameters in structural analysis and lattice dynamics studies.

Now the inelastic portion of the scattering has been separated from the total scattering. Although the phonon spectrum still cannot be deduced, other dy-

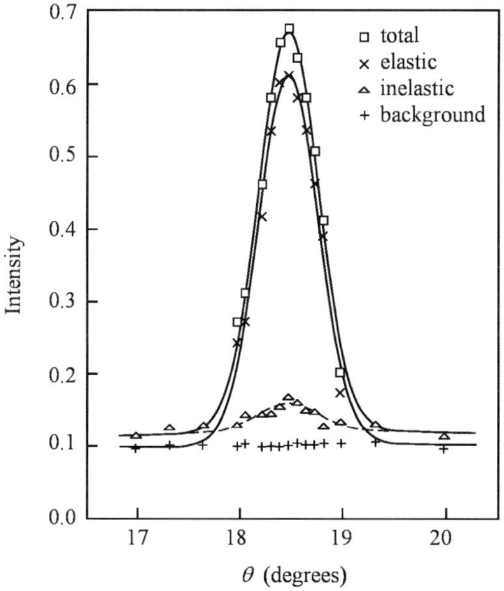

Fig. 6.14 RSMR intensity as a function of scattering angle near the Bragg angle for reflection from the single-crystal Si (4 0 0) plane.

Table 6.1 Elastically scattered fraction $\xi/\varepsilon(\nu_r)$ near the (4 0 0) Bragg reflection.

θ (°)	$\xi/\varepsilon(\nu_r)$
17.97	0.79(1)
18.30	0.898(9)
18.48	0.888(9)
18.64	0.916(8)

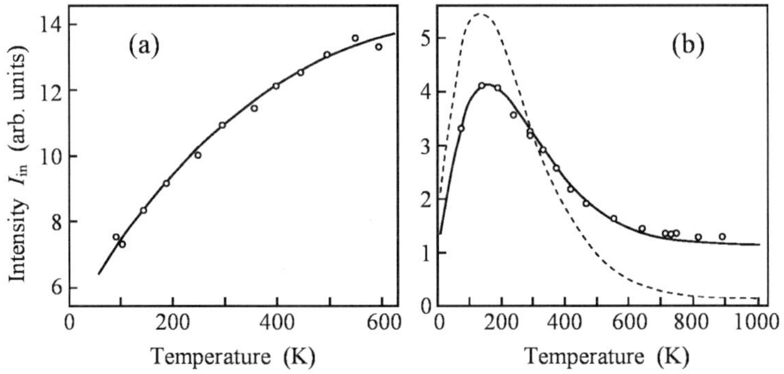

Fig. 6.15 Inelastic scattering intensity I_{in} as a function of temperature for γ-rays scattered from (a) the (4 4 4) plane of single-crystal Si and (b) the (10 0 0) plane of single-crystal KCl. In (b), assuming only the single-phonon process produced the dashed line, and multiple-phonon processes were required to yield a satisfactory fit (solid curve).

namic information can be extracted. Figure 6.15 shows inelastic scattering intensity I_{in} as a function of temperature for single-crystal Si and KCl [49, 50]. For the Si(4 4 4) reflection, the line in Fig. 6.15(a) is the calculated result using the single-phonon approximation, which agrees with the experiment. However, for the KCl(10 0 0) reflection, the single-phonon approximation produced a poor agreement as shown by the dashed line in Fig. 6.15(b), and only when multiple-phonon processes were included did the calculated result (solid line) yield a good fit. Figure 6.16 shows similar results, but the inelastic scattering intensity is plotted against the scan angle near θ_B. The solid lines in Fig. 6.16 were calculated using the single-phonon approximation. For the Si(8 0 0) reflection, when the scan angle is larger than 2.5°, the discrepancy grows, indicating that the single-phonon process is mainly concentrated near θ_B.

Other RSMR experimental methods have also been developed, e.g., using the transmission Laue method instead of Bragg scattering, and using amorphous solids or viscous liquids instead of single crystals or polycrystals. Rayleigh scattering of Mössbauer radiation has been recognized as a valuable method, especially

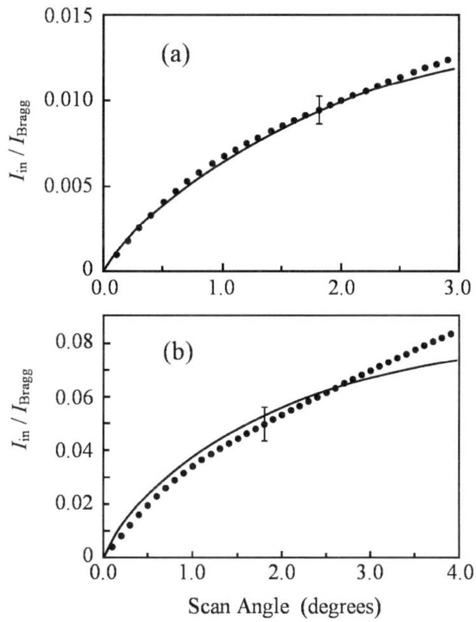

Fig. 6.16 Dependence of the ratio of the inelastic scattering intensity I_{in} to the Bragg intensity I_{Bragg} on the scan angle for reflections from (a) the (4 0 0) plane and (b) the (8 0 0) plane of single-crystal Si [48].

in surface sciences and for studying the structures and dynamics of large biological molecules.

6.4.3
Measuring Dynamic Parameters Using RSMR

For Rayleigh scattering, the Debye–Waller factor is

$$f_D = e^{-2W} = e^{-Q^2\langle u^2\rangle}.$$

For a Bragg reflection ($\mathbf{Q} = \mathbf{k}' - \mathbf{k}_0 = \boldsymbol{\tau}$), we have

$$W = B\left(\frac{\sin\theta_B}{\lambda}\right)^2 \tag{6.51}$$

where $B = 16\pi^2\langle u^2\rangle$ and $\lambda = 0.8602$ Å (for $E_\gamma = 14.4$ keV).

The following two experimental approaches can be used to obtain dynamic parameters. In the first approach, at a fixed temperature, W is measured for different values of $\sin\theta_B/\lambda$, which will give $\langle u^2\rangle$ and thus the Debye temperature θ_D. In the second approach, the scatterer temperature is changed and $\langle u^2\rangle$ is obtained as a function of temperature, which will also give θ_D.

6.4.3.1 The Fixed Temperature Approach

The incident γ-ray beam is not completely parallel but has a certain divergence. Also, the scatterer is not an ideal crystal, but may contain some defects. These imperfections cause the diffracted rays to be not completely concentrated at the Bragg angle θ_B, but having an angular distribution. Therefore, to calculate the actual diffraction intensity, we need to integrate Eq. (6.32) over a region within $\theta_B \pm \Delta\theta_B$. The integrated diffraction intensity is proportional to the elastic scattering intensity recorded by the detector within a specific time period, and it is given by [51]

$$I_{el}(\theta_B) = C|F(\theta_B)|^2 e^{-2W} \frac{1+\cos^2 2\theta_B}{\sin 2\theta_B} \tag{6.52}$$

where C is a constant. With the definition

$$\frac{I_{el}(\theta_B)\sin 2\theta_B}{|F(\theta_B)|^2(1+\cos^2 2\theta_B)} \equiv E(\theta_B)$$

and combining Eqs. (6.51) and (6.52), we get

$$\ln E(\theta_B) = \ln(Ce^{-2W}) = \ln C - 2B\left(\frac{\sin\theta_B}{\lambda}\right)^2. \tag{6.53}$$

The scattering form factor $F(\theta_B)$ has been tabulated and is readily available. From the experimental I_{in} data, the quantity $E(\theta_B)$ can be calculated, and when its logarithm is graphed against $(\sin\theta_B/\lambda)^2$, a linear relation is expected and the slope is just $-2B$.

This method has been used to study the Rayleigh scatterings from single-

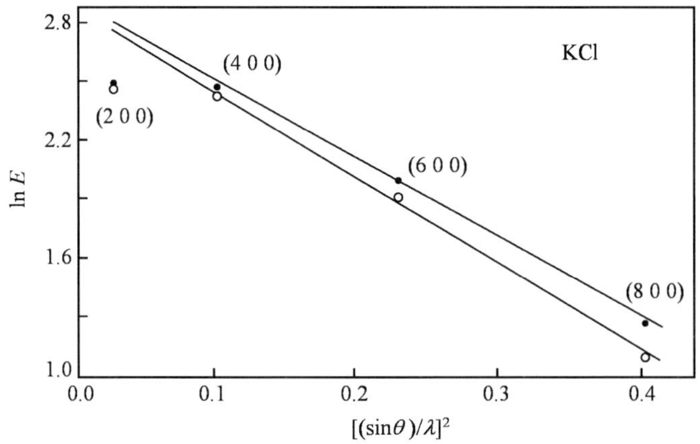

Fig. 6.17 Total scattering intensity (filled circles) and elastic scattering intensity (open circles) as functions of scattering angle for single-crystal KCl.

Table 6.2 Results of Debye temperature θ_D (K) for Al and KCl at room temperature. Under the RSMR method, values in column a were obtained from the total scattered intensities while those in b were obtained from the elastically scattered intensities.

	X-ray diffraction method [52]	RSMR method [51]		Neutron scattering method [53]
		a	b	
Al	390 ± 10	400 ± 14	387 ± 14	386 ± 10
	410 ± 9			
	397			
	389 ± 2			
KCl	240	213 ± 5	202 ± 5	–

crystal Al(1 1 1) and KCl(2 0 0) reflections (Fig. 6.17) [51]. The Debye temperatures θ_D derived from these experiments are listed in Table 6.2, where in column a under the RSMR method are the θ_D-values calculated from total scattered intensities and in column b are the θ_D-values from the elastically scattered intensities. One important observation is that the θ_D results from the elastically scattered intensities are always somewhat lower than the corresponding results from the total scattered intensities (which include inelastic scattering). The Debye temperature for aluminum has been measured by several authors using x-ray diffraction, and their results are slightly higher than the RSMR result. The x-ray data contained relatively large errors because of small number of experimental data points.

At each fixed temperature, a B-value can be measured, which leads to $\langle u^2 \rangle$ and f_D. Therefore, this approach allows absolute measurements of these parameters, and is a very useful method.

6.4.3.2 **The Variable Temperature Approach**

The scattering angle is now fixed at the Bragg angle θ_B. The elastically scattered intensity is measured as a function of temperature, and an experimental $f_D(T)$ curve is obtained. Using the explicit expression for $\langle u^2 \rangle$ based on the Debye model, Eq. (6.51) can be used to fit the experimental data. In this approach, more data points may be measured to reduce experimental uncertainty. Debye temperature values have been obtained using this approach of the RSMR method from the Al(1 1 1) and KCl(4 0 0) reflections as well as from a Ni crystal. The θ_D-values for Al and KCl using this approach are 387 and 202 K, respectively. Table 6.3 lists the θ_D-value for Ni from this method, along with the results from other methods. Note that the Ni θ_D-value from the RSMR method is smaller than that from the x-ray diffraction method.

Unfortunately, it is very difficult to obtain the phonon spectrum from the isolated inelastic portion of Rayleigh scattering. The Mössbauer effect does not

Table 6.3 Results of Debye temperature θ_D (K) for Ni at room temperature.

	X-ray diffraction method [54]	Specific heat method [54]	RSMR method [54]	Mössbauer absorption method [55]
Ni	417	441	406	413–437

seem to have any advantage, and there have been no reports of measurements of phonon spectra using RSMR.

6.4.4
RSMR and Anharmonic Effect

6.4.4.1 Using Strong Mössbauer Isotope Sources

In principle, RSMR is a very accurate method for studying the Debye–Waller factor f_D, but its application has been limited because the Mössbauer isotope sources are not strong enough. For the most common ^{57}Co source, its activity is seldom higher than 9.25×10^9 Bq (250 mCi) due to reasons such as self-absorption. Even when such a strong source is used, it would still take several months to complete the measurements for a sample [48]. There had been no reports of drastic improvement of accuracy in measuring f_D before the 1980s, when a strong ^{183}Ta source was successfully produced ($\sim 2.6 \times 10^{12}$ Bq) by placing a thin ^{181}Ta foil irradiated under a flux of 4×10^{14} neutrons cm^{-2} s^{-1} for a week [56]. A ^{183}Ta source has $E_\gamma = 46.48$ keV and a half-life of 5.1 days. Although ^{183}Ta has a short lifetime and a small maximum resonance cross-section σ_0, the estimated Mössbauer intensity of the 2.6×10^{12} Bq ^{183}Ta source is higher than a 3.7×10^9 Bq ^{57}Co source by a factor of 500. In addition, the self-absorption in the ^{183}Ta source is negligibly small. The natural linewidth of the 46.48 keV Mössbauer radiation is 2.5×10^{-6} eV, which is about four orders of magnitude smaller than the typical phonon energy and is suitable for separating the elastic and inelastic components in RSMR. Using such a strong source, the absorption spectrum and recoilless fraction f of metallic tungsten have been extensively studied [57], and the f-values are listed in Table 6.4. It can be seen that the accuracy of the f-values is better than 1%, a significant improvement over previous results. Fitting the experimental data using the Debye model gave $\theta_D = 336.5$ K. Also, the internal conversion coefficient was determined to be $\alpha = 8.76$. The first Mössbauer study of metallic W using the 46.48 keV radiation was in 1962 [58], in which $\theta_D = 320^{+70}_{-40}$ K and $\alpha = 11.0$ were deduced. The internal conversion coefficient in that study was obviously too high.

A special instrument known as QUEGS (quasi-elastic gamma-ray scattering) was designed [56] to be used for scattering experiments including RSMR. In addition to using a strong Mössbauer source, this instrument also has better specifications, such as angular resolution $\Delta\theta$ improved from 2.4° to 0.08° and mo-

Table 6.4 The f-values of metallic W at various temperatures (not corrected for thermal expansion).

Temperature (K)	$f(T)$
80	0.634(2)
297	0.299(1)
373	0.229(1)
469	0.155(1)
572	0.104(1)
621	0.0847(3)
663	0.0696(4)
770	0.0460(2)
869	0.0298(2)

mentum resolution from 0.340 to 0.011 Å^{-1}. Also, between the source and the scatterer is inserted a LiF(2 0 0) monochromator, which elastically scatters the 46.48 keV Mössbauer radiation with an almost 100% efficiency while greatly reducing other radiations.

6.4.4.2 Using Higher Temperature Measurements

The anharmonic effect is usually studied by accurate measurements of the Debye–Waller factor, as discussed in Chapter 5. However, this effect becomes significant only when higher order Bragg reflections are measured at high enough temperatures. Under such experimental conditions, the thermal diffuse scattering component is also very large. Therefore, neutron or x-ray scattering experiments give poor results for the anharmonic effect. The ability to separate different scattering components in RSMR therefore becomes very advantageous [59, 60].

When the anharmonic effect exists, f_D is the same as Eq. (5.16), except for scattering vector $\boldsymbol{Q}$ replacing the wave vector $\boldsymbol{k}$:

$$e^{-2W} = \exp\left\{-Q^2\langle u_Q^2\rangle - \frac{Q^4}{12}[3\langle u_Q^2\rangle^2 - \langle u_Q^4\rangle] + O(Q^6)\right\} \tag{6.54}$$

where $\langle u_Q^2\rangle$ is the atomic mean-square displacement in the $\boldsymbol{Q}$ direction and the quantity in brackets is known as the non-Gaussian term. The above expression has been discussed in detail in the literature [61, 62]. For cubic crystals at high temperatures ($T > \theta_D$), we have [63]

$$\langle u_Q^2\rangle = \frac{3\hbar^2}{Mk_B\theta_D^2}T + \gamma_2 T^2 + \gamma_3 T^3 \tag{6.55}$$

and

$$3\langle u_Q^2\rangle^2 - \langle u_Q^4\rangle = \gamma_4 T^3. \tag{6.56}$$

Because $V_3 \neq 0$ and $V_4 \neq 0$ in Eq. (4.12), γ_2 and γ_3 represent contributions from the isotropic anharmonic effect (including thermal expansion) and γ_4 represents the anisotropic anharmonic contribution depending on Miller indices. These anharmonic effect coefficients can be obtained by fitting the experimental data of Debye–Waller factor f_D as functions of Q and T. This is because the elastically scattered intensity is

$$I_{el}(\boldsymbol{Q}, T) = Ce^{-2W(\boldsymbol{Q},T)}$$

or

$$\ln I_{el}(\boldsymbol{Q}, T) = -2W(\boldsymbol{Q}, T) + C.$$

Using Eqs. (6.54), (6.55), and (6.56), we obtain

$$\begin{aligned} -\ln I_{el}(\boldsymbol{Q}, T) &= 2W(\boldsymbol{Q}, T) - C \\ &= Q^2 \frac{3\hbar^2}{Mk_B\theta_D^2} T + Q^2\gamma_2 T^2 + Q^2\gamma_3 T^3 + \frac{Q^4}{12}\gamma_4 T^3 - C. \end{aligned} \tag{6.57}$$

Consider the Cu(2 0 0) and Cu(4 0 0) reflections, with the corresponding scattering vectors $\boldsymbol{Q}_1$ and $\boldsymbol{Q}_2$. Suppose we now measure the Bragg scattering intensities at two temperatures T_0 and T from these two reflections: $I_{el}(\boldsymbol{Q}_1, T_0)$, $I_{el}(\boldsymbol{Q}_1, T)$, $I_{el}(\boldsymbol{Q}_2, T_0)$, $I_{el}(\boldsymbol{Q}_2, T)$. Let the non-Gaussian terms be

$$\gamma_4 T_0^3 \equiv D(T_0) \quad \text{and} \quad \gamma_4 T^3 \equiv D(T).$$

Substitute these into Eq. (6.57) and after simple rearrangements, we obtain

$$D(T) - D(T_0) = \frac{12}{Q_1^2 - Q_2^2}\left[\frac{1}{Q_1^2}\ln\frac{I_{el}(\boldsymbol{Q}_1, T_0)}{I_{el}(\boldsymbol{Q}_1, T)} - \frac{1}{Q_2^2}\ln\frac{I_{el}(\boldsymbol{Q}_2, T_0)}{I_{el}(\boldsymbol{Q}_2, T)}\right]. \tag{6.58}$$

On the other hand, from the definitions of $D(T_0)$ and $D(T)$, we get

$$D(T) - D(T_0) = \gamma_4(T^3 - T_0^3). \tag{6.59}$$

Suppose we measure $I_{el}(\boldsymbol{Q}_1, T_0)$ and $I_{el}(\boldsymbol{Q}_2, T_0)$ at $T_0 = 300$ K, and measure $I_{el}(\boldsymbol{Q}_1, T)$ and $I_{el}(\boldsymbol{Q}_2, T)$ at various higher temperatures T. Using the measured intensity values in Eq. (6.58), we can calculate $D(T) - D(T_0)$ and plot this quantity as a linear function of $T^3 - T_0^3$, whose slope is γ_4. The experimental results for Al [42], NaCl [59], and Zn [60] are shown in Fig. 6.18. Similar linear relations are also found for Cu [63, 64] and KCl [65]. For metallic Zn, the straight line does

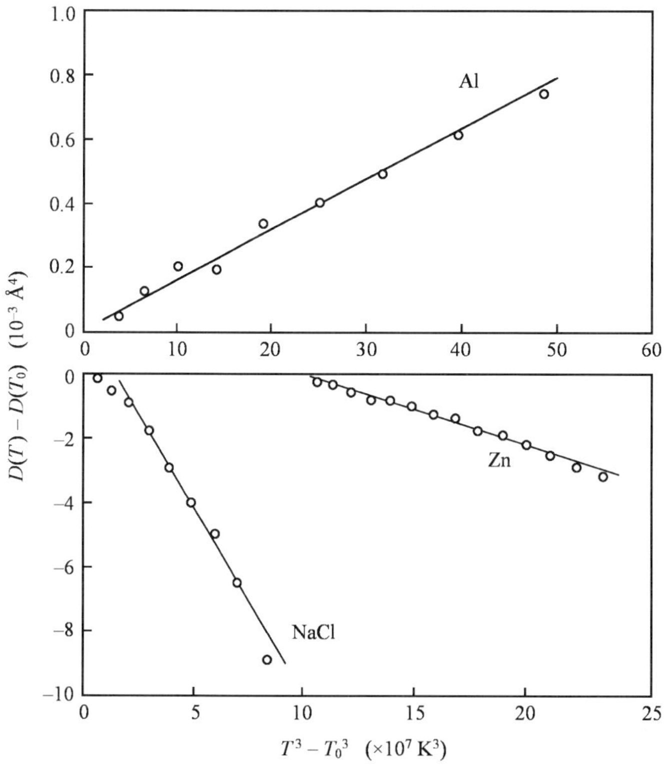

Fig. 6.18 Plots of the non-Gaussian term $[D(T) - D(T_0)]$ versus $[T^3 - {T_0}^3]$ for $T_0 = 300$ K.

not go through origin, indicating that the anharmonic effect is only appreciable at room temperature.

Based on the Rayleigh scattering data from a fixed plane at different temperatures, quantities related to the Debye–Waller factor e^{-2W} can be deduced. Figure 6.19 shows the results for Cu scatterers using different radiation sources. In Fig. 6.19(a), the calculated curves using the harmonic approximation and Morse anharmonic potential are included for the (2 0 0) reflection. The anharmonic coefficients derived from the fittings are listed in Table 6.5 and dynamic parameters for Cu obtained from different Mössbauer sources are listed in Table 6.6. It can be seen from Fig. 6.19 that the fitted curves agree with experiments very well, indicating that the measurements were made quite accurately.

However, discrepancies are often found among results reported by different authors, due to the fact that the anharmonic effect is relatively small and due to differences in experimental conditions and in single-crystal sample qualities. For example, the γ_4 values for NaCl reported by different authors show a considerable disagreement.

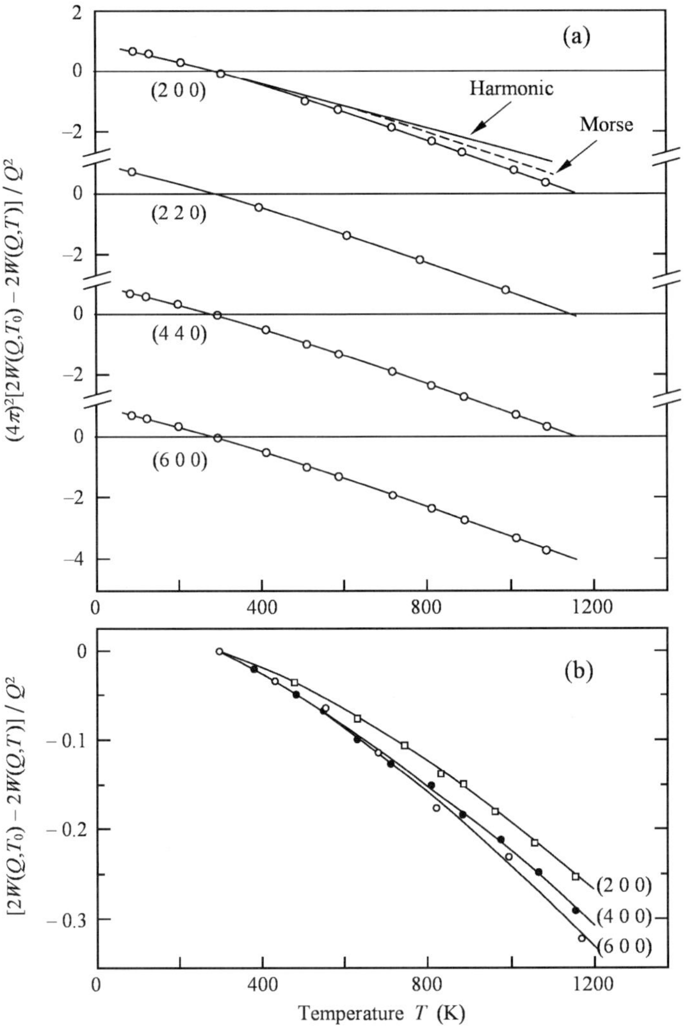

Fig. 6.19 Raleigh scattering results from Cu scatterers using Mössbauer radiation from (a) a ^{183}Ta source and (b) a ^{57}Co source.

Table 6.5 Dynamics parameters of Cu, Ag, and Pb [63].

Crystal	θ_D (K)	γ_2 (Å^2 K^{-2})	γ_3 (Å^2 K^{-3})	γ_4 (Å^4 K^{-3})
Cu	312(3)	$4.3(8) \times 10^{-9}$	$-2(8) \times 10^{-13}$	$6.0(8) \times 10^{-14}$
Ag	214(4)	$2(1) \times 10^{-9}$	$7.7(9) \times 10^{-12}$	$4.2(7) \times 10^{-13}$
Pb	83(10)	$-7(2) \times 10^{-8}$	$2.0(3) \times 10^{-10}$	$8(1) \times 10^{-12}$

Table 6.6 The coefficient γ_4 and Debye temperature θ_D for Cu.

	γ_4 (Å^4 K^{-3})	θ_D (K)
RSMR using ^{183}Ta source [63]	$6.0(8) \times 10^{-14}$	312
RSMR using ^{57}Co source [64]	$1.2(4) \times 10^{-13}$	320
Specific heat method [66]	–	315

References

1 G.N. Belozerski. *Mössbauer Studies of Surface Layers* (Elsevier, Amsterdam, 1993).

2 R.L. Mössbauer, F. Parat, and W. Hoppe. A solution of the phase problem in the structure determination of biological macromolecules. In *Mössbauer Spectroscopy II: The Exotic Side of the Method*, U. Gonser (Ed.), pp. 5–30 (Springer-Verlag, Berlin, 1981).

3 F.E. Wagner. Applications of Mössbauer scattering techniques. *J. de Physique (Colloque)* 37, C6-673–C6-689 (1976).

4 V.A. Belyakov. Diffraction of Mössbauer gamma rays in crystals. *Sov. Phys. Usp.* 18, 267–291 (1975) [Russian original: *Usp. Fiz. Nauk* 115, 553–601 (1975)].

5 B. Kolk. Studies of dynamical properties of solids with the Mössbauer effect. In *Dynamical Properties of Solids*, vol. 5, G.K. Horton and A.A. Maradudin (Eds.), pp. 1–328 (North-Holland, Amsterdam, 1984).

6 D.C. Champeney. The scattering of Mössbauer radiation by condensed matter. *Rep. Prog. Phys.* 42, 1017–1054 (1979).

7 U. van Bürck, G.V. Smirnov, R.L. Mössbauer, H.J. Maurus, and N.A. Semioschkina. Enhanced nuclear resonance scattering in dynamical diffraction of gamma rays. *J. Phys. C* 13, 4511–4529 (1980).

8 A.N. Artem'ev, V.V. Sklyarevskii, G.V. Smirnov, and E.P. Stepanov. Energy analysis of resonance γ rays diffracted by an α-$^{57}Fe_2O_3$ single crystal. *Sov. Phys. JETP* 36, 736–737 (1973) [Russian original: *ZhÉTF* 63, 1390–1392 (1972)].

9 E. Burkel. Introduction to x-ray scattering. *J. Phys.: Condens. Matter* 13, 7477–7498 (2001).

10 A. Storruste. The Rayleigh scattering of 0.41 MeV gamma-ray at various angles. *Proc. Phys. Soc. A* 63, 1197–1201 (1950).

11 P.B. Moon. The hard components of scattered gamma-rays. *Proc. Phys. Soc. A* 63, 1189–1196 (1950).

12 A.T. Forrester. On coherence properties of light waves. *Am. J. Phys.* 24, 192–196 (1956).

13 E.L. O'Neill and L.C. Bradley. Coherence properties of electromagnetic radiation. *Phys. Today* 14, June, 28–34 (1961).

14 P.J. Black and P.B. Moon. Resonant scattering of the 14-keV iron-57 γ-ray, and its interference with Rayleigh scattering. *Nature* (London) 188, 481–482 (1960).

15 P.J. Black and D.E. Evans. Comment on a paper by M.K.F. Wong, 'On the Mössbauer effect'. *Proc. Phys. Soc.* 86, 417–419 (1965).

16 M.K.F. Wong. On the Mössbauer effect. *Proc. Phys. Soc.* 85, 723–734 (1965).

17 S. Bernstein and E.C. Campbell. Nuclear anomalous dispersion in Fe^{57} by the methods of total reflection. *Phys. Rev.* 132, 1625–1633 (1963).

18 P.J. Black, D.E. Evans, and D.A. O'Connor. Interference between

Rayleigh and nuclear resonant scattering in crystals. *Proc. R. Soc. (London)* A270, 168–185 (1962).
19 P.J. Black, G. Longworth, and D.A. O'Connor. Interference between Rayleigh and nuclear resonant scattering in single crystals. *Proc. Phys. Soc.* 83, 925–936 (1964).
20 V.K. Voitovetskii, I.L. Korsunskii, A.I. Novikov, and Yu.F. Pazhin. Diffraction of resonance γ rays by nuclei and electrons in tin single crystals. *Sov. Phys. JETP* 27, 729–735 (1968) [Russian original: *ZhÉTF* 54, 1361–1373 (1968)].
21 V.K. Voitovetskii, I.L. Korsunskii, Yu.F. Pazhin, and R.S. Silakov. Resonant Bragg scattering of gamma rays by nuclei in high orders of reflection, and production of directed beams of pure Mössbauer radiation. *JETP Lett.* 12, 212–215 (1970) [Russian original: *ZhÉTF Pis. Red.* 12, 314–318 (1970)].
22 A.N. Artem'ev, I.P. Perstnev, V.V. Sklyarevskii, G.V. Smirnov, and E.P. Stepanov. Interference of nuclear and electron scatterings of 14.4 keV resonance γ-ray in Bragg reflection from an α-Fe_2O_3 crystal. *Sov. Phys. JETP* 37, 136–141 (1973) [Russian original: *ZhÉTF* 64, 261–272 (1973)].
23 R.M. Mirzababaev, V.V. Sklyarevskii, and G.V. Smirnov. Azimuthal dependence of the purely nuclear diffraction for 14.4 keV resonant gamma-rays. *Phys. Lett. A* 41, 349–350 (1972).
24 H. Bokemeyer, K. Wohlfahrt, E. Kankeleit, and D. Eckardt. Mössbauer conversion spectroscopy: measurements on the first excited states of $^{180,182}W$ and ^{145}Pm. *Z. Phys. A* 274, 305–318 (1975).
25 G.V. Smirnov, V.V. Sklyarevskii, A.N. Artem'ev, and R.A. Voscanyan. Anomalies in shape and width of the Mössbauer line measured in Bragg scattering of 14.4 keV γ-rays from the α-$^{57}Fe_2O_3$ perfect single crystal. *Phys. Lett. A* 32, 532–533 (1970).
26 A.M. Afanas'ev and Yu. Kagan. Suppression of inelastic channels in resonant nuclear scattering in crystal. *Sov. Phys. JETP* 21, 215–224 (1965) [Russian original: *ZhÉTF* 48, 327–341 (1965)].
27 Yu. Kagan, A.M. Afanas'ev, and I.P. Perstnev. Theory of resonance Bragg scattering of γ quanta by regular crystals. *Sov. Phys. JETP* 27, 819–824 (1968) [Russian original: *ZhÉTF* 54, 1530–1541 (1968)].
28 U. van Bürck. Coherent effects in resonant diffraction: theory. *Hyperfine Interactions* 27, 219–230 (1986).
29 V.V. Sklyarevskii, G.V. Smirnov, A.N. Artem'ev, R.M. Mirzababaev, and E.P. Stepanov. Suppression effect under conditions of purely nuclear Laue diffraction of resonant γ rays. *Sov. Phys. JETP* 37, 474–475 (1973) [Russian original: *ZhÉTF* 64, 934–936 (1973)].
30 G.V. Smirnov, V.V. Mostovoi, Yu.V. Shvyd'ko, V.N. Seleznev, and V.V. Rudenko. Suppression of a nuclear reaction in $Fe^{57}BO_3$ crystal. *Sov. Phys. JETP* 51, 603–609 (1980) [Russian original: *ZhÉTF* 78, 1196–1208 (1980)].
31 P.J. Black and I.P. Duerdoth. A direct observation of diffraction in nuclear resonant scattering. *Proc. Phys. Soc.* 84, 169–171 (1964).
32 G.V. Smirnov. Coherent effects in resonant diffraction: experiment. *Hyperfine Interactions* 27, 203–218 (1986).
33 G.V. Smirnov, V.V. Sklyarevskii, R.A. Voskanyan, and A.N. Artem'ev. Nuclear diffraction of resonant γ radiation by an antiferromagnetic crystal. *JETP Lett.* 9, 70–73 (1969) [Russian original: *ZhÉTF Pis. Red.* 9, 123–127 (1969)].
34 P.P. Kovalenko, V.G. Labushkin, A.K. Ovsepyan, É.R. Sarkisov, E.V. Smirnov, A.R. Prokopov, and V.N. Seleznev. Mössbauer diffraction determination of the magnetic structure of the Fe_3BO_6 crystal. *Sov. Phys. Solid State* 26, 1849–1851 (1984) [Russian original: *Fiz. Tverd. Tela* 26, 3068–3072 (1984)].
35 V.I. Mal'tsev, E.P. Naiden, S.M. Zhilyakov, R.P. Smolin, and L.M. Borisyuk. The magnetic structure of

Fe_3BO_6. *Sov. Phys. Crystallogr.* 21, 58–60 (1976) [Russian original: *Kristallografiya* 21, 113–117 (1975)].

36 J.P. Hannon and G.T. Trammell. Mössbauer diffraction: II. Dynamic theory of Mössbauer optics. *Phys. Rev.* 186, 306–325 (1969).

37 G.T. Trammell. Elastic scattering at resonance from bound nuclei. *Phys. Rev.* 126, 1045–1054 (1962).

38 M.K.F. Wong. Scattering amplitude for multipole mixtures in the Mössbauer effect. *Phys. Rev.* 149, 378–379 (1966).

39 R. Röhlsberger, O. Leupold, J. Metge, H.D. Rüter, W. Sturhahn, and E. Gerdau. Nuclear forward scattering of synchrotron radiation from unmagnetized α-^{57}Fe. *Hyperfine Interactions* 92, 1107–1112 (1994).

40 K.S. Singwi and A. Sjölander. Resonance absorption of nuclear gamma rays and dynamics of atomic motions. *Phys. Rev.* 120, 1093–1102 (1960).

41 E.B. Zholotoyabco and E.M. Eolin. *Coherent Rayleigh Scattering of Mössbauer Radiation* (Zhinachi, Riga, 1986) [in Russian].

42 G. Albanese and C. Ghezzi. Anharmonic contributions to elastic and inelastic scattering of x rays at Bragg reflections in aluminum. *Phys. Rev. B* 8, 1315–1323 (1973).

43 G. Albanese and A. Deriu. High energy resolution x-ray spectroscopy. *Rivista Nuovo Cimento* (Ser. 3) 2 (no. 9), 1–40 (1979).

44 G. Albanese. Rayleigh scattering of Mössbauer γ-rays in disordered systems. In *Applications of the Mössbauer Effect*, vol. 1, Yu.M. Kagan and I.S. Lyubutin (Eds.), pp. 63–81 (Gordon and Breach, New York, 1985).

45 V.I. Goldanskii and Yu.F. Krupyanskii. The study of biopolymer dynamics by Rayleigh scattering of Mössbauer radiation (RSMR). Glass-like model of protein and DNA. In *Applications of the Mössbauer Effect*, vol. 1, Yu.M. Kagan and I.S. Lyubutin (Eds.), pp. 83–112 (Gordon and Breach, New York, 1985).

46 D.A. O'Connor and N.M. Butt. The detection of the inelastic scattering of gamma rays at crystal diffraction maxima using the Mössbauer effect. *Phys. Lett.* 7, 233–235 (1963).

47 N.N. Lobanov, V.A. Bushuev, V.S. Zasimov, and R.N. Kuz'min. Elastic and inelastic scattering of Mössbauer γ quanta in a pyrolytic graphite crystal. *Sov. Phys. JETP* 49, 572–576 (1979) [Russian original: *ZhÉTF* 76, 1128–1135 (1979)].

48 K. Krec and W. Steiner. Investigation of a silicon single crystal by means of the diffraction of Mössbauer radiation. *Acta Crystallogr. A* 40, 459–465 (1984).

49 G. Albanese, C. Ghezzi, A. Merlini, and S. Pace. Determination of the thermal diffuse scattering at the Bragg reflection of Si and Al by means of the Mössbauer effect. *Phys. Rev. B* 5, 1746–1757 (1972).

50 G. Albanese, C. Ghezzi, and A. Merlini. Determination of the inelastic scattering at Bragg reflections of KCl by means of the Mössbauer effect: contribution of multiphonon-scattering terms. *Phys. Rev. B* 7, 65–72 (1973).

51 N.M. Butt and D.A. O'Connor. The determination of x-ray temperature factors for aluminium and potassium chloride single crystals using nuclear resonant radiation. *Proc. Phys. Soc.* 90, 247–252 (1967).

52 R.E. Dingle and E.H. Medlin. The x-ray Debye temperature of aluminum. *Acta Crystallogr. A* 28, 22–27 (1972).

53 D.L. McDonald. Neutron diffraction study of the Debye–Waller factor for aluminum. *Acta Crystallogr.* 23, 185–191 (1967).

54 D.K. Kaipov and S.M. Zholdasova. Determination of the Debye temperature of nickel by Rayleigh scattering of Mössbauer radiation. In *Applications of the Mössbauer Effect*, vol. 4, Yu.M. Kagan and I.S. Lyubutin (Eds.), pp. 1383–1385 (Gordon and Breach, New York, 1985).

55 F.E. Obenshain and H.H.F. Wegener. Mössbauer effect with Ni^{61}. *Phys. Rev.* 121, 1344–1349 (1961).

56 W.B. Yelon, G. Schupp, M.L. Crow, C. Holmes, and J.G. Mullen. A gamma-ray diffraction instrument for high-intensity Mössbauer sources. *Nucl. Instrum. Methods B* 14, 341–347 (1986).

57 B.R. Bullard, J.G. Mullen, and G. Schupp. Mössbauer line-shape parameters for ^{183}W and ^{191}Ir in metallic tungsten and iridium. *Phys. Rev. B* 43, 7405–7415 (1991).

58 O.I. Sumbaev, A.I. Smirnov, and V.S. Zykov. The Mössbauer effect on tungsten isotopes. *Sov. Phys. JETP* 15, 82–87 (1962) [Russian original: *ZhÉTF* 42, 115–123 (1962)].

59 N.M. Butt and G. Solt. Anharmonic non-Gaussian contribution to the Debye–Waller factor for NaCl. *Acta Crystallogr. A* 27, 238–243 (1971).

60 G. Albanese, A. Deriu, and C. Ghezzi. Anharmonic contributions to the Debye–Waller factor for zinc. *Acta Crystallogr. A* 32, 904–909 (1976).

61 A.A. Maradudin, P.A. Flinn, and C. Ghezzi. Anharmonic contributions to the Debye–Waller factor. *Phys. Rev.* 129, 2529–2547 (1963).

62 G.A. Wolfe and B. Goodman. Anharmonic contributions to the Debye–Waller factor. *Phys. Rev.* 178, 1171–1189 (1969).

63 J.T. Day, J.G. Mullen, and R.C. Shukla. Anharmonic contribution to the Debye–Waller factor for copper, silver, and lead. *Phys. Rev. B* 52, 168–176 (1995). Erratum: *Phys. Rev. B* 54, 15548 (1996).

64 C.J. Martin and D.A. O'Connor. Anharmonic contributions to Bragg diffraction: I. Copper and aluminium. *Acta Crystallogr. A* 34, 500–505 (1978).

65 C.J. Martin and D.A. O'Connor. Anharmonic contributions to Bragg diffraction: II. Alkali halides. *Acta Crystallogr. A* 34, 505–512 (1978).

66 J. de Launay. The theory of specific heats and lattice vibrations. In *Solid State Physics*, vol. 2, F. Seitz and D. Turnbull (Eds.), pp. 219–303 (Academic Press, New York, 1956).

7
Synchrotron Mössbauer Spectroscopy

It was in 1974 that the possibility of using synchrotron radiation (SR) as a source for Mössbauer measurements was proposed [1], but not until 1985 did a breakthrough take place when SR with an energy width of 10^{-8} eV at 14.413 keV was obtained and used to observe the transmission spectrum of stainless steel [2]. SR provides polarized pulsed radiation of high intensity, high collimation, and narrow beam. The only drawback is that SR is far from monochromatic. However, its energy can be adjusted and it can cover an energy range for a majority of Mössbauer transitions. Initially, the high intensity of the SR source was exploited for scattering experiments where conventional radiation sources could not provide adequate results. Soon after, it was realized that the pulsed nature of SR is most suitable for measuring time spectra – using a short SR pulse ($<10^{-10}$ s) to excite a nuclear ensemble to form a so-called exciton and observing its coherent decay at different time intervals. The third-generation synchrotron storage rings can give a pulse of 100 ps every 2–3 ns. Therefore, the method measuring time spectra is called time domain Mössbauer spectroscopy whereas the transmission method is referred to as energy domain Mössbauer spectroscopy. In the past two decades, significant progress has been made in synchrotron Mössbauer spectroscopy, especially in the time domain method, providing a direct and efficient approach to the study of the Mössbauer effect and hyperfine fields. There have emerged several new research areas which are not accessible with the conventional radiation sources. One of the exciting advances was the phonon DOS measured by SR sources in 1995. It had been known as soon as the Mössbauer effect was discovered that a phonon DOS could be measured by this effect, but because of technical difficulties it was not realized until SR became available.

At the present time, third-generation SR sources are in operation, such as those at the European Synchrotron Radiation Facility (ESRF) in Grenoble (France), the Advanced Photon Source (APS) in Argonne (USA), and the Super Photon ring (SPring-8) in Hyogo (Japan). The most distinct advantage of SR is its high brilliance (measured in photons s^{-1} eV^{-1} sr^{-1} mm^{-2}). The brilliance of a third-generation SR source is about 9 to 10 orders of magnitude higher than that of a rotating target x-ray generator, and about 12 orders of magnitude higher than that of a ^{57}Co source of 3.7×10^8 Bq. However, SR sources require an enormously large and costly facility, and will not be available in ordinary Mössbauer laborato-

Mössbauer Effect in Lattice Dynamics. Yi-Long Chen and De-Ping Yang

ISBN: 978-3-527-40712-5

ries. The time spectra are usually quite complicated. Therefore, the SR sources will not replace the conventional radiation sources. The fact that transmission Mössbauer spectroscopy is regarded as a "classical" method by some authors implies that synchrotron Mössbauer spectroscopy has opened a modern era of this research field. In this chapter, we briefly describe synchrotron Mössbauer spectroscopy and its possible applications in lattice dynamics. The reader may find more in a specialized volume devoted to the theory and experiments of this subject [3].

7.1 Synchrotron Radiation and Its Properties

When a charged particle undergoes a circular motion in a magnetic field, it radiates electromagnetic waves because of its large centripetal acceleration. It was discovered in 1948 that the radiation from electrons in a synchrotron accelerator is very unique (known as synchrotron radiation), having high intensity, narrow beam width, adjustable energy, and a broad energy spectrum. It was soon recognized that SR can serve as ideal radiation sources in the energy range of 10 eV to 100 keV for applications in all scientific research. High-energy electron synchrotrons for producing SR have been constructed in many countries around the world. Both classical and quantum mechanical theories of SR have been successfully developed and described in detail in textbooks [4, 5]. Here, we will simply quote the results to discuss some of the properties of SR.

7.1.1 The Angular Distribution of Radiation

If we consider the electron motion as nonrelativistic, the radiation has the dipole pattern, and its power distribution is

$$\frac{\mathrm{d}P}{\mathrm{d}\Omega} = \frac{e^2}{16\pi^2 \varepsilon_0 c^3} |\dot{\mathbf{v}}|^2 \sin^2 \psi \tag{7.1}$$

where ψ is the angle between the radiation direction unit vector $\boldsymbol{n}$ and the acceleration vector $\dot{\mathbf{v}}$. It is easy to see that the power is maximum in the directions perpendicular to acceleration, and it is zero along the acceleration. When the electron energy is very high, the motion must be treated as relativistic ($\beta = v/c \approx 1$), and after Lorentz transformation the dipole radiation in the rest frame of the electron is now concentrated in the direction of the electron velocity. Let x, y, and z be the axes of the laboratory reference frame, and the electron's orbit be in the y–z plane. When its velocity $\mathbf{v}$ and acceleration $\dot{\mathbf{v}}$ are in the z and y directions (Fig. 7.1), respectively, the angular distribution of radiation in terms of the observer's spherical coordinates θ and ϕ is

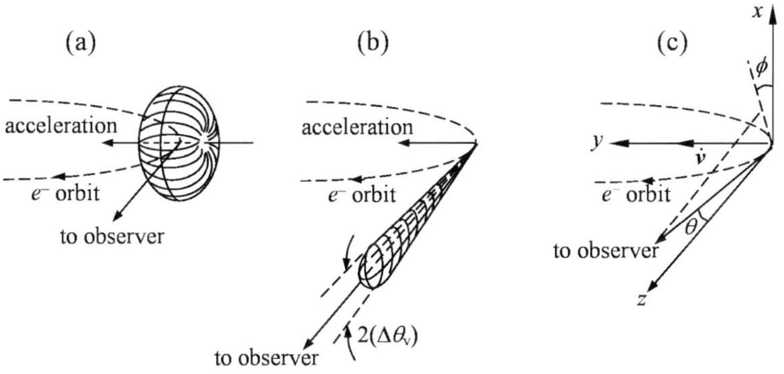

Fig. 7.1 Radiation distribution for an electron for (a) nonrelativistic motion and (b) relativistic motion in a circular orbit. (c) The laboratory frame.

$$\frac{\mathrm{d}P(\theta,\phi)}{\mathrm{d}\Omega}=\frac{P_0|\dot{\mathbf{v}}|^2}{(1-\beta\cos\theta)^3}\left[1-\frac{(1-\beta^2)\sin^2\theta\sin^2\phi}{(1-\beta\cos\theta)^2}\right] \tag{7.2}$$

where $P_0 = e^2/(16\pi^2\varepsilon_0 c^3)$. Now we analyze this angular distribution in two special planes, one horizontal (yz plane) and one vertical (xz plane) with respect to the electron orbit, and the angle θ will be written as θ_h and θ_v in these two respective planes.

1. If the observer is in the yz plane, $\phi = \pm\pi/2$, and Eq. (7.2) becomes

$$\frac{\mathrm{d}P(\theta_\mathrm{h},\phi)}{\mathrm{d}\Omega}=\begin{cases}\dfrac{P_0|\dot{\mathbf{v}}|^2}{(1-\beta)^3}\xrightarrow{\beta\to 1}\infty, & \text{when } \theta_\mathrm{h}=0\\ 0, & \text{when } \theta_\mathrm{h}=\pm\cos^{-1}\beta.\end{cases} \tag{7.3}$$

Therefore, when observed in the direction of the instantaneous velocity $\mathbf{v}$, the SR is limited within an angle of $\Delta\theta_\mathrm{h}$ centered at $\mathbf{v}$, which can be approximately expressed as

$$\Delta\theta_\mathrm{h} = 2\cos^{-1}\beta \approx 2\gamma^{-1} \tag{7.4}$$

where $\gamma = (1-\beta^2)^{-1/2} = E/(m_0c^2)$. For an orbiting electron with an energy of $E = 2$ GeV, this angle is very small, $\Delta\theta_\mathrm{h} = 1.7'$. As the β-value approaches 1, $\Delta\theta_\mathrm{h}$ becomes smaller, so the radiation along the $\mathbf{v}$ direction intensifies whereas the radiation in the opposite direction ($-z$) diminishes.

(2) If the observer is in the xz plane, $\phi = 0$ or π, and Eq. (7.2) becomes

$$\frac{\mathrm{d}P(\theta_\mathrm{v},\phi)}{\mathrm{d}\Omega}\begin{cases}=\dfrac{P_0|\dot{\mathbf{v}}|^2}{(1-\beta)^3}, & \text{when } \theta_\mathrm{v}=0\\ =P_0|\dot{\mathbf{v}}|^2, & \text{when } \theta_\mathrm{v}=\dfrac{\pi}{2}.\end{cases} \tag{7.5}$$

Although there does not exist a critical value of θ_v above which the radiation approaches zero, the ratio of the intensity at $\theta_v = 0$ to that at $\theta_v = \pi/2$ is $1/(1-\beta^2)^3$. When $\beta \to 1$, this ratio is extremely large, and the majority of radiation can still be considered [5] as distributed within the angle of

$$\Delta\theta_v \approx 2\gamma^{-1}. \tag{7.6}$$

In summary, SR is concentrated in a narrow cone in the z-direction, similar to a beam from a searchlight.

7.1.2
The Total Power of Radiation

Integrating Eq. (7.2) over all angles, we obtain the total power radiated by the electron:

$$P = \frac{1}{6\pi\varepsilon_0}\frac{e^2}{c^3}|\dot{\mathbf{v}}|^2\gamma^4. \tag{7.7}$$

As mentioned earlier, $\gamma \sim 10^4$; therefore, the radiation power from a relativistic electron is about 10^{16} times higher than that of a nonrelativistic electron as described in Eq. (7.1). With such a high power confined in a small cone of radiation, we can see why the photon density could be extremely high, reaching 10^{19} to 10^{20} photons s^{-1} mm^{-2} $mrad^{-2}$ $(0.1\%\ bw)^{-1}$.

7.1.3
The Frequency Distribution of Radiation

Synchrotron radiation is composed of pulses of duration $< 10^{-10}$ s with a period in the microsecond range or less. As we know, such a pulsed radiation series contain a wide spectrum of frequency components. When the electron velocity approaches the speed of light, the fundamental frequency ω_0 (orbiting frequency) is no longer the major frequency component, but its high-order harmonics $n\omega_0$. Let the pulse duration of the electron beam be τ'. The pulse duration as observed in the laboratory reference frame is then

$$\tau = (1 - \boldsymbol{\beta}\cdot\mathbf{n})\tau' \approx \tau'\gamma^{-2} = \frac{R}{c}\gamma^{-3}. \tag{7.8}$$

This pulse duration τ determines the maximum frequency of the radiation:

$$\omega_{max} \approx \frac{c}{R}\gamma^3. \tag{7.9}$$

When $E = 3$ GeV and $B = 0.8$ T, ω_{max} can be calculated to be 4.9×10^{18} rad s^{-1}, corresponding to a wavelength of 3.9 Å. Therefore, SR covers a wide range from

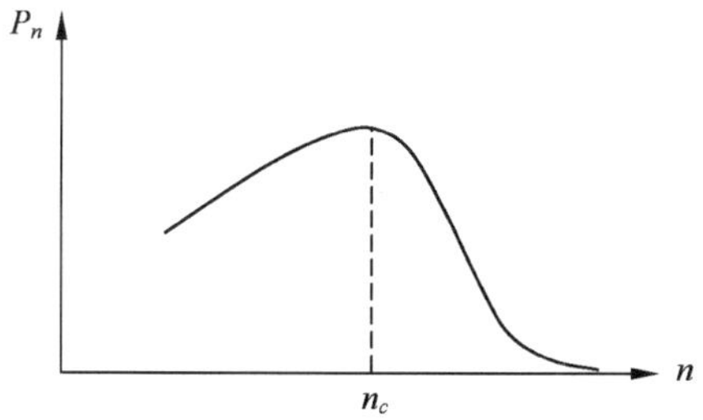

Fig. 7.2 Typical synchrotron radiation spectrum as a function of ω/ω_0.

radio frequency to very hard x-rays, as shown in a typical spectrum in Fig. 7.2. In reality, an electron beam in the synchrotron has a certain physical size and contains many electrons, which would make τ significantly larger and the resultant radiation is the superposition of the contributions from all the electrons in the beam.

7.1.4 Polarization

At a location far from the electron orbit, the electric field vector $\boldsymbol{E}$ is always in the direction determined by $\boldsymbol{n} \times (\boldsymbol{n} \times \dot{\boldsymbol{v}})$. Within the orbit plane, the vector $\boldsymbol{E}$ is collinear with $\dot{\boldsymbol{v}}$, and the radiation observed in this plane will be completely linearly σ-polarized. Above or below the orbit plane, the radiation contains elliptically polarized components. In general, the observed radiation intensity can be consid-

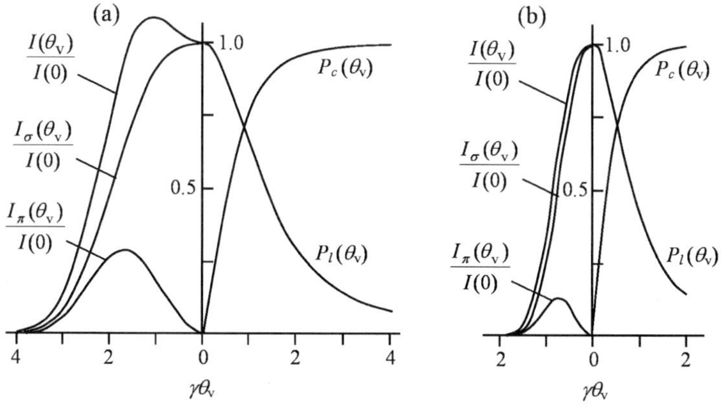

Fig. 7.3 Relative intensities (I, I_σ, I_π), linear polarization P_l, and circular polarization P_c as functions of the angle θ_v in the plane perpendicular to the orbit: (a) $\lambda = 10\lambda_c$ and (b) $\lambda = \lambda_c$, where λ_c is the wavelength corresponding to the maximum P_n in Fig. 7.2 and $\gamma = E/m_0c^2$.

ered as the sum of I_σ and I_π. We may define a degree of linear polarization P_l and a degree of circular polarization P_c [6]:

$$P_l = \frac{(I_\sigma - I_\pi)}{(I_\sigma + I_\pi)},$$
$$P_c = \pm\frac{2\sqrt{I_\sigma I_\pi}}{(I_\sigma + I_\pi)}. \qquad (7.10)$$

Figure 7.3 shows the quantities P_l and P_c as functions of the angle θ_v and the wavelength. Within the plane of the orbit, the radiation is 100% linearly polarized ($P_l = 1$ and $P_c = 0$). Deviating from this plane, both P_l and I decrease, whereas P_c increases slightly. When very far from this plane, the radiation is largely characterized by elliptical polarization ($P_l = 0$ and $P_c = 1$). When averaged over all angles, one finds that 75% of the radiation is linearly polarized.

7.2 Synchrotron Mössbauer Sources

The bandwidth of SR beams is too large for nuclear resonant scattering experiments. Even after going through a double-crystal pre-monochromator Si(1 1 1), its bandwidth can only be reduced to the order of eV. The nuclear resonant width is 4.66×10^{-9} eV in ^{57}Fe, so only about one part in a 10^8 of the incident SR is useful. This implies a serious problem of signal-to-noise ratio. Fortunately, ultra-high collimation of SR can be achieved, and provides monochromatic SR beams of bandwidth of meV, sub-meV, or μeV.

If a perfectly collimated and monochromatic SR beam is incident on a perfect crystal at a Bragg angle, the probability of reflection can be very large, close to unity. In practice, highly effective monochromators have been designed and good signal-to-noise ratios of 10^3 or better have been achieved [7].

To date, SR can be tuned to the Mössbauer transition energies not only for the most common isotope ^{57}Fe but also for others such as ^{83}Kr, ^{151}Eu, ^{119}Sn, ^{161}Dy, and ^{201}Hg.

7.2.1 The meV Bandwidth Sources

The desired bandwidth of a monochromator for nuclear resonant scattering experiments is within several μeV. This is broad enough to cover all hyperfine transitions of each Mössbauer nucleus in a sample, while the prompt background may be reduced to a manageable level. There are two types of monochromators, one based on the scattering by electrons and the other on resonant scattering by nuclei. The first type can offer a bandwidth within a few meV, and is described in this section. The second type can provide the desired bandwidth of a few μeV, and is discussed in the next section.

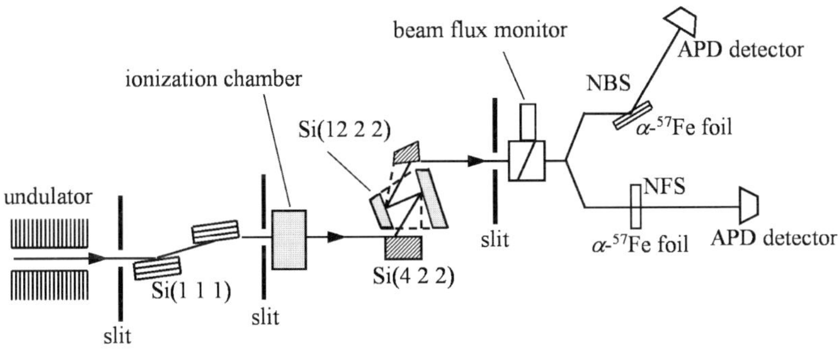

Fig. 7.4 Schematic representation of the geometry in nuclear Bragg scattering (NBS) and nuclear forward scattering (NFS) experiments using an SR source [11].

In combination with modern detectors, monochromators based on electron scattering have been improved so that the prompt rate can be reduced to a level that allows nuclear resonant scattering experiments to be successfully carried out. Using such SR sources, many recent lattice dynamics experiments as well as nuclear forward scattering experiments have been performed with acceptable signal-to-noise ratios. Therefore, SR with a few meV has become an important photon source for nuclear resonant scattering work.

The experimental setup of nuclear resonant scattering is shown Fig. 7.4. A beam of SR from an undulator is incident on a high-heat-load premonochromator, which consists of two symmetric Si(1 1 1) reflectors, narrowing the bandwidth to about 1–2 eV. The further reduction of the bandwidth to meV can be achieved by a high-resolution monochromator (HRM) [8]. Here an ionization chamber as the beam intensity monitor and some slits are also placed. Up to now, HRMs have been constructed by using two particular reflections. (1) A reflection with a Bragg angle near 90° [9]. Under this reflection the angular acceptance can be maximized for a given energy bandwidth. Several allowed reflections at 14.413 keV in Si are off the (10 6 4), (12 2 2), and (9 7 5) planes with Bragg angles of 90°, 77.5° and 80.4°, respectively. Figure 7.5(a) shows a pair of channel-cut Si(10 6 4) crystals arranged in a dispersive geometry. This HRM provides highly monochromatic, highly collimated, and high energy-resolution beams but with a low transmission. (2) An asymmetric Bragg reflection [10]. This also appreciably increases the angular acceptance. Asymmetric reflection means that the reflecting planes are not parallel to the physical surface of the crystal. In Fig. 7.4, the HRM using an asymmetric channel-cut Si(12 2 2) crystal nested within an asymmetric channel-cut Si(4 2 2) crystal provides a beam of a 6.7 meV bandwidth at 14.413 keV nuclear resonance for ^{57}Fe. With such a radiation source, the phonon DOS of α-Fe has been observed for the first time [7]. Figure 7.5(b) shows a modern HRM using extremely asymmetric angles on high-order reflections, for instance the Si(9 7 5) reflection [12, 13], which can reduce

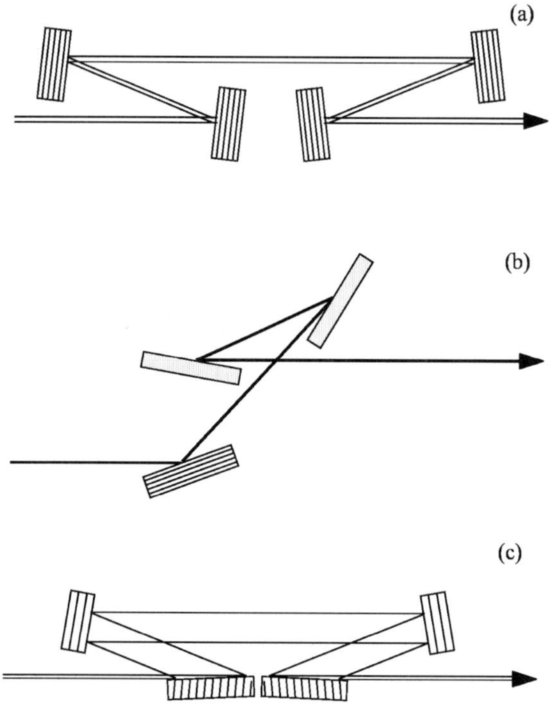

Fig. 7.5 Three versions of HRM with meV bandwidths. (a) Two pairs of symmetric channel-cut Si(10 6 4) high-order reflections in a dispersive geometry [9]. (b) An HRM optimized for energy resolution: two asymmetrically cut, high-order crystal reflections in Si(9 7 5) [13]. (c) A four-reflection version, similar to (b) [14].

the bandwidth to 0.8 meV with an efficiency of about 50%. To study lattice dynamics, the HRM must be tunable over a region sufficiently large for measuring phonon energies, i.e., a few hundred meV. This is achieved by mounting the crystal assembly on high-precision angular encoders (made by Kohzu Precision Co. Ltd), which can provide a minimum rotating step size of ∼0.012 μrad. This corresponds an energy step of 15 μeV. Temperature stability and monitoring are other important aspects; *e.g.*, a change of 13 mK on both crystals will produce an energy shift of about 1 meV [12].

Generally speaking, a meV bandwidth is about 10^6 times wider than the resonance linewidth of ^{57}Fe. A scatterer contains a large number of electrons in addition to the resonant nuclei, and those electrons will non-resonantly scatter all the SR in the meV band. Therefore, only 10^{-6} of the detected photon count is due to nuclear resonant scattering. The electron scattering process is prompt, whereas the nuclear resonant scattering is a time-delayed process since the typical lifetime of the nuclear isomeric state is long compared to the incident SR pulse (for ^{57}Fe, $\tau_0 \approx 141$ ns). Using this time difference, photons from non-resonant electron scattering can be in principle discriminated. Of course this requires a detector of good time response, with rise and fall times shorter than a nanosecond. This criterion is satisfied by the recently developed fast avalanche photodiode detectors (APD) with time resolutions of 0.1 to 1 ns, which can sustain an intense prompt scattering ($\sim 10^9$ photons s^{-1}) during the flash of SR and several nanoseconds later are able to detect a single delayed photon of nuclear scattering [15]. For a multi-element scatterer, a very good method is to find special reflections that the electronic Bragg reflections from different atoms may be cancelled, or even forbidden, such as the $(2n+1\ 2n+1\ 2n+1)$ plane in α-Fe_2O_3.

7.2.2 The μeV Bandwidth Sources

For producing a μeV bandwidth source, the first stage monochromator is also Si(1 1 1), followed by a HRM, usually making use of nuclear Bragg scattering. This combination reduces the bandwidth to within 10^{-6} to 10^{-8} eV, approaching the nuclear energy level's natural width. A single crystal of α-$^{57}Fe_2O_3$ or $^{57}FeBO_3$ can be used for such a monochromator [16, 17]. A film of hundreds of artificially structured nuclear multilayers can produce very strong Bragg scattering, while the electron Bragg scattering is relatively weak. For example, the nuclear periods of $25 \times [^{57}Fe(22\,Å)/Sc(11\,Å)/^{56}Fe(22\,Å)/Sc(11\,Å)]$ [18] and $25 \times [^{57}Fe(17\,Å)/Cr(10\,Å)]$ [19] multilayers have been designed. Structurally, the nuclear interplanar distance is twice the electronic interplanar distance, making the electron Bragg scattering much weaker than the nuclear Bragg scattering. Another type of monochromator is the grazing incident antireflection film (GIAR film) [20], for example, consisting of an $^{57}Fe_5B_4C$ layer on a Ta backing. Because of the grazing incident angle, the radiation's path in the coating is relatively long, resulting in a total reflection. This can be considered as an extreme case of interference. Using this method, the electronic reflectivity is reduced to 0.04 and the bandwidth of the source is reduced to 0.5×10^{-6} eV.

In order to protect the detectors from very intense prompt radiation, a new technique was developed, in which the ratio of prompt to delayed radiations is reduced before detection by polarization-selective optics [21, 22]. It has been pointed out that when x-rays undergo a 45° Bragg reflection, the π-polarized component is almost entirely eliminated [23]. When the 14.413 keV SR from the Si(1 1 1) monochromator is subjected to a 45° Bragg reflection from a Si(8 4 0) polarizer, the radiation has only the σ-polarized component remaining (Fig. 7.6). The beam is then directed perpendicularly to an Fe foil of 10.5 μm thickness (95% ^{57}Fe enrichment). An external magnetic field $\boldsymbol{B}$ is applied in the plane of the foil but making a 45° with the horizontal (i.e., the orbital plane of the electrons where the vector $\boldsymbol{E}$ lies). As has been shown for optics, when an optically active material is inserted between polarizer and analyzer crystals, part of the incident σ-polarized radiation is converted to a π-polarization component [24]. Here an Fe foil with an external magnetic field $\boldsymbol{B}$ acts like this material. There will be six allowed transitions in the foil ($\Delta m = 0$ and $\Delta m = \pm 1$), produc-

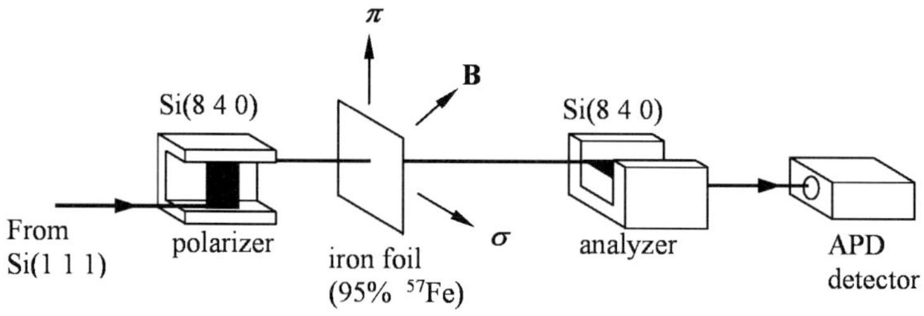

Fig. 7.6 Experimental setup of a polarizer, an Fe foil with magnetic field ***B***, an analyzer, and a detector, for suppressing radiation from non-resonant scattering.

ing σ- and π-polarization mixed components of the nuclear forward scattering with comparable intensities. Behind the Fe foil is an analyzer which is exactly the same as the polarizer except for a 90° rotation. In this orientation, the σ-component (instead of the π-component) and the prompt radiation caused by electron scattering can be almost completely suppressed, while the π-component is transmitted. To a very good approximation, only the resonant part of the radiation transmitted from the polarizer can have its polarization state modified. This is the principle of this assembly, converting σ-polarization to π-polarization and producing a 14.413 keV SR Mössbauer source of bandwidth 10^{-7} eV, with the electron scattering radiation suppressed to a fraction of 5.4×10^{-7}.

Now we discuss how to obtain a single-line source from the incident SR. Using pure Bragg scattering for making SR monochromatic usually involves hyperfine interactions, and as a result the spectrum of the "filtered" radiation contains several spectral components. If this radiation is used for recording a time spectrum, these components interfere with one another and produce "quantum beats." If a single-line monochromatic source is desired, an additional absorber may be added to filter out the unwanted spectral components. But because of the dynamic diffraction effect, the width of the resultant single line is several times wider than the absorber natural width. Also, this method causes significant intensity loss.

To circumvent these difficulties, studies have shown that when single-crystal $^{57}FeBO_3$ in an external magnetic field is heated to higher than its Neèl temperature T_N to eliminate magnetic hyperfine interaction, the Bragg reflection from its (3 3 3) plane will result in a single-line source of nearly the natural linewidth [25]. $FeBO_3$ is antiferromagnetic and the principal axis of its EFG is perpendicular to the hyperfine magnetic field. When the temperature approaches T_N, the magnetic hyperfine interaction becomes weaker and eventually disappears. As shown in Fig. 7.7, the low-energy line of the quadrupole doublet gradually loses its intensities due to destructive interference. When $T = 75.9$ °C, a single-line source of width $2.9\Gamma_n$ can be obtained in the Bragg angle direction.

An experimental setup using this method is shown in Fig. 7.8. A double-crystal Si(1 1 1) monochromator reduces the radiation bandwidth to 2.8 eV. A channel-

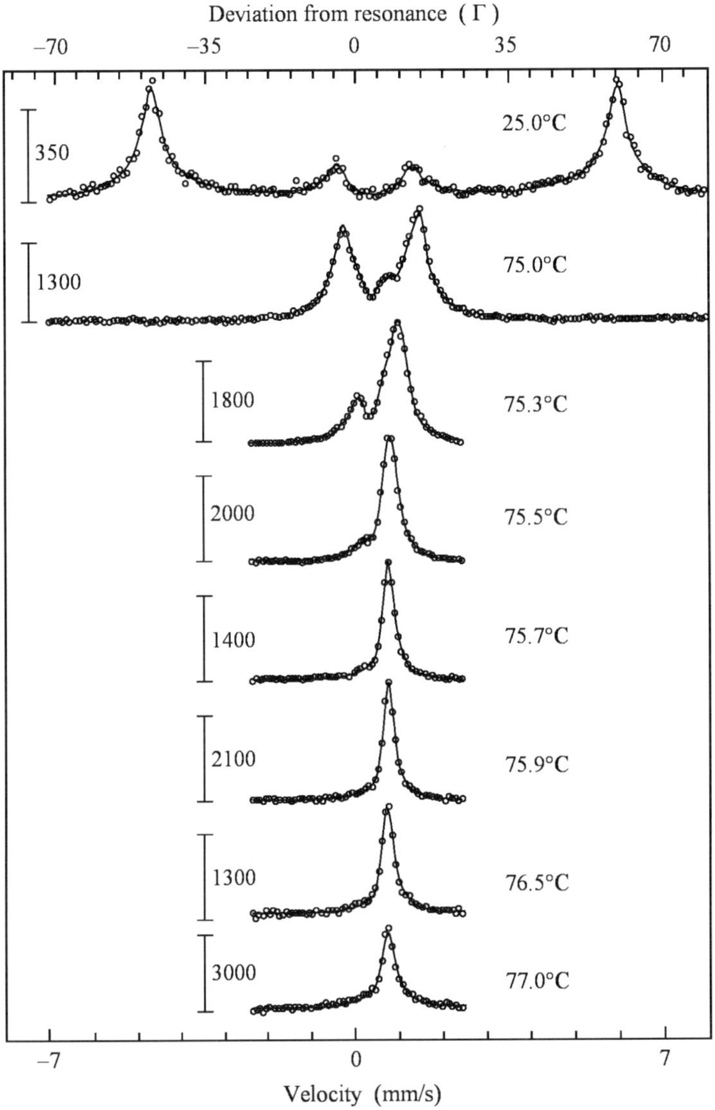

Fig. 7.7 Mössbauer diffraction spectra from a $^{57}FeBO_3$ crystal at different temperatures, using resonant γ-radiation from a $^{57}Co(Cr)$ source.

cut Si(8 4 0) polarizer reduces the π-polarized component from 1% to less than 10^{-4}% and the bandwidth to meV. Single-crystal $^{57}FeBO_3$, placed in an oven, has its (3 3 3) plane in the vertical orientation and an external magnetic field of 10 mT is also applied so that the crystal is magnetized in the vertical direction. The outcome is an extremely narrow ($10 \times 35\ \mu rad^2$), linearly polarized (electric field vec-

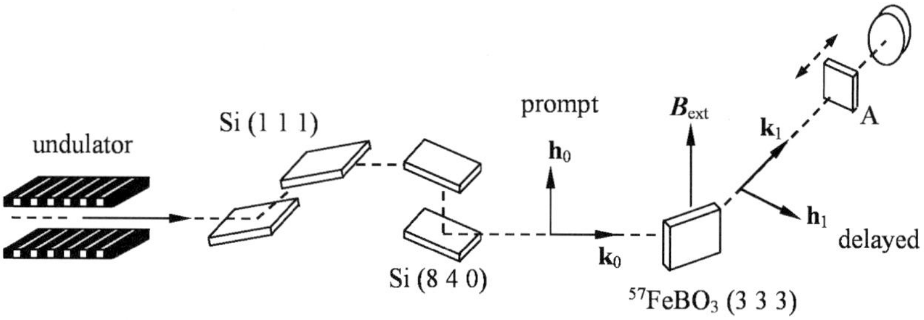

Fig. 7.8 Experimental setup for obtaining monochromization and polarization of ^{57}Fe Mössbauer synchrotron radiation. The ***h*** vectors represent the polarization directions, and the ***k*** vectors represent the propagation directions. The absorber A is an iron foil (95% ^{57}Fe) and D is an avalanche photodiode detector.

tor $\boldsymbol{E}$ in the vertical plane), completely recoilless, and highly intense Mössbauer source. The electron-scattered component is reduced to 10^{-10} of the original level. According to estimates, when the integrated current in the storage ring is 130 mA, the radiation within the above solid angle is equivalent to a ^{57}Co source of an enormous activity of 3.7×10^{13} Bq.

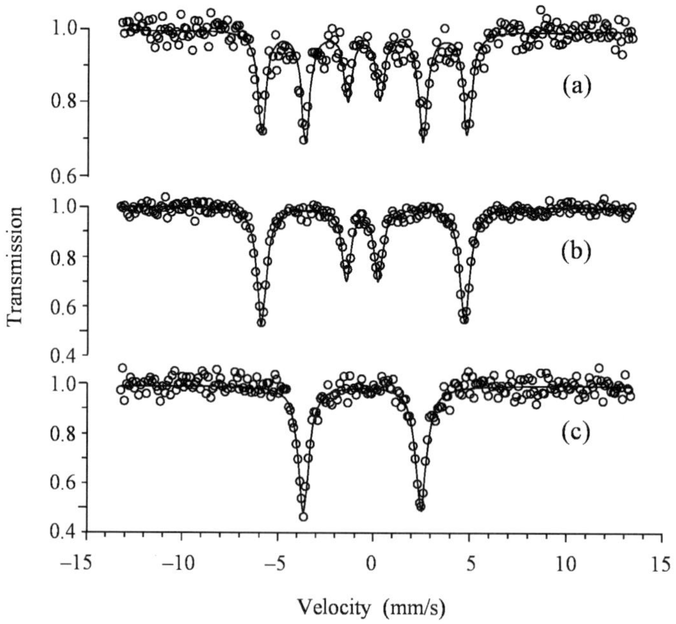

Fig. 7.9 Mössbauer transmission spectra of an ^{57}Fe foil of 1.3 μm thickness measured with SR Mössbauer source: (a) zero external magnetic field, (b) $\boldsymbol{B}_{ext} \perp \boldsymbol{k}'$ and $\boldsymbol{B}_{ext} \perp \boldsymbol{h}'$, and (c) $\boldsymbol{B}_{ext} \parallel \boldsymbol{h}'$.

The Mössbauer source discudded above has been tested using stainless steel and α-Fe absorbers (all with 95% ^{57}Fe enrichment). For stainless steel foils of 1 and 10 μm thicknesses, the resonance efficiencies are $\varepsilon = 70$ and 86%, respectively (ε is defined in Eq. (1.24)), which means that the radiation corresponds to a completely recoilless Mössbauer radiation, or $f_s = 1$. In order to verify its polarization state, the radiation is used to measure transmission spectra of an α-Fe foil of 1.3 μm thickness (Fig. 7.9). The spectrum in Fig. 7.9(a) is a sextet as expected from a nonmagnetized sample. When an external magnetic field of $B_{ext} = 10$ mT is applied in two different directions, the spectra in Figs. 7.9(b) and (c) show four and two lines, respectively, corresponding to the $\Delta m = \pm 1$ and $\Delta m = 0$ transitions. These results clearly confirm that this Mössbauer source is linearly polarized. It is also easy to see that the spectra in Figs. 7.9(b) and (c) resemble those in Fig. 2.24.

7.3 Time Domain Mössbauer Spectroscopy

The initial inception of time domain Mössbauer spectroscopy was in 1960, and it was called "time filtering" at that time. Although some interesting results were reported, research effort using this approach soon faced many difficulties. After SR became available, the excellent pulsed and periodic properties of SR revitalized this area of research, and the development in the last two decades has also warranted it a proper name: time domain Mössbauer spectroscopy (TDMS). Experimentally, the methodology uses nuclear Bragg scattering or forward scattering to observe the coherent decay at different times after the nuclear system has been excited. Mössbauer parameters (such as f_{LM} and δ_{SOD}) and hyperfine interactions can also be studied by TDMS, through the analysis of new phenomena such as speed-up effect of initial decay, dynamical beats, quantum beats, etc. We describe these phenomena in this section. The theoretical aspects of TDMS were derived from classical optics for isotopic radiation sources by Lynch et al. [26], and for SR by Kagan et al. [27] and Hannon and Trammell [28–30].

7.3.1 Nuclear Exciton

Most results of elastic nuclear resonant experiments, where the coherent effects are clearly revealed, can be easily understood if we assume that the Mössbauer nuclei in a sample are excited as a whole and consequently decay freely. The γ-ray emitted by an excited nucleus may be re-absorbed or scattered by other nuclei that are identical to the emitting nucleus. So, it is possible for a nuclear excitation to propagate elastically throughout the entire ensemble of nuclei. Each nucleus is no longer isolated, but interacts with others. Without this interaction, an excited nucleus would decay with the natural lifetime. But an interacting ensemble of nuclei behaves differently. Such a collective nuclear excitation phased in time by Mössbauer radiation or SR pulses is known as a nuclear exciton [27, 29], which is a spatial superposition of various excited hyperfine levels of all nuclei in the sample. The elastic decay of this exciton is characterized by speed-up and beat modulations of intensity, and exhibits a peculiar property, namely the emitted γ-rays exist predominantly in spatially coherent channels, i.e., mostly oriented in the forward or the Bragg direction.

On the other hand, inelastic decay of an exciton does not differ from the decay of an excited individual nucleus.

7.3.2
Enhancement of Coherent Channel

In the kinematical approximation, the resonant scattering amplitude of Mossbauer radiation by n nuclei in a sample can be expressed by a phased sum as

$$f^{\mathrm{N}} = \sum_{l=1}^{n} \mathrm{e}^{i\boldsymbol{s}\cdot\boldsymbol{r}_l} f_l^{\mathrm{N}}(\boldsymbol{k}_0, \boldsymbol{k}', \omega) \tag{7.11}$$

where $\boldsymbol{s} = \boldsymbol{k}' - \boldsymbol{k}_0$, f_l^{N} is just the scattering amplitude of an incident photon of energy $\hbar\omega$ by an atom l, and $\boldsymbol{k}_0$ and $\boldsymbol{k}'$ are wave vectors before and after scattering. If all nuclei in the sample are equivalent, we can factor f_l^{N} out of the summation. To find the scattering intensity, which is proportional to $|f_l^{\mathrm{N}}|^2$, we must calculate the product of the double sum in l and l'. It is convenient to calculate first the terms with $l = l'$, whose sum equals n, then those with $l \neq l'$. Therefore, one gets [31]

$$I(\omega, \boldsymbol{s}) \approx \left| \sum_{l}^{n} \mathrm{e}^{i\boldsymbol{s}\cdot\boldsymbol{r}_l} \right|^2 \tag{7.12a}$$

$$= n + \left(\sum_{l}^{n} \mathrm{e}^{-i\boldsymbol{s}\cdot\boldsymbol{r}_l} \right) \left(\sum_{l' \neq l}^{n} \mathrm{e}^{i\boldsymbol{s}\cdot\boldsymbol{r}_{l'}} \right). \tag{7.12b}$$

The double sum accounts for interference contributions from all pairs of nuclei l and l'. If there is no spatial correlation between atoms in the sample and $\boldsymbol{s} \neq 0$, the relative phases $\boldsymbol{s} \cdot \boldsymbol{r}_l$ are uniformly distributed over the interval 0–2π. It can be proved that the double sum in Eq. (7.12b) approaches zero provided that the number n is large enough. Hence, the scattering intensity will be proportional to n, and applies to all incoherent processes, such as internal conversion.

If we have $\boldsymbol{s} = 0$ or $\boldsymbol{\tau}$ (where $\boldsymbol{\tau}$ is a reciprocal-lattice vector), Eq. (7.12a) gives n^2, typical for spatially coherent scattering. Therefore the coherent radiative channel (NFS or NBS) is immensely enhanced relative to incoherent channels. An estimate shows that for the 14.4 keV resonance in ^{57}Fe the enhancement can be as high as 1000 [32]. The coherent constructive interference during the decay of a nuclear exciton determinates the physical origin of a strong enhancement in the radiative channel. This interesting effect has been considered by many researchers [27, 29–31, 33–36].

Note that in most nuclear scattering experiments, multiple scattering processes cannot be neglected, but must be treated by the dynamical theory. In such a case, the enhancement is often accompanied by a broadening of the frequency distribution of the scattered radiation.

7.3.3
Speed-Up of Initial Decay

We begin with some early results from nuclear forward scattering in conventional Mössbauer spectroscopy. Figure 7.10 shows a block diagram of a circuit for measuring the Mössbauer effect as a function of time (i.e., a time spectrum) using a ^{57}Co isotope source. It is a typical delay coincidence circuit. First, suppose the absorber A is temporarily removed from the apparatus. The 123 keV γ_1 signal sets the zero time at the formation of the Mössbauer energy levels, and this signal is used as the "start" pulse for the time-amplitude converter (TAC). After the decay process, the 14.4 keV γ_M is used as the "stop" pulse. The TAC output, a pulse whose amplitude is proportional to the delay between the γ_1 and γ_M signals, is to

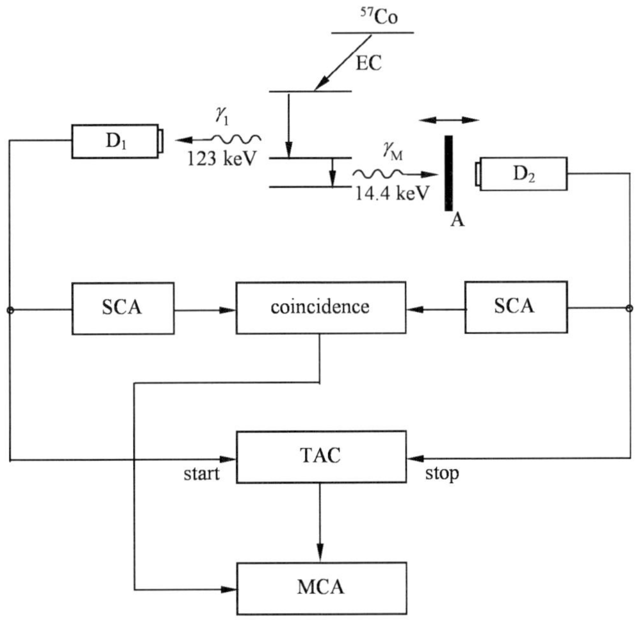

Fig. 7.10 Block diagram of the prompt delay coincidence detection circuit.

be stored in the multichannel analyzer (MCA). One single-channel analyzer (SCA), connected to each of the two detectors, is used for energy discrimination. The signal from the coincidence output is utilized as a gate pulse for the MCA. With no absorber, the standard circuit measures the lifetime, so the output should follow a simple exponential law:

$$I(t) = I_0 e^{-t/\tau_0}. \tag{7.13}$$

Now the absorber A containing ^{57}Fe is inserted and driven to oscillate. When its velocity is high, the radiation is slightly absorbed but the exponential time dependence in Eq. (7.13) is hardly affected. When the velocity of the absorber is such that it allows a resonance absorption, we observe immediately after $t = 0$ a higher decay rate than predicted by Eq. (7.13). This phenomenon is known as the speed-up of initial decay. At a later time, the decay rate is partially restored, resulting in the ringing pattern in an overall time spectrum, similar to the theoretical curves in Fig. 7.11.

Before describing this effect, we will start with a simpler situation, under which the emitted γ-spectrum presents an exact Lorentzian distribution. Considering the Mössbauer nucleus as a damped oscillator, we describe its radiation by an electromagnetic wave with an angular frequency ω_0, a speed $c = \omega/k$ propagating in the z-direction, and an exponentially decaying amplitude. The electric field of this wave is expressed as

$$E(z,t) = E_0 \exp\left[i(\omega_0 t - kz) - \frac{\Gamma}{2} t\right]. \tag{7.14}$$

Neglecting the kz term for the moment and Fourier transforming this function into the frequency domain, we obtain

$$E(\omega) = \frac{1}{2\pi}\int_{-\infty}^{\infty} E_0 \exp\left(i\omega_0 t - \frac{\Gamma}{2} t\right) e^{-i\omega t}\, dt = \frac{1}{2\pi i}\frac{E_0}{(\omega - \omega_0) - i\Gamma/2}. \tag{7.15}$$

The frequency dependence of the relative intensity is

$$I(\omega) \propto \frac{E_0^2}{4\pi^2}\frac{1}{(\omega - \omega_0)^2 + \Gamma^2/4} \tag{7.16}$$

which is the familiar Lorentzian distribution with a maximum intensity at $\omega = \omega_0$ and a FWHM of Γ.

If we detect photons in the time interval from 0 to t_m only, the upper limit in the above integral would be t_m, and the relative intensity would be [36]

$$I(\omega, t_m) \propto \frac{1 + e^{-\Gamma t_m} - 2e^{-\Gamma t_m/2}\cos[(\omega - \omega_0)t_m]}{(\omega - \omega_0)^2 + \Gamma^2/4}. \tag{7.17}$$

When $t_m \ll 1/\Gamma$, the above simplifies to

$$I(\omega, t_m) \propto \frac{2 - 2\cos[(\omega - \omega_0)t_m]}{(\omega - \omega_0)^2} = 4\left[\frac{\sin[(\omega - \omega_0)t_m/2]}{(\omega - \omega_0)}\right]^2. \tag{7.18}$$

This frequency distribution has a half-width of approximately $1/t_m$. Therefore, if t_m is much smaller than $1/\Gamma$, the spectral width of the emitted line will increase from the natural width Γ to a much larger value of $1/t_m$. The above also shows that an exponential decay corresponds to the Lorentzian frequency distribution. On the other hand, if the frequency distribution is no longer Lorentzian but like that in Eq. (7.18), the decay is also expected to deviate from the exponential decay with the natural lifetime τ_0.

In order to describe the speed-up effect, Lynch et al. [26] have applied classical optics theory to the transmission of radiation through a dispersive medium of an assembly of resonant atoms. First, the incident radiation is decomposed into frequency components, each of which gets absorbed and phase-shifted differently during propagation. Then, the time evolution of the outgoing radiation is obtained by Fourier transformation. They arrived at the following result for the relative intensity of the transmitted radiation:

$$I'(\omega, \tau) = e^{-\tau}\left|\sum_{n=0}^{\infty}\left(i\frac{4}{t_a}\frac{\omega - \omega_0}{\Gamma}\right)^n\left(\frac{t_a\tau}{4}\right)^{n/2} J_n(\sqrt{t_a\tau})\right|^2 \tag{7.19}$$

where t_a is the effective absorber thickness, $\tau \equiv t_m/\tau_0$, $\tau_0 = 141$ ns is the average lifetime of the excited state in ^{57}Fe, and J_n represents the Bessel function of the nth order. Not only does this formula correctly describe the experimental results of time dependence of resonantly transmitted γ-rays, it also marks the beginning of time domain Mössbauer spectroscopy. When $\omega = \omega_0$, this reduces to

$$I'(\omega_0, \tau) = e^{-\tau}[J_0(\sqrt{t_a\tau})]^2. \tag{7.20}$$

Figure 7.11 shows some of the graphs based on Eq. (7.20) which describes the main features of the experimental results. To observe significant effects of the speed-up of initial decay, the effective thickness t_a should be larger than 1. If we limit our attention to the region near $t_m \approx 0$ in Eq. (7.20), the high-order terms in the Taylor expansion of J_0 may be neglected and Eq. (7.20) becomes

$$I'(\omega_0, \tau) \approx e^{-\tau}[1 - (\sqrt{t_a\tau}/2)^2]^2 \approx \exp\left[-\tau\left(1 + \frac{t_a}{2}\right)\right]. \tag{7.21}$$

In this case, it is again approximately an exponential decay, except for the additional term $t_a/2$, which causes a faster decay process. When $t_a = 2$, the decaying

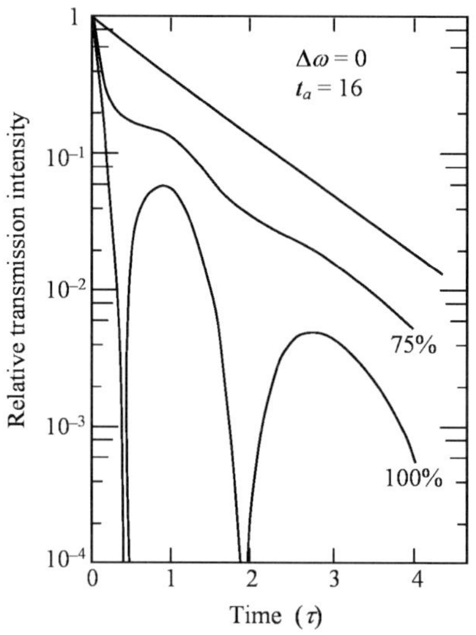

Fig. 7.11 Time dependence of radiation after transmission through a resonant filter, calculated according to Eq. (7.20), assuming that the radiation is 75 or 100% recoilless. The straight line represents an exponential decay for comparison. All curves are normalized to 1 at $t = 0$.

process would appear to be twice as fast. Since $t_a = a_m n_a f \sigma_0 d$, the speed-up effect clearly depends on the physical thickness of the absorber, the abundance of the resonance isotope (a_m), and the recoilless fraction. For example, if the number of resonance isotopes is decreased, the decay tends to have less ringing; if the medium does not contain the resonance isotope at all, the time spectrum curve becomes a straight line (Fig. 7.11). This interesting phenomenon was first experimentally observed by Lynch et al. in 1960.

The amount of broadening in the frequency distribution in Eq. (7.17) is dictated by how much shorter the measurement time t_m is compared with the average lifetime τ_0. The frequency distribution broadening can be quantitatively measured by transmission Mössbauer spectra [36], as shown in Fig. 7.12(a). In the experiment, both the source and absorber were stainless steel. Each spectrum was collected from the formation of the 14.4 keV state to a time t_m. As can be seen, the spectral line clearly becomes broader as t_m decreases. When $t_m = 150$ ns (t_m having the same value as τ_0, or $\tau = 1.00$), the half-width of the spectral line is about 0.5 mm s^{-1}, which is still about twice the width of a typical absorption line. The curves in Fig. 7.12(b) are calculated based on Eq. (7.18), and they agree with the experimental data quite well. Incidentally, the speed-up in decay and the broadening of spectral line are related by the time–energy uncertainty principle, i.e., $\Delta t \Delta E \sim \hbar$.

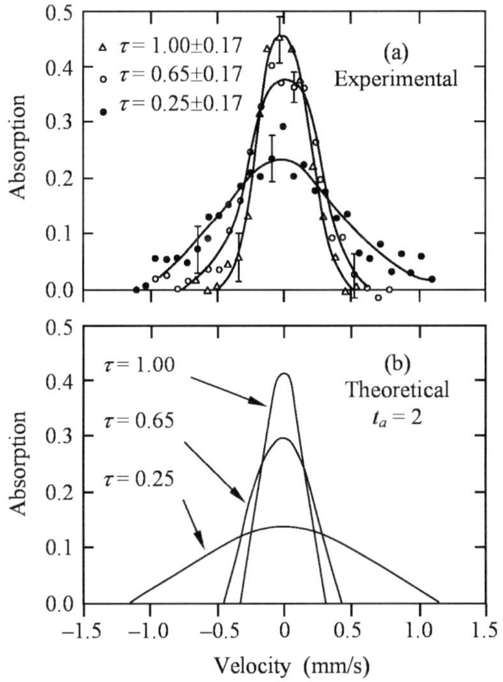

Fig. 7.12 (a) Mössbauer spectra of a stainless steel absorber observed at various delay times τ (relative to the lifetime τ_0) after the formation of the 14.4 keV excited state. (b) Theoretical curves calculated using $t_a = 2$ and the corresponding τ values.

7.3.4
Nuclear Forward Scattering of SR

Synchrotron radiation consists a series of sharp pulses, with a duration of about 10^{-10} s for each pulse and a separation of about 10^{-6} s between pulses. The duration time is very small compared to the lifetime of the Mössbauer level in ^{57}Fe ($\sim 10^{-7}$ s). Therefore, such a coherent SR flash causes a simultaneous excitation of nuclear ensemble in the sample. There also exists a time correlation. Due to the long lifetime of the excited states, the prompt radiation scattered by electrons (as a background) and the delayed radiation of nuclear resonant scattering are separated in time.

Nuclear forward scattering (NFS) experiments may be considering the time domain analog of conventional Mössbauer experiments. In the latter case, the recorded signal presents an incoherent sum of the spectral components of the transmitted radiation. By contract, in NFS the time response is a coherent sum of spectral components of the scattered radiation. In NFS there is only one coherent decay channel, and the nuclear exciton involves only one wave propagating in the sample, and hence no suppression of absorption is possible. The dynamical theory of scattering under these conditions is a familiar one in optics and leads to the concept of refractive index. The nuclear "refractive index" is usually represented by a 2×2 matrix. Its imaginary components are related to the absorption cross-section and its real components describe dispersion of the γ-radiation. If the incident beam is purely σ-polarized, the refractive index can be written as a complex scalar [37]

$$n(\omega) = 1 + \frac{\mu_r}{4k_0} \frac{\Gamma}{\hbar(\omega_0 - \omega) + i\Gamma/2} \tag{7.22}$$

where μ_r is defined in (1.16). Comparing this with Eq. (6.19), we find that the refractive index is connected to the forward scattering amplitude f^N by

$$n(\omega) = 1 + \frac{2\pi}{k_0^2} n f^N, \tag{7.23}$$

where n is the number density of resonance nuclei per unit volume. When two or more hyperfine lines are excited, the refractive index must be written as

$$n(\omega) = 1 + \frac{\mu_r}{4k_0} \sum_{j=1}^{n} \frac{\Gamma}{\hbar(\omega_0 - \omega_j) + i\Gamma/2}. \tag{7.24}$$

In principle, the scattering problem should be solved by a quantum mechanical method. For the sake of a semiquantitative discussion, we adopt a classical picture and introduce the electric field amplitude $A_0(\omega)$. The amplitude of the wave transmitted by a medium with a thickness d is then

$$A(\omega) = A_0(\omega) \exp[-in(\omega)k_0 d], \tag{7.25}$$

and the transmission is

$$T'(\omega) = \exp[-in(\omega)k_0 d]. \tag{7.26}$$

Assuming that the system under investigation is linear and time-invariant, for a given frequency response of the system, we can calculate the time response by a Fourier transform. Because we are only interested in the nuclear resonant part, only the second term in Eq. (7.22) is needed for inserting $n(\omega)$ into Eq. (7.26), and we have

$$T'(\omega) = \exp\left[-\frac{it_a/2}{2\hbar(\omega - \omega_0)/\Gamma + i}\right]. \tag{7.27}$$

Therefore, the time response function $R(t)$ of the sample is obtained by the reverse Fourier transform:

$$R(t) = \frac{c}{2\pi}\int_{-\infty}^{\infty} T'(\omega)\mathrm{e}^{-i\omega t}\,\mathrm{d}\omega \tag{7.28}$$

where c is a frequency-independent constant. After the integration is carried out [27], we obtain

$$R(t) = c\left[\delta(t) - \mathrm{e}^{-i\omega_0 t}\mathrm{e}^{-\tau/2}\left(\frac{t_\mathrm{a}}{2\tau_0}\right)\frac{J_1(\sqrt{t_\mathrm{a}\tau})}{\sqrt{t_\mathrm{a}\tau}}\theta(t)\right] \tag{7.29}$$

where

$$\theta(t) = \begin{cases} 1 & t > 0 \\ 0 & t < 0 \end{cases} \quad \text{and} \quad \tau = \frac{t}{\tau_0}.$$

The time-domain intensity for NFS is given by

$$I_\mathrm{fs}(t \geq 0) = |R(t \geq 0)|^2 = |c|^2\mathrm{e}^{-\tau}\left(\frac{t_\mathrm{a}}{2\tau_0}\right)^2\left[\frac{J_1(\sqrt{t_\mathrm{a}\tau})}{\sqrt{t_\mathrm{a}\tau}}\right]^2. \tag{7.30}$$

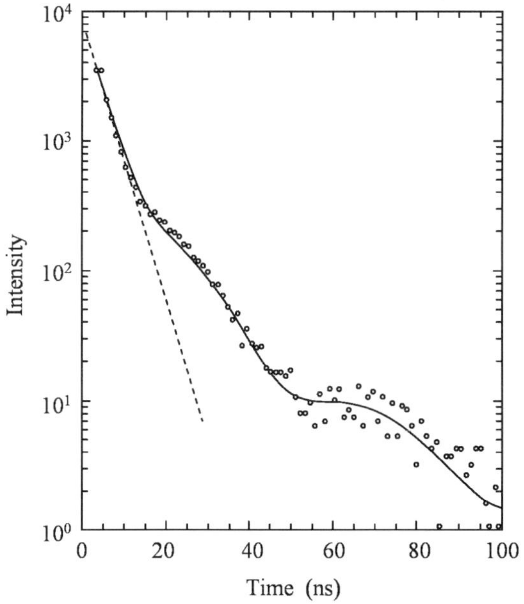

Fig. 7.13 Time response of ^{57}Fe/Sc/^{56}Fe/Sc nuclear multilayer at the nuclear Bragg angle. The solid line is a dynamical diffraction theory fit and the dashed line indicates the initial decay with a lifetime of 4 ns.

The quantity $|c|^2$ includes the incident intensity I_0 of SR and an attenuating factor due to photoelectric absorption.

This result can be similarly simplified near $\tau \approx 0$ by taking the appropriate approximations, and it becomes

$$I_{\mathrm{NFS}}(t) \propto t_{\mathrm{a}} \exp\left[-\tau\left(1+\frac{t_{\mathrm{a}}}{4}\right)\right]. \tag{7.31}$$

Comparing this with Eq. (7.21), we see that for the same t_{a}, the rate of decay using SR is slower than that using a radiation source. Figure 7.13 shows the time spectra from a nuclear multilayer film at the Bragg reflection [18]. The spectrum reveals the remarkable decay speed-up, with an initial decay equivalent to a lifetime of only 4 ns (the dashed line).

7.3.5
Dynamical Beat (DB)

In the previous section, we were only concerned with the decay characteristics during a short duration before the first zero of the Bessel function J_0 or J_1. Now we want to observe a complete time spectrum where the Bessel function J_0 or J_1 passes through zero several times. This requires a longer measurement time depending on the particular value of t_{a}. Such spectra exhibit the "ringing" pattern [38] as shown in Fig. 7.14. This type of intensity modulation is called dynamical beat (DB). Both decay speed-up and ringing are results of the coherent decay of a nuclear exciton. It can also be understood as an interference effect in coherent nuclear forward scattering or nuclear resonant scattering at the Bragg angle. Unlike the radiation source, the SR produces nuclear forward scattering with almost no background counts. The quality of a time domain spectrum depends only on the intensity of SR and the size of the time window.

In order to investigate a pure DB without disturbance by a quantum beat, materials with single-line absorption should be used. The time spectra of NFS from $(NH_4)_2Mg^{57}Fe(CN)_6$ powder samples of different t_{a} are shown in Fig. 7.14 where it can be seen that (1) the DB is aperiodic and the apparent periods increase with time and (2) the apparent periods decrease with increasing effective thickness t_{a}. These two characteristic features of DB are determined basically by the Bessel function J_1 with the argument $\sqrt{t_{\mathrm{a}}\tau}$. Note also that the spectra in the first 10 to 20 ns cannot be resolved because of detector overload and veto in the electronic circuit temporarily.

7.3.6
Quantum Beat (QB)

Beats can be easily observed in sound waves and radio waves, but they appear in optical waves, x-rays, or γ-rays only under certain particular conditions. Quantum effects may be completely ignored in long-wavelength cases, but must be consid-

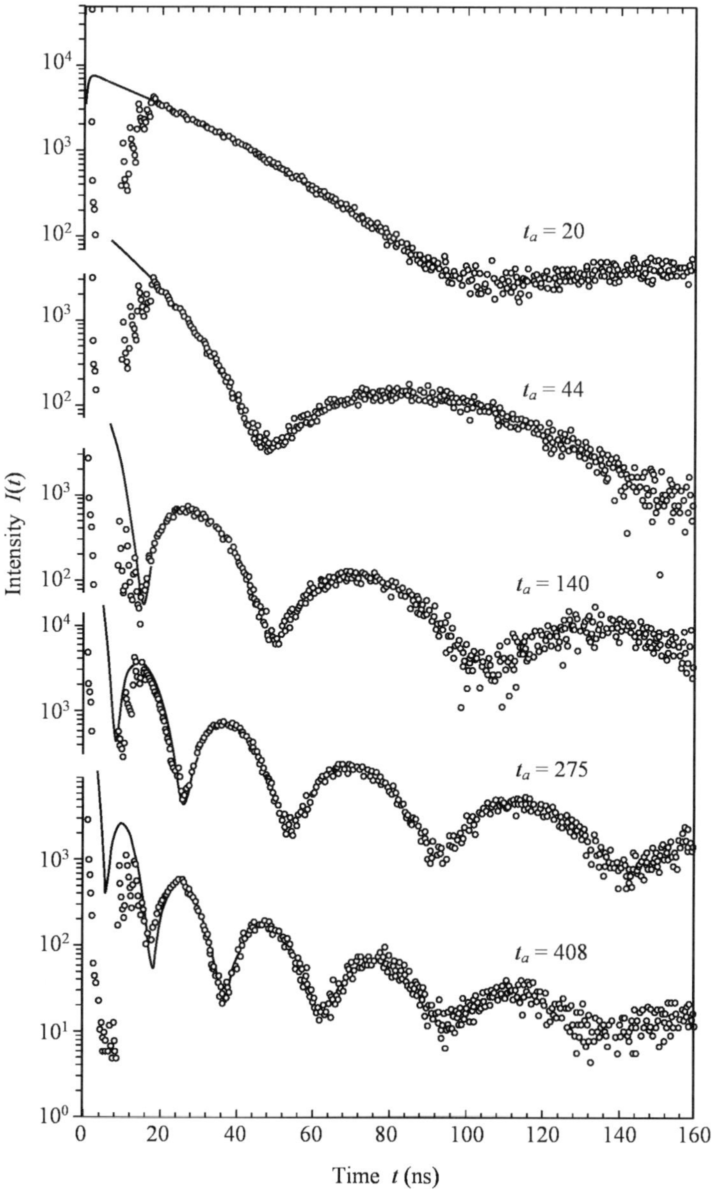

Fig. 7.14 Time evolution of NFS SR through $(NH_4)_2Mg^{57}Fe(CN)_6$ powder samples of different effective thicknesses t_a. The aperiodic modulation is the DB. The solid lines are fits using the NFS theory [38].

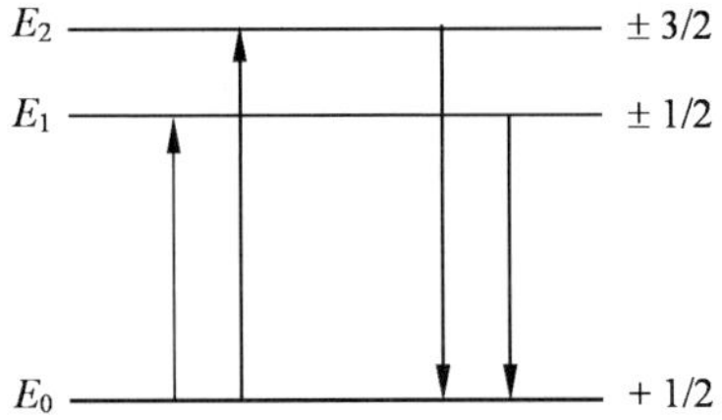

Fig. 7.15 Transitions involved in the phenomenon of quantum beats.

ered in the latter group, which may be the reason why quantum beat (QB) was so named. Quantum beats in optics were first observed in 1964. We now discuss the QB in γ-rays.

Suppose that SR pulses coherently excite an ensemble of ^{57}Fe nuclei in a sample from their ground state to quadrupole split energy levels E_1 and E_2 (Fig. 7.15). Two nuclear excitons with energies E_1 and E_2 are created by the above process. Of cause, they are also coherent and will interfere with each other. It can be shown [39] that the intensity of NFS is given by

$$\begin{aligned} I_{fs}(t) &\propto |N_1 e^{iE_1 t/\hbar} + N_2 e^{iE_2 t/\hbar}|^2 e^{-t/\tau_0} \\ &= (N_1^2 + N_2^2)e^{-t/\tau_0} + 2N_1 N_2 e^{-t/\tau_0} \cos\left(\frac{E_2 - E_1}{\hbar}\right)t. \end{aligned} \tag{7.32}$$

In our case, $N_1 = N_2 = N$ is proportional to the concentration of Mössbauer nuclei in the sample. Therefore, expression (7.32) becomes

$$I_{fs}(t) \propto N^2 e^{-it/\tau_0} \cos^2(\Omega t/2) \tag{7.33}$$

with $\Omega = (E_2 - E_1)/\hbar$. This indicates a periodic decay, known as QBs, an interference phenomenon in the time domain. As in the analysis of any other interference phenomenon, we have added the two amplitudes in Eq. (7.32) before taking the square to calculate the intensity, but the most important condition here is the simultaneous and instantaneous excitation of the nuclei in the absorber.

With the QB included, Eqs. (7.30) and (7.33) can be combined to give the total intensity of nuclear forward scattering

$$I_{fs}(t) \propto t_a^2 e^{-t/\tau_0} \left[\frac{J_1(\sqrt{t_a \tau})}{\sqrt{t_a \tau}}\right]^2 \cos^2(\Omega t/2), \tag{7.34}$$

which indicates that QB is periodic. By appropriate choice of the direction of a weak external magnetic field, only the two $\Delta m = 0$ transitions in ^{57}Fe metal foils are excited, and time spectra from such samples [37] are shown in Fig. 7.16, where the time windows are open before the first zero of J_1. Since both $\Delta m = 0$ transitions have the same partial effective thickness, the dynamical beats due to

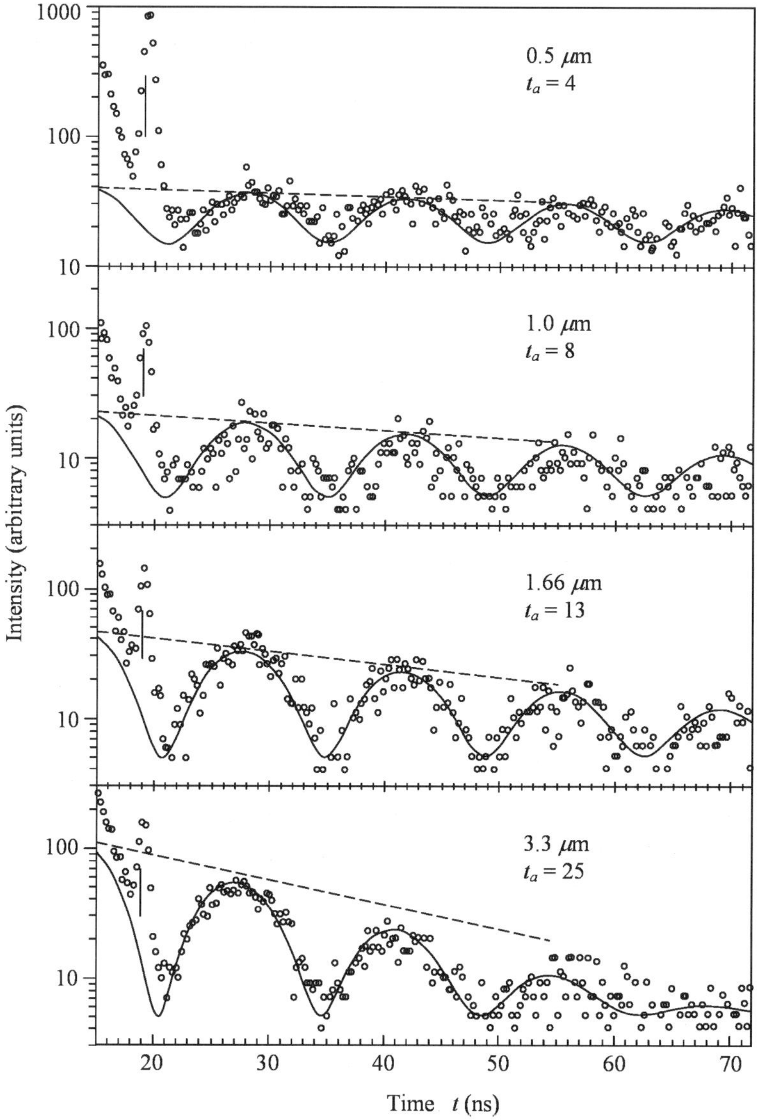

Fig. 7.16 Time evolution of NFS through ^{57}Fe metal foils of different effective thicknesses in a vertical magnetic field. Only the two $\Delta m = 0$ transitions were excited. The solid lines are computations based on Eq. (7.30). The dashed lines indicate the exponential decay of the envelope as calculated using Eq. (7.31).

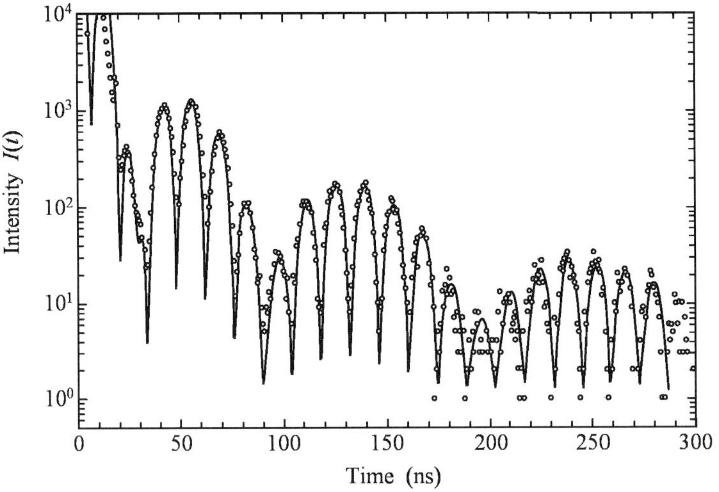

Fig. 7.17 Time evolution of NFS through a ^{57}Fe metal foil (~9 μm) at 4 K in a vertical magnetic field of 1 T [38]. Only the two $\Delta m = 0$ transitions were excited, with effective thickness $t_a \approx 75$ each. The DB is seen as an envelope modulation over the fast QB. The solid lines are fits using the NFS theory.

these transitions coincide. The fact that intensities of NFS increase with t_a is reflected in this figure. Increasing the time window or the effective thickness t_a, the DB and QB melt into hybrid forms of beating (Fig. 7.17).

If SR simultaneously excites more than two transitions, the interference pattern between the spectral lines of resonant scattering will be very complex. The nuclear Bragg scattering (NBS) of $FeBO_3$ can provide such an example [16, 41, 42]. A magnetic field is applied to the $FeBO_3$ crystal perpendicular to its scattering plane $(\boldsymbol{k}_0, \boldsymbol{k}')$ so that an internal magnetic field is parallel to $\boldsymbol{k}_0 + \boldsymbol{k}'$, resulting in only four $\Delta m = \pm 1$ transitions (Fig. 7.18).

The phase and intensities of these transitions are $+1$, $-1/3$, $+1/3$, and -1, respectively. Note that only transitions of the same polarization state interfere; hence we get the following simple time spectrum modulated by QBs:

$$I(t) \propto e^{-t/\tau_0} \left| \sin\left[\frac{1}{2}\Omega(1,6)t\right] - \frac{1}{3} e^{-i(\Delta\Omega)t} \sin\left[\frac{1}{2}\Omega(3,4)t\right] \right|^2 \tag{7.35}$$

where the first term describes the main features in the spectrum because it has a higher beat frequency corresponding to a period of 8.1 ns at room temperature. From this beat frequency, we calculate transition energy $\hbar\Omega(1,6) = 5.156 \times 10^{-7}$ eV, and therefore the magnetic hyperfine field $B_{\text{eff}} = 33.35 \pm 0.02$ T. Using QBs to deduce the hyperfine field value is a very accurate method, because we often observe more than one or two periods. As shown in Fig. 7.19, the first eight oscillations are very definite and the period can be precisely determined.

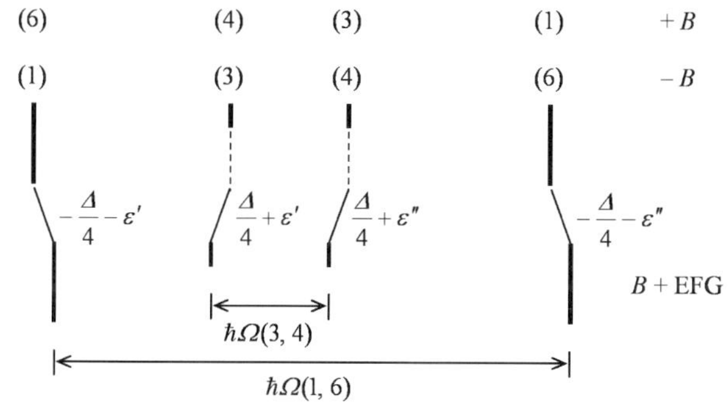

Fig. 7.18 Four $\Delta m = \pm 1$ spectral lines in $^{57}FeBO_3$ due to hyperfine splittings with a magnetic field B and an electric field gradient, where the ε shifts of the full Hamiltonian are added. The transitions are (1) $+1/2 \leftrightarrow +3/2$, (3) $+1/2 \leftrightarrow -1/2$, (4) $-1/2 \leftrightarrow +1/2$, and (6) $-1/2 \leftrightarrow -3/2$.

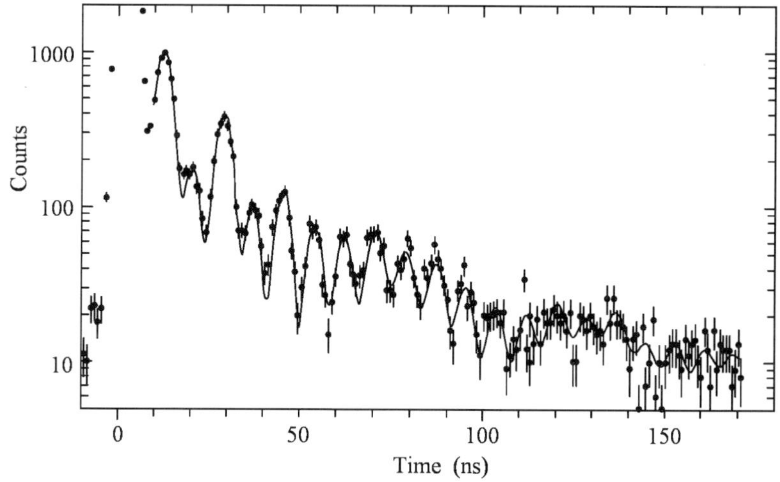

Fig. 7.19 Time spectrum (quantum beat) measured at the Bragg angle θ_B of $^{57}FeBO_3(1\ 1\ 1)$.

Usually, the frequencies of minor QBs are smaller and their intensities are also weaker, causing only small-amplitude modulations in the main oscillation. As a consequence, the overall spectrum seems somewhat ragged, but the hyperfine field measurement is not compromised.

Isomer shifts can also be measured with the forward scattering approach. To do this, a reference sample with a known isomer shift is attached to the sample under investigation. In such an experiment, one exciton can extend over these two samples. Two time spectra (with and without reference sample) are needed.

The isomer shift of the sample with respect to the reference sample is then obtained from a beat pattern of time spectra [43].

7.3.7
Distinctions between Time Domain and Energy Domain Methods

Energy domain Mössbauer spectroscopy is based on the method of resonance absorption. The transmitted counts of photons are measured as functions of their energies, i.e., an energy spectrum, which represents an incoherent sum of the spectral components of the transmitted radiation. In other words, the transmitted spectrum reflects the incoherent process of nuclear resonance absorption by individual nuclei.

By contrast, TDMS belongs to the scattering method. A scattering spectrum measured as a function of time is a coherent sum of the spectral components of the scattered radiation from nuclei collectively excited by an SR pulse. This leads to important interference effects in TDMS.

This is the fundamental distinction between these two methods, and consequently there are many theoretical and experimental differences between time domain and energy domain Mössbauer spectroscopies. The different characteristic features of time domain and energy domain spectra are demonstrated in Figs. 7.20 and 7.21 [31].

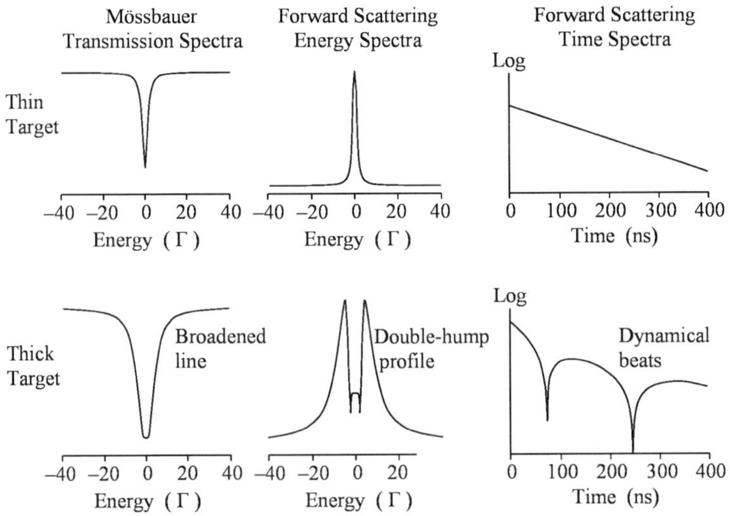

Fig. 7.20 Mössbauer transmission spectra (left column), and synchrotron radiation scattering spectra in the energy domain (middle column) and in the time domain (right column) for the case of a single resonance in ^{57}Fe-enriched stainless steel. Upper row: a thin target of 0.2 μm; lower row: a thick target of 3 μm.

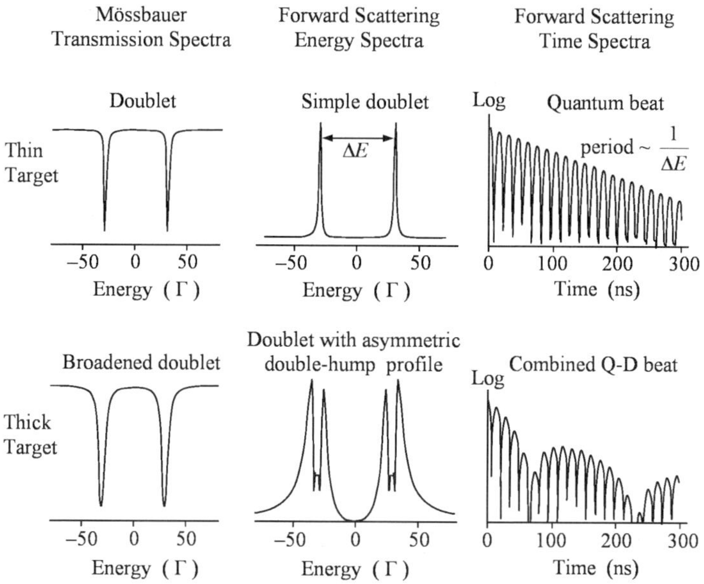

Fig. 7.21 Mössbauer transmission spectra (left column), and synchrotron radiation scattering spectra in the energy domain (middle column) and in the time domain (right column) for the case of a resonance doublet in ^{57}Fe-enriched stainless steel. Upper row: a thin target of 0.2 μm; lower row: a thick target of 3 μm.

7.3.8
Measurement of the Lamb–Mössbauer Factor

In the time domain, it is also through measuring t_a that the Lamb–Mössbauer factor f_{LM} is determined. Using nuclear forward scattering (NFS), t_a may be measured by the speed-up effect of coherent decay or the minimum positions of dynamical beats. We now look at each of these two methods.

1. The first example is single-crystal guanidinium nitroprusside (GNP), $(CN_3H_6)_2[Fe(CN)_5NO]$ [44]. With a relatively thin scatterer and the time window open before the first zero of J_1 function, the resultant NFS spectra are shown in Fig. 7.22. The values of t_a are deduced by fitting the experimental data with Eq. (7.30) or (7.31). If the number of Mössbauer nuclei per unit area is known, f_{LM} can be easily calculated. The experimental results are $f_{LM}^{(a)} = 0.122 \pm 0.010$ and $f_{LM}^{(c)} = 0.206 \pm 0.010$ for the orientations with the crystal a-axis and c-axis parallel to the incident SR beam, respectively. The corresponding Debye temperatures are 140 and 160 K. The difference in f_{LM} values along different crystal directions is due to the G-K effect. This method is applicable in either single-line resonant scattering or quadrupole doublet resonant scattering.

2. In the second example, a larger time window is open to observe as many periods of DB as possible so that t_a and f_{LM} can be deduced more accurately. Since

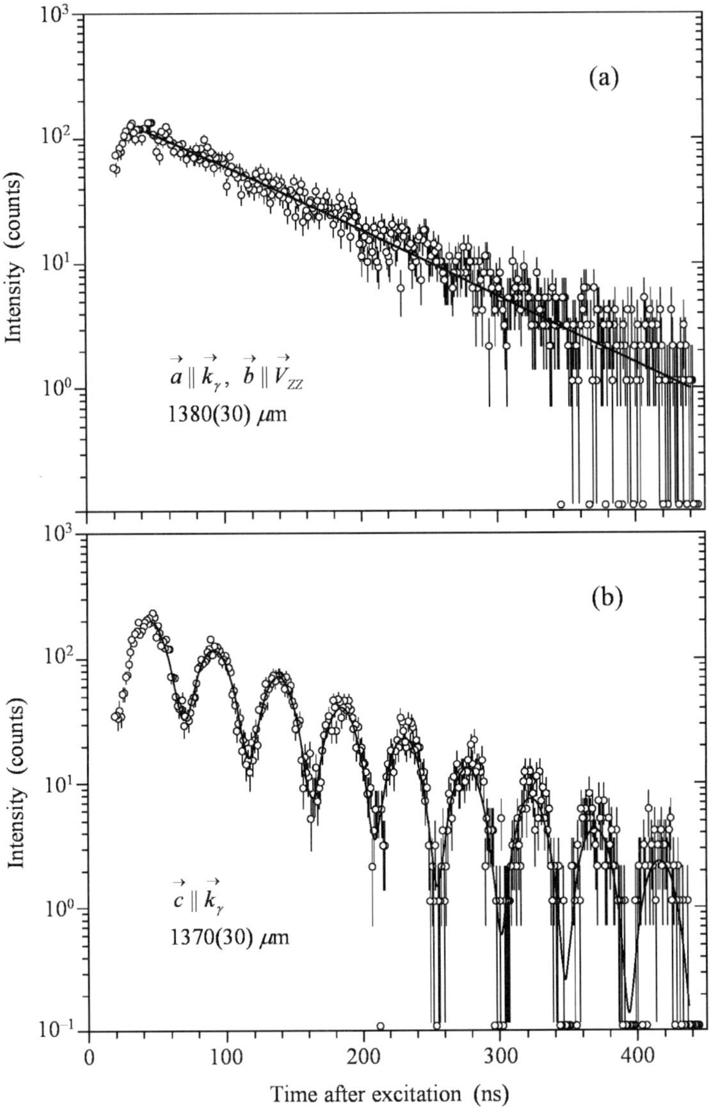

Fig. 7.22 NFS spectra of guanidinium nitroprusside single crystals recorded at room temperature with single-crystal orientations and thicknesses as indicated. The solid lines result from a least-squares fit.

$t_a = \sigma_0 f_{LM} n_a d$, if the scatterer is kept at the same temperature, the thickness d is the only variable in t_a. In this case the DB patterns have already been demonstrated in Fig. 7.15. On the contrary, suppose now we keep d constant but make the scatterer temperature the only variable, then the Lamb–Mössbauer factor f_{LM} and the magnetic hyperfine field (or Ω) will both change accordingly. The method involves measurements of a series of time spectra at different temperatures and the determination of the temperature dependence of both the magnetic hyperfine field from the QB periods and the Lamb–Mössbauer factor f_{LM} from the DB "periods." Considering only DB and using $t_a \propto f_{LM}$, Eq. (7.34) gives the relationship between the relative intensity $I(t)$ and the Lamb–Mössbauer factor f_{LM} as follows:

$$I(t, f_{LM}) \propto e^{-t/\tau_0} f_{LM} J_1^2(c\sqrt{f_{LM}}t) \tag{7.36}$$

where c is a constant. The intensity minima are just the zeroes of J_1, and by locating these minima the Lamb–Mössbauer factor f_{LM} can be determined. An excellent set of experiments has been carried out using a polycrystalline α-Fe foil of thickness (10.57 ± 0.13) μm and a 95% ^{57}Fe enrichment [45]. A magnetic field of 0.6 T was applied in the plane of the foil so that only the $\Delta m = 0$ transitions were allowed. Time spectra at many different temperatures were obtained using SR of bandwidth 10 meV. The f_{LM}-values at those temperatures are listed in Table 7.1, and selected spectra are shown in Fig. 7.23.

The first zero of J_1 occurs when $c\sqrt{f_{LM}}t = 3.83$. When temperature T increases, causing f_{LM} to decrease, the first intensity minimum in the time spec-

Table 7.1 The Lamb–Mössbauer factor f_{LM} and the splitting $\hbar(\delta\omega)$ of an Fe foil in the temperature range 9.7–1048 K, obtained from the ^{57}Fe NFS time spectra.

Temperature (K)	f_{LM}	$\frac{f_{LM}}{f_{LM}\ (9.7\ K)}$	$\hbar(\delta\omega)$ ($\times 10^{-9}$ eV)
9.7	0.890 ± 0.020	1	297.42 ± 0.31
50	0.886 ± 0.020	0.996 ± 0.003	297.24 ± 0.31
100	0.868 ± 0.019	0.976 ± 0.003	296.13 ± 0.31
150	0.850 ± 0.019	0.955 ± 0.003	295.44 ± 0.31
200	0.823 ± 0.018	0.925 ± 0.003	294.31 ± 0.31
250	0.796 ± 0.018	0.895 ± 0.003	292.71 ± 0.30
298	0.771 ± 0.017	0.866 ± 0.003	290.60 ± 0.30
348	0.739 ± 0.016	0.831 ± 0.003	287.40 ± 0.30
513	0.649 ± 0.015	0.730 ± 0.003	274.44 ± 0.32
693	0.526 ± 0.014	0.591 ± 0.004	252.33 ± 0.38
773	0.492 ± 0.012	0.553 ± 0.002	236.36 ± 0.31
873	0.430 ± 0.010	0.483 ± 0.002	209.67 ± 0.29
973	0.359 ± 0.010	0.403 ± 0.003	152.95 ± 0.34
1008	0.336 ± 0.012	0.377 ± 0.006	119.20 ± 0.46
1023	0.316 ± 0.012	0.355 ± 0.006	89.02 ± 0.67
1031	0.298 ± 0.009	0.335 ± 0.004	60.76 ± 0.50
1033	0.286 ± 0.008	0.322 ± 0.003	50.97 ± 0.57
1042	0.295 ± 0.010	0.332 ± 0.005	19.49 ± 1.4
1048	0.285 ± 0.008	0.321 ± 0.003	6.97 ± 0.60

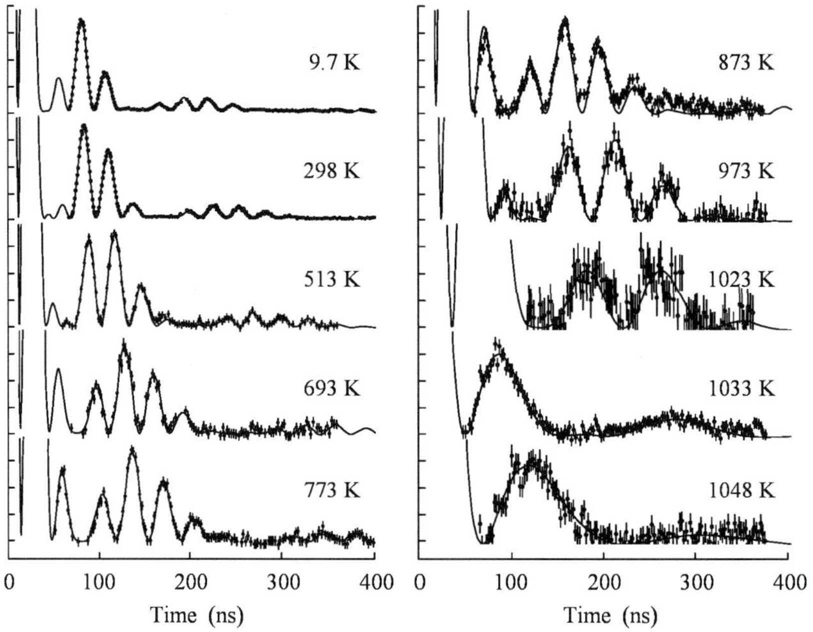

Fig. 7.23 Forward scattering time spectra of an ^{57}Fe foil at temperatures from 9.7 to 1048 K.

trum must then be shifted towards a later time, as seen in the experimental spectra. This fact can also be seen in Fig. 7.14, where the first minimum shifts to a later time as t_a decreases.

In Fig. 7.23, as the temperature increases, the QB frequency first decreases gradually, but when T is higher than 773 K the QB frequency decreases at a much faster rate, which reflects the temperature dependence of the magnetic hyperfine field B. When T approaches T_C, the magnetic hyperfine field decreases drastically and eventually disappears. For $T > T_C$, only DB remains. Comparing the two spectra at 693 and 1048 K, it is easy to see that the latter's f_{LM}-value is only one half of that of the former, but they have almost the same DB pattern. The reason is that the two $\Delta m = 0$ transitions become one after the magnetic hyperfine field breaks down, and the scattering intensity is therefore doubled, which happens to compensate the intensity loss due to the smaller f_{LM}. The relation between the magnetic hyperfine field B and temperature T is given in Fig. 7.24, which agrees well with previous results using other methods.

In either of the above two methods, measuring f_{LM} is not based on the height or area of the absorption spectral lines, therefore avoiding problems such as the saturation effect. In addition, since the samples are relatively thick compared to that in the transmission method, the error in the number of Mössbauer nuclei per unit area is also smaller. These factors improve the accuracy of the determination of f_{LM}. As listed in Table 7.1, the error in the absolute f_{LM}-value at room temperature is 2% and that in the relative f_{LM}-value is 0.4% (the relative f_{LM}-value

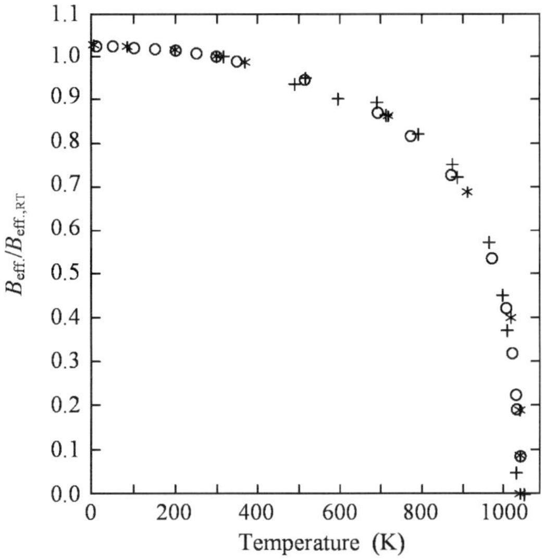

Fig. 7.24 Temperature dependence of effective internal magnetic field B_{eff} in α-Fe, normalized to the room temperature value. The results from nuclear forward scattering using SR (circles) are compared with transmission Mössbauer effect results (crosses and stars).

was at best given with an error of 1% in transmission Mössbauer spectroscopy with a radiation source).

7.4 Phonon Density of States

Soon after the Mössbauer effect was discovered, attempts were made to use it to measure the atomic vibration frequency distribution – phonon density of states (DOS). But it was not very successful for a long time because of several experimental difficulties; e.g., typical phonon energy transfers could not be reached with the conventional Doppler technique, the radiation sources were too weak to provide satisfactory statistical errors, etc. It was in 1995 that the phonon DOS $g(\omega)$ of α-Fe was first measured by incoherent inelastic nuclear resonant scattering using SR [7, 46, 47]. The experiments were performed at high-brilliance undulator sources with energy resolutions in the range of 6 meV. Since then the technique has made appreciable progress and nowadays phonon DOS are routinely recorded with sub-meV resolution.

This is a new technology and has several advantages. First, the cross-section of resonant scattering is usually large, which guarantees high counting rates. In addition, SR has high brilliance and narrow beams, especially suitable for studying those thin films and biological samples which may be of a small size or with a low content of the Mössbauer isotope. Second, when the lifetime of the excited nucleus is much longer than the SR pulse duration (~50 ps), the nuclear resonant scattering process can be separated from electronic scattering by counting only the delayed products such as atomic K-fluorescence photons after the disappearance of the prompt radiation and electronic scattering [48]. This leads to excellent signal-to-noise ratios (S/N ≈ 103) [46]. The noise level is basically determined by the detector and by the associated electronics. The high S/N ratio allows one to discriminate the multi-phonon contributions against the measured data. Third, incoherent inelastic nuclear resonant scattering directly offers the phonon/vibrational DOS regardless whether the material is single crystal, polycrystalline, or amorphous. However, such scattering is only sensitive to the vibrations of Mössbauer atoms; i.e., this technique provides a partial density of states. In addition, high precision and short experimental time are also important advantages.

7.4.1 Inelastic Nuclear Resonant Scattering

The theoretical basis for extracting lattice dynamics from Mössbauer measurements was given at the beginning of the 1960s by Singwi and Sjölander [49] and by Visscher [50].

A nucleus excited by resonance absorption of γ-rays may decay via one of the two mechanisms: radioactive decay or internal conversion with its subsequent fluorescence radiation. The relative probabilities of the two mechanisms are $1/(1+\alpha)$ and $\alpha/(1+\alpha)$, respectively, where α is the internal conversion coefficient. For most Mössbauer isotopes, $\alpha > 1$ and the dominating mechanism is internal conversion, an incoherent decay process. Thus, the total yield of the delayed K-fluorescence photons is given by

$$I(E) = I_0 n'_a \eta_k \frac{\alpha_k}{1+\alpha} \sigma(E) \tag{7.37}$$

where I_0 is the incident photon flux, n'_a the effective area density of the nuclei, η_k the K-fluorescence yield, and α_k the partial internal conversion coefficient. Also in Eq. (7.37), $\sigma(E)$ is the cross-section for nuclear resonant absorption of a photon with energy E:

$$\sigma(E) = \frac{\pi}{2} \sigma_0 \Gamma S(E - E_0) \tag{7.38}$$

where σ_0 is given by Eq. (1.12), and $S(E)$ is the normalized absorption probability per unit energy interval due to phonons. $S(E)$ is also $\boldsymbol{k}$-dependent in the general case of an anisotropic lattice. According to Singwi and Sjölander [49], it can be represented by

$$S(\boldsymbol{k}_0, E) = \frac{1}{2\pi} \operatorname{Re} \int_0^{\infty} \mathrm{d}\tau e^{-iE\tau - \Gamma|\tau|/2} F(\boldsymbol{k}_0, \tau) \tag{7.39}$$

with

$$F(\boldsymbol{k}_0, \tau) = \langle \mathrm{e}^{-i\boldsymbol{k}_0 \cdot \boldsymbol{u}(0)} \mathrm{e}^{i\boldsymbol{k}_0 \cdot \boldsymbol{u}(t)} \rangle_T \tag{7.40}$$

where Re indicates the real part of the integral, $\boldsymbol{k}_0$ is the wave vector of the incident γ-ray, E is the difference between the energy of the γ-ray and the resonance energy of the nucleus, $\tau = t/\hbar$, t is the time, Γ is the natural width of the excited nuclear state, and F is the time-dependent correlation function, which describes the correlation between the displacements $\boldsymbol{u}$ of the same nucleus at different moments of time. Using displacement $\boldsymbol{u}$ in Eq. (5.18), it can be proved that

$$F(\boldsymbol{k}_0, \tau) = f_{\mathrm{LM}}(\boldsymbol{k}_0) \exp\left[\frac{E_\mathrm{R}}{N} \sum_s \frac{(\boldsymbol{h} \cdot \boldsymbol{e}_s)^2}{\hbar\omega_s} (\langle n_s + 1 \rangle \mathrm{e}^{i\omega_s \tau} + \langle n_s \rangle \mathrm{e}^{-i\omega_s \tau})\right] \tag{7.41}$$

where

$$f_{\mathrm{LM}}(\boldsymbol{k}_0) = \exp\left[-\frac{E_\mathrm{R}}{N} \sum_s \frac{(\boldsymbol{h} \cdot \boldsymbol{e}_s)^2}{\hbar\omega_s} \langle 2n_s + 1 \rangle\right], \tag{7.42}$$

where $\boldsymbol{h} = \boldsymbol{k}_0/k$, and $f_{\mathrm{LM}}(\boldsymbol{k}_0)$ is the angular dependent Lamb–Mössbauer factor. When the sample is a cubic Bravais crystal, Eq. (4.130) gives $(\boldsymbol{h} \cdot \boldsymbol{e}_s)^2 = 1/3$. When this is substituted into (7.42), it becomes identical to (1.81) and the Lamb–Mössbauer factor $f_{\mathrm{LM}}(\boldsymbol{k}_0)$ is identical to the recoilless fraction f. If the harmonic lattice model is valid, we can treat the time-dependent correlation function $F(\boldsymbol{k}_0, \tau)$ for small displacements $\boldsymbol{u}$ as done in Section 4.6.1, namely to expand the exponent of (7.40) in a power series. As a consequence, the phonon absorption probability is written as a sum of the elastic and inelastic components:

$$S(\boldsymbol{k}_0, E) = f_{\mathrm{LM}}(\boldsymbol{k}_0)\left[S_0(\boldsymbol{k}_0, E) + \sum_{n=1} S_n(\boldsymbol{k}_0, E)\right]. \tag{7.43}$$

The $n = 0$ term describes elastic nuclear absorption without phonon creation or annihilation, and it can be written as

$$S_0(\boldsymbol{k}_0, E) = \int \frac{\mathrm{d}\tau}{2\pi} \exp(-iE\tau - \Gamma|\tau|/2) = \frac{\Gamma}{2\pi} \frac{1}{E^2 + \Gamma^2/4} = \mathscr{L}(E) \tag{7.44}$$

where $\mathscr{L}(E)$ is a Lorentzian centered at zero energy, i.e., $E - E_0 = 0$ (see Eq. (1.13)). The first term S_1 describes a single-phonon nuclear inelastic absorption. By analogy, the $n = 1$ term can be calculated:

$$S_1(\boldsymbol{k}_0, E) = \frac{E_\mathrm{R}}{N} \sum_s \frac{(\boldsymbol{h} \cdot \boldsymbol{e}_s)^2}{\hbar\omega_s} [\langle n_s + 1 \rangle \mathscr{L}(E - \hbar\omega_s) + \langle n_s \rangle \mathscr{L}(E + \hbar\omega_s)]. \tag{7.45}$$

Since the width Γ, being orders of magnitude smaller than typical phonon energies, can be neglected in Eqs. (7.43) and (7.44), then

$$\lim_{\Gamma\to 0} S_0(\boldsymbol{k}_0, E) = \delta(E) \tag{7.46}$$

$$\lim_{\Gamma\to 0} S_1(\boldsymbol{k}, E) = \frac{E_R}{N}\sum_s \frac{(\boldsymbol{h}\cdot\mathbf{e}_s)^2}{\hbar\omega_s}[\langle n_s+1\rangle\delta(E-\hbar\omega_s) + \langle n_s\rangle\delta(E+\hbar\omega_s)] \tag{7.47}$$

where $\delta(E)$ is the Dirac δ-function. As can be seen, the part after the summation sign in formula (7.47) is identical to that in (4.137) for inelastic neutron scattering.

In our discussion the energy scale is chosen relative to the resonant nuclear transition energy E_0. Hence, SR with energy larger $(E > 0)$ or smaller $(E < 0)$ than E_0 will be inelastically absorbed by creation or annihilation of phonons. The $\boldsymbol{k}_0$-dependence of $S(\boldsymbol{k}_0, E)$ can be dropped if a cubic crystal or a polycrystalline sample is used and, for $E > 0$, the second term in Eq. (7.47) is zero, so

$$S_1(E) = \frac{E_R}{N}\sum_s \frac{\langle n_s+1\rangle}{\hbar\omega_s}\delta(E-\hbar\omega_s). \tag{7.48}$$

The higher order $S_n(E)$ terms are given by successive convolutions with the single-phonon term:

$$S_n(E) = \frac{1}{n} S_{n-1}(E) \otimes S_1(E). \tag{7.49}$$

According to the definition of phonon DOS in Eq. (4.95), Eq. (7.48) becomes

$$S_1(E) = \frac{E_R g(E)}{E(1-e^{-\beta E})} = \frac{E_R}{2E} g(E) \coth\left(\frac{\beta E}{2} + 1\right) \tag{7.50}$$

where we have used the fact that the mean value of $(\boldsymbol{h}\cdot\mathbf{e}_s)^2$ over all the modes in a cubic lattice is $1/3$. The energy spectrum of inelastic nuclear absorption satisfies the detailed balance condition, which means that for any particular energy the ratio of phonon creation and phonon annihilation probabilities is given by a factor $e^{\beta|E|}$. So

$$S_1(E) = e^{\beta E} S_1(-E). \tag{7.51}$$

Because this is independent of the material, one can use both wings of the phonon spectrum to get the partial phonon DOS, $g(E)$. The result is

$$g(E) = \frac{E}{E_R}(S_1(E) + S_1(-E)) \tanh\frac{\beta E}{2} \quad \text{with } E \geq 0 \tag{7.52}$$

where $S_1(-E)$ has a similar expression to (7.50) (see Eq. 4.139). In general, $g(E)$ is $\boldsymbol{k}_0$-dependent and takes the following form [51, 52]:

$$g(\boldsymbol{h}, E) = V_a \sum_j \int \frac{d\boldsymbol{q}}{(2\pi)^3} \delta(E - \hbar\omega_j(\boldsymbol{q})) |\boldsymbol{h} \cdot \boldsymbol{e}_j(\boldsymbol{q})|^2 \qquad (7.53)$$

where V_a is the volume of the unit cell.

7.4.2
Measurement of DOS in Solids

A typical experimental setup for incoherent inelastic nuclear resonant scattering is shown in Fig. 7.25. The incident SR beam, reduced down to the meV bandwidth, is energy-tunable within a range to cover a particular phonon spectrum. For the resonance nucleus ^{57}Fe, it is favorable to record the K-fluorescence photons of 6.4 keV following internal conversion as a product of incoherent absorption. However, if the nuclear transition energy is below the K-edge, K-fluorescence is not possible, as in the case of ^{119}Sn. Since the L-fluorescence photons have energies often too low to be efficiently detected, nuclear resonant fluorescence has to be used then. The product of incoherent absorption does not form a collimated beam, but rather is emitted isotropically. In order to collect sufficient number of counts, the first detector is situated at a distance of about 1 mm from the sample, covering about a quarter of a complete sphere [53]. The second detector records the nuclear forward scattering, and is situated far away from the sample to reduce the contribution from incoherent scattering. A sharp peak will be recorded by the second detector, which gives the instrumental function of the high-resolution monochromator and precisely determines the energy position of nuclear resonance. This function is necessary for subsequent data processing.

Resonant nuclei in a sample provide a very accurate energy reference with natural width resolution of the nuclear level (~neV). The energy transfer is determined as a difference between the incident energy and the nuclear transition energy. Experimentally, this difference is between the elastic peak in the incoherent spectrum and the peak in the forward scattering spectrum. Therefore, the ana-

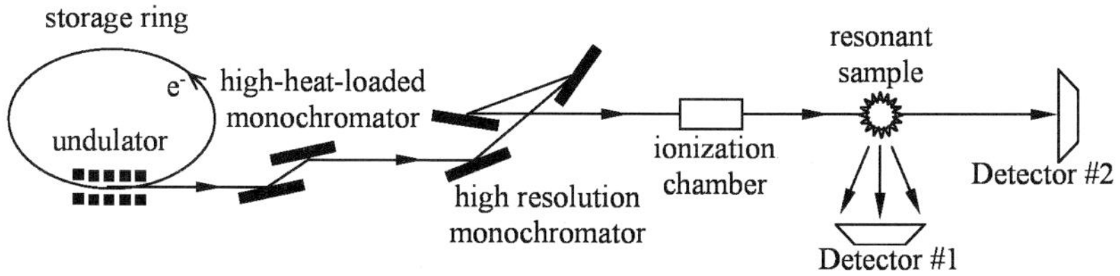

Fig. 7.25 Experimental setup for measurements of inelastic nuclear scattering with SR.

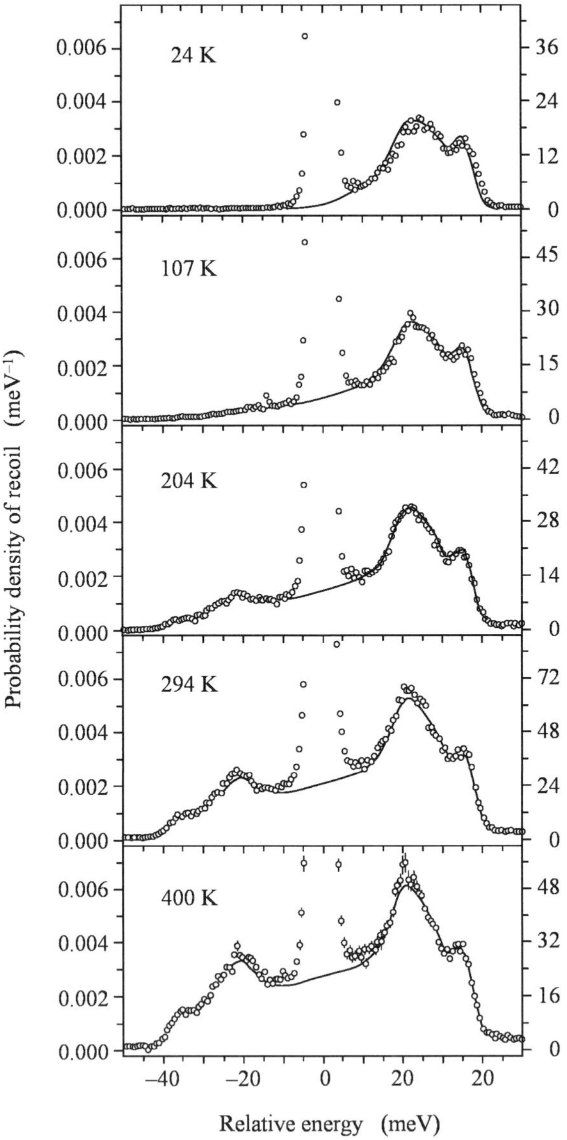

Fig. 7.26 Energy spectra of inelastic nuclear absorption of synchrotron radiation by α-^{57}Fe at various temperatures [53, 54]. Solid lines are calculations according to Eqs. (7.43), (7.48), and (7.49), based on the results of neutron scattering at room temperature and convoluted with the instrumental function of the monochromator.

lyzer part of the traditional inelastic scattering setup is omitted. If the incident energy is off-resonance, excitation of nuclei may be assisted by creation ($E > 0$) or annihilation ($E < 0$) of phonons in the sample. In other words, resonance excitation takes place if the incident energy plus the energy exchanged with a particular vibrational mode equals the resonance energy.

As mentioned in Eq. (7.45), the phonon annihilation probability is proportional to $\langle n_s \rangle$, the Bose occupation number, while the phonon creation probability is proportional to $\langle n_s + 1 \rangle$. This means that the incident photon may gain energy only from an existing phonon, whereas it may lose energy to an existing phonon or for the creation of a new phonon. Hence, the total yield of the delayed fluorescence photons gives a direct measure of the phonon DOS. As an example of experimental results, the temperature dependence of nuclear inelastic absorption in α-Fe is shown in Fig. 7.26.

At high temperatures $\langle n_s \rangle \approx \langle n_s + 1 \rangle$, so the spectra of inelastic absorption are somewhat symmetric. At low temperatures many low-energy phonons are suppressed, $\langle n_s \rangle$ approaches zero, and $\langle n_s + 1 \rangle$ approaches unity. Therefore, the spectra become very asymmetric. When $T = 400$ K, the thermal energy $k_B T = 34$ meV, the occupation is relatively high for all phonon states, and the energy spectrum is only slightly asymmetric. At $T = 24$ K, $k_B T = 2$ meV, the low-energy phonon states (e.g., below 10 meV) are mostly unoccupied. However, inelastic absorption with an energy transfer is still possible, because the recoil may excite phonons even in a "frozen" crystal.

It should be noted that the instrumental resolution is constant and determined with high precision, which allows one to extract the phonon DOS with an accuracy within a few percent.

7.4.3 Extraction of Lamb–Mössbauer Factor, SOD Shift, and Force Constant

The phonon DOS, Lamb–Mössbauer factor f_{LM}, second-order Doppler shift δ_{SOD}, and force constant Φ are all obtainable only after tedious data analysis. First, we discuss Lipkin's sum rules [55], which give various moments of the measured spectra. As has already been proved, the sum rules provide a very useful tool to treat the inelastic nuclear absorption data, because they simplify the normalization of the spectra by decomposing $S(E)$ into the multi-phonon contributions, and their various moments provide model-independent information on lattice dynamics (a similar case is in Section 4.4.3). The first three moments are

$$\int ES(\boldsymbol{k}_0, E)\,\mathrm{d}E = E_R, \tag{7.54}$$

$$\int (E - E_R)^2 S(\boldsymbol{k}_0, E)\,\mathrm{d}E = 4E_R \overline{T}_k, \tag{7.55}$$

$$\int (E - E_R)^3 S(\boldsymbol{k}_0, E)\,\mathrm{d}E = \frac{E_R}{M} \hbar^2 \bar{\Phi}_k \tag{7.56}$$

where $\overline{T}_k$ is the mean kinetic energy in the $\boldsymbol{k}_0$-direction, and $\bar{\Phi}_k$ is the mean force constant experienced by the resonant nuclei in the $\boldsymbol{k}_0$-direction. The details of Eq. (7.54) are elaborated in Chapter 1.

The central task in data analysis is to separate precisely the elastic part from the inelastic part in the measured spectrum. Only after this can we get $S(E)$, $S_1(E)$, and $S_n(E)$ necessary to calculate lattice dynamics parameters. However, here one faces some serious problems. Due to the saturation effect, the area of the elastic peak is not proportional to the Lamb–Mössbauer factor f_{LM}. Another problem arises from the energy dependence of extinction of the incident radiation in a thick sample. Off nuclear resonance, the incident beam is only slightly weakened by electronic absorption. At resonance ($E = 0$), an additional strong Mössbauer absorption takes place. As a result, the elastically scattered intensity is reduced in height by an essentially unknown factor. In contrast, the nuclear forward scattered radiation at $E = 0$ may be scattered by the electrons into the detector and increase the elastic scattering intensity. After the elastic scattering is removed, a procedure to normalize the measured energy spectrum may be used [46], which provides an accurate determination of f_{LM} and partial DOS.

There are several approaches to the removal of the measured elastic peak. In Ref. [54] the inelastic scattering spectrum is separated through interpolating the experimental data in elastic peak region of about ± 8 meV using $S(E) \propto E(1 - \mathrm{e}^{-\beta E})^{-1}$, which results from the relation $g(|E|) \propto E^2$, valid for small energies.

We discuss in detail the following alternative approach [56]. The experimentally measured intensity spectrum is not $I(E)$, but rather a convolution of the normalized instrument resolution function of the monochromator $R(E)$ with a modified function $S_{\mathrm{exp}}(E)$:

$$I_{\mathrm{exp}}(E) = R(E) \otimes a S_{\mathrm{exp}}(E) \tag{7.57}$$

with

$$S_{\mathrm{exp}}(E) = f_{\mathrm{LM}} c \delta(E) + f_{\mathrm{LM}} \sum_{n=1} S_n(E) \tag{7.58}$$

where a is a normalization constant, and the factor c ($\neq 1$) takes into account the extinction of the incident beam mentioned just above and serves to restore the proper height of the central elastic peak. Using this, $S(E)$ can be written as

$$S(E) = S_{\mathrm{exp}}(E) + f_{\mathrm{LM}}(1 - c)\delta(E). \tag{7.59}$$

Processing experimental data, our aim is to determine the three parameters c, a, and f_{LM}. First, the constant a can be derived from Lipkin's sum rule of the first moment:

$$
\begin{aligned}
A &= \int E I_{\exp}(E)\,\mathrm{d}E \\
&\approx a \int E S_{\exp}(E)\,\mathrm{d}E = a \int E \left[f_{\mathrm{LM}} c\delta(E) + f_{\mathrm{LM}} \sum_{n=1} S_n(E) \right] \mathrm{d}E \\
&= a \int f_{\mathrm{LM}} \sum_{n=1} S_n(E)\,\mathrm{d}E = a E_{\mathrm{R}}
\end{aligned}
\tag{7.60}
$$

where, to a good approximation, $R(E)$ is assumed to be a symmetric function. The part of first moment of $R(E)$ is of the order of 1% of E_{R}, so the slight asymmetry of $R(E)$ may be ignored in (7.60). In the integral over $\mathrm{d}E$, the contributions of all terms that are even in E cancel because of the multiplication by E. It is the same for elastic scattering; so this factor a does not influence the result. To obtain the value of A, and hence the value of a, a numerical integration of $I_{\exp}(E)$ weighted by E must be done. Finally, the measured intensity spectrum is normalized as follows:

$$
\int \frac{1}{a} I_{\exp}(E)\,\mathrm{d}E = \int R(E) \otimes S_{\exp}(E)\,\mathrm{d}E = c f_{\mathrm{LM}} + (1 - f_{\mathrm{LM}}),
\tag{7.61}
$$

which means that the numerical integration of the normalized experimental data gives the sum of elastic and inelastic scattering parts, and c and f_{LM} are correlated.

The factor a may be found by an iterative procedure. Starting with a reasonable trial value for a, we can calculate the phonon DOS according to Eq. (7.52). However, we confine our attention to the low-energy region of the phonon DOS, which has a Debye behavior, i.e., it should follow the E^2 law. When this procedure was applied to the analysis of spectra from metmyoglobin [56], $c = 0.9840$ gives the best agreement with an E^2 dependence. After c is obtained, the inelastic spectrum is separated from the elastic component. Now $S_{\exp}(E)$ is substituted into Eq. (7.59), to restore $S(E)$ to its proper form.

Although the value of f_{LM} may be determined together with c due to (7.61), it would be in general found by some other independent approach. In fact, f_{LM} can be solved from (7.61) as

$$
f_{\mathrm{LM}} = 1 - \frac{1}{a} \int I'_{\exp}(E)\,\mathrm{d}E
\tag{7.62}
$$

with $I'_{\exp}(E)$ being the measured spectrum with the elastic peak removed. This expression shows that the f_{LM}-value is determined without requiring specific knowledge about isotope abundance, shape or thickness of the sample, resonant cross-section, hyperfine fields, and so on. This distinguishing feature reduces systematic error, ingeniously avoids the saturation effect, and leads to the very precise determination of f_{LM}-values, as can be seen in Table 7.2.

Table 7.2 The Lamb–Mössbauer factor f_{LM} and δ_{SOD} of various compounds. The values were obtained by inelastic nuclear resonant absorption with ^{57}Fe, ^{119}Sn, or ^{151}Eu as resonant nuclei [57].

Compound	f_{LM}	δ_{SOD} (Γ)	Ref.
Fe metal (bcc), foil	0.805(3)	−2.47(4)	46
	0.791(15)	−2.50(13)	54
	0.80(1)		58
	0.796(2)	−2.49(2)	13
Stainless steel, $Fe_{55}Cr_{25}Ni_{20}$, foil	0.742(10)	−2.41(4)	Evaluated from [46]
	0.76(5)		58
Fe metal, nanocrystalline powder	0.726(5)	−2.62(12)	Evaluated from [59]
Fe_3Al, foil	0.743(3)	−2.46(2)	Evaluated from [60]
Fe_2Tb, Laves phase, film	0.679(3)	−2.39(2)	61
$Fe_{67}Tb_{33}$, amorphous film	0.595(5)	−2.39(3)	61
$SrFeO_3$, powder	0.811(10)	−2.57(4)	Evaluated from [57]
$FeBO_3$, single crystal	0.81(3)		62
Fe_2O_3, powder	0.793(4)	−2.56(4)	63
$[Fe(bpp)_2][BF_4]$, polycrystalline	0.10(5)		58
α-Sn(500 Å)/InSb(001)	0.14(2)		64
β-Sn, foil	0.042(6)		65
SnO_2, powder	0.628(9)	−0.357(6)	66

With the normalized spectrum, one can easily calculate the different moments. From the second moment (7.55), δ_{SOD} is obtained from measurements with directions of the incident radiation along orthogonal axes [52]:

$$\delta_{SOD} = -E_0 \frac{\langle v^2 \rangle}{2c^2} = -\frac{\langle E \rangle_{2x} + \langle E \rangle_{2y} + \langle E \rangle_{2z} - 3E_R}{2E_0} \tag{7.63}$$

where $\langle E \rangle_j = \int E^j S(E)\, dE$ and E_0 is the nuclear transition energy. Note that the extracted δ_{SOD} by this way is separated from isomer shift δ_{IS}. The average force constant $\bar{\Phi}_k$ projected on the direction of the incident radiation is given straightforwardly by (7.56):

$$\bar{\Phi}_k = \frac{E_0^2}{2\hbar^2 c^2 E_R^2} [\langle E \rangle_3(\boldsymbol{k}) - 3E_R \langle E \rangle_2(\boldsymbol{k}) + 2E_R^2]. \tag{7.64}$$

The f_{LM}- and δ_{SOD}-values of α-Fe measured by this method at various temperatures are compared with those by conventional Mössbauer spectroscopy in Figs. 7.27 and 7.28, respectively.

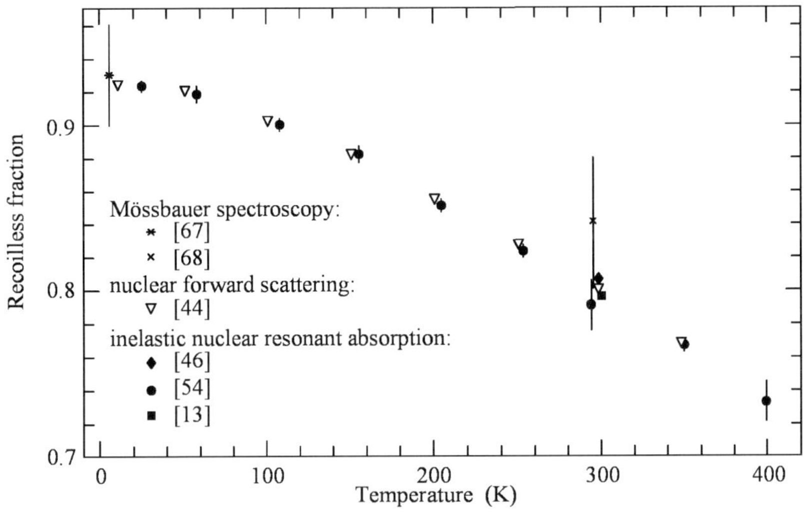

Fig. 7.27 Recoilless fraction or Lamb–Mössbauer factor of iron metal (bcc) versus temperature. The samples were polycrystalline iron foils at ambient pressure. The employed methods comprise Mössbauer spectroscopy, nuclear forward scattering, and inelastic nuclear resonant absorption.

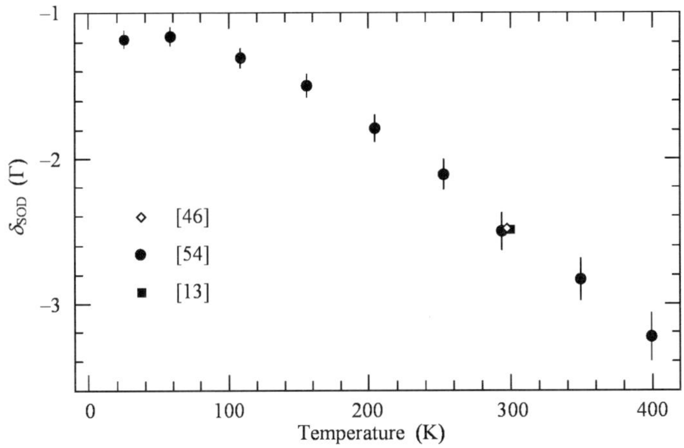

Fig. 7.28 SOD shift versus temperature of polycrystalline iron foils (bcc) at ambient pressure.

To get the phonon DOS, we have to extract the single-phonon contribution $S_1(E)$ convoluted with $R(E)$. Using the convolution theorem, Eq. (7.49) has a very simple form in the Fourier space:

$$\mathscr{F}\{S_n(E)\} = (1/n)\mathscr{F}\{S_{n-1}(E)\} \cdot \mathscr{F}\{S_1(E)\} \tag{7.65}$$

where $\mathscr{F}$ indicates the Fourier transform. This recursive relation has the closed solution

$$\mathscr{F}\{S_n(E)\} = (1/n!)[\mathscr{F}\{S_1(E)\}]^n. \tag{7.66}$$

Thus, taking the Fourier transform of Eq. (7.59) and summing up the multi-phonon contributions, one obtains

$$\mathscr{F}\{S(E)\} = f_{\mathrm{LM}} \exp[\mathscr{F}\{S_1(E)\}]. \tag{7.67}$$

Now, the single-phonon contribution can be solved from this expression,

$$S_1(E) = \int \mathrm{d}te^{-iEt} \ln[\mathscr{F}\{S(E)\}/f_{\mathrm{LM}}]. \tag{7.68}$$

This method is known as the Fourier-logarithm decomposition [69]. We need to also correct for the influence of $R(E)$, which can be done by simultaneously multiplying the numerator and the denominator in the above expression by $a\mathscr{F}\{R(E)\}$:

$$\begin{aligned} S_1(E) &= \int \mathrm{d}te^{-iEt} \ln\{[a\mathscr{F}\{S(E) \otimes R(E)\}]/[af\mathscr{F}\{R(E)\}]\} \\ &= \int \mathrm{d}te^{-iEt} \ln\{1 + [\mathscr{F}\{I'_{\mathrm{exp}}(E)\}]/[af\mathscr{F}\{R(E)\}]\}, \end{aligned} \tag{7.69}$$

by which one can finally calculate the phonon DOS using the measured inelastic scattering spectrum. As an example, the measured phonon DOS of α-Fe is shown in Fig. 7.29.

Now we discuss an important issue in the determination of phonon DOS, i.e., the necessity of the Mössbauer effect. As can be seen from the above procedure, what we want are only the inelastic contributions, while the central elastic peak due to the Mössbauer effect must be removed. Since the incident radiation in the recoilless process does not interact with phonons, is it possible to use a resonant absorption other than that in a Mössbauer nucleus? Here, a Mössbauer transition, regarded and used as an energy analyzer, is an extremely precise energy reference [48]. However, all nuclear transitions are excellent energy references, not just the Mössbauer transitions. The "analyzer" referred here is not the same as the analyzer used in Mössbauer Rayleigh scattering. Let us discuss this problem briefly.

It is important that resonant absorption or scattering is used because of the large cross-sections to provide high counting rates. Furthermore, since the cross-section is proportional to λ^2, low-energy transitions (e.g., $<$200 keV) are more favorable. In order to observe delayed products of the decays, the lifetime of the nuclear excited state must not be too short; otherwise it would be difficult to dis-

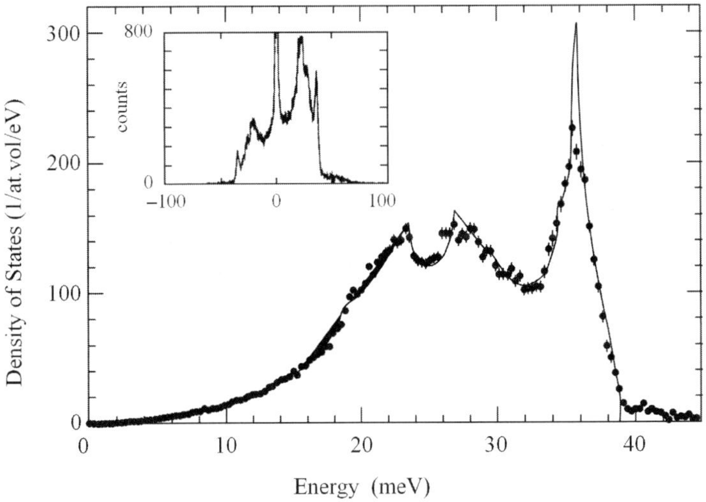

Fig. 7.29 Phonon density of states for α-Fe as measured in inelastic nuclear resonant scattering with 920 μeV energy resolution [13] (circles) and as reconstructed from neutron results (solid line). The inset shows the raw data from inelastic nuclear resonant scattering.

criminate against the electronic scattering which causes serious background. It so happens that the isotopes satisfying these conditions are mostly Mössbauer isotopes. In additional, the relative contribution from the multi-phonon terms is $(-\ln f)^n/n!$ [49]. If f_{LM} approaches 1, the sum in Eq. (7.43) converges quickly, and it would be sufficient to take only the first few leading terms, e.g., taking up to the three-phonon term and neglecting all higher order terms. Therefore, having a large f_{LM} is very important for the precise separation of the single-phonon term from the rest. It seems that the Mössbauer effect plays a pivotal role, and the methodology is called "phonon-assisted Mössbauer effect." Although the SR energy is completely tunable for resonance excitations of all Mössbauer isotopes, only a few of them have been used in experiments, such as ^{57}Fe, ^{119}Sn, ^{169}Tm, ^{181}Ta, ^{151}Eu, ^{161}Dy, and ^{83}Kr. The aspect of excellent energy resolution of Mössbauer effect is not exploited here, because we are not measuring hyperfine interactions, but phonon DOS in the meV range.

7.5
Synchrotron Methods versus Conventional Methods

Synchrotron Mössbauer spectroscopy has attracted a significant amount of attention from researchers and become a well-established methodology in the last ten years. Here we make a comprehensive comparison between synchrotron Mössbauer spectroscopy and the conventional Mössbauer spectroscopy.

Table 7.3 Comparison between a modern SR source and a ^{57}Co source with 10 mCi activity [63].

Radiation property	SR	^{57}Co source
Relevant spectral flux (ph s^{-1} eV^{-1})	2.5×10^{12}	2.5×10^{9}
Brightness (ph s^{-1} eV^{-1} sr^{-1})	2.8×10^{22}	2.5×10^{12}
Brilliance (ph s^{-1} eV^{-1} sr^{-1} mm^{-2})	2.8×10^{22}	2.8×10^{10}
Typical beam size (mm^2)	1×1	10×10
Energy resolution (eV)	Variable	4.66×10^{-9}
Energy range (eV)	3.7×10^{-6}	$\approx 1 \times 10^{-4}$
Polarization	100% linear	Unpolarized

Table 7.3 lists some of the typical properties (such as intensity, resolution, size, and polarization) of the γ-rays from SR and a ^{57}Co source. As an additional feature of SR, its pulsed γ-rays have excellent properties (each pulse duration < 100 ps, and variable periods between a few and hundreds of nanoseconds), which is uniquely suitable for measuring time spectra.

Synchrotron radiation provides extremely strong and narrow photon beams, which facilitates spectral measurement under special experimental conditions, such as high temperature, high pressure, high magnetic field, and working with small samples ($\sim$1 mm^2) or nanostructured thin films. In biological samples, the concentration of Mössbauer isotope is usually too low for conventional Mössbauer spectroscopy, but SR should be strong enough to produce detectable signals.

Results from synchrotron Mössbauer spectroscopy have usually higher accuracy than results from the conventional methods. Whether using nuclear forward scattering or inelastic nuclear resonant scattering, the experimental values of recoilless fraction f have typically three significant figures, as shown in Tables 7.1 and 7.2. The main reason for better accuracy is the absence of the saturation effect in synchrotron Mössbauer methods. In conversion Mössbauer experiments, one must measure the height or the area of spectral peaks, whose uncertainties are usually higher that a few percent. This is because the "thin absorber approximation" is usually not satisfied, causing the saturation effect. Scattering experiments require only relatively short measurement time, ranging from a few minutes to a few hours. The conventional Mössbauer spectroscopy may require up to hundreds of hours. The hyperfine parameters measured from synchrotron experiments have comparable accuracies to those from conventional experiments. In quantum beat experiments, enough data accumulation is required in the chosen time window in order to yield satisfactory measurements of the periods.

The most important contribution from synchrotron Mössbauer spectroscopy is its ability of measuring phonon DOS directly. So far, this cannot be achieved by conventional Mössbauer spectroscopy. Although phonon DOS may be deduced from inelastic neutron scattering (see Section 4.6.1) by extracting force constants from fitting dispersion curves, inelastic nuclear resonant scattering using SR can provide the phonon DOS directly, independent of the dispersion relations, and hence is a model-independent method. This direct method is also much more accurate (within a few percent), much better than that which neutron scattering methods can provide. In nuclear resonant scattering, the Mössbauer nuclei serve directly as analyzers. But in neutron scattering (Fig. 4.24), the scattered neutrons must be diffracted by an "analyzer crystal," which obviously introduces added uncertainty. Inelastic nuclear resonant scattering using SR can also allow us to measure partial density of states (PDOS), which is remarkable because PDOS is very difficult to obtain using other methods. For studying vibrations of Fe atoms in various ferrous and ferric compounds, iron PDOS contains much needed information. For studying Fe as impurities, PDOS can be used to elucidate local vibration modes. The availability of experimental PDOS is particularly significant, because PDOS can now be calculated using first-principles methods (see Fig. 4.27), and be compared with experimental results.

However, synchrotron Mössbauer spectroscopy suffers from several shortcomings. Because time domain experiments are based on the interference phenomenon, the corresponding spectra are very complex, whereas conventional Mössbauer spectra provide certain direct visual information. If two or more hyperfine fields are involved, time spectra may be severely modulated. To alleviate these difficulties, new experimental procedures are being developed [70, 71], such as the time-integrated nuclear forward scattering method using SR, where an absorber mounted on a Mössbauer drive is inserted between the sample and the detector. The scattered intensity as a function of v is then measured, which in principle is similar to an energy spectrum from conventional Mössbauer spectroscopy. This is an interesting concept, but it cannot be fully implemented until all technical problems are resolved. The second major drawback of synchrotron Mössbauer spectroscopy is obviously the high expense involved in constructing and maintaining such large centralized synchrotron facilities and their unavailability to local individual users.

In summary, synchrotron Mössbauer spectroscopy is an important supplement of the conventional Mössbauer spectroscopy. The former will never completely replace the latter, but help to solve problems that the conventional methods cannot study or unable to provide satisfactory results.

We now compare experimental results from three materials, α-Fe, Fe_2O_3, and $(CN_3H_6)_2[Fe(CN)_5NO]$, each of which has been investigated by synchrotron Mössbauer spectroscopy and by conventional Mössbauer spectroscopy.

α-Fe has been extensively investigated by the conventional and synchrotron Mössbauer methods, as well as by inelastic neutron scattering. For the recoilless fraction values of α-Fe, we can compare data in Table 5.3 (using conventional Mössbauer) with data in Table 7.1 (using SR nuclear forward scattering time spectra), and with data in Table 7.2 (using inelastic nuclear resonant absorption). From the above three methods, the recoilless fraction values are $f = 0.78, 0.77$, and 0.80, respectively, which obviously agree with one another very well. Temperature dependence of the recoilless fraction or the Lamb–Mössbauer factor is shown in Fig. 7.27, which contains data from Mössbauer, SR nuclear forward scattering, and SR nuclear resonant absorption. All data points follow exactly the same trend and are quite consistent. The density of phonon states in Fig. 7.29 (using SR nuclear resonance scattering) compares very well with that in Fig. 4.19 (using neutron scattering). The polarization effects of an external magnetic field as shown in the transmission spectra using conventional ^{57}Co radiation (Fig. 2.24) resemble their SR counterparts (Fig. 7.9). Furthermore, the effective magnetic hyperfine field as plotted in Fig. 7.24 is a compilation of results from both transmission Mössbauer and SR nuclear forward scattering.

Fe_2O_3 powder is another example for which comparisons can be made. Its recoilless fraction f from the traditional Mössbauer results is 0.66, as listed in Table 5.3, and the Lamb–Mössbauer factor f_{LM} from inelastic nuclear resonant absorption using SR is 0.793, as in Table 7.2. Strictly, these two values are not expected to be equal. Nevertheless, we list both of them here because they are comparable.

One more example is the remarkable anisotropic vibrational mean-square displacement of guanidinium nitroprusside, $(CN_3H_6)_2[Fe(CN)_5NO]$, whose SR nuclear forward scattering spectra are presented in Fig. 7.22 with the corresponding Lamb–Mössbauer factors $f_{\mathrm{LM}}^{(a)} = 0.122 \pm 0.010$ and $f_{\mathrm{LM}}^{(c)} = 0.174 \pm 0.002$ for the orientations with the single-crystal a-axis and c-axis parallel to the incident SR beam, respectively [44]. An investigation of single-crystal guanidinium nitroprusside using Mössbauer line broadening yielded $f_{\mathrm{LM}}^{(a)} = 0.118 \pm 0.003$, $f_{\mathrm{LM}}^{(b)} = 0.206 \pm 0.010$, and $f_{\mathrm{LM}}^{(c)} = 0.198 \pm 0.002$ for the principal crystal directions [72]. Another study using Mössbauer saturation and polarization effects provided $f_{\mathrm{LM}}^{(a)} = 0.118 \pm 0.008$, $f_{\mathrm{LM}}^{(b)} = 0.174 \pm 0.008$, and $f_{\mathrm{LM}}^{(c)} = 0.202 \pm 0.008$ as the Lamb–Mössbauer factors [73]. These three sets of data from experiments using synchrotron and conventional Mössbauer methods are again consistent with one another.

References

1 S.L. Ruby. Mössbauer experiments without conventional sources. *J. de Physique (Colloque)* 35, C6-209–C6-211 (1974).

2 E. Gerdau, R. Rüffer, H. Winkler, W. Tolksdorf, C.P. Klages, and J.P. Hannon. Nuclear Bragg diffraction of synchrotron radiation in yttrium iron garnet. *Phys. Rev. Lett.* 54, 835–838 (1985).

3 E. Gerdau and H. de Waard (Eds.). Nuclear resonant scattering of synchrotron radiation. *Hyperfine Interactions*, 123–125, (1999/2000).

4 L.D. Landau and E.M. Lifshitz. *The Classical Theory of Fields* (Pergamon Press, New York, 1975).

5 J.D. Jackson. *Classical Electrodynamics*, 2nd edn (Wiley, New York, 1975).

6 G.N. Kulipanov and A.N. Skrinskii. Utilization of synchrotron radiation: current status and prospects. *Soviet Physics Uspekhi* 20, 559–586 (1977) [Russian original: *Uspekhi Fiz. Nauk* 122, 369–418 (1977)].

7 M. Seto, Y. Yoda, S. Kikuta, X.W. Zhang, and M. Ando. Observation of nuclear resonant scattering accompanied by phonon excitation using synchrotron radiation. *Phys. Rev. Lett.* 74, 3828–3831 (1995).

8 T.M. Mooney, T.S. Toellner, W. Sturhahn, E.E. Alp and S.D. Shastri. High-resolution, large-angular-

acceptance monochromator for hard X rays. *Nucl. Instrum. Methods A* 347, 348–351 (1994).

9 G. Faigel, D.P. Siddons, J.B. Hastings, P.E. Haustein, J.R. Grover, J.P. Remeika, and A.S. Cooper. New approach to the study of nuclear Bragg scattering of synchrotro radiation. *Phys. Rev. Lett.* 58, 2699–2701 (1987).

10 T. Ishikawa, Y. Yoda, K. Izumi, C.K. Suzuki, X.W. Zhang, M. Ando, and S. Kikuta. Construction of a precision diffractometer for nuclear Bragg scattering at the photon factory. *Rev. Sci. Instrum.* 63, 1015–1018 (1992).

11 M. Seto, Y. Yoda, S. Kikuta, X.W. Zhang, and M. Ando. Temperature dependence of phonon energy spectra with nuclear resonant scattering of synchrotron radiation. *Nuovo Cimento D* 18, 381–384 (1996).

12 T.S. Toellner, M.Y. Hu, W. Sturhahn, K. Quast and E.E. Alp. Inelastic nuclear resonant scattering with sub-meV energy resolution. *Appl. Phys. Lett.* 71, 2112–2114 (1997).

13 A.I. Chumakov and W. Sturhahn. Experimental aspects of inelastic nuclear resonance scattering. *Hyperfine Interactions* 123/124, 781–808 (1999).

14 T.S. Toellner. Monochromatization of synchrotron radiation for nuclear resonant scattering experiments. *Hyperfine Interactions* 125, 3–28 (2000).

15 A.Q.R. Baron. Report on the X-ray efficiency and time response of a 1 cm^2 reach through avalanche diode. *Nucl. Instrum. Methods A* 352, 665–667 (1995).

16 U. van Bürck, R.L. Mössbauer, E. Gerdau, R. Rüffer, R. Hollatz, G.V. Smirnov, and J.P. Hannon. Nuclear Bragg scattering of synchrotron radiation with strong speedup of coherent decay, measured on antiferromagnetic $^{57}FeBO_3$. *Phys. Rev. Lett.* 59, 355–358 (1987).

17 G. Faigel, D.P. Siddons, J.B. Hastings, P.E. Haustein, J.R. Grover, and L.E. Berman. Observation of the full time evolution of the nuclear collective-decay mode in crystalline $^{57}Fe_2O_3$ excited by synchrotron radiation. *Phys. Rev. Lett.* 61, 2794–2796 (1988).

18 A.L. Chumakov, G.V. Smirnov, A.Q.R. Baron, J. Arthur, D.E. Brown, S.L. Ruby, G.S. Brown, and N.N. Salashchenko. Resonant diffraction of synchrotron radiation by a nuclear multilayer. *Phys. Rev. Lett.* 71, 2489–2492 (1993).

19 T.S. Toellner, W. Sturhahn, R. Röhlsberger, E.E. Alp, C.H. Sowers, and E.E. Fullerton. Observation of pure nuclear diffraction from a Fe/Cr antiferromagnetic multilayer. *Phys. Rev. Lett.* 74, 3475–3478 (1995).

20 R. Röhlsberger, E. Gerdau, M. Harsdorff, O. Leupold, E. Lüken, J. Metge, R. Rüffer, H.D. Rüter, W. Sturhahn, and E. Witthoff. Broad-band nuclear resonant filters for synchrotron radiation: a new source for nuclear diffraction experiments. *Europhys. Lett.* 18, 561–566 (1992).

21 D.P. Siddons, M. Hart, Y. Amemiya, and J.B. Hastings. X-ray optical activity and the Faraday effect in cobalt and its compounds. *Phys. Rev. Lett.* 64, 1967–1970 (1990).

22 T.S. Toellner, E.E. Alp, W. Sturhahn, T.M. Mooney, X. Zhang, M. Ando, Y. Yoda, and S. Kikuta. Polarizer/analyzer filter for nuclear resonant scattering of synchrotron radiation. *Appl. Phys. Lett.* 67, 1993–1995 (1995).

23 M. Hart. X-ray polarization phenomena. *Phil. Mag. B* 38, 41–56 (1978).

24 E.E. Alp, W. Sturhahn, T.S. Toellner. Polarizer–analyzer optics. *Hyperfine Interactions* 125, 45–68 (2000).

25 G.V. Smirnov. Synchrotron Mössbauer source of ^{57}Fe radiation. *Hyperfine Interactions* 125, 91–112 (2000).

26 F.J. Lynch, R.E. Holland, and M. Hamermesh. Time dependence of resonantly filtered gamma rays from Fe^{57}. *Phys. Rev.* 120, 513–520 (1960).

27 Yu. Kagan, A.M. Afanas'ev, and V.G. Kohn. On excitation of isomeric nuclear states in a crystal by synchrotron radiation. *J. Phys. C* 12, 615–631 (1979).

28 G.T. Trammell and J.P. Hannon. Quantum beats from nuclei excited by synchrotron pulses. *Phys. Rev. B* 18, 165–172 (1978). Erratum. *Phys. Rev. B* 19, 3835–3836 (1979).

29 J.P. Hannon and G.T. Trammell. Coherent excitations of nuclei in crystals by synchrotron radiation pulses. *Physica B* 159, 161–167 (1989).

30 J.P. Hannon and G.T. Trammell. Coherent γ-ray optics. *Hyperfine Interactions* 123/124, 127–274 (1999).

31 G.V. Smirnov. General properties of nuclear resonant scattering. *Hyperfine Interactions* 123/124, 31–77 (1999).

32 G.T. Trammell and J.P. Hannon. Comment on '*N* dependence of coherent radiation from crystal'. *Phys. Rev. Lett.* 61, 653 (1988).

33 G.T. Trammell. Gamma-ray diffraction by resonant nuclei. In *Chemical Effects of Nuclear Transformations* (Proceedings of the 1960 Symposium), pp. 75–85 (International Atomic Energy Agency, Vienna, 1961).

34 Yu. Kagan and A.M. Afanas'ev. Coherence effects during nuclear resonant interaction of gamma quanta in perfect crystals. In *Mössbauer Spectroscopy and Its Applications* (Proceedings of the 1971 Symposium), pp. 143–167 (International Atomic Energy Agency, Vienna, 1972).

35 D.F. Zaretskii and V.V. Lomonosov. Spontaneous emission of gamma quanta from crystals. *Sov. Phys. JETP* 21, 243–246 (1965) [Russian original: *Zh. Eksp. Teor. Fiz.* 48, 368–374 (1965)].

36 C.S. Wu, Y.K. Lee, N. Benczer-Koller, and P. Simms. Frequency distribution of resonance line versus delay time. *Phys. Rev. Lett.* 5, 432–435 (1960).

37 U. van Bürck, D.P. Siddons, J.B. Hastings, U. Bergmann, and R. Hollatz. Nuclear forward scattering of synchrotron radiation. *Phys. Rev. B* 46, 6027–6211 (1992).

38 U. van Bürck. Coherent pulse propagation through resonant media. *Hyperfine Interactions* 123/124, 483–509 (1999).

39 J. Odeurs, R. Coussement, C. L'abbé, G. Neyens, G.R. Hoy, E.E. Alp, W. Sturhahn, T.S. Toellner, and C. Johnson. Time-integrated energy domain measurements with synchrotron radiation. *Hyperfine Interactions* 113, 455–463 (1998).

40 Yu.V. Shvyd'ko, U. van Bürck, W. Potzel, P. Schindelmann, E. Gerdau, O. Leupold, J. Metge, H.D. Rüter, and G.V. Smirnov. Hybrid beat in nuclear forward scattering of synchrotron radiation. *Phys. Rev. B* 57, 3552–3561 (1998).

41 R. Hollatz, R. Rüffer, and E. Gerdau. Determination of hyperfine parameters using quantum beats. *Hyperfine Interactions* 42, 1141–1144 (1988).

42 R. Rüffer, E. Gerdau, M. Grote, R. Hollatz, R. Röhlsberger, H.D. Rüter, and W. Sturhahn. Nuclear Bragg diffraction using synchrotron radiation: a new method for hyperfine spectroscopy. *Nucl. Instrum. Methods A* 303, 495–502 (1991).

43 E.E. Alp, W. Sturhahn, and T.S. Toellner. Synchrotron Mössbauer spectroscopy of powder samples. *Nucl. Instrum. Methods B* 97, 526–529 (1995).

44 H. Grünsteudel, V. Rusanov, H. Winkler, W. Meyer-Klaucke, and A.X. Trautwein. Mössbauer spectroscopy in the time domain applied to the study of single-crystalline guanidinium nitroprusside. *Hyperfine Interactions* 122, 345–351 (1999).

45 U. Bergmann, S.D. Shastri, D.P. Siddons, B.W. Batterman, and J.B. Hastings. Temperature dependence of nuclear forward scattering of synchrotron radiation in α-^{57}Fe. *Phys. Rev. B* 50, 5957–5961 (1994).

46 W. Sturhahn, T.S. Toellner, E.E. Alp, X. Zhang, M. Ando, Y. Yoda, S. Kikuta, M. Seto, C.W. Kimball, and B. Dabrowski. Phonon density of states measured by inelastic nuclear resonant scattering. *Phys. Rev. Lett.* 74, 3832–3835 (1995).

47 A.I. Chumakov, R. Rüffer, H. Grünsteudel, H.F. Grünsteudel, G. Grübel, J. Metge, O. Leupold, and H.A. Goodwin. Energy dependence of nuclear recoil measured with incoherent nuclear scattering of synchrotron radiation. *Europhys. Lett.* 30, 427–432 (1995).

48 E.E. Alp, W. Sturhahn, and T.S. Toellner. Lattice dynamics and inelastic nuclear resonant x-ray scattering. *J. Phys.: Condens. Matter* 13, 7645–7658 (2001).

49 K.S. Singwi and A. Sjölander. Resonance absorption of nuclear gamma rays and dynamics of atomic motions. *Phys. Rev.* 120, 1093–1102 (1960).

50 W.M. Visscher. Study of lattice vibrations by resonance absorption of nuclear gamma rays. *Ann. Phys.* 9, 194–210 (1960).

51 V.G. Kohn, A.I. Chumakov, and R. Rüffer. Nuclear resonant inelastic absorption of synchrotron radiation in an anisotropic single crystal. *Phys. Rev. B* 58, 8437–8444 (1998).

52 V.G. Kohn and A.I. Chumakov. DOS: evaluation of phonon density of states from nuclear resonant inelastic absorption. *Hyperfine Interactions* 125, 205–221 (2000).

53 R. Rüffer and A.I. Chumakov. Nuclear inelastic scattering. *Hyperfine Interactions* 128, 255–272 (2000).

54 A.I. Chumakov, R. Rüffer, A.Q.R. Baron, H. Grünsteudel, and H.F. Grünsteudel. Temperature dependence of nuclear inelastic absorption of synchrotron radiation in α-^{57}Fe. *Phys. Rev. B* 54, R9596–R9599 (1996).

55 H.J. Lipkin. Mössbauer sum rules for use with synchrotron sources. *Hyperfine Interactions* 123/124, 349–366 (1999).

56 K. Achterhold, C. Keppler, A. Ostermann, U. van Bürck, W. Sturhahn, E.E. Alp, and F.G. Parak. Vibrational dynamics of myoglobin determined by the phonon-assisted Mössbauer effect. *Phys. Rev. E* 65, 051916-1–051916-13 (2002).

57 W. Sturhahn and A. Chumakov. Lamb–Mössbauer factor and second-order Doppler shift from inelastic nuclear resonant absorption. *Hyperfine Interactions* 123/124, 809–824 (1999).

58 A.I. Chumakov, J. Metge, A.Q.R. Baron, R. Rüffer, Yu.V. Shvyd'ko, H. Grünsteudel, and H.F. Grünsteudel. Radiation trapping in nuclear resonant scattering of x rays. *Phys. Rev. B* 56, R8455–R8458 (1997).

59 B. Fultz, C.C. Ahn, E.E. Alp, W. Sturhahn, and T.S. Toellner. Phonons in nanocrystalline ^{57}Fe. *Phys. Rev. Lett.* 79, 937–940 (1997).

60 B. Fultz, T.A. Stephens, W. Sturhahn, T.S. Toellner, and E.E. Alp. Local chemical environments and the phonon partial densities of states of ^{57}Fe in $^{57}Fe_3Al$. *Phys. Rev. Lett.* 80, 3304–3307 (1998).

61 W. Keune and W. Sturhahn. Inelastic nuclear resonant absorption of synchrotron radiation in thin firms and multilayers. *Hyperfine Interactions* 123/124, 847–861 (1999).

62 A.I. Chumakov, R. Rüffer, A.Q.R. Baron, H. Grünsteudel, H.F. Grünsteudel, and V.G. Kohn. Anisotropic inelastic nuclear absorption. *Phys. Rev. B* 56, 10758–10761 (1997).

63 W. Sturhahn, E.E. Alp, T.S. Toellner, P. Hession, M. Hu, and J. Sutter. Introduction to nuclear resonant scattering with synchrotron radiation. *Hyperfine Interactions* 113, 47–58 (1998).

64 B. Roldan Cuenya, W. Keune, W. Sturhahn, T.S. Toellner, and M.Y. Hu. Structure and vibrational dynamics of interfacial Sn layers in Sn/Si multilayers. *Phys. Rev. B* 64, 235321-1–235321-12 (2001).

65 A. Barla, R. Rüffer, A.I. Chumakov, J. Metge, J. Plessel, and M.M. Abd-Elmeguid. Direct determination of the phonon density of states in β-Sn. *Phys. Rev. B* 61, R14881–R14884 (2000).

66 M.Y. Hu, T.S. Toellner, W. Sturhahn, P.M. Hession, J.P. Sutter, and E.E. Alp. A high-resolution monochromator for

inelastic nuclear resonant scattering experiments using ^{119}Sn. *Nucl. Instrum. Methods A* 430, 271–276 (1999).

67 T.A. Kovats and J.C. Walker. Mössbauer absorption in Fe^{57} in metallic iron from the Curie point to the γ–δ transition. *Phys. Rev.* 181, 610–618 (1969).

68 H. Vogel, H. Spiering, W. Irler, U. Volland, and G. Ritter. The Lamb–Mössbauer factor of metal iron foils at 4.2 K. *J. de Physique (Colloque)* 40, C2-676 (1979).

69 D.W. Johnson and J.C.H. Spence. Determination of the single-scattering probability distribution from plural-scattering data. *J. Phys. D* 7, 771–780 (1974).

70 C. L'abbé, R. Callens and J. Odeurs. Time-integrated synchrotron Mössbauer spectroscopy. *Hyperfine Interactions* 135, 275–294 (2001).

71 I. Serdons, S. Nasu, R. Callens, R. Coussement, T. Kawakami, J. Ladrière, S. Morimoto, T. Ono, K. Vyvey, T. Yamada, Y. Yoda, and J. Odeurs. Isomer shift determination in Eu compounds using stroboscopic detection of synchrotron radiation. *Phys. Rev. B* 70, 014109-1–014109-5 (2004).

72 V. Rusanov, H. Winkler, C. Ober, and A.X. Trautwein. Mössbauer study of the vibrational anisotropy and of the light-induced population of metastable states in single-crystal guanidinium nitroprusside. *Eur. Phys. J.* B12, 191–198 (1999).

73 V. Rusanov, S. Stankov, V. Angelov, N. Koop, H. Winkler, and A.X. Trautwein. On the theory of the polarization effects and its experimental test by angle-dependent Mössbauer spectroscopy with guanidimium nitroprusside $(CN_3H_6)_2[Fe(CN)_5NO]$ (GNP) single crystals. *Nucl. Instrum. Methods B* 207, 205–218 (2003).

8
Mössbauer Impurity Atoms (I)

In most cases, the Mössbauer atom appears as an impurity; even in the α-Fe lattice, ^{57}Fe is an isotopic impurity. Therefore, studying impurity atoms is an important part of Mössbauer spectroscopy.

Among the methods for studying the dynamics of impurity atoms, the Mössbauer effect has many special characteristics. In addition to high energy resolution, its isotope selectivity gives the Mössbauer method a unique advantage, making it the best means for obtaining information on impurity–host and host–host force constants. Moreover, only the Mössbauer method allows the investigation of "isolated" impurity atoms at extremely low concentrations. Neutron inelastic scattering and heat capacity are not sensitive methods for studying impurity atoms, and do not provide observable effects unless the impurity concentration is larger than 1 at.%. On the other hand, good Mössbauer spectra can be obtained from alloys where the Mössbauer atom concentrations are as low as 10^{-4} to 10^{-2} at.% [1, 2].

The presence of impurity atoms destroys the translational symmetry of the lattice and complicates theoretical treatments. The prevalent approach to solving the equations of motion is to use the Green's functions and take advantage of the symmetry around the impurity atom as much as possible to simplify the calculations. Maradudin and other authors pioneered the theoretical calculations, but in 1968 Mannheim proposed a relatively simple and practical model, which is discussed here in detail. Also, we limit our attention to substitutional impurities of low concentrations (<0.1 at.%) in this chapter, because the interactions among them may be neglected and the theoretical analysis is much simplified. The studies of impurities with higher concentrations are discussed in Chapter 9.

8.1
Theory of Substitutional Impurity Atom Vibrations

8.1.1
The General Method

First we introduce the Green's function for the perfect Bravais lattice. In such a solid, Eq. (4.28) can be written as

Mössbauer Effect in Lattice Dynamics. Yi-Long Chen and De-Ping Yang

ISBN: 978-3-527-40712-5

$$\sum_{\beta,l'}[\Phi_{\alpha\beta}(l,l') - \omega^2 M_0 \delta_{\alpha\beta}\delta_{ll'}]u_\beta(l') = 0. \tag{8.1}$$

For calculating the thermal average of a physical quantity such as $\langle u^2(0)\rangle$, it is very convenient to use the retarded Green's function, defined as [3–6]

$$G_{\alpha\beta}(ll', t-t') = -\frac{i}{\hbar}\theta(t-t')\langle[u_\alpha(l,t)u_\beta(l',t')]\rangle, \tag{8.2}$$

where

$$\theta(t-t') = \begin{cases} 1, & \text{when } t > t' \\ 0, & \text{when } t < t' \end{cases} \tag{8.3}$$

is the step function, and $\langle\ldots\rangle$ represents thermal averaging. We may Fourier-transform the Green's functions from the time domain to the frequency domain:

$$G_{\alpha\beta}(ll', \omega \pm i\varepsilon) = \frac{1}{2\pi}\int_{-\infty}^{\infty} G_{\alpha\beta}(ll', t-t')\mathrm{e}^{i(\omega \pm i\varepsilon)(t-t')}\,\mathrm{d}(t-t')$$

where $\varepsilon \to +0$, and it is often convenient to assign $t' = 0$.

For a harmonic lattice, the Green's functions in Eq. (8.2) satisfy the equations of motion in Eq. (8.1), as shown in Appendix F.1 and Ref. [6]:

$$\sum_{l'',\gamma}[M_0\omega^2\delta_{\alpha\gamma}\delta_{ll''} - \Phi_{\alpha\gamma}(l,l'')]G_{\gamma\beta}(l''l', \omega) = \delta_{\alpha\beta}\delta_{ll'}. \tag{8.4}$$

This may be written in the following matrix form:

$$\boldsymbol{L}_0\boldsymbol{G} = \boldsymbol{I} \tag{8.5}$$

where $\boldsymbol{I}$ is the unit matrix and $\boldsymbol{L}_0 = \boldsymbol{M}_0\omega^2 - \boldsymbol{\Phi}$. For a perfect lattice of N atoms, the solution to Eq. (8.5) is

$$G_{\alpha\beta}(ll', \omega + i\varepsilon) = \frac{1}{NM_0}\sum_{\boldsymbol{k},j}\frac{\mathbf{e}_\alpha(\boldsymbol{k}j)\cdot\mathbf{e}_\beta(\boldsymbol{k}j)}{\omega^2 - \omega_j^2(\boldsymbol{k}) + i\varepsilon}e^{i\boldsymbol{k}\cdot(\boldsymbol{l}-\boldsymbol{l}')}. \tag{8.6}$$

To understand the physical meanings of these Green's functions, let us look at the static Green's function $G_{\alpha\beta}(ll', \omega = 0)$ as an example. It is the displacement of atom l in the α-direction when a unit force is applied on atom l' in the β-direction. From Eq. (8.5), it is easy to see that the Green's functions are reciprocals of the force constants in $\boldsymbol{\Phi}$. Because of this fact, the lattice Green's functions have the same symmetry properties as the force constants (see Appendix F.3).

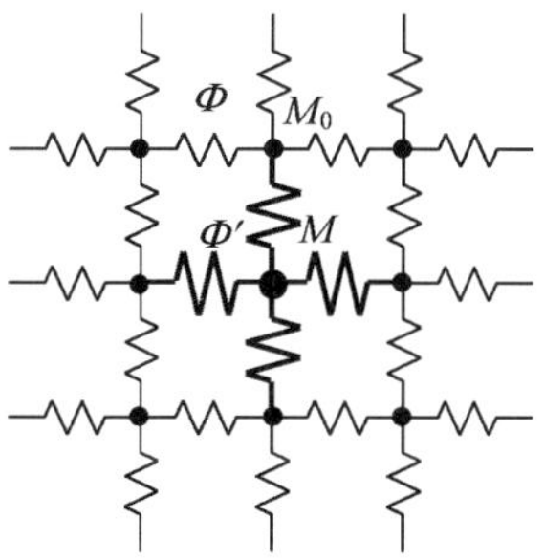

Fig. 8.1 Formation of an impurity as a result of atom M substituting M_0.

Suppose that an atom of mass M replaces an atom of mass M_0 in a perfect lattice, forming a substitutional impurity (Fig. 8.1). The substitution not only causes a change in this atom's mass but also alters the interatomic interactions. However, these changes are localized within a region involving the impurity atom and its nearest neighbors. For studying the dynamics of the impurity atom, we may consider the combination of the impurity atom and its nearest neighbor as a new "molecule." The regions occupied by such molecules are known as the impurity space. The Hamiltonian of a harmonic lattice containing an isolated impurity atom is [7–9]

$$\mathscr{H} = \mathscr{H}_0 + \Delta\mathscr{H} \tag{8.7}$$

where $\mathscr{H}_0$ is the Hamiltonian of the perfect lattice with atomic mass M_0 and force constant tensor $\mathbf{\Phi}$

$$\mathscr{H}_0 = \sum_l \frac{p_l^2}{2M_0} + \frac{1}{2}\sum_{\substack{\alpha,\beta \\ l,l'}} \Phi_{\alpha\beta}(l,l')u_\alpha(l)u_\beta(l') \tag{8.8}$$

and $\Delta\mathscr{H}$ contains only the contributions from the atoms in the impurity space

$$\Delta\mathscr{H} = \sum_l \frac{p_l^2}{2}\left(\frac{1}{M_l} - \frac{1}{M_0}\right) + \frac{1}{2}\sum_{\substack{\alpha,\beta \\ l,l'}} [\Phi'_{\alpha\beta}(l,l') - \Phi_{\alpha\beta}(l,l')]u_\alpha(l)u_\beta(l'). \tag{8.9}$$

We now introduce a new matrix for the perturbation term

$$\boldsymbol{U} = \boldsymbol{L}_0 - \boldsymbol{L} \tag{8.10}$$

where $\boldsymbol{L}_0$ corresponds to the perfect host lattice and $\boldsymbol{L}$ to the impurity lattice. If each impurity atom interacts with z nearest neighbors, $\boldsymbol{U}$ is a $3n \times 3n$ matrix, where $n = z + 1$. Therefore, $\boldsymbol{U}$ is completely localized within the impurity space (U-space) and n is not a large number. The matrix elements are

$$U_{\alpha\beta}(l,l') = (M_0 - M_l)\omega^2\delta_{ll'}\delta_{\alpha\beta} + [\Phi'_{\alpha\beta}(l,l') - \Phi_{\alpha\beta}(l,l')]. \quad (8.11)$$

The equation of motion for the impurity atom is

$$\sum_{\beta,l'}[M_0\omega^2\delta_{\alpha\beta}\delta_{ll'} - \Phi_{\alpha\beta}(l,l')]u_\beta(l') = \sum_{\beta,l'} U_{\alpha\beta}(l,l')u_\beta(l'). \quad (8.12)$$

This inhomogeneous equation is satisfied by [10]

$$u_\alpha(l) = \sum_{\substack{\beta,\gamma \\ l',l''}} G_{\alpha\beta}(ll',\omega)U_{\beta\gamma}(l',l'')u_\gamma(l''), \quad (8.13)$$

in which the summation over four different indices is tedious. We will take an alternative approach, introducing the impurity lattice Green's function $G'_{\alpha\beta}(ll',\omega)$ and utilizing its relation with $G_{\alpha\beta}$ (the Dyson equation) to obtain $\langle u_\alpha{}^2(0)\rangle$ and the recoilless fraction f.

For a lattice with impurity atoms, a relationship similar to Eq. (8.5) can be written (see Eq. (F.19) in Appendix F) as

$$\boldsymbol{G}' = (\boldsymbol{M}\omega^2 - \boldsymbol{\Phi}')^{-1}\boldsymbol{I} = \boldsymbol{M}^{-1/2}[\omega^2\boldsymbol{I} - \boldsymbol{D}]^{-1}\boldsymbol{M}^{-1/2} \quad (8.14)$$

where $\boldsymbol{M}$ is a $3N \times 3N$ diagonal matrix in which three of the elements are M and the rest are M_0. The only nondiagonal matrix is $\boldsymbol{D} = \boldsymbol{M}^{-1/2}\boldsymbol{\Phi}'\boldsymbol{M}^{-1/2}$, but can be diagonalized by a unitary matrix $\boldsymbol{B}$ with elements $B_a(l,s)$ (see Eq. (4.29)). Then the Green's function of the perturbed lattice can be expressed as [11, 12]

$$G'_{\alpha\beta}(ll',\omega) = \frac{1}{[M_l M_{l'}]^{1/2}}\sum_s \frac{B^*_\alpha(l,s)B_\beta(l',s)}{\omega^2 - \omega_s^2}. \quad (8.15)$$

Using Eqs. (8.5) and (8.10), we see that

$$\boldsymbol{U} = \boldsymbol{L}_0 - \boldsymbol{L} = \boldsymbol{G}^{-1} - \boldsymbol{G}'^{-1},$$

thus $\boldsymbol{G}'$ and $\boldsymbol{G}$ are related through

$$\boldsymbol{G}' = \boldsymbol{G} + \boldsymbol{G}\boldsymbol{U}\boldsymbol{G}' \quad (8.16)$$

which is known as the Dyson equation. Here, both $\boldsymbol{G}'$ and $\boldsymbol{G}$ are $3N \times 3N$ matrices. To study effects of impurities on a variety of physical phenomena, only elements of $\boldsymbol{G}'$ in the impurity space are needed. For this reason, we partition each of the matrices as follows [13–15]:

$$\boldsymbol{U} = \left[\begin{array}{c|c} \boldsymbol{U} & 0 \\ \hline 0 & 0 \end{array}\right] \quad (8.17)$$

$$G = \left[\begin{array}{c:c} g & G_{12} \\ \hdashline G_{21} & G_{22} \end{array}\right] \tag{8.18}$$

$$G' = \left[\begin{array}{c:c} G_1{}' & G_{12}{}' \\ \hdashline G_{21}{}' & G_{22}{}' \end{array}\right] \tag{8.19}$$

where $\boldsymbol{g}$ and $\boldsymbol{G}_1{}'$ are both $3n \times 3n$ matrices, formed by the U-space matrix elements of $\boldsymbol{G}$ and $\boldsymbol{G}'$, respectively. Equation (8.16) now becomes four matrix equations of lower dimensions, and we are interested in the first one:

$$\boldsymbol{G}_1{}' = \boldsymbol{g} + \boldsymbol{g}\boldsymbol{U}\boldsymbol{G}_1{}'. \tag{8.20}$$

Once the Green's function matrix $\boldsymbol{G}_1{}'$ is obtained, we can calculate all the dynamic parameters, including the impurity atom's vibration frequency, mean-square displacement $\langle u^2(0)\rangle$, the recoilless fraction f, the Debye temperature, etc. We will still use the general form of the Dyson equation (Eq. (8.16)) with the understanding that each matrix is evaluated only in the U-space.

For an impurity atom at the origin in a cubic host, it is only necessary to evaluate the x-direction mean-square displacement $\langle u_x{}^2(0)\rangle$. According to Refs. [11, 16, 17] or Eq. (F.28) in Appendix F, it can be written as

$$\begin{aligned}\langle u_x^2(0)\rangle &= \lim_{\substack{t\to 0\\ \varepsilon\to 0}} \langle u_x(0,t)u_x(0,0)\rangle \\ &= \lim_{\varepsilon\to 0}\left[-\frac{\hbar}{\pi}\int_0^\infty \coth\left(\frac{\beta\hbar\omega}{2}\right) \text{Im } G'_{xx}(00,\omega+i\varepsilon)\,d\omega\right] \\ &= \lim_{\varepsilon\to 0}\frac{i\hbar}{2\pi}\int_0^\infty \coth\left(\frac{\beta\hbar\omega}{2}\right)[G'_{xx}(00,\omega+i\varepsilon) - G'_{xx}(00,\omega-i\varepsilon)]\,d\omega\end{aligned} \tag{8.21}$$

where $\langle\ldots\rangle$ represents thermal averaging and Im represents the imaginary part of G'_{xx}. Analogous to the case of a perfect lattice, the impurity mean-square displacement can be expressed as

$$\langle u_x^2(0)\rangle = \frac{\hbar}{2M}\int_0^\infty \frac{1}{\omega}\coth\left(\frac{\beta\hbar\omega}{2}\right) g'(\omega)\,d\omega. \tag{8.22}$$

In general, $g'(\omega)$ is the modified vibrational DOS for the impure lattice, or partial DOS. The function $g'(\omega)$ referring to the impurity atom is often called the impurity dynamic response function:

$$\begin{aligned} g'(\omega) &= -\frac{2M\omega}{\pi}\text{ Im } G'_{xx}(00,\omega+i0) \\ &= \frac{iM\omega}{\pi}[G'_{xx}(00,\omega+i0) - G'_{xx}(00,\omega-i0)]. \end{aligned} \tag{8.23}$$

Therefore, the main task in studying $\langle u^2(0)\rangle$ is to obtain the impurity Green's functions through Eq. (8.20), in which M_0 and M are known, the host Green's functions can be calculated, and the only variables are the impurity–host and host–host force constants.

In addition to the recoilless fraction, the second-order Doppler effect can also be evaluated. This is because the mean-square velocity is calculated using Green's functions as [15]

$$\langle v_\alpha^2(0)\rangle = \frac{\hbar}{\pi}\int_0^\infty \coth\left(\frac{\beta\hbar\omega}{2}\right) \operatorname{Im} G'_{\alpha\alpha}(00, \omega^2 - i0)\omega^2 \, d\omega. \tag{8.24}$$

8.1.2
Mass Defect Approximation

We now discuss isotopic impurities in a cubic Bravais lattice, which is the simplest type of substitutional impurity without any force constant changes. The general characteristics of the impurity vibrations can be obtained by studying such a simple model. An obvious example is ^{57}Fe in metallic iron, for which Eq. (8.11) becomes

$$U_{\alpha\beta}(l, l') = \eta M_0 \omega^2 \delta_{ll'} \delta_{\alpha\beta} \tag{8.25}$$

where $\eta = (M_0 - M)/M_0$ and for metallic iron $\eta = 1.78\%$.

In the impurity space $\boldsymbol{gU}$ is a 3×3 matrix, whose elements are

$$(\boldsymbol{gU})_{\alpha\beta, ll'} = \delta_{ll'} \delta_{\alpha\beta} \eta M_0 \omega^2 G_{\alpha\beta}(00, \omega). \tag{8.26}$$

Using Eq. (8.20), we can easily get

$$G'_{\alpha\alpha}(00, \omega) = \frac{G_{\alpha\alpha}(00, \omega)}{1 - \eta M_0 \omega^2 G_{\alpha\alpha}(00, \omega)} \tag{8.27}$$

where the Green's function for the perfect host lattice is [14, 18]

$$G_{\alpha\alpha}(00, \omega) = \frac{1}{NM_0} \sum_{kj} \frac{\mathbf{e}_\alpha(\boldsymbol{k}j) \cdot \mathbf{e}^*_\beta(\boldsymbol{k}j)}{\omega^2 - \omega_j^2(\boldsymbol{k})} \delta_{\alpha\beta} = \frac{1}{3NM_0} \sum_{kj} \frac{1}{\omega^2 - \omega_j^2(\boldsymbol{k})}. \tag{8.28}$$

The frequency of the impurity vibration is determined by the following equation:

$$1 - \eta M_0 \omega^2 G_{\alpha\alpha}(00, \omega) = 0 \tag{8.29}$$

which corresponds to a singular point in Eq. (8.20) [14]. We now consider the details in two different situations.

8.1.2.1 **Resonance Modes**

In this case, the impurity vibration frequency ω is within the band from 0 to ω_{m}, where ω_{m} is the highest frequency of the host lattice vibration. Generally, $G'_{\alpha\alpha}$ is a complex function, and we need to find its imaginary part before calculating $\langle u^2(0)\rangle$ of the impurity vibration. Changing ω^2 to $(\omega^2 + i\varepsilon)$ in Eq. (8.28), and using Eqs. (F.19) and (4.96), we obtain

$$G_{\alpha\alpha}(00, \omega + i\varepsilon) = \frac{1}{3NM_0}\sum_s \frac{1}{\omega^2 - \omega_s^2 + i\varepsilon}$$

$$= \frac{1}{M_0} p \int_0^{\omega_{\mathrm{m}}} \frac{g(\omega')}{\omega'^2} \mathrm{d}\omega' - i\frac{\pi}{M_0}\frac{g(\omega)}{2\omega} \tag{8.30}$$

where $g(\omega)$ is the phonon DOS of the host lattice.

Substituting Eq. (8.30) into (8.27), the recoilless fraction f can be derived from Eq. (8.22):

$$-\ln f = k^2\langle u^2(0)\rangle = \frac{k^2\hbar}{2M}\int_0^{\omega_{\mathrm{m}}} \frac{g'(\omega)}{\omega} \coth\frac{\beta\hbar\omega}{2} \mathrm{d}\omega \tag{8.31}$$

where

$$g'(\omega) = \frac{\frac{M}{M_0} g(\omega)}{\left[1 - \eta\omega^2 p \int_0^{\omega_{\mathrm{m}}} \frac{g(\omega')}{\omega^2 - \omega'^2} \mathrm{d}\omega'\right]^2 + \left[\frac{\pi\eta\omega g(\omega)}{2}\right]^2}. \tag{8.32}$$

Calculations by many authors [3, 9, 11, 14, 15, 19] all arrived at Eq. (8.31). Frequencies of vibrations with very large amplitudes are solutions of the following equation:

$$1 - \eta\omega_{\mathrm{r}}^2 p \int_0^{\omega_{\mathrm{m}}} \frac{g(\omega')}{\omega_{\mathrm{r}}^2 - \omega'^2} \mathrm{d}\omega' = 0. \tag{8.33}$$

In this case, the denominator of Eq. (8.32) shows the resonance characteristics, i.e., the frequency of the impurity atom ω_{r} resonates with one of the host modes. Those modes with frequencies that satisfy (8.33) are known as resonance modes.

8.1.2.2 **Localized Modes**

When $\omega > \omega_{\mathrm{m}}$, we have localized modes. In this case, the Green's function $G_{\alpha\alpha}(00, \omega + i\varepsilon)$ is a positive real number. It can be seen from Eq. (8.29) that a necessary condition for $\omega > \omega_{\mathrm{m}}$ is $\eta > 0$ (impurity atoms lighter than the host ones). In fact, η should be larger than a critical value η_{cr}. When $\omega = \omega_{\mathrm{L}}$, $G'_{\alpha\alpha}$ has a singular point because the denominator in Eq. (8.27) becomes zero. The corresponding normal modes (ω_L) are known as local modes. In such a mode, vibration amplitude decreases exponentially as a function of distance from the impurity atom, hence the name. Obviously, the frequencies of local modes must satisfy Eq. (8.29):

$$1 - \eta M_0 \omega_L^2 G_{\alpha\alpha}(00, \omega) = 0. \tag{8.34}$$

According to (8.28), for a cubic lattice the above equation can be written as

$$1 = \frac{\eta \omega_L^2}{3N} \sum_{kj} \frac{1}{\omega^2 - \omega_j^2(\boldsymbol{k})} \tag{8.35}$$

or

$$1 = \eta \omega_L^2 \int_0^{\omega_m} \frac{g(\omega')\,d\omega'}{\omega^2 - \omega'^2}. \tag{8.36}$$

In order to determine the changes in $\langle u^2(0) \rangle$ caused by the existence of local modes, it is instructive to look at the behavior of $G'_{\alpha\alpha}(00, \omega)$ near ω_L. We now replace the denominator of Eq. (8.27) by its Taylor expansion near $\omega = \omega_L$ [20], in which the first term automatically vanishes and the second is the lowest order term (see Eq. (F.26) in Appendix F):

$$\begin{aligned} \operatorname{Im} G'_{\alpha\alpha}(00, \omega + i\varepsilon) \\ &= \pi \frac{G_{\alpha\alpha}(00, \omega)}{\dfrac{d}{d\omega}[\eta M_0 \omega^2 G_{\alpha\alpha}(00, \omega)]} \delta(\omega - \omega_L) \\ &= -\frac{\pi}{2\eta M_0 \omega_L} \left[\frac{\eta \omega_L^4}{3N} \sum_{kj} \frac{1}{[\omega^2 - \omega_j^2(\boldsymbol{k})]^2} - 1 \right]^{-1} \delta(\omega - \omega_L). \end{aligned} \tag{8.37}$$

Substituting Eq. (8.37) into (8.21) gives an expression for the recoilless fraction:

$$-\ln f = \frac{k^2 \hbar}{2M} \frac{1 - \eta}{\eta \omega_L} \frac{\coth\left[\dfrac{1}{2}\beta\hbar\omega_L\right]}{\eta \omega_L^4 \displaystyle\int_0^{\omega_m} \frac{g(\omega')}{(\omega_L^2 - \omega'^2)^2} d\omega' - 1}. \tag{8.38}$$

For isotopic impurities or in cases where the force constants are approximately the same, the vibration of the impurity atom depends strongly on the η-value, and the amplitude increases as M decreases. When η is larger than the critical value η_{cr} (where $\eta_{cr} > 0$), the above vibrations will be mostly in discrete local modes. The lighter the impurity atom M is, the more the modes are localized, and the frequencies are higher. In general, it is difficult to excite these high-frequency modes by the recoil energy after the nucleus absorbs a photon. Consequently, the recoilless fraction increases, which is an important fact. The opposite happens as the ratio M/M_0 increases. Low-frequency resonance modes gradually dominate the vibrations, and f becomes more temperature-dependent.

If we use the Debye model to approximate the impurity vibrations, then the following simple relationship between θ'_D of the impurity and θ_D of the host can be derived [21, 22]:

$$\theta'_D = \theta_D\sqrt{\frac{M_0}{M}}. \tag{8.39}$$

This is to be expected because when the vibrations of both the impurity and the host follow the same model, $\omega \sim 1/\sqrt{M}$. However, Eq. (8.39) would only be valid for $T = 0$.

8.2 The Mannheim Model

In the Mannheim model, in addition to the impurity atom mass, the impurity–host force constant changes are also considered. However, only the nearest neighbor central forces are taken into account for either the host or the impurity–host system, and the anharmonic effects are neglected. This model was first developed for fcc and bcc lattices using group theory [23], followed by derivations using an alternative method [17], and later it was applied to the diamond structure [2]. The most significant contribution of the Mannheim model is that it has derived a simple and analytical expression for $\langle u^2(00)\rangle$ or f, and it has been in practical use because it agrees well with experimental results [24]. Considering only the nearest neighbor central force seems to be somewhat a crude model, but the Mannheim model is the only practical one available for Mössbauer spectroscopy.

We now use the fcc lattice to illustrate the essentials of this model. Suppose the impurity atom is at the origin with 12 host nearest neighbors. The symmetry point group of the impurity site is still O_h (or group O plus a central inversion i), and Eq. (8.20) will involve a 39×39 matrix. The irreducible representations for an impurity site having symmetry O_h in the fcc lattice are

$$\Gamma_{\text{fcc}} = A_{1g} \oplus A_{2g} \oplus 2E_g \oplus 2F_{1g} \oplus 2F_{2g} \oplus A_{2u} \oplus E_u \oplus 4F_{1u} \oplus 2F_{2u} \tag{8.40}$$

where each irreducible representation corresponds to one particular normal mode in the lattice. It can be shown that only in representation F_{1u} is the impurity atom involved in the lattice vibration. F_{1u} is a three-dimensional representation and appears four times. Therefore, in order to describe the vibration of the impurity atom, four three-dimensional basis vectors are needed.

When studying small vibrations of a lattice or a molecule, it is customary to use a displacement from the equilibrium position as the basis of an irreducible representation. However, in order to simplify calculations, it is necessary to introduce a set of orthonormal symmetry coordinates (linear combination of displacements) as the bases of $4F_{1u}$ [25], such that the matrices $\boldsymbol{G}$ and $\boldsymbol{U}$ in Eq. (8.16) are block-diagonalized. Owing to cubic symmetry, we only need to consider the x-components of the following four symmetry coordinates (see Appendix G):

$$\begin{aligned}
S_0 &= u_x(000) \\
2\sqrt{2}S_1 &= u_x(110) + u_x(\bar{1}\bar{1}0) + u_x(101) + u_x(\bar{1}0\bar{1}) \\
&\quad + u_x(1\bar{1}0) + u_x(\bar{1}10) + u_x(\bar{1}01) + u_x(10\bar{1}) \\
2\sqrt{2}S_2 &= u_y(110) + u_y(\bar{1}\bar{1}0) + u_z(101) + u_z(\bar{1}0\bar{1}) \\
&\quad - u_y(1\bar{1}0) - u_y(\bar{1}10) - u_z(\bar{1}01) - u_z(10\bar{1}) \\
2S_3 &= u_x(011) + u_x(0\bar{1}\bar{1}) + u_x(01\bar{1}) + u_x(0\bar{1}1)
\end{aligned} \tag{8.41}$$

where $u_\alpha(xyz)$ represents the α-direction unit displacement of the atom located at xyz.

The Green's function for the host lattice is translation invariant, and only depends on the relative position between atoms, as shown in Eq. (8.6). Therefore

$$G_{\alpha\beta}(l, l') = G_{\alpha\beta}(l - l', 0) = G_{\alpha\beta}(l - l') \tag{8.42}$$

and other symmetry properties of Green's functions are detailed in Appendix F.3.

The following shorthand notations will be used to avoid lengthy writing of the results:

$$U_{\alpha\beta}(110) = U_{\alpha\beta}(110, 000), \quad U_{\alpha\beta}(0, 0) = U_{\alpha\beta}(000, 000), \tag{8.43}$$

and

$$\begin{aligned}
g_0 &= G_{xx}(000), \quad g_1 = G_{xx}(110), \\
g_2 &= G_{xy}(110), \quad g_3 = G_{xx}(011), \\
A &= g_0 + G_{xx}(020) + G_{xx}(200) + G_{xx}(220) \\
B &= G_{xy}(220) + 2G_{xy}(211) \\
C &= g_3 + G_{xx}(211) \\
D &= G_{xx}(200) + G_{xx}(020) - G_{xx}(220) - g_0 \\
E &= G_{xy}(112) - g_2 \\
F &= g_0 + 2G_{xx}(020) + G_{xx}(022) \\
H &= g_1 + G_{xx}(121) \\
K &= G_{xy}(211).
\end{aligned} \tag{8.44}$$

We now calculate the matrix elements of $\boldsymbol{G}$ and $\boldsymbol{U}$ in the space spanned by the basis vectors S_0, S_1, S_2, and S_3 defined in (8.41):

$$\langle S_0|\boldsymbol{G}|S_0\rangle = \langle u_x(000)|\boldsymbol{G}|u_x(000)\rangle = G_{xx}(000) = g_0, \tag{8.45a}$$

$$\begin{aligned}\langle S_0|\boldsymbol{G}|S_1\rangle &= \left\langle u_x(000)|\boldsymbol{G}|\frac{1}{2\sqrt{2}}[u_x(110) + u_x(\bar{1}\bar{1}0) \right.\\ &\qquad \left. + u_x(101) + u_x(\bar{1}0\bar{1}) + \cdots]\right\rangle \\ &= \frac{1}{2\sqrt{2}}[G_{xx}(110) + G_{xx}(\bar{1}\bar{1}0) + G_{xx}(101) + G_{xx}(\bar{1}0\bar{1}) + \cdots] \\ &= \frac{8}{2\sqrt{2}}G_{xx}(110) = 2\sqrt{2}g_1. \end{aligned} \tag{8.45b}$$

In the above derivations, we have used the relation $G_{\alpha\beta}(l-l') = \langle u_\alpha(l')|\boldsymbol{G}|u_\beta(l)\rangle$ and the symmetry properties of Green's functions (Appendix F.3). Similarly, we have

$$\begin{aligned}\langle S_0|\boldsymbol{G}|S_2\rangle &= \left\langle u_x(000)|\boldsymbol{G}|\frac{1}{2\sqrt{2}}[u_y(110) + \cdots]\right\rangle \\ &= \frac{8}{2\sqrt{2}}G_{xy}(110) = 2\sqrt{2}g_2, \end{aligned} \tag{8.45c}$$

$$\begin{aligned}\langle S_0|\boldsymbol{G}|S_3\rangle &= \left\langle u_x(000)|\boldsymbol{G}|\frac{1}{2}[u_x(011) + \cdots]\right\rangle \\ &= \frac{4}{2}G_{xx}(011) = 2g_3. \end{aligned} \tag{8.45d}$$

The calculations of some of the other matrix elements are very tedious and would be unrevealing to be reproduced here. For example, $\langle S_2|\boldsymbol{G}|S_2\rangle = A + 2C$, which is actually simplified from a sum of 64 terms. Finally, the following matrix elements are obtained:

$$\boldsymbol{G}_{F_{1u}} = \begin{bmatrix} g_0 & 2\sqrt{2}g_1 & 2\sqrt{2}g_2 & 2g_3 \\ 2\sqrt{2}g_1 & A+2C & B & 2\sqrt{2}H \\ 2\sqrt{2}g_2 & B & 2E-D & 2\sqrt{2}K \\ 2g_3 & 2\sqrt{2}H & 2\sqrt{2}K & F \end{bmatrix} \tag{8.46}$$

$$U_{F_{1u}} = \begin{bmatrix} U_{xx}(00) & 2\sqrt{2}U_{xx}(110) & 2\sqrt{2}U_{xy}(110) & 2U_{xx}(011) \\ 2\sqrt{2}U_{xx}(110) & -U_{xx}(110) & -U_{xy}(110) & 0 \\ 2\sqrt{2}U_{xy}(110) & -U_{xy}(110) & -U_{xx}(110) & 0 \\ 2U_{xx}(011) & 0 & 0 & -U_{xx}(011) \end{bmatrix}. \tag{8.47}$$

Next, we inverse the matrix $(\boldsymbol{I} - \boldsymbol{GU})$ to evaluate the impurity Green's function $\boldsymbol{G}'$. However, in order to simplify the derivation, we have to utilize the relations between the Green's functions of the host lattice. The nearest neighbor forces are given by Eq. (F.38) in Appendix F as follows:

$$\sum_{\beta,l} \Phi_{\alpha\beta}(0,l)G_{\alpha'\beta}(ll',\omega) = -\delta_{\alpha\alpha'}\delta_{0l'} + M_0\omega^2 G_{\alpha\alpha'}(0l',\omega). \tag{8.48}$$

Since we are only interested in the x-direction motion of the central atom in an fcc lattice, the relevant equations derived from Eq. (8.48) are

$$\Phi_{xx}(0,0)g_0 + 8\Phi_{xx}(110)g_1 + 8\Phi_{xy}(110)g_2 + 4\Phi_{xx}(011)g_3 = -1 + M_0\omega^2 g_0, \tag{8.49a}$$

$$\Phi_{xx}(0,0)g_1 + \Phi_{xx}(110)(A+2C) + \Phi_{xy}(110)B + 2\Phi_{xx}(011)H = M_0\omega^2 g_1, \tag{8.49b}$$

$$\Phi_{xx}(0,0)g_2 + \Phi_{xx}(110)B + \Phi_{xy}(110)(2E-D) + 2\Phi_{xx}(011)K = M_0\omega^2 g_2, \tag{8.49c}$$

$$\Phi_{xx}(0,0)g_3 + 4\Phi_{xx}(110)H + 4\Phi_{xy}(110)K + \Phi_{xx}(011)F = M_0\omega^2 g_3. \tag{8.49d}$$

According to the central force approximation in Appendix E, some useful expressions for our calculation are

$$8\Phi_{xx}(110) = 8\Phi_{xy}(110) = -\Phi_{xx}(0,0), \tag{8.50}$$

$$\Phi_{xx}(011) = 0, \tag{8.51}$$

$$8U_{xx}(110) = 8U_{xy}(110) = U_{xx}(0,0) - \eta M_0\omega^2, \tag{8.52}$$

$$U_{xx}(011) = 0. \tag{8.53}$$

Now we are able to simplify Eqs. (8.46), (8.47), and (8.49) to expressions that contain only the parameters M_0, $\Phi_{xx}(0,0)$, η, λ, and g_0, where $\lambda = 1 - \Phi'_{xx}(0,0)/\Phi_{xx}(0,0)$. After somewhat lengthy calculations, one would arrive at the following inverse matrix:

$$(\boldsymbol{I} - \boldsymbol{GU})^{-1} = \frac{1}{\Delta}\begin{bmatrix} 1+\lambda(a_1+a_2) & -\frac{\lambda}{2\sqrt{2}}(1+a_0) & -\frac{\lambda}{2\sqrt{2}}(1+a_0) & 0 \\ -2\sqrt{2}b_1 & 1-\lambda+b_0+c_2 & -c_1 & 0 \\ -2\sqrt{2}b_2 & c_2 & 1-\lambda+b_0+c_1 & 0 \\ -2b_3 & -\frac{c_3}{\sqrt{2}} & -\frac{c_3}{\sqrt{2}} & \Delta \end{bmatrix} \tag{8.54}$$

where $a_i = -M_0\omega^2 g_i$, $b_i = -M_0\omega^2(\eta - \lambda)g_i$, and $c_i = -\lambda M_0\omega^2(1-\eta)g_i$. Δ is the determinant $|\boldsymbol{I} - \boldsymbol{GU}|$, which is given by

$$\Delta = (1-\lambda)(1-\eta)[1-\rho(\omega^2)S(\omega^2)] \tag{8.55}$$

where

$$\rho(\omega^2) = \frac{\eta}{1-\eta} + \frac{\omega^2}{\mu(+2)}\frac{\lambda}{1-\lambda}, \tag{8.56}$$

and

$$S(\omega^2) = -1 + M_0\omega^2 G_{xx}(00). \tag{8.57}$$

We notice that $\Delta = 0$ is exactly the condition for having a resonance mode or a localized mode:

$$1 - \rho(\omega^2)S(\omega^2) = 0. \tag{8.58}$$

We now solve for $\boldsymbol{G}'$ using the Dyson equation. Multiplying the first row of (8.54) by the first column of matrix (8.46) and making use of Eqs. (8.49a) and (8.50), we obtain

$$G'_{xx}(00,\omega) = \frac{1 + S(\omega^2)\left[1 - \rho(\omega^2) + \dfrac{\eta}{1-\eta}\right]}{M_0\omega^2(1-\eta)[1-\rho(\omega^2)S(\omega^2)]}. \tag{8.59}$$

It is easy to show that, in the case of mass defect approximation only ($\lambda = 0$), Eqs. (8.59) and (8.58) reduce exactly to Eqs. (8.27) and (8.29). Now, substituting Eq. (8.59) with the complex function $S(\omega^2 + i\varepsilon)$ in it into Eq. (8.23), we get the vibrational DOS function for an impurity atom from the unperturbed phonon DOS as follows:

$$g'(\omega) = \frac{M_0}{M}\frac{g(\omega)}{[1-\rho(\omega^2)S(\omega^2)]^2 + \left[\dfrac{1}{2}\pi\omega\rho(\omega^2)g(\omega)\right]^2} + \frac{M_0}{M}\frac{\delta(\omega-\omega_L)}{\dfrac{M_0}{M} - [1+\rho(\omega^2)]^2 + \omega_L^4[\rho(\omega^2)]^2\displaystyle\int_0^\infty \frac{g(\omega')\,d\omega'}{(\omega_L^2 - \omega'^2)^2}}. \tag{8.60}$$

Here, the first term is the contribution from the resonance modes whose frequencies lie in the range of the normal modes of the host lattice. The second term is the contribution from the localized mode ($\omega_L > \omega_{max}$), which exists if the mass of an impurity atom is sufficiently light and/or the binding of an impurity atom to the host lattice is sufficiently strong.

Equations (8.58) through (8.60) are the results of the Mannheim model, which has been successfully applied to the following four types of impurity–host systems: simple cubic, face-centered cubic, body-centered cubic, and the diamond structure [2]. The response function $g'(\omega)$ is calculated from the $g(\omega)$ of the host lattice, which is either known or not difficult to obtain for many host systems. For a cubic lattice, the Mannheim model is still the best suitable method for calculating the values of $\langle u^2 \rangle$ and $\langle v^2 \rangle$ for the impurity atom based on the measurements of the recoilless fraction f and the second-order Doppler shift.

The method highlighted above is not the simplest; there are several other approaches that are slightly less cumbersome. The results for the function $\boldsymbol{G}'$ given by different authors [2, 12, 23] seem to be different at first glance, but one can easily verify that they are all identical to Eq. (8.59). The Green's function for the host introduced by some authors [23] differs from Eq. (8.7) by a negative sign, causing also a negative sign in $\boldsymbol{G}'$. The definitions of $\boldsymbol{G}$ and $\boldsymbol{U}$ here are consistent with Refs. [2, 13].

At the present time direct observation of the vibrational DOS for an impurity atom is possible by using inelastic nuclear resonant scattering of synchrotron radiation, as discussed in Chapter 7. Figures 8.2, 8.3, and 8.4 show the measured and the calculated vibrational DOS for impurity ^{57}Fe in Al [26], Cu [26], and Cr [27], whose unperturbed phonon DOS are given correspondingly. From these figures one can find a good agreement between experimental results and the theoretical curves of the Mannheim model. For Fe in Al, the measured vibrational DOS shows that the vibrational modes are in resonance with host normal modes.

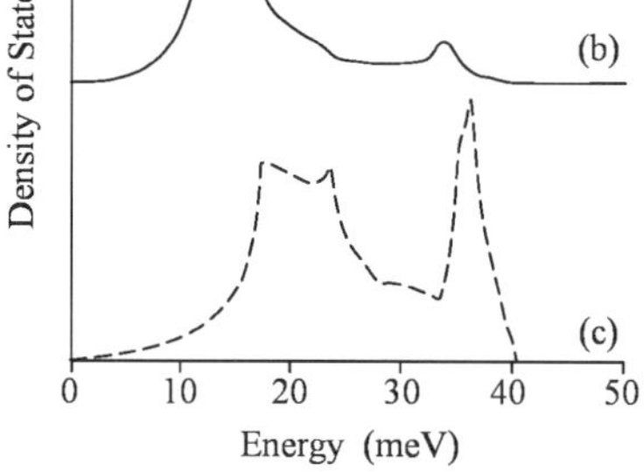

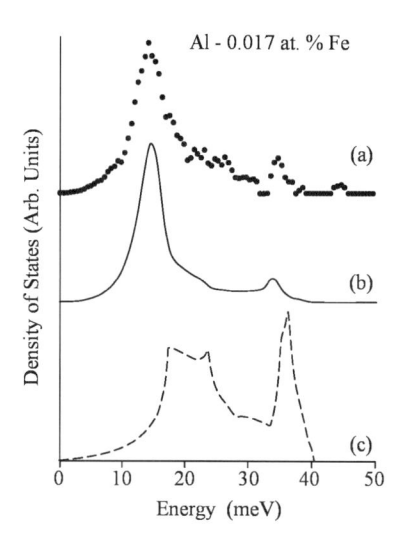

Fig. 8.2 (a) Vibrational DOS of ^{57}Fe in Al–0.017 at.% Fe measured using inelastic nuclear resonance scattering. (b) Vibrational DOS of Fe atom in Al calculated on the basis of the Mannheim model with $\Phi/\Phi' = 0.94$ [23]. (c) Unperturbed phonon DOS of Al.

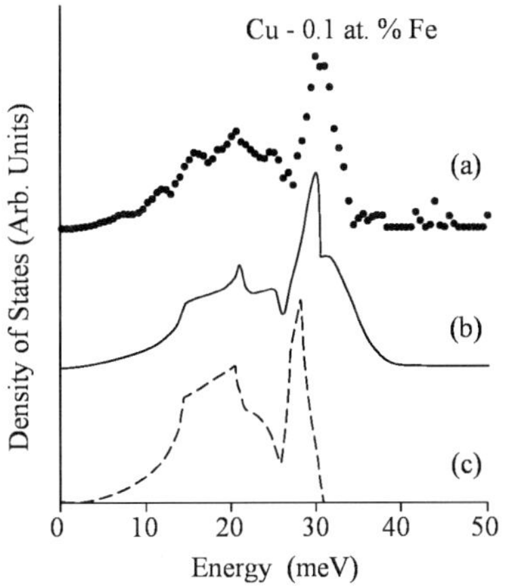

Fig. 8.3 (a) Vibrational DOS of ^{57}Fe in Cu–0.1 at.% Fe measured using inelastic nuclear resonance scattering. (b) Vibrational DOS of Fe atom in Cu calculated on the basis of the Mannheim model with $\Phi/\Phi' = 0.79$ [23]. (c) Unperturbed phonon DOS of Cu.

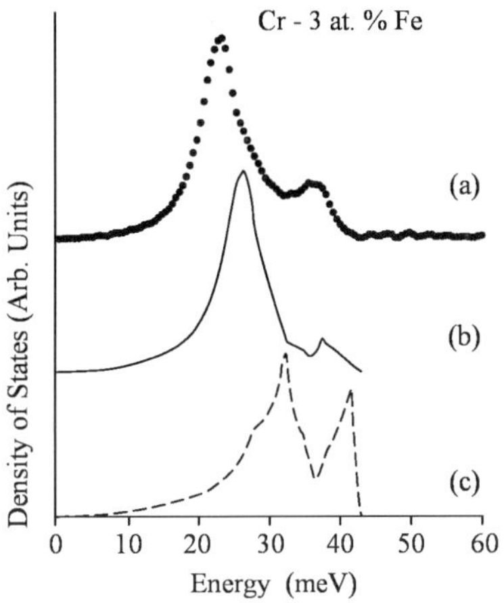

Fig. 8.4 (a) Vibrational DOS of ^{57}Fe in $^{57}Fe_{0.03}Cr_{0.97}$ alloy film measured using inelastic nuclear resonance scattering [27]. (b) Vibrational DOS of Fe atom calculated on the basis of the Mannheim model with $\Phi/\Phi' = 1.25$ [28]. (c) Unperturbed phonon DOS of Cr.

In the case of Fe in Cu, besides the resonant modes, a peak interpreted as being the localized mode predicted by the Mannheim model was observed. The Mannheim model assumes harmonic forces only. Therefore, the localized modes appear as the Dirac function. Anharmonic contributions to the interatomic forces, as well as other phonon interactions, are expected to broaden these sharp lines into narrow frequency bands, as is observed in Fig. 8.3. As for Fe in Cr, the agreement between the measured and calculated DOS is poor because the latter was taken for a bulk sample, not a film. So, one may get a better agreement by choosing an appropriate ratio Φ'/Φ.

8.3
Impurity Site Moments

In Chapter 4, we discussed how to describe lattice dynamics using frequency moments $\mu(n)$ instead of using the response function $g(\omega)$. For an isolated substitutional impurity, the corresponding impurity site moments are defined as

$$\mu'(n) = \int_0^\infty \omega^n g'(\omega)\, d\omega \quad \text{and} \quad \mu'(0) = 1. \tag{8.61}$$

Since both integrands are site-dependent, these moments must also be site-dependent.

Based on the concept of weighted mean frequencies in Ref. [29], we might give another definition of the nth site moment for a compositional disordered solid in the form

$$\mu'(l,n) = \sum_{s=1}^{\infty} |B_\alpha(l,s)|^2 \omega_s^n \tag{8.62}$$

where $B_\alpha(l,s)$ are elements of a unitary matrix given in Section 4.1.4. For a perfect cubic lattice the normal mode s is replaced by $(\boldsymbol{k}j)$, and from Eq. (4.38) one gets

$$|B_\alpha(l,s)|^2 = \frac{|\mathbf{e}_\alpha(\boldsymbol{k}j)|^2}{N} = \frac{1}{3N}. \tag{8.63}$$

Therefore, the moment defined by (8.62) reduces to the usual expression (4.105):

$$\mu_\alpha(l,n) = \frac{1}{3N}\sum_{\boldsymbol{k},j} \omega_j^n(\boldsymbol{k}),$$

and in this case the indices l and α can be omitted. Therefore, the definition in (8.62) is equivalent to (8.61).

Again according to Ref. [29] and using expression (8.62), $\langle u^2\rangle$ and $\langle v^2\rangle$ of the impurity atom at $l = 0$ in a cubic host can be expressed, for high temperatures $(T > \theta_D/2)$, as

$$\begin{aligned}
\langle u^2\rangle &= \frac{k_B T}{M}\left[\mu'(-2) + \frac{1}{12}\left(\frac{\hbar}{k_B T}\right)^2 - \frac{1}{720}\left(\frac{\hbar}{k_B T}\right)^4 \mu'(+2) + \cdots\right],\\
\langle v^2\rangle &= \frac{3k_B T}{M}\left[1 + \frac{1}{12}\left(\frac{\hbar}{k_B T}\right)^2 \mu'(+2) - \frac{1}{720}\left(\frac{\hbar}{k_B T}\right)^4 \mu'(+4) + \cdots\right].
\end{aligned} \tag{8.64}$$

And for $T \to 0$ as

$$\langle u^2 \rangle = \frac{\hbar}{2M}\mu'(-1), \quad \langle v^2 \rangle = \frac{3\hbar}{2M}\mu'(+1). \tag{8.65}$$

If the prime is removed from μ' in (8.65), the expressions are also valid, but only for the perfect lattice.

It has been pointed out [12, 29] that, for any harmonic cubic Bravais lattice with central or noncentral neighbor forces, the moment $\mu(+2)$ is given by

$$\mu(+2) = \frac{\Phi_{xx}(0,0)}{M_0} \tag{8.66}$$

where

$$\Phi_{xx}(0,0) = -\sum_{l \neq 0} \Phi_{xx}(0,l).$$

For a substitutional impurity atom, a similar relation can be written:

$$\mu'(+2) = \frac{\Phi'_{xx}(0,0)}{M} \tag{8.67}$$

where

$$\Phi'_{xx}(0,0) = -\sum_{l \neq 0} \Phi'_{xx}(0,l).$$

Generally speaking, contributions to $\Phi_{xx}(0,0)$ can be made by up to the sixth or seventh nearest neighbors, but for cubic lattices such as fcc or bcc, summing up just the nearest neighbors would be sufficient and the result is

$$\frac{\Phi_{xx}(0,0)}{M_0} = \frac{1}{2}\omega_{\mathrm{m}}^2$$

where ω_{m} is the maximum frequency of lattice vibration [10]. Good agreements have been obtained between this approximation and $\mu(+2)$, as shown in Table 8.1. Discrepancies between $(1/2){\omega_{\mathrm{m}}}^2$ and $\mu(+2)$ are only significant for a small number of lattices. In Eq. (8.56), $(1/2){\omega_{\mathrm{m}}}^2$ has been replaced with $\mu(+2)$ so that the expression is more general.

Incidentally, because of the anharmonic effect, Φ', Φ, and Φ'/Φ all depend on temperature; thus both $g'(\omega)$ and $g(\omega)$ as well as the frequency moments are all functions of temperature. For ^{57}Fe in Cu, for example, Φ and $\mu(+2)$ vary about 2% for every 100 K temperature change [1].

Table 8.1 The parameters $\mu(+2)$ and β_n of cubic lattices [12].

Lattice		T[a] (K)	$\mu(+2)$ (10^{26} rad^2 s^{-2})	$\frac{1}{2}\omega_m^2$ (10^{26} rad^2 s^{-2})	β_{-2}	β_{-1}	β_{+1}	β_{+4}
fcc	Al	80	16.91	18.73	0.556	0.842	1.046	0.759
	Ni	RT	15.29	15.54	0.603	0.865	1.038	0.797
	Cu	RT	10.10	10.79	0.559	0.848	1.042	0.779
	Kr	10	0.44	0.45	0.555	0.851	1.041	0.782
	Pd	RT	8.30	9.46	0.509	0.830	1.046	0.765
	Ag	RT	4.85	5.22	0.524	0.833	1.047	0.763
	Xe	10	0.33	0.33	0.558	0.852	1.041	0.784
	Pt	90	5.85	6.75	0.506	0.823	1.050	0.747
	Au	RT	3.45	4.32	0.443	0.788	1.063	0.695
	Pb	100	0.92	1.01	0.491	0.801	1.061	0.702
bcc	Na	90	2.84	2.88	0.454	0.791	1.057	0.747
	Cr	RT	21.90	18.21	0.691	0.910	1.022	0.879
	Fe	RT	18.02	16.99	0.599	0.870	1.034	0.818
	Rb	120	0.40	0.43	0.382	0.753	1.068	0.712
	Nb	RT	8.28	8.51	0.534	0.851	1.037	0.816
	Mo	RT	14.43	13.01	0.674	0.904	1.024	0.865
	Ta	RT	5.24	5.22	0.617	0.880	1.030	0.845
	W	RT	9.83	9.10	0.658	0.894	1.028	0.842

[a] RT, room temperature.

Since the relation between $\mu(+2)$ and the host restoring force, Eq. (8.66), does not depend on the specific lattice model, we may introduce a dimensionless ratio β_n for relating $\mu(+2)$ to other frequency moments:

$$\beta_n = \frac{[\mu(+2)]^{n/2}}{\mu(n)}. \tag{8.68}$$

Measuring frequency moments of the various orders can provide the impurity–host force constant ratios Φ'/Φ as an important parameter. In particular, low-temperature Φ'/Φ ratios can only be obtained through analysis of these moments. Since the Mössbauer effect has the distinctive advantage in obtaining the low-temperature data, $\mu(-1)$ and $\mu(-2)$ can be calculated from accurately measured f-values. As for $\mu(+1)$, the result is usually not as good because the second-order Doppler shift at low temperatures is not as pronounced.

We now summarize the results of $\mu'(n)$ derived from several different lattice models, especially the Mannheim model. Details can also be found in a few good review articles [2, 12, 28].

8.3.1 The Einstein Model

The simplest model for lattice vibrations is of course the Einstein model. The lattice is treated as independent oscillators (atoms) with the same frequency $\omega_E = (\Phi/M_0)^{1/2}$. If there is a substitutional impurity of mass M and the new force constant Φ', the frequency becomes ω'_E, and

$$\frac{\mu'_E(n)}{\mu_E(n)} = \left(\frac{\omega'_E}{\omega_E}\right)^n = \left(\frac{M_0}{M}\right)^{n/2}\left(\frac{\Phi'}{\Phi}\right)^{n/2}. \tag{8.69}$$

8.3.2 The Einstein–Debye Model

Using the Debye model, the frequency moments can be expressed in terms of Debye temperatures as in Eq. (4.109):

$$\mu(n) = \frac{3}{n+3}\left(\frac{k_B}{\hbar}\right)^n [\theta_D(n)]^n \quad \text{and} \quad \mu'(n) = \frac{3}{n+3}\left(\frac{k_B}{\hbar}\right)^n [\theta'_D(n)]^n. \tag{8.70}$$

Substituting these into Eq. (8.69), we obtain

$$\theta'_D(n) = \theta_D(n)\left(\frac{M_0}{M}\right)^{1/2}\left(\frac{\gamma'}{\gamma}\right)^{1/2} \tag{8.71}$$

where the force constant ratio is written as γ'/γ, and called the Einstein–Debye force constant ratio. Equation (8.71) is the result of combining the two models, and hence known as the Einstein–Debye model.

8.3.3 The Maradudin–Flinn Model

Using the fcc lattice as an example and taking the central force approximation, the following expression is obtained [7]:

$$\frac{\mu'(-2)}{\mu(-2)} = \frac{M}{M_0}\left[1 + \frac{(1-\Phi'/\Phi)}{\mu(+2)\mu(-2)} + \frac{5}{4}\frac{(1-\Phi'/\Phi)^2}{\mu(+2)\mu(-2)} + \cdots\right]. \tag{8.72}$$

This model is most suitable when the impurity only causes a small change in the force constant.

8.3.4 The Visscher Model

Considering only the interactions among nearest neighbors in a simple cubic lattice [30] and using the Visscher model [31], the following result has been obtained for high temperatures:

$$\frac{\mu'(-2)}{\mu(-2)} = \frac{M}{M_0}\left[1 - 0.675\left(1 - \frac{\Phi}{\Phi'}\right)\right] \tag{8.73}$$

which is valid for all values of Φ/Φ', not just small force constant changes.

8.3.5 The Mannheim Model

1. Even moments. Detailed analyses of two extreme situations (ω very high and very low) can be found in Ref. [12]. When ω is very high

$$\frac{\mu'(+2)}{\mu(+2)} = \frac{M_0}{M}\left(\frac{\Phi'}{\Phi}\right) \tag{8.74}$$

which is essentially the ratio of (8.67) to (8.66), consistent with the Einstein model for $n = 2$. When ω is very low, another important relation can be derived:

$$\frac{\mu'(-2)}{\mu(-2)} = \frac{M}{M_0}\left[1 - \beta_{-2}\left(1 - \frac{\Phi}{\Phi'}\right)\right] \tag{8.75}$$

where

$$\beta_{-2} = \frac{1}{\mu(+2)\mu(-2)} = \frac{5}{9}\left(\frac{\theta_D(-2)}{\theta_D(+2)}\right)^2.$$

This is valid for a wide temperature range, and is not the same as from the Einstein model because $\beta_{-2} \neq 1$. In order to compare this with Eq. (8.72), we write Φ/Φ' as $\Phi/\Phi' = [1 - (1 - \Phi'/\Phi)]^{-1}$ and expand it as a polynomial of the small quantity $(1 - \Phi'/\Phi)$:

$$\frac{\mu'(-2)}{\mu(-2)} = \frac{M}{M_0}\left[1 + \beta_{-2}\left(1 - \frac{\Phi'}{\Phi}\right) + \beta_{-2}\left(1 - \frac{\Phi'}{\Phi}\right)^2 + \cdots\right]. \tag{8.76}$$

This differs from Eq. (8.72) in the third term by a factor of 5/4, which is much larger than β_{-2}, limiting the Maradudin–Flinn model to cases with $\Phi' \approx \Phi$.

Using Eq. (8.70), we may replace the moments $\mu'(-2)$, $\mu(-2)$, and $\mu(+2)$ in Eq. (8.75) with $\theta'_D(-2)$, $\theta_D(-2)$, and $\theta_D(+2)$, respectively, and obtain

$$\left[\frac{\theta_D(-2)}{\theta'_D(-2)}\right]^2 = \frac{M}{M_0}\left[1 - \frac{5}{9}\left(\frac{\theta_D(-2)}{\theta_D(+2)}\right)^2\left(1 - \frac{\Phi}{\Phi'}\right)\right]. \tag{8.77}$$

The Debye temperature $\theta_D(+2)$ of the host lattice can be determined either by the heat capacity method or by neutron scattering, which usually give consistent re-

Table 8.2 Frequency moments $\omega(n) = [\mu(n)]^{1/n}$ ($\times 10^{13}$ rad s^{-1}) from dispersion relations and heat capacity data [28].

Metal (temp. (K))	Method[a]	$\omega(+4)$	$\omega(+2)$	$\omega(+1)$	$\omega(-1)$	$\omega(-2)$	$\omega_D(-3)$	Ref.
Cu (296)	NS	3.38	3.18	3.05	2.69	2.37	4.30	12, 32
	NS	3.37	3.18	3.04	2.69	2.37	4.36	33
	NS	3.39	3.20	3.07	2.72	2.39	4.34	34
Cu (80)	NS	3.45	3.29	3.11	2.75	2.42	4.47	12, 36
	NS	3.41	3.21	3.10	2.76	2.43	4.52	34
	HC	3.38(4)	3.21(3)	–	2.72(2)	2.43(2)	–	12, 37
	HC	–	–	–	–	–	4.49	35
	HC	3.41(2)	3.24(1)	–	2.76(1)	2.43(1)	4.52(1)	42
Al (80)	NS	4.41	4.11	3.93	3.47	3.07	5.77	12, 39
	NS	4.43	4.13	–	3.47	3.06	5.66	39
	HC	4.27(4)	4.06(2)	–	3.39(3)	3.09(3)	–	37
	HC	–	–	–	–	–	5.61	35
	HC	4.36(13)	4.08(1)	–	3.45(1)	3.05(1)	5.63(1)	39
Al (300)	NS	4.34	4.03	–	3.35	2.94	5.37	39
Pt (90)	NS	2.60	2.42	2.30	1.99	1.72	3.05	12, 40
	HC	2.80(2)	2.52(1)	–	1.99(2)	1.72(2)	–	41
	HC	–	–	–	–	–	3.07	35
	HC	2.93(1)	2.60(5)	2.41(4)	2.01(3)	1.73(2)	3.12(3)	42
V (296)	XDS	3.68	3.54	3.44	3.15	2.81	5.13	12, 43
	HC	4.35(3)	4.13(3)	3.98(3)	3.49(3)	2.98(2)	5.22(4)	38

[a] NS, neutron scattering; HC, heat capacity; XDS, x-ray diffuse scattering.

sults (Table 8.2), while $\theta'_D(-2)$ and $\theta_D(-2)$ are obtained by fitting the f versus temperature curve. Finally, we can calculate the force constant ratio Φ/Φ' from Eq. (8.77). One example of the application of the Mannheim model is the studies of ^{119}Sn impurities in host lattices of Si, Ge, and α-Sn using Mössbauer spectroscopy [44]. The results are listed in Table 8.3. A comparison of Φ/Φ' with the Einstein–Debye force constant ratio γ/γ' shows that they do not agree. However,

Table 8.3 Mössbauer effect results of ^{119}Sn impurities in host lattices of Si, Ge, and α-Sn [2, 44].

	$\theta_D'(-2)$	$\frac{\Phi_{xx}(0,0)}{\Phi'_{xx}(0,0)}$	$\frac{\gamma}{\gamma'}$	f(300 K) (Mannheim model)	f(300 K) (experimental)
Si	223(4)	1.92(15)	1.31(6)	0.332(14)	0.34(3)
Ge	191(4)	2.51(30)	1.48(7)	0.226(20)	0.22(3)
α-Sn	161(3)	1	1	0.125(13)	0.13(1)

the Mannheim model force constant ratio gives an f-value in each case that agrees very well with the experimental result.

2. Odd moments. While the even moments describe the impurity atom motion that resembles the classical vibration, the odd moments depict the zero-point motion of the impurity atoms. No analytical expressions of $\mu(\pm 1)$ are available from the Mannheim model, but for $0.25 \le M/M_0 \le 4$ and $0.2 \le \Phi_{xx}(0,0)/\Phi'_{xx}(0,0) \le 5$, two semi-empirical formulas have been derived:

$$\frac{\mu'(-1)}{\mu(-1)} \approx \left(\frac{M}{M_0}\right)^{1/2+a} \left\{1 - \left(\frac{M}{M_0}\right)^{-2a} \beta_4 \left[1 - \left(\frac{\Phi_{xx}(0,0)}{\Phi'_{xx}(0,0)}\right)^{1/2}\right]\right\}, \tag{8.78}$$

$$\frac{\mu'(+1)}{\mu(+1)} \approx \left(\frac{M}{M_0}\right)^{1/2+b} \left(\frac{\Phi'_{xx}(0,0)}{\Phi_{xx}(0,0)}\right)^{(1/2)[1-b(M/M_0)]}. \tag{8.79}$$

For Cu as the host lattice,

$$a = \frac{1}{2}\left(\frac{1}{\sqrt{\beta_{-1}}} - 1\right) = 0.043 \quad \text{and} \quad b = \frac{1}{2}(\beta_1 - 1) = 0.021.$$

Values of β_{-1}, β_1, and β_4 can be found in Table 8.1.

Figure 8.5 shows plots of $\mu'(-1)/\mu(-1)$ and $\mu'(+1)/\mu(+1)$ against Φ/Φ' for various parameters of M/M_0, all within the ranges indicated above. The black circles in the graphs represent calculated values using Eq. (8.60) to evaluate $g'(\omega)$ from known room temperature $g(\omega)$ for Cu, followed by using Eq. (8.61) to evaluate $\mu(\pm 1)$ and $\mu'(\pm 1)$. The solid curves in Fig. 8.5 are best fits using formulas (8.78) and (8.79), whereas the dashed curves are the Einstein model results [12].

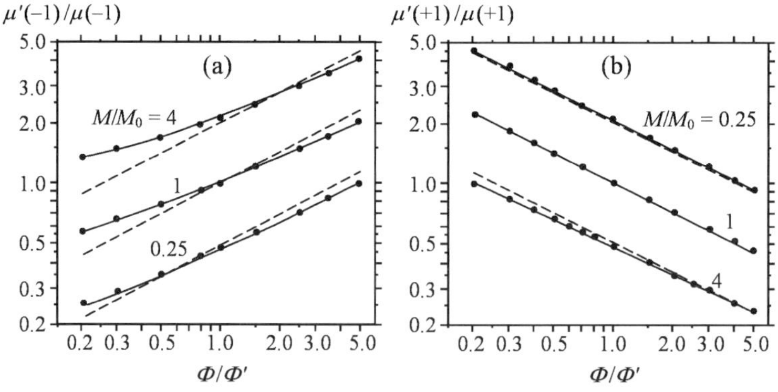

Fig. 8.5 Theoretical predictions of (a) $\mu'(-1)/\mu(-1)$ and (b) $\mu'(+1)/\mu(+1)$ as functions of the force constant ratio Φ/Φ'. The solid lines represent the results from Eqs. (8.78) and (8.79), while the dashed lines represent the results based on the Einstein model.

3. The McMillan ratio. It was found that an impurity's McMillan ratio $\langle\omega\rangle/\langle\omega^{-1}\rangle$ is also an important parameter of lattice dynamics. It is approximately a constant, and can be deduced from Mössbauer experiments. Using the Mannheim model, the impurity McMillan ratio may be expressed in terms of frequency moments. In the high-temperature limit [12]

$$\left(\frac{\langle\omega\rangle}{\langle\omega^{-1}\rangle}\right)_{\text{HT}} = \frac{1}{\mu'(-2)} = \frac{\Phi_{xx}(0,0)}{M}\left(\frac{1}{\beta_{-2}} - 1 + \frac{\Phi_{xx}(0,0)}{\Phi'_{xx}(0,0)}\right)^{-1}, \tag{8.80}$$

and in the low temperature limit

$$\left(\frac{\langle\omega\rangle}{\langle\omega^{-1}\rangle}\right)_{\text{LT}} = \frac{\mu'(+1)}{\mu'(-1)} \approx \frac{\Phi_{xx}(0,0)}{M}\left(\frac{\Phi'_{xx}(0,0)}{\Phi_{xx}(0,0)}\right)^{1/2}\frac{\beta_{-1}}{\beta_{+1}}\left\{1-\beta_4\left[1-\left(\frac{\Phi_{xx}(0,0)}{\Phi'_{xx}(0,0)}\right)^{1/2}\right]\right\}^{-1}. \tag{8.81}$$

In summary, analysis of the f-values and their temperature dependence for ^{57}Fe, ^{119}Sn, and ^{197}Au in fcc and bcc lattices shows that the Mannheim model can adequately describe experimental results for most cases when the force constant ratio and mass ratio fall into the following ranges: $0.65 \leq \Phi_{xx}(0,0)/\Phi'_{xx}(0,0) \leq 2.6$ and $0.3 \leq M/M_0 \leq 3.5$. In host lattices of more massive atoms, noncentral forces become appreciable; in some cases, the presence of the impurity atom causes host atoms to deviate from their original equilibrium positions, and it would not be sufficient to merely consider the nearest neighbors.

8.4 Examples of Mössbauer Studies of ^{57}Fe, ^{119}Sn, and ^{197}Au Impurities

As substitutional impurity atoms in Mössbauer studies, ^{57}Fe is the most commonly used isotope, followed by ^{119}Sn. To investigate the effects of heavy impurities ($M/M_0 > 1$), ^{197}Au is also a suitable choice. In this section, we discuss the experimental results using these three isotopes.

8.4.1 ^{57}Fe Impurity Atoms

The Mannheim model is very successful in describing the dynamic properties of ^{57}Fe impurities, because only a very low impurity concentration (10^{-4} to 10^{-2}) is required for obtaining a good spectrum [1], and thus the impurity atoms are iso-

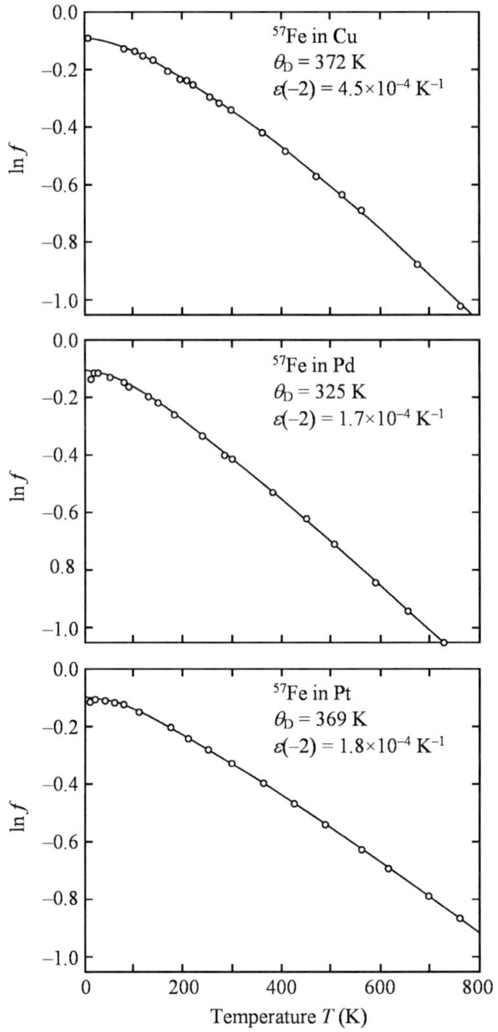

Fig. 8.6 Recoilless fractions f of ^{57}Fe in single-crystal Cu, Pd, and Pt as functions of T. The solid lines are Debye functions, corrected for the anharmonic effect $\varepsilon(-2)$.

lated from each other. Mössbauer spectra of ^{57}Fe in most of the cubic host lattices such as Ag, Al, Au, Cr, Cu, Ir, Mo, Nb, Ni, Pd, Pt, Rh, Ta, V, and W have been investigated. Figure 8.6 shows how the recoilless fraction f varies with temperature for ^{57}Fe in Cu, Pd, and Pt [28, 45]. The parameters such as f, $\theta'_D(n)$, and force constant ratio Φ/Φ' for ^{57}Fe in various cubic hosts are listed in Tables 8.4 and 8.5.

Table 8.4 Recoilless fraction $f(T)$, $\theta_D'(-2)$, and $\theta_D(-2)$ for ^{57}Fe impurities in cubic metals [12, 24, 25, 46, 47].[a]

Host	T[b] (K)	$f(T)$	Best value $f(T)$	$\theta_D'(-2)$ (K)	$\theta_D(-2)$ (K)
Ag	RT	*0.64(4)*			
		0.52(3)			
		0.58(3)	–		211
Al	RT	*0.54(2)*			
		0.52(5)			
		0.50(5)	0.50(5)		405
Au	RT	*0.589(14)*			
		0.62(5)			
		0.583(10)	0.583(10)		164
Cr	RT	0.76(2)			
		0.792(9)	0.790(9)		
Cu	RT	0.710(14)			
		0.727(16)			
		0.725(34)			
		0.710(10)			
		0.709(6)			
		0.703(7)			
		0.710(6)	0.709(5)	372(3)	317(10)
	4	0.917(19)			
		0.910(7)			
		0.911(6)	0.911(5)		
Ir	RT	*0.807(25)*			
		0.79(3)			
		0.812(5)	0.812(5)		
	4	0.914(5)	0.914(5)		
Mo	RT	0.78(5)			
		0.76(3)			
		0.77(1)			
		0.753(8)			
		0.773(11)	0.763(11)		
	4	0.907(10)			
		0.885(11)	0.907(10)		
Nb	RT	0.63(3)			
		0.659(8)			
		0.660(10)			
		0.644(4)	0.648(14)		226
	4	*0.881(6)*			
		0.846(10)	0.881(10)		

Host	T[b] (K)	$f(T)$	Best value $f(T)$	$\theta_D'(-2)$ (K)	$\theta_D(-2)$ (K)
Ni	RT	0.80(1)			
		0.81(5)	0.80(1)	505(10)	401
Pd	RT	0.652(36)			
		0.652(15)			
		0.661(6)			
		0.657(24)	0.659(4)	325(3)	257
	4	*0.813(13)*			
	13	*0.875(15)*			
	20	0.891(6)	0.891(10)		
Pt	RT	0.729(25)			
		0.723(36)			
		0.729(16)			
		0.723(8)	0.725(7)	369(3)	231
	4	*0.85(5)*			
	12	*0.897(10)*			
	20	0.905(8)	0.905(8)		
Rh	RT	0.785(17)			
		0.783(25)			
		0.78(10)			
		0.781(5)	0.781(5)		255
	4	0.875(18)			
		0.910(6)	0.906(6)		
Ta	RT	*0.77(4)*			
		0.76(3)			
		0.704(8)	0.704(8)		235
V	RT	*0.55(3)*			
		0.76(3)			
		0.547(24)			
		0.76(1)			
		0.70(1)	–		
	4	*0.913(10)*	–		
W	RT	*0.86(3)*			
		0.86(5)			
		0.797(9)	0.797(9)		263
	4	0.916	0.916(13)		

[a] Values in italics may be less reliable.
[b] RT, room temperature.

Table 8.5 Force constant ratio Φ/Φ' calculated from temperature dependence of recoilless fraction $f(T')$ in ^{57}Fe Mössbauer effect experiments, from neutron dispersion and heat capacity data [12].[a]

Host	$\frac{M_0}{M}$	Using Mössbauer and neutron dispersion data			Using neutron dispersion data	Using heat capacity data	
		From $f(T')$, $g(\omega)$ at T_0, and Eq. (8.57)	Impurity T' (K)[b]	Host T_0 (K)[b]	From $\frac{\mu'(-1)}{\mu(-1)}$ $T' \approx 4$ K $T_0 \approx 4$ K	From $\frac{\mu'(-2)}{\mu(-2)}$ $T' \approx 296$ K $T_0 \approx 150$ K	From $\frac{\mu'(-1)}{\mu(-1)}$ $T' \approx 4$ K $T_0 \approx 150$ K
Al	0.47	1.6(3)	RT	80	–	1.6(6)	–
Au	3.46	1.49(10)	RT	RT	–	1.59(8)	–
Cr	0.91	1.43(10)	RT	RT	–	1.39(20)	–
Cu	1.12	0.76(1)	–	–	–	–	–
		0.80(10)	80	80	0.81(13)	0.80(7)	0.80(4)
		0.82(3)	RT	RT	–	–	–
		0.87(3)	471	473	–	–	–
		0.91(3)	677	673	–	–	–
Ir	3.38	–	–	–	–	2.15(16)	2.56(40)
Mo	1.69	2.25(15)	RT	RT	*2.49(64)*	2.31(20)	2.55(76)
Nb	1.63	1.63(9)	RT	RT	*2.10(50)*	1.67(10)	2.17(55)
Ni	1.03	0.80(10)	RT	RT	–	–	–
Pd	1.87	1.72(2)	–	–	–	–	–
		1.72(2)	126	120	1.96(50)	1.78(9)	1.94(70)
		1.78(5)	RT	RT	–	–	–
		1.71(4)	655	673	–	–	–
Pt	3.43	1.60(25)	78–110	90	1.97(40)	1.73(7)	2.16(53)
Rh	1.81	–	–	–	–	1.84(10)	2.32(40)
Ta	3.18	1.84(8)	RT	RT	*2.5(7)*	1.94(15)	2.7(8)
W	3.23	2.42(18)	RT	RT	*2.6(10)*	2.53(21)	2.9(12)

[a] Values in italics may be less reliable.
[b] RT, room temperature.

8.4.2
^{119}Sn Impurity Atoms

Dynamic parameters such as f, Φ'/Φ, and $\theta_D(-2)$ from several investigations of ^{119}Sn impurities in Ag, Al, Au, Pb, Pd, Pt, Si, Ge, α-Sn, Cu, and Rh are summarized in Table 8.6. In all experiments, the ^{119}Sn concentration was higher than 1 at.%, and thus the ^{119}Sn atoms cannot be treated as isolated impurities. However, these studies have shown that for such high impurity concentrations, the Mann-

Table 8.6 Dynamic parameters Φ'/Φ, f, $\theta'_D(-2)$, and $\theta_D(-2)$ for ^{119}Sn impurities in several host lattices [2, 12, 24, 44, 47].

Host	$\frac{M}{M_0}$	$\frac{\Phi'_{xx}(0,0)}{\Phi_{xx}(0,0)}$	f Room temp.	f 4 K	$\theta'_D(-2)$ (K)	$\theta_D(-2)$ (K)
Ag	1.10	0.81(13)	0.27(1)	0.80(2)	190(8)	211
Al	4.41	0.49	0.14(2)		153(6)	405
Au	0.60	0.54	0.18(5)	0.85(2)	180(8)	164
Pb	0.57	1.64	0.016(15)	0.80(6)		88
Pd	1.12	1.08(28)	0.48(5)		262(20)	272
Pt	0.61	0.32	0.44(5)		212(9)	236
Si	4.24	0.52	0.34(3)		223(4)	526
Ge	1.64	0.40	0.22(3)		191(4)	297
α-Sn	1.00	1.00	0.13(1)		161(3)	169(7)
Cu	1.87	0.70(10)	0.30(3)		206(10)	314
Rh	1.15	–	0.44(5)		248(20)	235

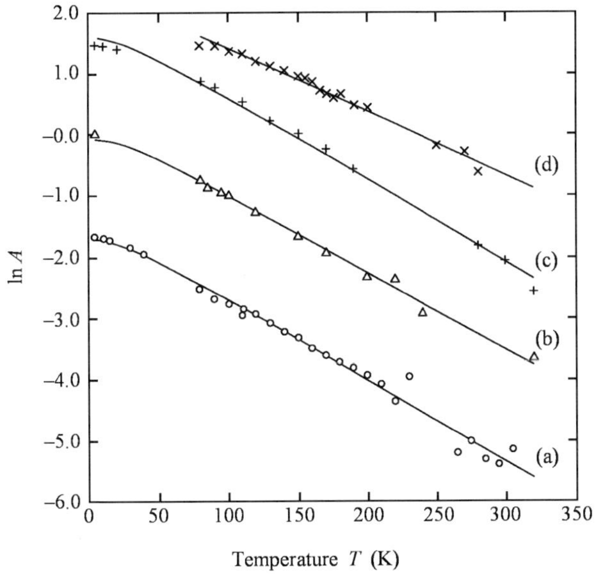

Fig. 8.7 Natural logarithm of the integrated intensity as a function of temperature for ^{119}Sn in Pb at concentrations of (a) 1.3%, (b) 1.6%, (c) 3.1%, and (d) 5.7%.

heim model can still give satisfactory results. For example, implantation of ^{119}Sn at a concentration of 10^{13} atoms cm^{-2} and ^{119}In at 10^{11} atoms cm^{-2} in Si and Ge resulted in the same recoilless fraction f to within 2% [24].

Figure 8.7 shows recent experimental results of ^{119}Sn impurities in crystalline Pb [48]. When the atomic concentration is varied between 1.3 and 3.1%, θ_D remains unchanged at 116 K; only when the concentration is raised to 5.7% does the curve show a lower slope.

^{119}Sn belongs to group IV in the periodic table and has the same valence electronic structure as Si and Ge. Therefore, ^{119}Sn is an ideal impurity in Si and Ge for Mössbauer studies, and has many valuable applications. Table 8.6 gives the results of such investigations. For ^{119}Sn in a Si or Ge host lattice, the impurity response function $g'(\omega)$ differs from the host response function $g(\omega)$ significantly [20], with the former having a much increased low-frequency peak. On

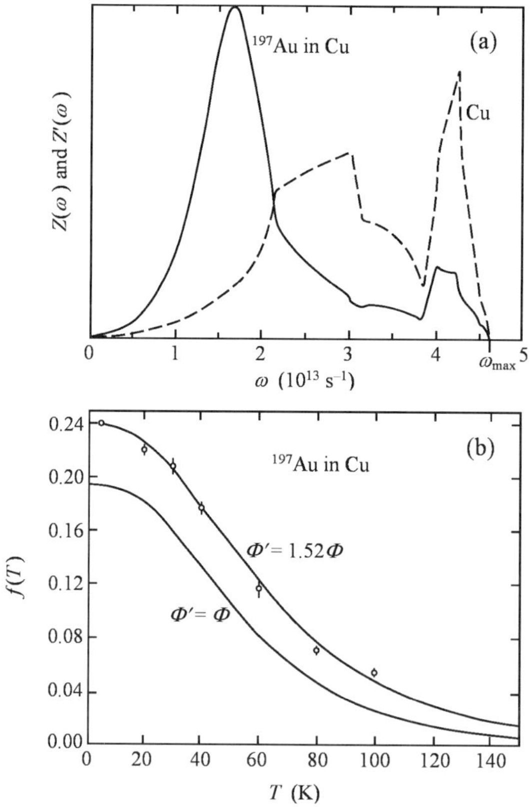

Fig. 8.8 (a) Vibration DOS $g'(\omega)$ of ^{197}Au impurities in Cu and $g(\omega)$ of pure Cu. (b) Experimental and calculated recoilless fraction $f(T)$ as a function of temperature for ^{197}Au in Cu.

Table 8.7 Values of the impurity–host force constant ratio $\Phi'_{xx}(0,0)/\Phi_{xx}(0,0)$ for ^{197}Au in host lattices of Cu and Ag [49].

Host	Au atomic concentration (%)	$\frac{M}{M_0}$	$\frac{\Phi'_{xx}(0,0)}{\Phi_{xx}(0,0)}$
Cu	2	3.2	1.51(4)
			1.57(5)
			1.63(5)
			1.72(6)
Ag	5	1.83	1.54(11)
			1.34(10)
			1.39(10)

the other hand, very few changes were observed in the force constant ratio $\Phi'_{xx}(0,0)/\Phi_{xx}(0,0)$ and the Debye temperature θ'_{D}.

8.4.3
^{197}Au Impurity Atoms

Since the ^{197}Au Mössbauer transition energy (77.3 keV) is relatively high, reasonable values of recoilless fraction f can only be observed at or below the liquid nitrogen temperature. There have been extensive studies of ^{197}Au impurities in Cu and Ag. Figure 8.8(a) shows the dynamic response functions $g(\omega)$ and $g'(\omega)$ for Cu host and Au impurities in Cu. In $g'(\omega)$, the resonant modes broaden due to the effect of a large mass difference ($M/M_0 = 3.10$) exceeding the opposite effect of an increased force constant ratio ($\Phi'/\Phi = 1.52$). Figure 8.8(b) shows data points and fitted curves of f versus temperature [17, 49]. Table 8.7 lists the values of the force constant ratio $\Phi'_{xx}(0,0)/\Phi_{xx}(0,0)$ obtained from fitting the $f(T)$ curves using the Mannheim model.

8.5
Interstitial Impurity Atoms

When the foreign atoms are located at interstitial positions, the lattice vibrations become extremely complicated. It is very difficult to obtain a theoretical impurity DOS $g'(\omega)$ in terms of the host DOS $g(\omega)$. But experimentally when the Mössbauer effect is used for studying the dynamic properties of lattices with interstitial impurities, parameters such as $\theta'_{\mathrm{D}}(n)$ and Φ'/Φ can be obtained by fitting the recoilless fraction f and second-order Doppler shift δ_{SOD} using an approximate $g'(\omega)$ based on the Debye model. To date, Mössbauer studies of interstitial impurity atoms have yielded the following general conclusions: (1) the f- and θ_D-values for interstitial impurities are very different from those for substitutional impurities; (2) the s-electron density for an interstitial impurity also differs significantly from that for a substutional impurity; and (3) the electric field gradient at the nucleus of an interstitial impurity atom is usually large and has a certain distribution, causing severe broadening of the spectral lines.

We now discuss two examples of Mössbauer effect studies of interstitial impurities.

8.5.1
^{57}Fe Impurities in Au

In this study, a source was produced by electroplating ^{57}Co into single-crystal gold chips of 99.995% purity followed by heating and quenching [50]. When a natural iron foil was used as the absorber, the Mössbauer spectrum has a single line as shown in Fig. 8.9(a). The recoilless fractions f_s were measured using the "wide black absorber" technique, for the quenched and annealed samples. The f_s-values and other pertinent parameters are listed in Table 8.8.

In Fig. 8.9, the single line spectrum (Fig. 8.9(a)) demonstrates that ^{57}Fe is at a cubic site (the host lattice of Au is fcc) and leads us to believe that the impurities in the quenched source occupy the substitutional positions. The linewidth $\Gamma = 0.235$ mm s^{-1} is slightly larger than the typical value. Similar line broadening has been observed in several other fcc hosts where the nearest neighbor distance is increased. The emission spectrum of the annealed source shows an additional broadened doublet, whose centroid was shifted towards lower energy, characteristic of an increase in the s-electron density at the Fe nucleus. This indi-

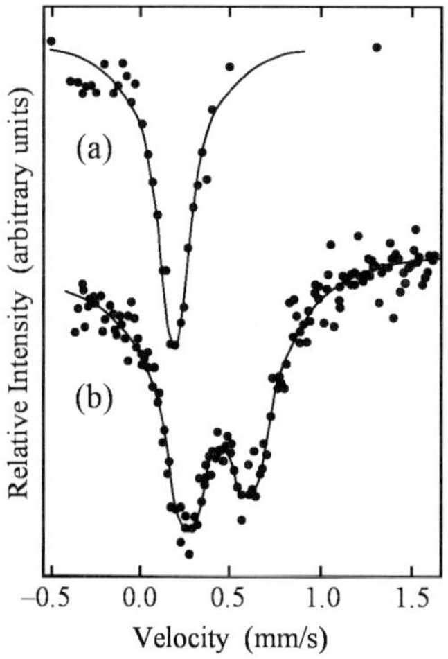

Fig. 8.9 Room temperature Mössbauer spectra of ^{57}Fe in Au. (a) 90% of the Fe impurities are in substitutional sites (using a natural iron foil absorber). (b) 25% of Fe impurities are in substitutional sites and the rest in interstitial sites (using a nitroprusside absorber).

Table 8.8 Dynamic and Mössbauer effect parameters for ^{57}Fe in Au [50].

Heat treatment	f_s (296 K)	$\theta'_D(-2)$ (K)	$\Phi'_{xx}(0, 0)$ / $\Phi_{xx}(0, 0)$	Γ (mm s^{-1})	δ_{IS} (mm s^{-1}) (rel. to α-Fe)	ΔE_Q (mm s^{-1})
Quenched	0.583(7)	282(5)	~0.9	0.235(5)	−0.635(5)	~0
Annealed	0.73(1)	366(10)	~1.3	0.40(2)	−0.34(1)	0.40(1)

cates that some Fe atoms in the annealed sample now occupy the interstitial sites. Since both f_s and $\theta'_D(-2)$ increase after annealing, the mean-square displacement $\langle u^2 \rangle$ must have decreased, and the restoring force on the Fe atom is stronger than the Au–Au interaction. Because the atomic radius of Fe is smaller than that of Au, if the Fe impurities were in substitutional sites, they would be less strongly bonded to the lattice than the original Au atoms, and would not result in a higher recoilless fraction f_s. This is similar to the results from ^{57}Co in an indium crystal, where the experimental value for f_s at 297 K is 0.7 but the theoretical estimate using a substitutional model was between 0.15 and 0.2. Therefore, the ^{57}Co atoms are in interstitial positions. Also, it is found that the quadrupole-split doublet does not have significant broadening, indicating that all interstitial positions have an identical environment.

8.5.2
^{57}Fe Impurities in Diamond

Diamond possesses many excellent physical and chemical properties, some of which are very unique. Being a very hard material due to the strong carbon–carbon covalent bonds, the amplitudes of the atomic vibrations in diamond are very small and the vibration frequencies are as high as 10^{14} Hz. Its Debye temperature, $\theta_D = 2230$ K, is also the highest of all known materials. These peculiar properties have stimulated the interest of many Mössbauer spectroscopists. There have been several reports of making a source or an absorber by replacing a carbon atom with a Mössbauer isotope, and the largest recoilless fractions have been observed [51–55].

Figure 8.10 shows a typical emission spectrum of ^{57}Co in diamond, which is a superposition of a singlet with 20% intensity and an asymmetric quadrupole-split doublet with 80% intensity. The singlet is due to ^{57}Co in the high-symmetry (HS) substitutional sites while the doublet is due to ^{57}Co in low-symmetry (LS) interstitial sites. These interstitial sites have a large average EFG of $V_{zz} = 1.2 \times 10^{18}$ V cm^{-2} resulting in $\Delta E_Q = 2.57$ mm s^{-1}, and have also a certain distribution of EFG values.

Table 8.9 lists values of f_s and θ'_D for ^{57}Co in diamond. These are the largest

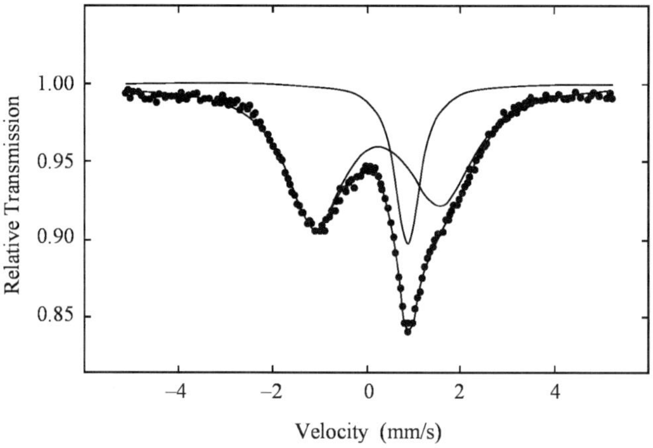

Fig. 8.10 Mössbauer emission spectrum of ^{57}Co in diamond, with both the source and a $Na_4Fe(CN)_6 \cdot 10H_2O$ absorber at room temperature. The ^{57}Co ions were hot-implanted into the diamond target at 830 K [54].

recoilless fractions and Debye temperatures ever observed in Mössbauer spectroscopy at room temperature. The Debye temperature θ'_D may also be estimated using Eq. (8.71). Assuming $\Phi'_{xx}(0,0) \approx \Phi_{xx}(0,0)$ and using $\theta_D = 2230$ K, we find $\theta'_D = 1023$ K, which is somewhat lower than the experimental value of 1300 K, indicating that $\Phi'_{xx}(0,0)$ is actually larger than $\Phi_{xx}(0,0)$.

There is a report of Mössbauer studies of nanophase diamond (NPD) films, which contain diamond-like sp^3 bonds concentrated into nodules of 20 to 30 nm in diameter and have many important solid-state properties [55]. Mössbauer spectra from ^{57}Fe-implanted NPD showed similar results for the interstitial sites, and the corresponding values of f and θ'_D are also included in Table 8.9.

Investigation of other properties such as the types of interstitial sites, local electronic configuration of Fe impurities, and distribution of EFG is usually very challenging [55–57], and sometimes requires a strong external magnetic field.

Table 8.9 Dynamic parameters from Mössbauer studies of ^{57}Co in diamond.

		f_s	θ'_D (K)		Ref.
Single-crystal diamond	HS	0.97(1)	1300(150)[a]	800(100)[b]	54
	LS	0.71(3)	550(50)[a]	450(50)[b]	
NPD films		0.69	523		55

[a] Derived from Mössbauer measurements between 4 and 100 K.
[b] Derived from Mössbauer measurements between 300 and 1100 K.

References

1 S.S. Cohen, R.H. Nussbaum, and D.G. Howard. Determination of an unambiguous parameter for the impurity–lattice interaction. *Phys. Rev. B* 12, 4095–4101 (1975).

2 J.W. Petersen, O.H. Nielsen, G. Weyer, E. Antoncik, and S. Damgaard. Lattice dynamics of substitutional ^{119m}Sn in silicon, germanium, and α-tin. *Phys. Rev. B* 21, 4292–4305 (1980). Erratum. *Phys. Rev. B* 22, 3135 (1980).

3 R.J. Elliott and D.W. Taylor. Theory of correlations and scattering of lattice vibrations by defects using double-time Green's functions. *Proc. Phys. Soc.* 83, 189–197 (1964).

4 D.W. Taylor. Dynamics of impurities in crystals. In *Dynamical Properties of Solids*, vol. 2, G.K. Horton and A.A. Maradudin (Eds.), pp. 285–384 (North-Holland, Amsterdam, 1975).

5 A.L. Fetter and J.D. Walecka. *Quantum Theory of Many-Particle Systems* (McGraw-Hill, New York, 1971).

6 A.A. Maradudin, E.W. Montroll, and G.H. Weiss. *Theory of Lattice Dynamics in the Harmonic Approximation* (Solid State Physics Supplement 3) (Academic Press, New York, 1963).

7 A.A. Maradudin and P.A. Flinn. Debye–Waller factor for Mössbauer resonant impurity atoms. *Phys. Rev.* 126, 2059–2071 (1962).

8 A.A. Maradudin, P.A. Flinn, and S. Ruby. Velocity shift of the Mössbauer resonance. *Phys. Rev.* 126, 9–23 (1962).

9 A.A. Maradudin. Phonons and lattice imperfections. In *Phonons and Phonon Interactions*, T.A. Bak (Ed.), pp. 424–504 (Benjamin, New York, 1964).

10 P.D. Mannheim. Influence of force-constant changes on the lattice dynamics of cubic crystals with point defects. *Phys. Rev.* 165, 1011–1018 (1968).

11 H. Böttger. *Principles of the Theory of Lattice Dynamics*, p. 100 (Physik Verlag, Weinheim, 1983).

12 J.M. Grow, D.G. Howard, R.H. Nussbaum, and M. Takeo. Frequency moments of cubic metals and substitutional impurities: a critical review of impurity–host force-constant changes from Mössbauer data. *Phys. Rev. B* 17, 15–39 (1978).

13 K. Lakatos and J.A. Krumhansl. Effect of force-constant changes on the incoherent neutron scattering from cubic crystals with point defects. *Phys. Rev.* 175, 841–858 (1968).

14 A.A. Maradudin. Some effects of point defects on the vibrations of crystal lattice. *Rep. Prog. Phys.* 28, 331–380 (1965).

15 A.A. Maradudin. Theoretical and experimental aspects of the effects of point defects and disorder on the vibrations of crystals. In *Solid State Physics*, vol. 18, F. Seitz and D. Turnbull (Eds.), pp. 273–420 (Academic Press, New York, 1966).

16 D.N. Zubarev. Double-time Green functions in statistical physics. *Sov. Phys. Usp.* 3, 320–345 (1960) [Russian original: *Usp. Fiz. Nauk* 71, 71–116 (1960)].

17 P.H. Dederichs and R. Zeller. Dynamical properties of point defects in metals. In *Point Defects in Metals II*, p. 9 (Springer-Verlag, Berlin, 1980).

18 A.A. Maradudin, P.A. Flinn, and S. Ruby. Velocity shift of the Mössbauer resonance. *Phys. Rev.* 126, 9–23 (1962).

19 A.A. Maradudin. Lattice dynamical aspects of the resonance absorption of gamma rays by nuclei bound in a crystal. *Rev. Mod. Phys.* 36, 417–432 (1964).

20 O.H. Nielsen. Lattice dynamics of substitutional ^{119m}Sn in silicon, germanium, and α-tin using an adiabatic bond-charge model. *Phys. Rev. B* 25, 1225–1240 (1982).

21 H.J. Lipkin. Some simple features of the Mössbauer effect: III. The f-factor for an impurity source in a crystal. *Ann. Phys.* 23, 28–37 (1963).

22 V.G. Bhide. *Mössbauer Effect and its Applications*, p. 97 (Tata McGraw-Hill, New Delhi, 1973).

23 P.D. Mannheim and A. Simopoulos. Influence of force-constant changes and localized modes on the V:Fe^{57} Mössbauer system. *Phys. Rev.* 165, 845–849 (1968). P.D. Mannheim and S.S. Cohen. Force-constant changes and the crystal impurity problem. *Phys. Rev. B* 4, 3748–3756 (1971). P.D. Mannheim. Localized modes and cell-model limit in the crystal impurity problem. *Phys. Rev. B* 5, 745–749 (1972).

24 H. Muramatsu, T. Miura, and H. Nakahara. Mössbauer spectroscopy of ^{119}Sn from implantations of radioactive ^{119}Sn in metals. *Phys. Rev. B* 42, 43–55 (1990).

25 K. Dettmann and W. Ludwig. Lokalizierte Schwingungszustände in kubischen Kristallen mit Punktdefkten. *Phys. Kondens. Materie* 2, 241–261 (1964).

26 M. Seto, Y. Kobayashi, S. Kitao, R. Haruki, T. Mitsui, Y. Yoda, S. Nasu, and S. Kikuta. Local vibrational densities of states of dilute Fe atoms in Al and Cu metals. *Phys. Rev. B* 61, 11420–11424 (2000).

27 T. Ruckert, W. Keune, W. Sturhahn, M.Y. Hu, J.P. Sutter, T.S. Toellner, and E.E. Alp. Phonon density of states in epitaxial Fe/Cr(001) superlattices. *Hyperfine Interactions* 126, 363–366 (2000).

28 D.G. Howard and R.H. Nussbaum. Force-constant changes and localized modes for iron impurities in cubic metals from Mössbauer-fraction measurements. *Phys. Rev. B* 9, 794–801 (1974).

29 R.M. Housley and F. Hess. Analysis of Debye–Waller-factor and Mössbauer-thermal-shift measurements: I. General theory. *Phys. Rev.* 146, 517–526 (1966).

30 K. Ohashi and K. Kobayashi. Effective Debye temperature. *J. Phys. F* 5, 1466–1474 (1975).

31 W.M. Visscher. Resonance absorption of gammas by impurity nuclei in crystals. *Phys. Rev.* 129, 28–36 (1963).

32 A.P. Miiller and B.N. Brockhouse. Crystal dynamics and electronic specific heats of palladium and copper. *Can. J. Phys.* 49, 704–723 (1971).

33 E.C. Svensson, B.N. Brockhouse, and J.M. Rowe. Crystal dynamics of copper. *Phys. Rev.* 155, 619–632 (1967).

34 R.M. Nicklow, G. Gilat, H.G. Smith, L.J. Raubenheimer, and M.K. Wilkinson. Phonon frequencies in copper at 49 and 298 °K. *Phys. Rev.* 164, 922–928 (1967).

35 N.E. Phillips. Low-temperature heat capacity of metals. In *CRC Critical Reviews in Solid State Sciences*, vol. 2, D.E. Schuele and R.W. Hoffman (Eds.), pp. 467–553 (Chemical Rubber Co., Cleveland, 1971).

36 G. Nilsson and S. Rolandson. Lattice dynamics of copper at 80 K. *Phys. Rev. B* 7, 2393–2400 (1973).

37 W.F. Giauque and P.F. Meads. The heat capacities and entropies of aluminum and copper from 15 to 300 °K. *J. Am. Chem. Soc.* 63, 1897–1901 (1941).

38 L.S. Salter. The temperature variation of the scattering properties of crystals. *Adv. Phys.* 14, 1–37 (1965).

39 G. Gilat and R.M. Nicklow. Normal vibrations in aluminum and derived thermodynamic properties. *Phys. Rev.* 143, 487–494 (1966).

40 D.H. Dutton, B.N. Brockhouse, and A.P. Miiller. Crystal dynamics of platinum by inelastic neutron scattering. *Can. J. Phys.* 50, 2915–2927 (1972).

41 G.T. Furukawa, M.L. Reilly, and J.S. Gallagher. Critical analysis of heat-capacity data and evaluation of thermodynamic properties of ruthenium, rhodium, palladium, iridium, and platinum from 0 to 300 K. A survey of the literature data on osmium. *J. Phys. Chem. Ref. Data* 3, 197 (1974).

42 J.L. Feldman and G.K. Horton. Critical analysis of the thermodynamic data for Pt and a prediction of $\Theta_{DW}(T)$. *Phys. Rev.* 137, A1106–A1108 (1965).

43 R. Colella and B.W. Batterman. X-ray determination of phonon dispersion in vanadium. *Phys. Rev. B* 1, 3913–3921 (1970).

44 O.H. Nielsen. Lattice dynamics of substitutional ^{199m}Sn in covalent semiconductors. In *Recent Developments in Condensed Matter Physics*, vol. 3, J.T. Devreese, L.F. Lemmens, V.E. van Doren, and J. van Royen (Eds.), pp. 157–163 (Plenum Press, New York, 1981).

45 R.H. Nussbaum, D.G. Howard, W.L. Nees, and C.F. Steen. Lattice-dynamical properties of Fe^{57} impurity atoms in Pt, Pd, and Cu from precision measurements of Mössbauer fractions. *Phys. Rev.* 173, 653–663 (1968).

46 S.K. Roy and N. Kundu. Anharmonic effects on the dynamics of Fe impurities in metallic hosts. *Phys. Lett. A* 160, 279–286 (1991). S.K. Roy and N. Kundu. Dynamical properties of ^{57}Fe impurities in different metallic solids from anharmonic recoilless fractions. *J. Phys. F: Met. Phys.* 17, 1051–1064 (1987).

47 H. Andreasen, S. Damgaard, J.W. Petersen, and G. Weyer. Isomer shifts and force constants of substitutional ^{119}Sn impurity atoms in FCC metals. *J. Phys. F* 13, 2077–2088 (1983).

48 S.N. Dickson, J.G. Mullen, and R.D. Taylor. Temperature dependence of the Mössbauer effect on Sn-119 doped metallic lead. *Hyperfine Interactions* 93, 1445–1451 (1994).

49 J.F. Prince, L.D. Roberts, and D.J. Erickson. Impurity–host force-constant changes from ^{197}Au Mössbauer recoilless-fraction measurements for dilute Cu(Au) and Ag(Au) alloys. *Phys. Rev. B* 13, 24–33 (1976).

50 C.F. Steen, D.G. Howard, and R.H. Nussbaum. Evidence for stable interstitial and substitutional sites of cobalt in gold from Mössbauer studies. *Solid State Commun.* 9, 865–869 (1971).

51 B.D. Sawicka, J.A. Sawicki, and H. de Waard. Mössbauer effect evidence of high internal pressure at iron atoms implanted in diamond. *Phys. Lett. A* 85, 303–307 (1981).

52 J.A. Sawicki and B.D. Sawicka. Implantation induced diamond to amorphous-carbon transition studied by conversion electron Mössbauer spectroscopy. *Nucl. Instrum. Methods* 194, 465–469 (1982).

53 M. de Potter and G. Langouche. Mössbauer study of the amorphization process in diamond. *Z. Phys. B* 53, 89–93 (1983).

54 J.A. Sawicki and B.D. Sawicka. Properties of ^{57}Fe hot-implanted into diamond crystals studied by Mössbauer emission spectroscopy between 4 and 300 K. *Nucl. Instrum. Methods B* 46, 38–45 (1990).

55 T.W. Sinor, J.D. Standifird, F. Davanloo, K.N. Taylor, C. Hong, J.J. Carroll, and C.B. Collins. Mössbauer effect measurement of the recoil-free fraction for ^{57}Fe implanted in a nanophase diamond film. *Appl. Phys. Lett.* 64, 1221–1223 (1994).

56 L.B. Kvashnina and M.A. Krivoglaz. Mössbauer spectra in crystals containing defects. *Phys. Metals Metallogr.* 23, 1–12 (1967) [Russian original: *Fiz. Metal. Metalloved.* 23, 1–14 (1967)].

57 V.A. Lagunov, V.I. Polozenko, and V.A. Stepanov. Influence of plastic deformation on the Mössbauer effect in Armco iron. *Sov. Phys. Solid State* 12, 2480–2482 (1971) [Russian original: *Fiz. Tverd. Tela* 12, 3064–3066 (1970)].

9
Mössbauer Impurity Atoms (II)

The previous chapter was devoted to the studies of Mössbauer impurities of dilute concentrations (<0.1 at.%). In this chapter, we discuss experimental results from Mössbauer effect studies of dynamics of several systems where the Mössbauer isotope concentration is larger than 0.1 at.%. These systems include metals, alloys, as well as amorphous, molecular crystalline, and low-dimensional materials (surface, multilayer, nanocrystals, etc.). It is often the case that Fe is one of the major constituents of the material, but ^{57}Fe is still an "isotopic impurity." The natural abundance for most of the Mössbauer isotopes is not 100%. It is for this reason that these systems are grouped together in this chapter.

Because we are now dealing with high-concentration impurities, the Mössbauer atoms cannot be treated as isolated impurities and the Mannheim model is no longer applicable. In most cases, anharmonic effects must be included because the harmonic approximation is no longer adequate. There are generally two approaches to understanding the Mössbauer spectra from these systems. If the dynamic response function $g'(\omega)$ of the Mössbauer atom vibrations is unknown, the frequency moments and Eq. (5.35) are utilized to fit the spectra to obtain the characteristic temperature $\theta_D(-2)$ and the parameter $\varepsilon(-2)$. If the material's $g'(\omega)$ is available, the analysis would then be very straightforward.

9.1
Metals and Alloys

9.1.1
Metals

The dynamic properties of metals of Mössbauer atoms have been thoroughly studied by many researchers (some results are listed in Table 9.1). For other metals, the Mössbauer scattering method may be applied (see Chapter 6).

As an example, we discuss zinc and its alloys, which were first extensively studied using the Mössbauer effect in the 1980s [8, 12–19].

The Mössbauer radiation used are the 93.3 keV γ-rays from ^{67}Zn, whose decay scheme is shown in Fig. 9.1. The most prominent characteristic of this source is

Mössbauer Effect in Lattice Dynamics. Yi-Long Chen and De-Ping Yang

ISBN: 978-3-527-40712-5

Table 9.1 Recoilless fractions and Debye temperatures in some metals.

Metal	f	θ_D (K)			
		Mössbauer effect	Neutron diffraction	X-ray diffraction	Specific heat [4]
α-Fe	0.93(3) at 4.2 K [1]	400 (30) [2]	420 [3]		460
β-Sn	0.40(2) at 100 K [5, 6]	140 [5]			170
Au	0.189 at 4.2 K [7]	168 [7]			180
Zn	$f_\perp = 6.4 \times 10^{-3}$ at 4.2 K [8]	$\theta_\perp = 242(10)$ [8]		254 [9]	250
	$f_\parallel = 2.6 \times 10^{-4}$	$\theta_\parallel = 149(20)$		169	
Ni	0.09(1) at 80 K [10]	413 [10], 406 [11]			440

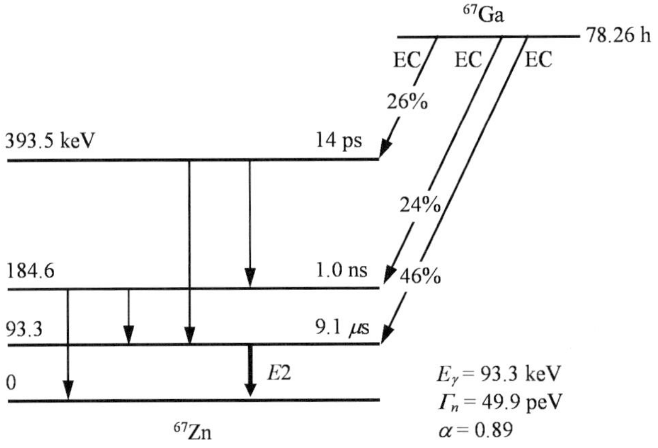

Fig. 9.1 Decay scheme of ^{67}Ga. The γ-ray energy E_γ, linewidth Γ_n, and internal conversion factor α are listed.

the long life of the excited state (9.1 μs), resulting in an extremely narrow energy level width ($\Gamma_n = 49.9 \times 10^{-12}$ eV) and consequently a very high energy resolution. It has been reported [12] that an energy resolution of 1.3×10^{-18} has been obtained using a single-crystal ^{67}Ga/ZnO source and a ^{67}Zn-enriched powder ZnO sample. In order to record a spectrum of such high resolution, the apparatus must be isolated from the slightest mechanical vibrations. In the γ-ray direction, a vibration velocity of about 0.16 μm s^{-1} would completely destroy the Mössbauer effect. Furthermore, due to the spin 5/2 of its ground state as shown in Fig. 9.2, hyperfine interactions would split the ^{67}Zn energy levels such that a ^{67}Zn Mössbauer spectrum would appear much more complicated than the ^{57}Fe or ^{119}Sn spectrum.

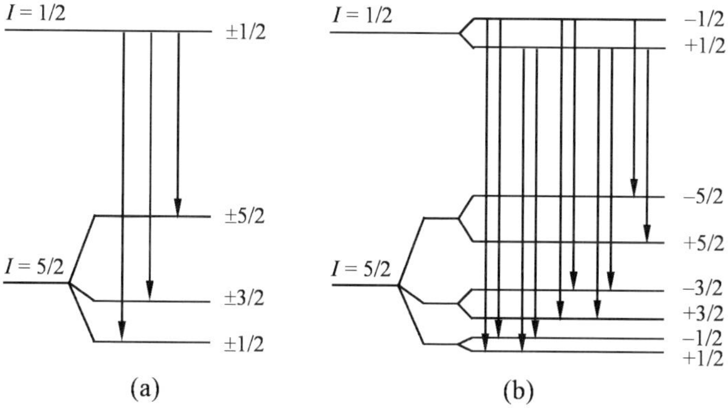

Fig. 9.2 Hyperfine splittings of ^{67}Zn: (a) pure quadrupole splitting and (b) a combination of electric quadrupole and magnetic dipole interactions.

Metallic Zn has a close-packed hexagonal lattice, with a relatively large c/a ratio of 1.861, from which a large anisotropy is expected in the mean-square displacement of the atomic vibrations [15].

For a polycrystalline Zn sample, a ^{67}Zn Mössbauer spectrum should be composed of three lines of equal intensity due to quadrupole splitting. However, because of the G–K effect, the intensities of the three lines are not exactly equal, as shown in Fig. 9.3.

In order to study the anisotropic recoilless fraction $(f_{\parallel}, f_{\perp})$ and atomic mean-square displacement $(\langle u_{\parallel}{}^2\rangle, \langle u_{\perp}{}^2\rangle)$, Mössbauer emission spectra of a single-

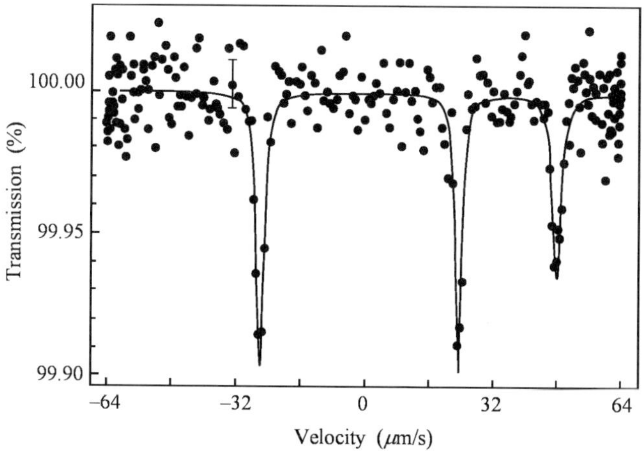

Fig. 9.3 Mössbauer spectrum of a polycrystalline ^{67}Zn metal absorber at 4.2 K, using a ^{67}Ga/Cu source [13].

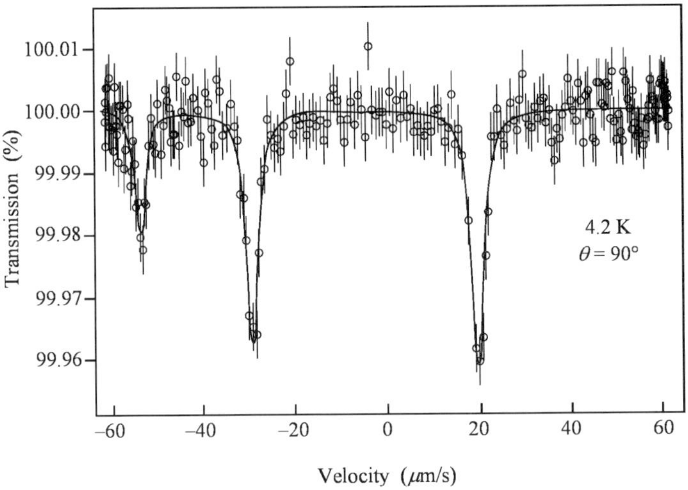

Fig. 9.4 Mössbauer emission spectrum of a ^{67}Ga/Zn single-crystal source at 4.2 K with its c-axis perpendicular to the 93.3 keV γ-ray direction. The absorber is β'-brass, also at 4.2 K.

crystal ^{67}Ga/Zn source have been measured [8]. The absorber used was β'-brass, a Cu–Zn alloy with a Zn content of 49.2 at.% (^{67}Zn-enriched to 91.9%). Let θ be the angle between the c-axis and the γ-ray direction. Mössbauer spectra at several given temperatures (4.2, 20.8, and 47 K) were obtained for θ-values of 90°, 75°, 60°, and 55°. A typical spectrum ($\theta = 90°$, $T = 4.2$ K) is shown in Fig. 9.4.

According to Eq. (5.44), a linear relation should exist between $\ln f(\theta)$ and $\cos^2\theta$ at a given temperature. Therefore, after background correction, we may calculate $\langle u_\perp{}^2\rangle$ and $\langle u_\parallel{}^2\rangle$, as well as $f_\perp$ and $f_\parallel$, from the total area of the three peaks. The results are listed in Tables 9.1 and 9.2, which show that the anisotropy in the recoilless fraction is enormous. At $T = 4.2$ K, $f_\perp/f_\parallel = 25$, and at $T = 47$ K,

Table 9.2 Experimental and theoretical ^{67}Zn recoilless fraction f, experimental center shift δ, and theoretical δ_{SOD} values for metallic Zn. Both δ and δ_{SOD} are relative to the respective value at 4.2 K [18].

T (K)	$f_\perp$ (%)		$f_\parallel$ (%)		δ (μm s^{-1})	δ_{SOD} (μm s^{-1})
	exp.	theor.	exp.	theor.	exp.	theor.
4.2	$1.07^{+0.13}_{-0.12}$	1.19	$0.043^{+0.088}_{-0.030}$	0.032	0.0	0.0
20.8	$0.80^{+0.18}_{-0.15}$	1.08	$0.098^{+0.53}_{-0.08}$	0.022	0.37(8)	0.267
47	$0.40^{+0.10}_{-0.05}$	0.62	$0.00018^{+0.0067}_{-0.00017}$	0.0016	4.46(19)	4.38

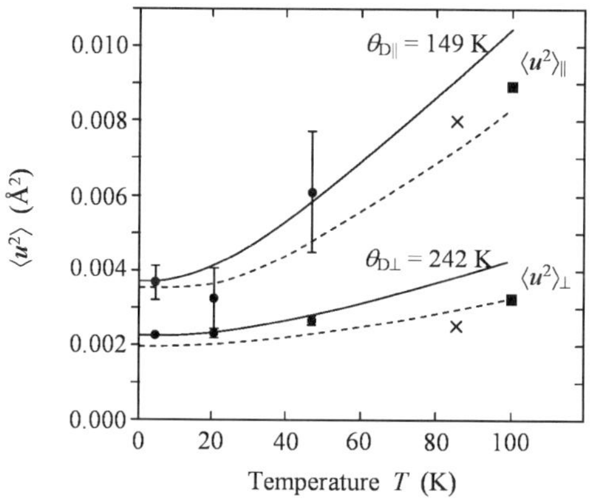

Fig. 9.5 Mean-square atomic displacements in zinc, parallel and perpendicular to the *c*-axis. Circles: from Mössbauer experiments; squares and crosses: from x-ray experiments. Solid lines: fits to Mössbauer data by the Debye model; dashed lines: results of calculations based on the modified axially symmetric (MAS) model.

$f_\perp/f_\parallel$ becomes as large as about 2.2×10^3, which means that the Mössbauer effect can hardly be observed in the direction $\theta = 0$.

When the Debye model is used to fit the results of $\langle u_\perp{}^2\rangle$ and $\langle u_\parallel{}^2\rangle$ at various temperatures (Fig. 9.5), Debye temperature values of $\theta_{D\perp} = 240$ K and $\theta_{D\parallel} = 149$ K are deduced [8].

9.1.2 Alloys

9.1.2.1 The β-Ti(Fe) Alloy [20]

This is an example of obtaining a solid's dynamics information via measuring the second-order Doppler shift δ_{SOD}. This alloy contains 9.3% Fe atoms. It has been revealed that when Fe atoms are doped into a Ti crystal to form a substitutional solid solution, they substantially stabilize its high-temperature bcc phase. We may understand this effect by considering the difference between the force constants $\Phi_{\text{Ti-Fe}}$ and $\Phi_{\text{Ti-Ti}}$, or the difference between the bond energies $E_{\text{Ti-Fe}}$ and $E_{\text{Ti-Ti}}$. Suppose that

$$E_{\text{Ti-Fe}} > E_{\text{Fe-Fe}} \quad \text{and} \quad E_{\text{Ti-Fe}} > E_{\text{Ti-Ti}}, \tag{9.1}$$

i.e., the attractive forces between like atoms (Ti–Ti or Fe–Fe) are weaker than those between unlike atoms (Ti–Fe). If this is the case, there would be no Ti-rich or Fe-rich regions, but rather a tendency to form certain ordering so that each Fe atom would coordinate with as many Ti atoms as possible. In other words, the tendency is to have as many Ti–Fe pairs as possible, not Ti–Ti. The above hypoth-

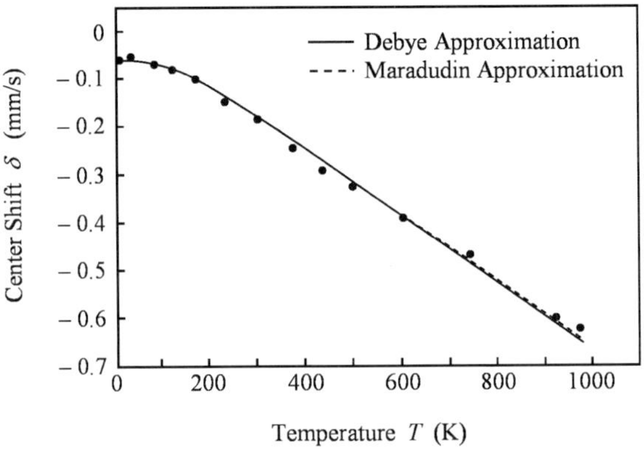

Fig. 9.6 Center shift δ in ^{57}Fe Mössbauer spectra of a β-Ti(Fe) alloy (Ti + 9.3 at.% Fe), as a function of temperature *T*.

esis can be verified by measuring δ_{SOD} as a function of temperature and calculating the force constant ratio $\Phi_{\text{Ti-Fe}}/\Phi_{\text{Ti-Ti}}$.

If we apply the Debye model for atomic vibrations by substituting the expression for $\langle v^2 \rangle$ in Eq. (5.4) into (5.54) and use α-Fe as the reference absorber, we would obtain

$$\delta = \delta_{IS} - \delta_{SOD}^{\alpha\text{-Fe}} - \frac{9k_B\theta_D}{2Mc^2}\left[\frac{1}{8} + \left(\frac{T}{\theta_D}\right)^4 \int_0^{\theta_D/T} \frac{x^3}{e^x - 1}\,dx\right] \tag{9.2}$$

where $\delta_{SOD}^{\alpha\text{-Fe}} = -0.229$ mm s^{-1} is the second-order Doppler shift of α-Fe at room temperature and δ_{IS} is the isomer shift (relative to α-Fe) of the Fe atoms in the β-Ti(Fe) alloy.

When Eq. (9.2) is used to fit the data of δ-values measured at different temperatures, as shown in Fig. 9.6, the parameters θ_D and δ_{IS} are determined to be $\theta_D = 497(26)$ K and $\delta_{IS} = -0.154$ mm s^{-1}. The negative sign indicates that the s-electron density at the Fe nuclei in the alloy is higher than that in α-Fe, which is due to electron transfer from Ti to Fe [21].

Let us use the Einstein–Debye formula for calculating the difference in the force constants. Substituting the known value of $\theta_D(\text{Ti}) = 420$ K and the above θ_D-value, along with the mass values of Ti and Fe, into Eq. (8.71), we obtain the ratio

$$\frac{\Phi_{\text{Ti–Fe}}}{\Phi_{\text{Ti–Ti}}} = \frac{M_{\text{Fe}}}{M_{\text{Ti}}}\left(\frac{\theta_D}{\theta_D(\text{Ti})}\right)^2 = 1.66 > 1, \tag{9.3}$$

which is consistent with the above hypothesis. This confirms that the Ti–Fe atoms are more tightly bonded than Ti–Ti, and therefore the Fe atoms in β-Ti(Fe) alloy stabilize its high-temperature bcc lattice.

9.1.2.2 **Cu–Zn Alloy (Brass)**

In this example, the densities of states (DOS) $g_\alpha(\omega)$ and $g_{\beta'}(\omega)$ for the α- and β'-phases are both known, and a detailed comparison between theoretical and experimental results can therefore be easily carried out.

The Cu–Zn system has many different phases, and since it is an important industrial material, it has always been the focus of theoretical and experimental research. Amongst the experimental methods, x-ray diffraction and neutron scattering are difficult to apply because the atomic structure factors of Cu and Zn are almost equal and their neutron scattering wavelengths are similar. Therefore, high-resolution ^{67}Zn Mössbauer spectroscopy is ideally suited for both the α- and β'-phases of this system, which are solid solutions of fcc and bcc structures, respectively.

Figure 9.7 shows the ^{67}Zn Mössbauer spectra from the α- and β'-phases of the Cu–Zn alloy [22, 23]. With Zn atoms being nonmagnetic and both phases having cubic symmetry in their structures, we would expect each of their spectra to have a single-line absorption with no magnetic or quadrupole splitting. However, the experimental Mössbauer spectrum of the α-phase shows four absorption lines. When the Zn content is varied in the range 4.3 to 24.6%, there are appreciable changes in the intensities of these lines but almost no changes in their positions, indicating that the four lines are not due to quadrupole interactions. They must be due to four different configurations in the α-phase, each having a different

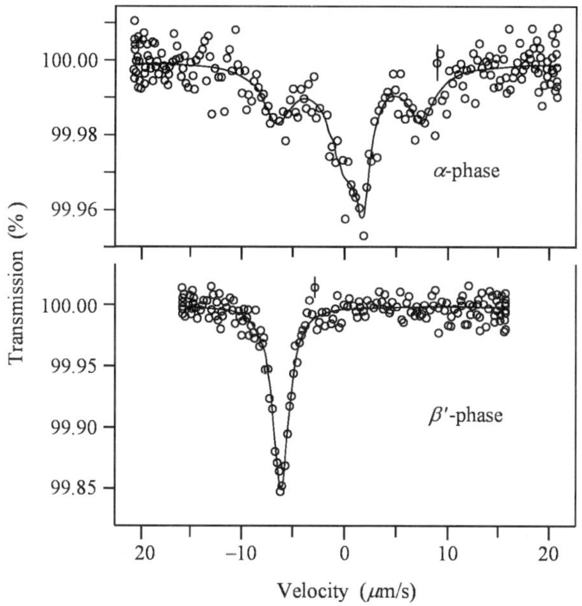

Fig. 9.7 ^{67}Zn Mössbauer spectra of α-phase (15.9 at.% Zn) and β'-phase (49.2 at.% Zn) Cu–Zn alloys at 4.2 K, using a ^{67}Ga/Cu source.

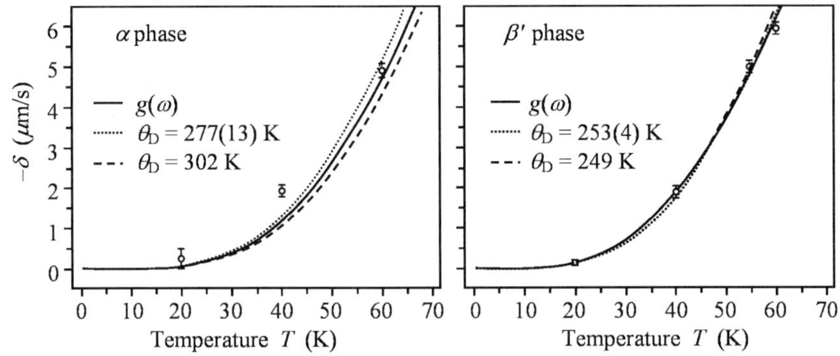

Fig. 9.8 ^{67}Zn isomer shift δ as a function of temperature T in α-phase and β'-phase Cu–Zn brass. Solid curves: calculations of the second-order Doppler shift using the phonon distribution $g(\omega)$ derived from inelastic neutron scattering data. Dotted curves: best fits to the Debye model. Dashed curve for α-phase: using $\theta_D = 302$ K derived from specific heat data. Dashed curve for β'-phase: using $\theta_D = 249$ K from data for f-factor (see Fig. 9.9).

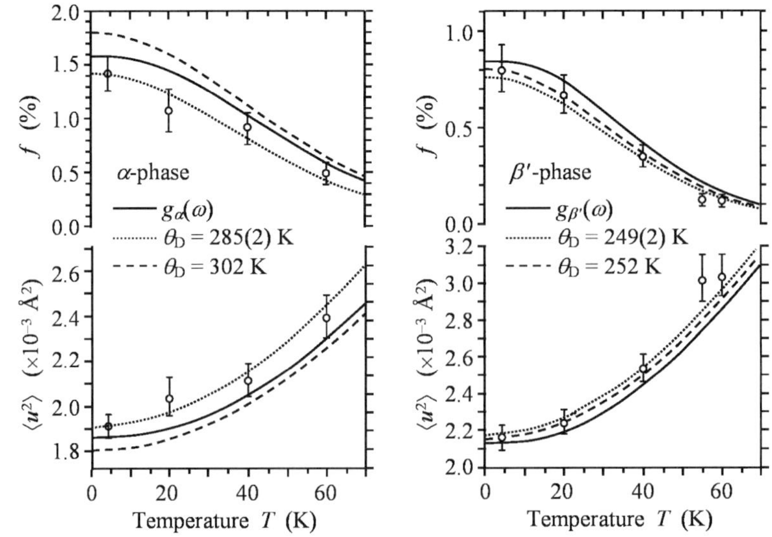

Fig. 9.9 Recoilless fraction f and mean-square displacement $\langle u^2 \rangle$ as functions of temperature T. Solid curves: calculated results based on the phonon distributions in Fig. 9.10. Dotted curves: best fits using the Debye model. Dashed curve for α-phase: based on $\theta_D = 302$ K derived from specific heat data. Dashed curve for β'-phase: based on $\theta_D = 252$ K derived from second-order Doppler shift data.

s-electron density at the Zn nuclei. Therefore, contrary to the belief that Zn has the binomial distribution in the α-phase of the Cu–Zn system, there must be a short-range order in the structure [24]. The existence of short-range order in the α-phase was therefore first verified unequivocally by the Mössbauer effect. Subsequent neutron scattering experiments also supported this conclusion [25]. From the viewpoint of lattice dynamics, the β'-phase is more interesting. At temperatures lower than 725 K, the β'-phase of $Cu_{0.5}Zn_{0.5}$ has the ordered CsCl structure, with eight Cu atoms surrounding one Zn atom, or vice versa, having a cubic symmetry. Since Zn and Cu atoms have similar mass values, substituting Cu with Zn causes very little change in the force constant [18].

How δ_{SOD} and recoilless fractions of α- and β'-phases vary with temperature are shown in Figs. 9.8 and 9.9 [22, 23]. The positions of the four subspectral lines exhibited the same temperature dependence within experimental uncertainty.

The following two conclusions can be drawn from the experiments:

1. Either using the DOS (as shown in Fig. 9.10) or applying the Debye model, there is a good agreement between theory and experiments. However, recoilless fractions derived from the

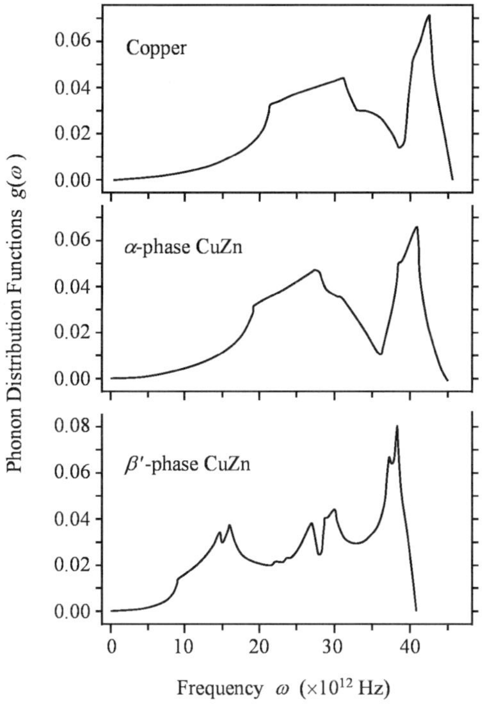

Fig. 9.10 Frequency distribution functions $g_{Cu}(\omega)$, $g_{\alpha}(\omega)$, and $g_{\beta'}(\omega)$ for metallic Cu, α-phase Cu–Zn, and β'-phase Cu–Zn [23].

DOS are somewhat higher than the experimental values. Analyses have shown that the phonon DOS may have a systematic deviation.

2. For the α-phase, there is an excellent agreement between $\theta_D = (285 \pm 2)$ K from the f measurements and $\theta_D = (277 \pm 13)$ K from the δ_{SOD} measurements. For the β'-phase, the corresponding values are $\theta_D = (249 \pm 2)$ K and $\theta_D = (252 \pm 2)$ K, about 30 K lower than the θ_D-values for the α-phase. The agreement between the θ_D-values from recoilless fraction and second-order Doppler shift measurements shows that the Debye model is a very good approximation for describing the lattice dynamics of this system.

9.2 Amorphous Solids

In a dilute gas, there is almost no interaction between the individual molecules, and because the molecular positions in space are constantly changing, they are in a completely disordered state. The main structure of an amorphous solid is a long-range disordered state, but the atomic arrangement may not be completely random and "short-range order" may exist, which has been revealed by a large amount of diffraction data. Typical diffraction patterns from amorphous materials consist of relatively wide haloes and diffuse rings, rather than the characteristic sharp points and lines from crystalline samples. A diffuse diffraction pattern indicates that the relative positions of atoms in an amorphous material are distributed in a certain range, unlike the completely random positions in a gas that would not even give diffuse rings in its diffraction pattern. So far, there is still lacking a single consistent description or definition for "amorphous materials." In contrast with a crystalline material or a gas, the structure of an amorphous material may be depicted as in Fig. 9.11(b). Based on the experimental results

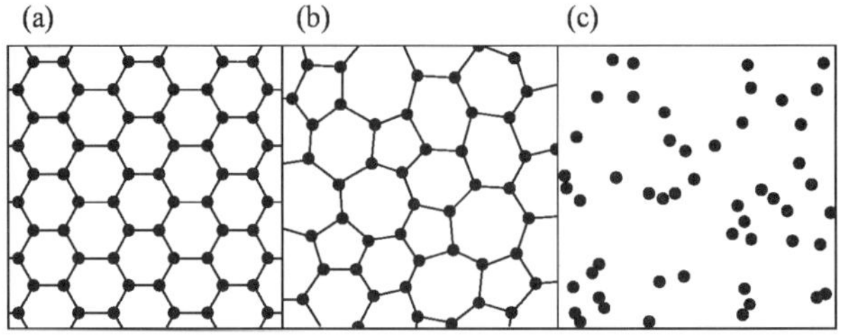

Fig. 9.11 Schematic representations of atomic arrangements in (a) a crystal, (b) an amorphous solid, and (c) a gas.

available to date, the general notion is that atoms or molecules in an amorphous material have no spatial periodicity and no translational symmetry, and thus long-range order is destroyed. Due to the local atomic interactions, certain characteristics of ordered atomic constitution and structural arrangement are still maintained in small regions of nanometer size, namely, short-range order.

The two most common types of short-range order are topographic short-range order (TSRO) and chemical short-range order (CSRO). TSRO refers to the situation where the relative atomic positions have certain order in a small region, while CSRO refers to order in the arrangement of different atoms. If the nearest neighbor positions of an atom are occupied by a different kind of atom, as in a transition metal–metalloid amorphous material, the transition metal atoms and metalloid atoms form nearest neighbors of each other because of the strong heteroatomic interaction.

An amorphous material is sometimes referred to as in a "glassy state" or as a "solidified liquid." But the liquid and amorphous structures have fundamental differences. We may imagine that a high pressure is applied to a liquid so that the atoms are compressed next to one another until the repulsive potential appears. The atoms cannot have diffusions or displacements beyond the typical interatomic distance, but are only allowed to execute thermal motions about their equilibrium positions. Such a material would be a solid similar to an amorphous material.

The physics of amorphous materials is an important and active research area. Compared with the studies of crystalline materials, research on amorphous materials is still under development. Numerous books are available, offering systematic descriptions of this area of research. Here, we focus on how the Mössbauer effect can be used for investigating the dynamics of amorphous materials.

Figure 9.12 shows Mössbauer spectra of an amorphous ferromagnet $Fe_{75}P_{15}C_{10}$ at two different temperatures, each composed of six broad absorption peaks. The broadening is similar to the wide diffuse rings in the diffraction patterns, except Mössbauer peak broadening here is a result of the distributions of magnetic hyperfine fields, EFGs, and s-electron densities at the Mössbauer nuclei. However, it is not immediately obvious which of these hyperfine interactions, or a combination, is responsible for the broadening. References [27, 28] give a method of separating the magnetic dipole interaction from the electric quadrupole interaction. The method uses an external magnetic field of a suitable radio frequency applied to the amorphous sample, causing a collapse of the magnetic hyperfine interaction [29]. This makes the average magnetic hyperfine field at the Mössbauer nuclei approach zero, leaving a pure quadrupole split spectrum. Therefore, based on this Mössbauer spectrum with the collapsed magnetic interactions, one may measure the distribution of EFG in the amorphous sample. For example, the ΔE_Q values in amorphous alloys $Fe_{74}Si_{10}B_{16}$ and $Fe_{40}Ni_{40}P_{14}B_6$ are distributed in the range from 0 to 1 mm s^{-1} while the most probable value occurs at $(\Delta E_Q)_{max} = 0.5$ mm s^{-1} [27, 28]. Using this method, the Mössbauer effect is regarded as an effective method for revealing the short-range order in amorphous solids.

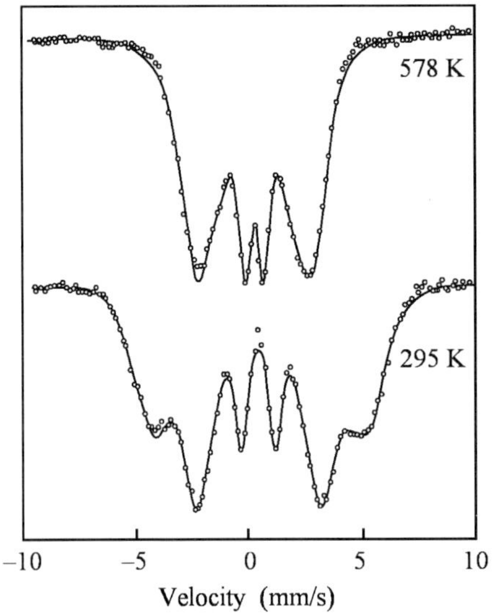

Fig. 9.12 Mössbauer spectra of an amorphous alloy $Fe_{75}P_{15}C_{10}$ [26].

Regardless how the hyperfine interactions are distributed, the total area can always be easily evaluated. The harmonic approximation is obviously applicable to the atomic vibrations in amorphous solids, so the recoilless fraction is still Eq. (5.13), except for the lack of an exact analytical expression for $\langle u^2 \rangle$. The Debye model could be used as an approximation, which must be based, of course, on the short-range order.

9.2.1
The Alloy YFe_2 [30]

The crystalline and amorphous phases of the alloy YFe_2 are denoted by c-YFe_2 and a-YFe_2, respectively. We now compare their lattice dynamics parameters.

For a-YFe_2 at $T > 58$ K, the magnetic hyperfine field disappears and only a certain distribution of quadrupole splitting exists. For c-YFe_2, the spectrum is a superposition of two sextets, corresponding to two different sublattices. Figure 9.13 shows the temperature dependences of the recoilless fraction f and center shift δ. Fitting each curve according to the Debye model gives $\theta_D = (350 \pm 10)$ K for c-YFe_2 and $\theta_D = (280 \pm 10)$ K for a-YFe_2. These results indicate that the Debye temperature of the amorphous state is lower than that of the crystalline state, namely $\theta_D{}^a/\theta_D{}^c = 0.80$. Similar results have also been observed in other amorphous alloys (Table 9.3).

The lower θ_D-values in amorphous alloys can be explained by the considerable changes in the vibrational density of states (VDOS) in comparison with that of the corresponding crystalline enhancement in the low-frequency region and a

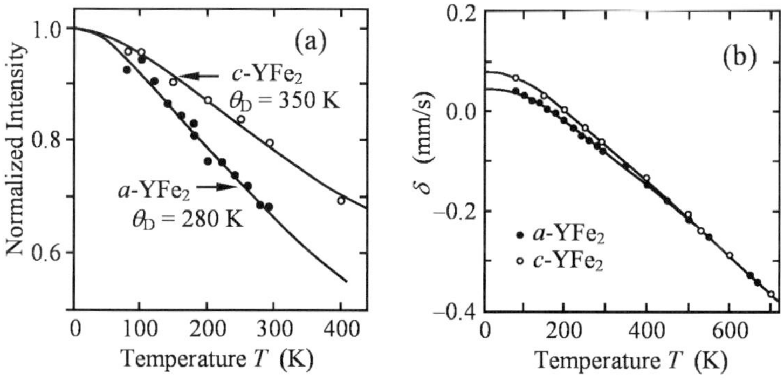

Fig. 9.13 (a) ^{57}Fe Mössbauer absorption intensity and (b) center shift as functions of temperature in amorphous a-YFe_2 and crystalline c-YFe_2.

Table 9.3 The $\theta_D{}^a/\theta_D{}^c$ ratios of several amorphous–crystalline alloy systems.

Alloy system	$\theta_D{}^a/\theta_D{}^c$ ratio
YFe_2	0.80
$Pd_{80}Si_{20}$	0.75
$Zr_{9.5}Fe_{90.5}$	~0.79
$Fe_{80}B_{20}$	~0.73
$Fe_{40}Ni_{40}P_{14}B_6$	0.87

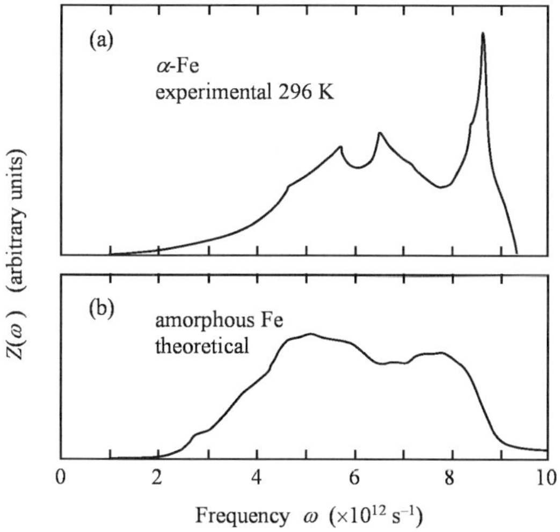

Fig. 9.14 Phonon DOS $g(\omega)$: (a) experimental result for α-Fe at 296 K; (b) theoretical result for amorphous Fe [32].

Table 9.4 Values of $\theta_D(n)$ for α-Fe and amorphous Fe.

	$\theta_D(-2)$ (K)	$\theta_D(-1)$ (K)	$\theta_D(+1)$ (K)	$\theta_D(+2)$ (K)
α-Fe	390	370	375	376
Amorphous Fe	351	345	354	361

softening in the high-frequency end [31], as clearly indicated by the DOS curves of α-Fe and its amorphous state, shown in Fig. 9.14. Calculations using the frequency moments [32] also provided results (Table 9.4) that show the same trend of $\theta_D{}^a/\theta_D{}^c < 1$.

9.2.2 The Alloy $Fe_{80}B_{20}$

We choose $Fe_{80}B_{20}$ as a second example, because the Fe–B amorphous alloys have been extensively investigated by various experimental methods including transmission Mössbauer spectroscopy. Recently, the Fe partial VDOS of $Fe_{80}B_{20}$ in both amorphous and crystalline phases were measured by the phonon-assisted Mössbauer effect [33], as shown in Fig. 9.15. The partial VDOS for the crystalline phase consists of two maxima at about 26 and 36 meV, typical of the bcc structure. This result for the amorphous phase is intriguing because of an excess vibrational density of states in the low-energy range of 4 to 21 meV, compared to

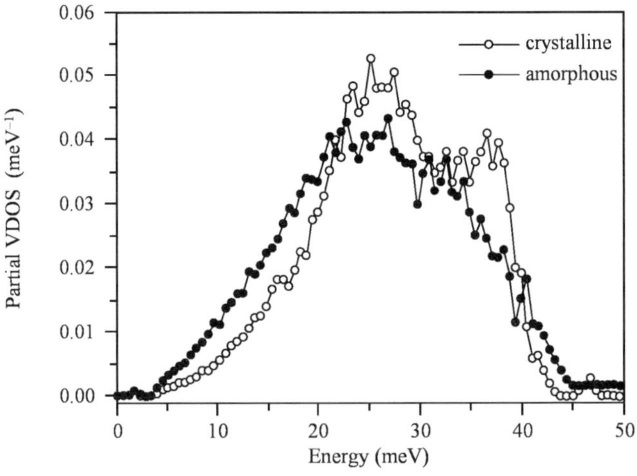

Fig. 9.15 Fe partial vibrational densities of states in amorphous and crystalline $Fe_{80}B_{20}$, as obtained using the phonon-assisted Mössbauer effect.

the usual Debye law ($g(\omega) \propto \omega^2$). Such an excess is now termed a "boson peak" and has been observed not only in a large number of amorphous materials but also in some disordered crystals [34, 35]. The origin of the boson peak is an issue that has attracted a lot of experimental and theoretical activities. A number of models for its explanation have been proposed [36–41], but it is still far from being completely understood [42, 43]. The existence of a boson peak is qualitatively consistent with the decrease in Debye temperature, as mentioned above.

9.3 Molecular Crystals

If a crystal is formed by van der Waals forces between individual atoms or finite molecules whose internal structure is covalently bonded, it is referred to as a typical molecular crystal. Many pure substances of nonmetals, nonionic oxides, and most solid organometallic compounds belong to the category of molecular crystals. The Mössbauer effect has often been applied in the studies of molecular crystals, in particular, those with complex ions such as $[Fe(CN)_6]^{4-}$, $[Ni(CN)_4]^{2-}$, $[Fe(C_5H_7O_3)_3]^{3+}$, and $[Fe(H_2O)_6]^{3+}$.

In general, intermolecular forces are weak compared to forces in covalent bonds. Therefore, the normal modes of vibration in a molecular crystal are usually separated into two different groups: the intermolecular modes related to vibrations of the molecular center of mass and the intramolecular modes of much higher frequencies within the molecule. Although the Born–von Karman theory can be used to calculate the intermolecular modes provided that the molecular framework may be treated as a rigid one, a promising theoretical method to study the structure and dynamics is the first-principles calculation based on the density-functional theory (DFT). This method has been quite successful in several examples [44].

9.3.1 The Concept of Effective Vibrating Mass M_{eff} [45]

In molecular crystals, the Mössbauer nucleus is usually at the center of the molecule. If the molecule is an ideal rigid body, the mass value in the expressions for f and δ_{SOD} would just be the molecular weight. In reality, the molecule is not entirely rigid due to some degree of internal vibration; the mass value may be replaced by an effective vibrating mass M_{eff}. Its upper limit is the molecular weight and lower limit is the mass of the Mössbauer atom (57 u for ^{57}Fe); the latter corresponds to the situation of Fe atoms as monatomic vibrating entities in a solid. In the high-temperature limit and after the substitution of M_{eff} for M, Eqs. (5.14) and (5.52) become, respectively,

$$\ln f = -\frac{3E_\gamma^2}{M_{\text{eff}}c^2 k_B \theta_M^2} T = \ln A(T) + \text{const}, \tag{9.4}$$

$$\delta_{SOD} = -\frac{3}{2}\frac{k_B T}{M_{eff} c} + \text{const.} \tag{9.5}$$

Because of the introduction of the effective vibrating mass M_{eff}, the corresponding Debye temperature θ_D will be affected and it is therefore replaced by the so-called Mössbauer lattice temperature θ_M. Equation (9.5) indicates that the effective vibrating mass M_{eff} can be deduced from the slope of a linear relation between δ and T:

$$M_{eff} = -\frac{3k_B}{2c}\left[\frac{d\delta_{SOD}}{dT}\right]^{-1} = -4.1601 \times 10^{-2}\left[\frac{d\delta_{SOD}}{dT}\right]^{-1} \tag{9.6}$$

where the numerical coefficient is such that it would give M_{eff} in u if δ_{SOD} is measured in mm s^{-1} and T in K. Substituting Eq. (9.6) into (9.4), we obtain the following formulas for determining the Mössbauer lattice temperatures for ^{57}Fe and ^{119}Sn from experimental data:

$$\theta_M(^{57}\text{Fe}) = 4.3202 \times 10^2\left[\frac{d\delta_{SOD}/dT}{d\ln A/dT}\right]^{1/2}, \tag{9.7}$$

$$\theta_M(^{119}\text{Sn}) = 7.1564 \times 10^2\left[\frac{d\delta_{SOD}/dT}{d\ln A/dT}\right]^{1/2}. \tag{9.8}$$

As soon as M_{eff} and θ_M are evaluated, the recoilless fraction f can be calculated using Eq. (9.4). For ^{57}Fe,

$$-\ln f = 7.75369 \times 10^3 \frac{T}{M_{eff}\theta_M^2}. \tag{9.9}$$

In such an approximation the dynamics of molecular crystals is described by parameters f, M_{eff}, and θ_M. These results have confirmed the feasibility of separating external from internal vibrations and have allowed a measure of molecular rigidity. Here is an example of such studies. Ferrocene, $Fe(C_5H_5)_2$, and its derivatives [46, 47] are compounds with an Fe atom sandwiched between two C_5H_5 rings. Their Mössbauer spectra are relatively simple, each having a doublet due to quadrupole splitting. Figure 9.16(a) shows ln f for ferrocene and Fig. 9.16(b) shows the center shift δ for one of its derivatives, both as functions of temperature T. The Mössbauer parameters deduced from the data are listed in Table 9.5.

As can be seen from these graphs, both ln f and δ are indeed linearly dependent on T in the range from 100 to 300 K. The values of the effective vibrating mass M_{eff} are about twice that of ^{57}Fe, but only 2/3 of the molecular weight (187 u), indicating that the molecule is not entirely rigid. Taking the slope data from rows 3 and 4 in Table 9.5 and using Eq. (9.7), Mössbauer lattice temperatures have been calculated to be $\theta_M = 93$ and 117 K for ferrocene and bis(3-methylpentadienyl)iron, respectively. Using Eq. (9.9), the recoilless fraction at

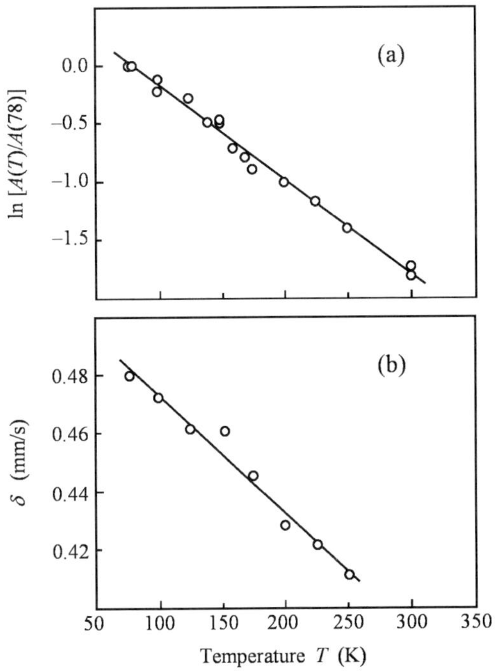

Fig. 9.16 (a) Temperature dependence of the recoilless fraction f for ferrocene and (b) temperature dependence of the center shift δ for bis(3-methylpentadienyl)iron.

Table 9.5 Lattice dynamics parameters for ferrocene and its derivative bis(3-methylpentadienyl)iron. The δ_{IS} values are with respect to that of α-Fe at 295 K.

	Ferrocene	**Bis(3-methylpentadienyl)iron**
δ_{IS} at 78 K (mm s^{-1})	0.542(9)	0.482(5)
ΔE_Q at 78 K (mm s^{-1})	2.452(12)	1.255(41)
$-d\delta/dT$ ($\times 10^{-4}$ mm s^{-1} K^{-1})	3.74	4.06
$-d \ln A/dT$ ($\times 10^{-3}$ K^{-1})	8.09(22)	5.49(41)
M_{eff} (u)	111(8)	103(8)
θ_M (K)	93	117
f at 295 K	0.092	0.109

room temperature $f = 0.092$ is calculated, which is consistent with the result of $f = 0.08$ as given in Ref. [48].

The deviation of the M_{eff} value from 57 u reflects the covalency of the bonding force between the Fe atom and its neighbors. Here, we present a recent experimental result which defines the meaning of the effective mass more precisely. In the temperature range of 15 to 550 K, the measured central shifts of iron(III) octahedral 16*a* and tetrahedral 24*d* sites in $Dy_3Fe_5O_{12}$ give an effective mass of 57 u [49]. It is evident that the ^{57}Fe atoms are indeed totally ionically bonded to their neighboring oxygen dianions. The effective mass is then expected to be exactly 57 u in the absence of covalency.

9.3.2 Vibrational DOS in Molecular Crystals

9.3.2.1 The Mode Composition Factor $e^2(l, j)$

For a molecular crystal, Eq. (7.53) is reduced to

$$g(\boldsymbol{h}, E) = \sum_j \delta(E - \hbar\omega_j)|\boldsymbol{h} \cdot \mathbf{e}(j)|^2, \tag{9.10}$$

and for a particular mode with ω_j, it is reduced to

$$g(\omega_j, \boldsymbol{h}) = |\boldsymbol{h} \cdot \mathbf{e}(j)|^2. \tag{9.11}$$

This means the probability that a vibrating mode j takes place is determined by $|\mathbf{e}(j)|^2$. On the other hand, the eigenvector $\mathbf{e}(l, j)$ of the lth atom vibrating in mode j within a molecule is related to the atomic displacement $\boldsymbol{u}(l, j)$ by [50]

$$\boldsymbol{u}(l, j) = \frac{1}{\sqrt{M_l}} \left(\frac{E(j)}{\hbar\omega_j^2} \right)^{1/2} \mathbf{e}(l, j). \tag{9.12}$$

Using this relation and the normalization properties of the vector $\mathbf{e}(l, j)$, one easily gets

$$e^2(r, j) = \frac{u^2(r, j) M_r}{\sum_{l=1}^{N} u^2(l, j) M_l} \tag{9.13}$$

where $u^2(r, j)$ is the mean-square displacement of the Mössbauer atom of mass M_r. As can be seen, both the numerator and the denominator in the right-hand side of the above expression represent the kinetic energy of the Mössbauer atoms. Therefore, the quantity $e^2(l, j)$ is called the mode composition factor [51]. The vibrating spectrum in a molecular crystal usually consists of discrete peaks. The factor $e^2(l, j)$ offering direct information on mode character is determined by the area fraction under each peak. We consider two simple cases. First, when the inter- and intramolecular vibrations are completely decoupled, the mean-square displacement in the acoustic mode is the same for all atoms, i.e., a translational mode. Therefore, for each acoustic mode the factor is

$$e_{\mathrm{ac}}^2 = \frac{M_r}{M_\Sigma} \tag{9.14}$$

where M_Σ is the total molecular mass. Second, consider a molecule that can be grouped into two rigid molecular fragments with masses M_1 and M_2. Let the Mössbauer atom be located in the first fragment. Assuming the vibrational motion involves no translation of the center of mass, the kinetic energy associated with the motion of the first fragment must be a fraction M_2/M_Σ of the total kinetic energy. A sub-fraction of M_r/M_2 is then associated with the motion of the Mössbauer atom. Hence, one obtains a composition factor for the stretching mode [51, 52]:

$$e_{\mathrm{str}}^2 = \frac{M_r}{M_1}\frac{M_2}{M_\Sigma}. \tag{9.15}$$

Measuring the factor e_{str}^2 permits one to calculate the fragment mass M_1:

$$M_1 = \frac{M_r M_\Sigma}{M_r + M_\Sigma e_{\mathrm{str}}^2}. \tag{9.16}$$

The simplest such fragment is one that contains only the Mössbauer atom and nothing else. In this case, the composition factor of this stretching mode reaches its maximum value

$$e_{\mathrm{max}}^2 = \frac{M_2}{M_\Sigma}. \tag{9.17}$$

The composition factor may be estimated in simple cases as done above, but in general it is determined experimentally.

9.3.2.2 An Example

We take the hexacyanoferrate(II) compound $(NH_4)_2MgFe(CN)_6$ [52] as an example of application of inelastic nuclear resonant scattering in molecular dynamics. The free complex ion $[Fe(CN)_6]^{4-}$ has octahedral symmetry. As mentioned in Chapter 8, the Fe atom vibrates in two threefold-degenerate F_{1u} modes only (Fig. 9.17). The measured Fe partial VDOS is illustrated in Fig. 9.18 where the energies and the widths of the peaks were found by a fit with Gaussian distributions (with the exception of the region of acoustic modes). The spectral parameters are listed in Table 9.6.

First, we consider the assignment of the three acoustic modes. According to Eq. (9.14), the total area under the DOS curve due to the three acoustic modes is equal to $M_{\mathrm{Fe}}/M_\Sigma = 0.21$ (where $M_\Sigma = 273$ u). The integral over the phonon DOS reaches this value at 12.7 meV. An acoustic peak occurs at ~8 meV. The peaks at higher energies come from intramolecular vibrations. The largest peak at 74.3 meV has a composition factor of $e^2 = 1.04$, which is higher than the maximum value of $e_{\mathrm{max}}^2 = 0.79$ according to Eq. (9.17). Due to the threefold degener-

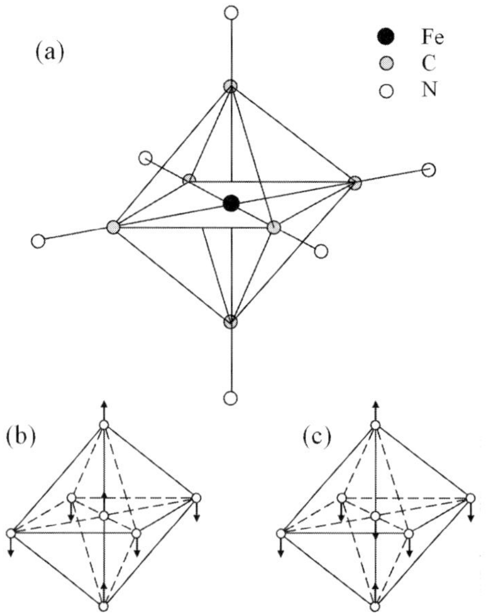

Fig. 9.17 (a) Fe surroundings in hexacyanoferrate (II) complexes. (b, c) The two vibrational F_{1u} modes (each with threefold degeneracy) of the ideal octahedron that involve motion of the central Fe.

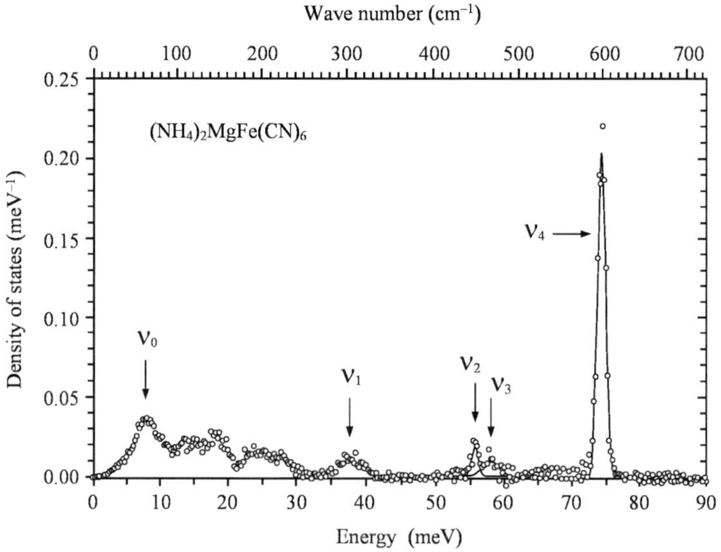

Fig. 9.18 Fe VDOS in hexacyanoferrate (II) at 30 K. Solid lines represent the fits of frequency distributions for several indicated modes with Gaussian functions.

Table 9.6 Composition of iron partial VDOS in $(NH_4)_2MgFe(CN)_6$ at 30 K.

Vibrational mode	ν_0	ν_1	ν_2	ν_3	ν_4
Frequency (meV)	0–12.5	37.9(6)	55.7(3)	58.0(6)	74.3(3)
Frequency (cm^{-1})	0–101	306(5)	449(2)	468(5)	600(2)
Width (FWHM) (meV)	–	3.8(5)	1.1(2)	3.0(5)	1.4(2)
Mean force constant ($N\ m^{-1}$)	15.7(5)	315(10)	681(8)	738(15)	1213(10)
Composition factor e^2		0.19(3)	0.23(3)		1.04(3)
Vibration amplitude (Å) at room temperature	0.176(1)	0.017(1)	0.013(1)		0.024(1)
Assignment	Acoustic	–	Stretching and bending		Degenerate stretching

acy, this peak may consist of three degenerate modes with $e^2 = 0.35$ for each stretching mode. Equation (9.16) then gives a fragment mass $M_1 = 102$ u. This unequivocally points to the fragment as the Fe atom and two CN groups (total 109 u), as shown in Fig. 9.17(b). Therefore, this mode can be identified as the threefold-degenerate stretching of this fragment against the rest of the molecule. Of course, one expects the other stretching mode where the fragment is composed of the Fe atom and four CN groups (Fig. 9.17(c)). But the estimated $e^2{}_{str}$ does not allow an unambiguous assignment that is consistent with the experimental data. The other three peaks have low factors e^2 (Table 9.6). Finally, two peaks with higher factors e^2 are assigned to the second F_{1u} modes. These results are basically consistent with that from normal mode analyses, even though not all modes are pure Fe–C stretching motion, but involve both Fe–C stretching and Fe–C–N bending components [53].

This is the simplest example. At the present time, the phonon-assisted Mössbauer effect has been successfully used to study the dynamics of macromolecules such as the protein myoglobin [54]. Furthermore, the first-principles calculation method has begun to simulate spectral measurements from molecular crystals.

9.4 Low-Dimensional Systems

9.4.1 Thin Films

In conventional Mössbauer spectroscopy, the conversion electron method is used for thin-film studies because of its high surface sensitivity. However, it seldom provides reliable data of dynamics parameters, including recoilless fraction f. Investigating the vibrational properties of thin films is particularly difficult because neither neutron inelastic scattering nor x-ray inelastic scattering is feasible for such a small amount of material in thin films. Infrared or visible light scattering can only provide a small part of the vibrational DOS. Recently, this problem has been successfully solved by x-ray inelastic scattering using SR.

A variety of interesting and useful phenomena occur under glancing incidence of x-rays on a flat surface of a material. These phenomena include total external reflection, interference fringes from layers on the substrate, and the formation of evanescent and standing waves [55, 56]. A valuable feature of these phenomena is the ability to enhance the output x-ray flux considerably by the interference effects. All of these phenomena have been utilized in the studies of surfaces, interfaces, thin films, and layered materials.

To understand the phenomena, let us first assume that x-rays, as electromagnetic traveling plane waves, impinge on a flat material under a glancing angle θ. The incident and reflected waves will be coherent and interfere to generate a standing wave with planes of maximum intensity parallel to the surface and with a spatial period $(\lambda/2)\sin\theta$ [55]. This describes the distance between the first antinode and successive antinodes above the surface. In the case of total reflection, the first antinode of the standing wave coincides with the surface, where the intensity may be up to four times the incident one.

This method becomes especially effective when the thin film under study is deposited on a high-reflection substrate, i.e., a material with a high electron density. Figure 9.19(a) shows an example of an FeB film on a Pd substrate where the critical angle of the air–FeB surface is θ_c(FeB). A synchrotron x-ray beam is allowed to penetrate into the FeB film and is strongly reflected by the Pd layer. If the reflection angle θ is higher than θ_c(FeB), the x-ray beam, already reflected from the Pd layer, is totally reflected back into the FeB film from the FeB–air interface, and is therefore trapped in this film which acts as a waveguide [57]. The x-ray inside the thin film is said to resonantly excite a guide mode, and as a result the electromagnetic field intensity within the thin film is strongly enhanced. Such a resonance always leads to a standing wave with a periodicity equal to an integer fraction of the thin film thickness – one of the conditions for resonance – and also with the antinode planes parallel to the interfaces. This type of resonance can be experimentally detected by deep minima in a reflection curve and by a sharp maximum in the secondary effect yield (e.g., fluorescence photons), shown in the left and middle columns in Fig. 9.19, respectively. The arrow in each reflectivity curve indicates the first-order resonance. The degree of field enhancement, the widths of the intensity peaks, and the number of peaks depend on the layer thickness. The intensity enhancement is most pronounced if the film under investigation is sandwiched between two highly reflecting layers (Fig. 9.19(b)).

In a recent study, this effect has been utilized in inelastic nuclear resonant scattering (INRS) to measure the vibrational DOS of a thin film. The experimental results of a 13 nm ^{57}Fe film on a 20 nm thick Pd layer [58] are illustrated in Fig.

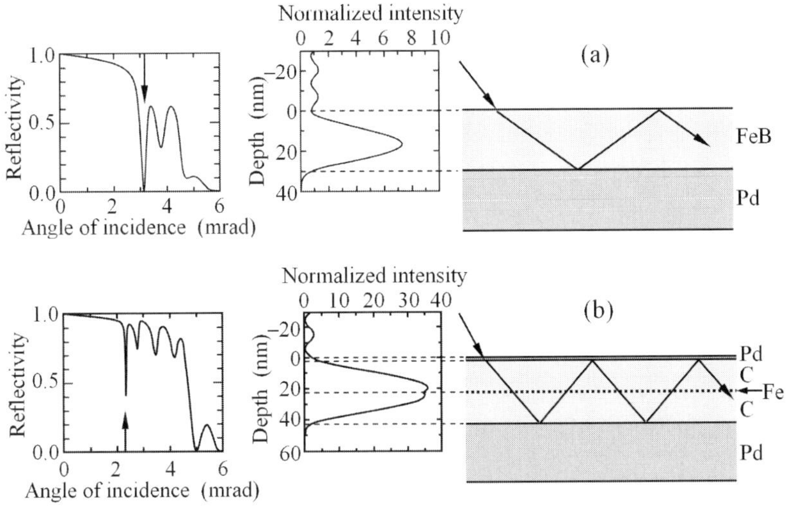

Fig. 9.19 Thin-film interference effects to enhance fluorescence signals: (a) inside a thin film of FeB on a total reflecting Pd substrate and (b) inside a Pd/C/Pd sandwich structure that acts as an x-ray waveguide. The graphs also show the specular reflectivity of the structure (left column) and the normalized field intensity as a function of depth (middle column).

9.20, where a guide mode of first order is excited and the average fluorescence yield inside the film reaches 6 times the incident SR intensity. At this peak, a maximum counting rate of 30 s^{-1} was observed which is comparable to that obtained from the bulk material.

The recorded vibrational DOS of thin films of thickness 13 and 28 nm are shown in Figs. 9.21(a) and (b). For comparison, the DOS of bulk α-Fe, obtained from a 10 μm thick foil under the same experimental conditions, is shown Fig. 9.21(c). The most obvious feature is that the longitudinal peak at 35 meV is broadened in comparison with the peak in bulk Fe. Much of this type of broadening may have been caused by the short phonon lifetimes in the thin film.

Assuming that each phonon is broadened in energy as a damped harmonic oscillator, each intensity at energy E' of the experimental DOS curve is convoluted with the characteristic spectrum of a damped harmonic oscillator function $D(E', E)$ [58, 59]:

$$g(E') = \int_0^\infty D(E', E) g(E)\, \mathrm{d}E \tag{9.18}$$

where

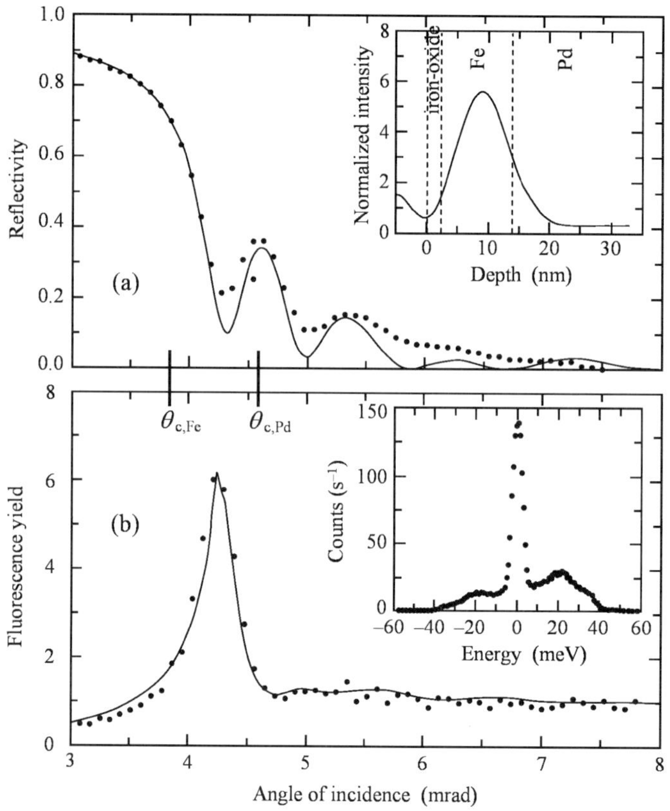

Fig. 9.20 (a) Angular dependence of reflectivity of 13 nm Fe on Pd. A guided mode is excited at an angle of 4.2 mrad. The inset shows the depth dependence of intensity inside the layer. (b) Angular dependence of the yield of delayed fluorescence from the Fe film. The inset shows the phonon spectrum recorded at an incident angle of 4.2 mrad.

$$D(E', E) = \frac{1}{\pi Q E'} \frac{1}{(E'/E - E/E')^2 + 1/Q^2} \tag{9.19}$$

and $g(E)$ is the phonon DOS of the bulk α-Fe. In the convolution, the quality factor Q is the sole parameter used to fit the data. The Q-values for the 13 and 28 nm Fe films were determined to be 13 ± 1 and 25 ± 2, respectively.

Using the glancing incidence technique, other thin films such as FeC_2Ni on Pd [60] as well as $FeBO_3$ and Fe islands on a W(110) surface [58] have been studied.

Because of the development of strong synchrotron x-ray sources, one may now directly record the VDOS in thin films without using the above technique, but relatively long measurement times are required. The Fe partial VDOS of Tb/Fe multilayers obtained with the direct method [61] is plotted in Fig. 9.22, where the VDOS of a bulk Fe and a 175 Å thick a-$Tb_{33}Fe_{67}$ amorphous alloy film are

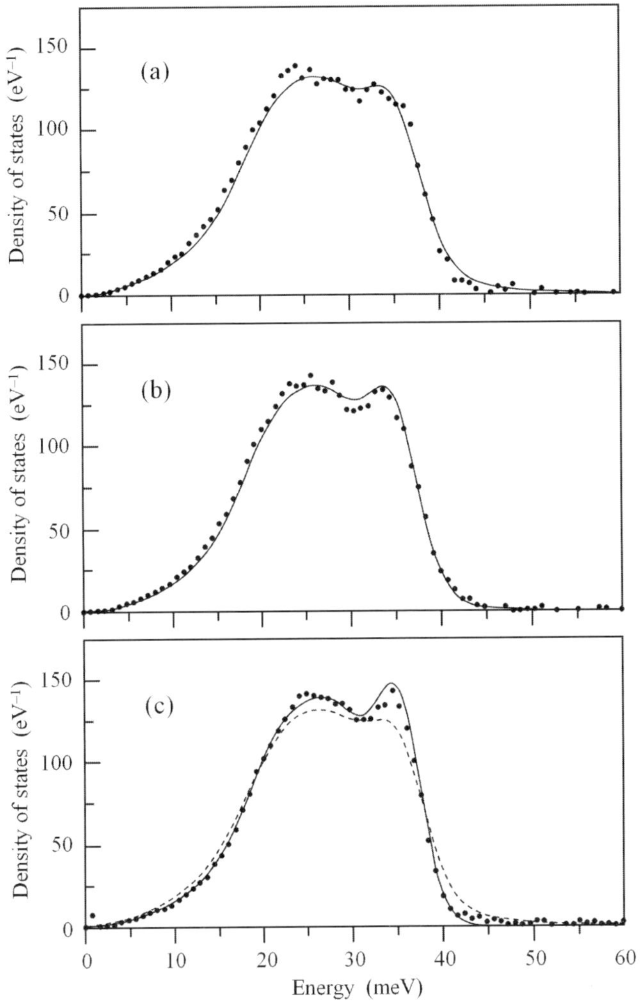

Fig. 9.21 (a) DOS of 13 nm Fe film and (b) DOS of 28 nm Fe film, in which the solid lines are calculated by Eq. (9.16). (c) DOS of bulk bcc α-Fe. The solid lines are calculated with force constant from inelastic neutron scattering and the dotted line in (c) corresponds to the solid line in (a).

also given for comparison. Distinct differences in the VDOS have been observed by varying the thickness of Fe or Tb. For the thicker Fe layers (samples C, D, and F), the VDOS exhibit phonon peaks at 36 meV (longitudinal phonons) as well as at 23 and 28 meV (transverse phonons), all typical of bulk bcc α-Fe. For the thinner Fe layers (samples A, B, and E), the VDOS have a broad and featureless peak and they resemble that observed in the amorphous alloy films $Tb_{1-x}Fe_x$ [62].

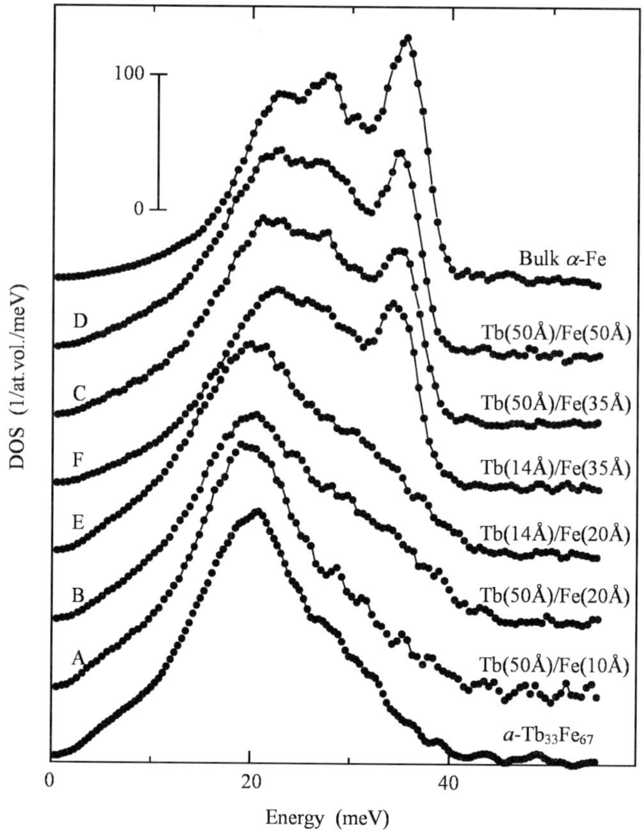

Fig. 9.22 Fe partial VDOS of the Tb/Fe samples (thickness values in Å in parentheses). The VDOS of a bulk Fe sample (top) and a 175 Å thick a-$Tb_{33}Fe_{67}$ amorphous film (bottom) are also shown for comparison.

9.4.2 Nanocrystals

Nanocrystalline materials, often defined as materials composed of crystallites smaller than 100 nm, have attracted much interest in recent years. Their unusual physical and chemical properties are related to the finite-size effect, quantum-size effect, and large surface-to-volume ratio. A detailed study of the dynamics of nanocrystals is one of the best ways to understand their properties. Very recently, inelastic neutron scattering [63–66] and INRS of γ-rays [67, 68], as well as theoretical calculations [69–71], have provided evidence of two distinct differences in the phonon DOS of materials in nanocrystalline and bulk phases. One such feature is an enhancement of phonon DOS at low energies. The second is

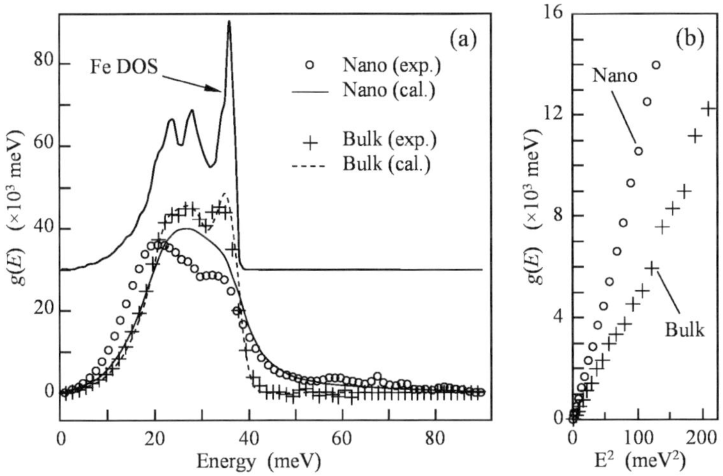

Fig. 9.23 (a) Phonon DOS of bcc Fe in bulk and nanocrystalline phases; dotted and solid curves are calculated as described in the text. (b) Enlargement of the low-energy part of phonon DOS curves in (a).

the broadening of phonon DOS in the relatively high-energy region. These features are not well understood and they continue to motivate more extensive investigations.

Figure 9.23(a) depicts the measured phonon DOS in nanocrystalline Fe by INRS of γ-rays [67]. For comparison, the phonon DOS of bulk bcc Fe obtained under the same conditions is overlaid in this figure, where the dashed curve was calculated in the same way as for the solid line in Fig. 9.21(c). The low-energy part of the phonon DOS curve is plotted in Fig. 9.23(b), where it is clear that the nanocrystalline DOS lies above that from bulk bcc Fe by a factor of about 2. The DOS below 15 meV can obviously be fitted to the Debye law $g(\omega) = aE^2$, where constant a dependents on the crystallite size. Such a quadratic dependence on energy has been confirmed also by other experiments [64, 68]. However, the functional dependence of the low-energy modes is controversially reported in the literature. The power-law exponent was found to be 1 [65, 69, 70], ~1.5 [71], and 2 [64, 67, 68]. At the present time, only some of theoretical calculations are fully consistent with the experimental data.

A broadening of peaks in DOS of nanocrystalline Fe is particularly evident for the longitudinal mode at 36 meV, and additional intensities above the high-frequency cutoff extends to about 50 meV. Due to the negligible background counts, INRS is a suitable method for studying the features at the high-energy region and especially the additional intensities. The broadening of DOS again can be described by a damped harmonic oscillator model with $Q = 5$, which produced the solid line for the nanocrystalline Fe in Fig. 9.23(a).

References

1 H. Vogel, H. Spiering, W. Irler, U. Volland, and G. Ritter. The Lamb–Mössbauer factor of metal iron foils at 4.2 K. *J. de Physique (Colloque)* 40, C2-676 (1979).

2 R.S. Preston, S.S. Hanna, and J. Heberle. Mössbauer effect in metallic iron. *Phys. Rev.* 128, 2207–2218 (1962).

3 B.N. Brockhouse, H.E. Abou-Helal, and E.D. Hallman. Lattice vibrations in iron at 296 °K. *Solid State Commun.* 5, 211–216 (1967).

4 G. Burns. *Solid State Physics* (Academic Press, New York, 1985).

5 C. Hohenemser. Measurement of the Mössbauer recoilless fraction in β-Sn for 1.3 to 370 °K. *Phys. Rev.* 139, A185–A196 (1965).

6 A. Barla, R. Rüffer, A.I. Chumakov, J. Metge, J. Plessel, and M.M. Abd-Elmeguid. Direct determination of the phonon density of states in β-Sn. *Phys. Rev. B* 61, R14881–R14884 (2000).

7 D.J. Erickson, L.D. Roberts, J.W. Burton, and J.O. Thomson. Precision determinations of the Mössbauer recoilless fraction for metallic gold in the temperature range $4.2 \leq T \leq 100$ °K. *Phys. Rev. B* 3, 2180–2186 (1971).

8 W. Potzel, W. Adlassnig, U. Närger, Th. Obenhuber, K. Riski, and G.M. Kalvius. Temperature dependence of hyperfine interactions and of anisotropy of recoil-free fractions: a Mössbauer study of the 93.3-keV resonance of ^{67}Zn in single crystals of zinc metal. *Phys. Rev. B* 30, 4980–4988 (1984).

9 E.F. Skelton and J.L. Katz. Examination of the thermal variation of the mean square atomic displacements in zinc and evaluation of the associated Debye temperature. *Phys. Rev.* 171, 801–808 (1968).

10 F.E. Obenshain and H.H.F. Wegener. Mössbauer effect with Ni^{61}. *Phys. Rev.* 121, 1344–1349 (1961).

11 D.K. Kaipov and S.M. Zholdasova. Determination of the Debye temperature of nickel by Rayleigh scattering of Mössbauer radiation. In *Applications of the Mössbauer Effect*, vol. 4, Yu.M. Kagan and I.S. Lyubutin (Eds.), pp. 1383–1385 (Gordon and Breach, New York, 1985).

12 W. Potzel. Recent ^{67}Zn-experiments. *Hyperfine Interactions* 40, 171–182 (1988).

13 W. Potzel, A. Forster, and G.M. Kalvius. The quadrupole interaction in zinc metal. *J. de Physique (Colloque)* 40, C2-29–30 (1979).

14 T. Katila and K. Riski. ^{67}Zn Mössbauer spectroscopy. *Hyperfine Interactions* 13, 119–148 (1983).

15 W. Potzel, U. Närger, Th. Obenhuber, J. Zpnkert, W. Adlassnig, and G.M. Kalvius. Anisotropy of the Lamb–Mössbauer factor in zinc metal. *Phys. Lett. A* 98, 295–298 (1983).

16 S.P. Tewari and P. Silotia. The effect of crystal anisotropy on the Lamb Mössbauer recoilless fraction and second-order Doppler shift in zinc. *J. Phys.: Condens. Matter* 1, 5165–5170 (1989).

17 M. Steiner, M. Köfferlein, W. Potzel, H. Karzel, W. Schiessl, G.M. Kalvius, D.W. Mitchell, N. Sahoo, H.H. Klauss, T.P. Das, R.S. Feigelson, and G. Schmidt. Investigation of electronic structure and anisotropy of the Lamb–Mössbauer factor in ZnF_2 single crystals. *Hyperfine Interactions* 93, 1453–1458 (1994).

18 M. Köfferlein, W. Potzel, M. Steiner, H. Karzel, W. Schiessl, and G.M. Kalvius. Applications of lattice-dynamic models to ^{67}Zn Lamb–Mössbauer factors and second-order Doppler shifts. *Phys. Rev. B* 52, 13332–13340 (1995).

19 Th. Obenhuber, W. Adlassnig, J. Zänkert, U. Närger, W. Potzel, and G.M. Kalvius. High resolution Mössbauer spectroscopy with ^{67}Zn in metallic systems. *Hyperfine Interactions* 33, 69–88 (1987).

20 E. Galvão da Silva and U. Gonser. Determination of relative force constant in β-Ti(Fe) alloys by means of the Mössbauer second-order

Doppler shift. *Appl. Phys. A* 27, 89–94 (1982).

21 D.A. Papaconstantopoulos. Electronic structure of TiFe. *Phys. Rev. B* 11, 4801–4807 (1975).

22 M. Peter, W. Potzel, M. Steiner, C. Schäffer, H. Karzel, W. Schiessl, G.M. Kalvius, and U. Gonser. ^{67}Zn Mössbauer study of lattice-dynamical effects and hyperfine interactions in Cu–Zn alloys. *Hyperfine Interactions* 69, 463–466 (1991).

23 M. Peter, W. Potzel, M. Steiner, C. Schäfer, H. Karzel, W. Schiessl, G.M. Kalvius, and U. Gonser. Lattice-dynamical effects and hyperfine interactions in Cu–Zn alloys. *Phys. Rev. B* 47, 753–762 (1993).

24 Th. Obenhuber, W. Adlassnig, U. Närger, J. Zänkert, W. Potzel, and G.M. Kalvius. Observation of short-range order in α-brass by ^{67}Zn Mössbauer spectroscopy. *Europhys. Lett.* 3, 989–994 (1987).

25 L. Reinhard, B. Schönfeld, G. Kostorz, and W. Bührer. Short-range order in α-brass. *Phys. Rev. B* 41, 1727–1734 (1990).

26 C.L. Chien and R. Hasegawa. Mössbauer study of amorphous $Fe_{75}P_{15}C_{10}$. *J. de Physique (Colloque)* 37, C6-759–761 (1976).

27 M. Kopcewicz, H.-G. Wagner, and U. Gonser. A direct determination of the quadrupole splitting in ferromagnetic amorphous metals. *Solid State Commun.* 48, 531–533 (1983).

28 M. Kopcewicz. The rf stimulation of amorphous metals. *Hyperfine Interactions* 40, 77–88 (1988).

29 L. Pfeiffer. The effects of radio frequency fields on ferromagnetic Mössbauer absorbers. In *Mössbauer Effect Methodology*, vol. 7, I.J. Gruverman (Ed.), pp. 263–298 (Plenum Press, New York, 1971).

30 Y. Nishihara, T. Katayama, and S. Ogawa. Mössbauer study of amorphous and crystalline YFe_2. *J. Phys. Soc. Japan* 51, 2487–2492 (1982).

31 M.H. Cohen, J. Singh, and F. Yonezawa. Elementary excitations of topologically disordered systems. *Solid State Commun.* 36, 923–926 (1980).

32 R. Yamamoto, K. Haga, T. Mihara, and M. Doyama. The vibrational states in a realistic model of amorphous iron. *J. Phys. F* 10, 1389–1399 (1980).

33 A. Gupta, P. Shah, N.P. Lalla, B.A. Dasannacharya, T. Harami, Y. Yoda, M. Seto, M. Yabashi, and S. Kikuta. Vibrational dynamics of some amorphous and quasicrystalline alloys. *Mater. Sci. Eng.* A304–306, 731–734 (2001).

34 S.N. Taraskin,Y.L. Lok, G. Natarajan, and S.R. Elliott. Origin of the boson peak in systems with lattice disorder. *Phys. Rev. Lett.* 86, 1255–1258 (2001).

35 J.J. Tu and A.J. Sievers. Experimental study of Raman-active two-level systems and the boson peak in LaF_3-doped fluorite mixed crystals. *Phys. Rev. B* 66, 094206 (2002).

36 U. Buchenau, Yu.M. Galperin, V.I. Gurevich, D.A. Parshin, M.A. Ramos, and H.R. Schober. Interaction of soft modes and sound waves in glasses. *Phys. Rev. B* 46, 2798–2808 (1992).

37 W. Schirmacher, G. Diezemann, C. Ganter. Harmonic vibrational excitations in disordered solids and the "boson peak". *Phys. Rev. Lett.* 81, 136–139 (1998).

38 S.N. Taraskin and S.R. Elliott. Anharmonicity and localization of atomic vibrations in vitreous silica. *Phys. Rev. B* 59, 8572–8585 (1999).

39 Y. Inamura, M. Arai, T. Otomo, N. Kitamura, and U. Buchenau. Density dependence of the boson peak of vitreous silica. *Physica B* 284–288, 1157–1158 (2000).

40 D.A. Parshin and C. Laermans. Interaction of quasilocal harmonic modes and boson peak in glasses. *Phys. Rev. B* 63, 132203 (2001).

41 V.L. Gurevich, D.A. Parshin, and H.R. Schober. Anharmonicity, vibrational instability, and boson peak in glasses. *Phys. Rev. B* 67, 094203 (2003).

42 F. Finkemeier and W. von Niessen. Boson peak in amorphous silica: a numerical study. *Phys. Rev. B* 63,

235204 (2001); *Phys. Rev. B* 66, 087202 (2002).

43 S.M. Nakhmanson, D.A. Drabold, and N. Mousseau. Comment on "Boson peak in amorphous silicon: a numerical study". *Phys. Rev. B* 66, 087201 (2002).

44 H. Paulsen, H. Winkler, A.X. Trautwein, H. Grünsteudel, V. Rusanov, and H. Toftlund. Measurement and simulation of nuclear inelastic-scattering spectra of molecular crystals. *Phys. Rev. B* 59, 975–984 (1999).

45 R.H. Herber. Structure, bonding, and the Mössbauer lattice temperature. In *Chemical Mössbauer Spectroscopy*, R.H. Herber (Ed.), pp. 199–216 (Plenum Press, New York, 1984).

46 R.D. Ernst, D.R. Wilson, and R.H. Herber. Bonding, hyperfine interactions, and lattice dynamics of bis(pentadienyl)iron compounds. *J. Am. Chem. Soc.* 106, 1646–1650 (1984).

47 Y.L. Chen, B.F. Xu, L. Zhang, and P.Z. Hu. Mössbauer effect study of iron decahydro-decaborate chelates containing ferrocenyl group. *Hyperfine Interactions* 53, 305–310 (1990).

48 F.L. Zhang, F. Yi, Y.L. Chen, and B.F. Xu. Determination of the optimum thickness of an absorber in Mössbauer spectroscopy. *J. Wuhan University (Natural Sci. Ed.)* 43, 348–352 (1997) [in Chinese].

49 G.J. Long, D. Hautot, F. Grandjean, D.T. Morelli, and G.P. Meisner. Reply to "Comment on 'Mössbauer effect study of filled antimonide skutterudites'". *Phys. Rev. B* 62, 6829–6831 (2000).

50 B.T.M. Willis and A.W. Pryor. *Thermal Vibrations in Crystallography*, p. 28 (Cambridge University Press, London, 1975).

51 J.T. Sage, C. Paxson, G.R.A. Wyllie, W. Sturhahn, S.M. Durbin, P.M. Champion, E.E. Alp, and W.R. Scheidt. Nuclear resonance vibrational spectroscopy of a protein active-site mimic. *J. Phys. C* 13, 7707–7722 (2001).

52 A.I. Chumakov, R. Rüffer, O. Leupold, and I. Sergueev. Insight to dynamics of molecules with nuclear inelastic scattering. *Struct. Chem.* 14, 109–119 (2003).

53 O. Zakharieva-Pencheva and V.A. Dementiev. Calculation of IR- and Raman-active vibrations of hexacyano complexes of Fe^{II}, Co^{III}, Mn^{II}, Cr^{III}, V^{II}. *J. Molec. Struct.* 90, 241–248 (1982).

54 K. Achterhold, C. Keppler, A. Ostermann, U. van Bürck, W. Sturhahn, E.E. Alp, and F.G. Parak. Vibrational dynamics of myoglobin determined by the phonon-assisted Mössbauer effect. *Phys. Rev. E* 65, 051916 (2002).

55 M.J. Bedzyk, G.M. Bommarito, and J.S. Schildkraut. X-ray standing waves at a reflecting mirror surface. *Phys. Rev. Lett.* 62, 1376–1379 (1989).

56 B.N. Dev, A.K. Das, S. Dev, D.W. Schubert, M. Stamm, and G. Materlik. Resonance enhancement of x rays in layered materials: application to surface enrichment in polymer blends. *Phys. Rev. B* 61, 8462–8468 (2000).

57 Y.P. Feng, S.K. Sinha, H.W. Deckman, J.B. Hastings, and D.P. Siddons. X-ray flux enhancement in thin-film waveguides using resonant beam couplers. *Phys. Rev. Lett.* 71, 537–540 (1993).

58 R. Röhlsberger. Vibrational spectroscopy of thin films and nanostructures by inelastic nuclear resonant scattering. *J. Phys.: Condens. Matter* 13, 7659–7677 (2001).

59 R. Röhlsberger, W. Sturhahn, T.S. Toellner, K.W. Quast, P. Hession, M. Hu, J. Sutter, and E.E. Alp. Phonon damping in thin films of Fe. *J. Appl. Phys.* 86, 584–587 (1999).

60 R. Röhlsberger, W. Sturhahn, T.S. Toellner, K.W. Quast, E.E. Alp, A. Bernhard, J. Metge, R. Rüffer, and E. Burkel. Vibrational density of states of thin films measured by inelastic scattering of synchrotron radiation. *Physica B* 263–264, 581–583 (1999).

61 T. Ruckert, W. Keune, W. Sturhahn, and E.E. Alp. Phonon density of

states in Tb/Fe multilayers. *J Magn. Magn. Mater.* 240, 562–564 (2002).

62 W. Keune and W. Sturhahn. Inelastic nuclear resonant absorption of synchrotron radiation in thin firms and multilayers. *Hyperfine Interactions* 123/124, 847–861 (1999).

63 B. Fultz, J.L. Robertson, T.A. Stephens, L.J. Nagel, and S. Spooner. Phonon density of states of nanocrystalline Fe prepared by high-energy ball milling. *J. Appl. Phys.* 79, 8318–8322 (1996).

64 H. Frase, B. Fultz, and J.L. Robertson. Phonons in nano-crystalline Ni_3Fe. *Phys. Rev. B* 57, 898–905 (1998).

65 U. Stuhr, H. Wipf, K.H. Andersen, and H. Hahn. Low-frequency modes in nanocrystalline Pd. *Phys. Rev. Lett.* 81, 1449–1452 (1998).

66 E. Bonetti, L. Pasquini, E. Sampaolesi, A. Deriu, and G. Cicognani. Vibrational density of states of nanocrystalline iron and nickel. *J. Appl. Phys.* 88, 4571–4575 (2000).

67 B. Fultz, C.C. Ahn, E.E. Alp, W. Sturhahn, and T.S. Toellner. Phonon in nanocrystalline ^{57}Fe. *Phys. Rev. Lett.* 79, 937–940 (1997).

68 L. Pasquini, A. Barla, A.I. Chumakov, O. Leupold, R. Rüffer, A. Deriu, and E. Bonetti. Size and oxidation effects on the vibrational properties of nanocrystalline α-Fe. *Phys. Rev. B* 66, 073410-1–073410-4 (2002).

69 A. Kara and T.S. Rahman. Vibrational properties of metallic nanocrystals. *Phys. Rev. Lett.* 81, 1453–1456 (1998).

70 D.Y. Sun, X.G. Gong, and X.Q. Wang. Soft and hard shells in metallic nanocrystals. *Phys. Rev. B* 63, 193412-1–193412-4 (2001).

71 P.M. Derlet, R. Meyer, L.J. Lewis, U. Stuhr, and H. van Swygenhoven. Low-frequency vibrational properties of nanocrystalline materials. *Phys. Rev. Lett.* 87, 205501-1–205501-4 (2001).

Appendices

Appendix A
Fractional Intensity $\varepsilon(v)$ and Area $A(t_a)$

Let Γ_s and Γ_a be the linewidths of the source and the absorber, respectively. They are often unequal, i.e., $\xi = \Gamma_s/\Gamma_a \neq 1$. Therefore, the fractional absorption intensity of γ-rays is the convolution between the Lorentzian emission spectrum and the Lorentzian absorption spectrum:

$$\varepsilon(S) = \frac{I(\infty) - I(v)}{I(\infty) - I_b}$$
$$= \frac{1}{2\pi}\int_{-\infty}^{\infty} \frac{f_s\Gamma_s}{(E - E_0 + S)^2 + \Gamma_s^2/4}$$
$$\times \left\{1 - \exp\left[-\frac{t_a\Gamma_a^2/4}{(E - E_0')^2 + \Gamma_a^2/4}\right]\right\} dE \qquad \text{(A.1)}$$

where $S = (v/c)E_0$, v is the Doppler velocity of the source, and E_0 and E'_0 are energy peak positions of the emission and absorption lines, respectively.

Introducing two dimensionless variables

$$x = \frac{2(E - E_0')}{\Gamma_a}, \quad y = \frac{2(S + E_0' - E_0)}{\Gamma_a}, \qquad \text{(A.2)}$$

we can rewrite integral (A.1) as

$$\varepsilon(y) = \frac{f_s\xi}{\pi}\int_{-\infty}^{\infty} \frac{1 - \exp[-t_a/(1 + x^2)]}{\xi^2 + (x + y)^2} dx$$
$$= f_s - \frac{f_s\xi}{\pi}\int_{-\infty}^{\infty} \frac{\exp[-t_a/(1 + x^2)]}{\xi^2 + (x + y)^2} dx. \qquad \text{(A.3)}$$

Substitution of

$$x = \tan\frac{\varphi}{2} \qquad \text{(A.4)}$$

Mössbauer Effect in Lattice Dynamics. Yi-Long Chen and De-Ping Yang

ISBN: 978-3-527-40712-5

into (A.3) then gives

$$\frac{\varepsilon(y)}{f_s} = 1 - \frac{\xi}{\pi} \frac{e^{-t_a/2}}{\xi^2 + y^2 + 1} \times \int_{-\pi}^{\pi} \frac{\exp[-(t_a/2)\cos\varphi]\, d\varphi}{1 + \dfrac{\xi^2 + y^2 - 1}{\xi^2 + y^2 + 1}\cos\varphi + \dfrac{2y}{\xi^2 + y^2 + 1}\sin\varphi}. \tag{A.5}$$

Let us introduce two more variables, ρ and θ, so that

$$\left.\begin{aligned} \frac{\xi^2 + y^2 - 1}{\xi^2 + y^2 + 1} &= \rho\cos\theta \\ \frac{2y}{\xi^2 + y^2 + 1} &= \rho\sin\theta \end{aligned}\right. \quad \left(0 < \theta < \frac{\pi}{2}\right), \tag{A.6}$$

and

$$\rho^2 = 1 - \frac{4\xi^2}{(\xi^2 + y^2 + 1)^2},$$

$$\theta = \cos^{-1}\left[\frac{\xi^2 + y^2 - 1}{\sqrt{(\xi^2 + y^2 + 1)^2 - 4\xi^2}}\right].$$

Therefore, the expression in (A.5) is simplified to

$$\frac{\varepsilon(y)}{f_s} = 1 - \frac{e^{-t_a/2}\sqrt{1-\rho^2}}{2\pi} \int_{-\pi}^{\pi} \frac{\exp[-(t_a/2)\cos\varphi]}{1 + \rho\cos(\varphi - \theta)} d\varphi. \tag{A.7}$$

In this integral, the denominator is a periodic function of φ with period 2π and can be expanded into a Fourier series

$$\frac{1}{1 + \rho\cos(\varphi - \theta)} = \sum_{n=-\infty}^{\infty} a_n(\rho, \theta) e^{in\varphi} \tag{A.8}$$

where

$$a_n(\rho, \theta) = \frac{e^{-in\theta}}{\sqrt{1-\rho^2}} \left(\frac{\sqrt{1-\rho^2} - 1}{\rho}\right)^{|n|}. \tag{A.9}$$

Using this Fourier series, (A.7) becomes

$$\frac{\varepsilon(y)}{f_s} = 1 - \frac{e^{-t_a/2}}{2\pi} \sum_{n=-\infty}^{\infty} e^{-in\theta} \left(\frac{\sqrt{1-\rho^2} - 1}{\rho}\right)^{|n|} \int_{-\pi}^{\pi} e^{-(t_a/2)\cos\varphi + in\varphi}\, d\varphi. \tag{A.10}$$

The integral happens to match the definition of the modified Bessel function I_n, and the expression is then written as

$$\frac{\varepsilon(\gamma)}{f_s} = 1 - e^{-t_a/2}\left[\sum_{n=-\infty}^{\infty} e^{-in\theta}\left(\frac{\sqrt{1-\rho^2}-1}{\rho}\right)^{|n|} I_n\left(-\frac{t_a}{2}\right)\right]. \tag{A.11}$$

Using the properties of these Bessel functions, $I_n(x) = I_{-n}(x)$ and $I_n(-x) = (-1)^n I_n(x)$, and sorting out terms containing $+n$ and $-n$, we finally get

$$\begin{aligned}\frac{\varepsilon(\gamma)}{f_s} = 1 - e^{-t_a/2}\Bigg\{ I_0\left(\frac{t_a}{2}\right) + \sum_{n=1}^{\infty}\Bigg[&\left(-\frac{\sqrt{1-\rho^2}-1}{\rho}e^{i\theta}\right)^n \\ &+\left(-\frac{\sqrt{1-\rho^2}-1}{\rho}e^{-i\theta}\right)^n\Bigg] I_n\left(\frac{t_a}{2}\right)\Bigg\}.\end{aligned} \tag{A.12}$$

This expression was obtained by Capaccioli et al. (*Nucl. Instrum. Methods B* 101, 280 (1995)).

Now we continue to treat the above expression by using

$$-\frac{\sqrt{1-\rho^2}-1}{\rho} = e^{-\beta} \quad (\beta > 0) \tag{A.13}$$

where the condition $\beta > 0$ is to guarantee that the sum in (A.12) converges. Thus, the formula in (A.13) can be compactly written as

$$\frac{\varepsilon(\gamma)}{f_s} = 1 - e^{-t_a/2} I_0\left(\frac{t_a}{2}\right) - 2e^{-t_a/2}\sum_{n=1}^{\infty}(e^{-n\beta}\cos n\theta) I_n\left(\frac{t_a}{2}\right). \tag{A.14}$$

If $\xi = 1$, this formula reduces to the result obtained by Ruby and Hicks (*Rev. Sci. Instrum.* 33, 27–30 (1962)). When resonance absorption takes place, $v_r = c(E_0 - E'_0)/E_0$, $\gamma = 0$, and $\theta = 0$, we obtain the following analytical formula for the line shape:

$$\frac{\varepsilon(\gamma)}{f_s} = 1 - e^{-t_a/2} I_0\left(\frac{t_a}{2}\right) - 2e^{-t_a/2}\sum_{n=1}^{\infty}\left(\frac{\xi-1}{\xi+1}\right)^n I_n\left(\frac{t_a}{2}\right), \tag{A.15}$$

which is explicitly expressed in terms of Γ_s and Γ_a through their ratio ξ. In the special case of $\Gamma_s = \Gamma_a$, i.e., $\xi = 1$, the above formula further simplifies to

$$\varepsilon(v_r) = f_s\left[1 - e^{-t_a/2} I_0\left(\frac{t_a}{2}\right)\right], \tag{A.16}$$

which is a well-known result.

Now we turn to the consideration of the area under an absorption line, which is given by the integral

$$A(t_a) = \int_{-\infty}^{\infty} \varepsilon(S)\,dS. \tag{A.17}$$

The following is based on the derivation by Williams and Brooks (*Nucl. Instrum. Methods* 128, 363 (1975)). Substituting (A.3) into (A.17) yields

$$\begin{aligned} A(t_a) &= \int_{-\infty}^{\infty} \frac{f_s \xi}{\pi} \left\{ \int_{-\infty}^{\infty} \frac{1 - \exp[-t_a/(1+x^2)]}{\xi^2 + (x+y)^2} dx \right\} dS \\ &= f_s \frac{\Gamma_a}{2} \int_{-\infty}^{\infty} \left\{ 1 - \exp\left[-\frac{t_a}{1+x^2} \right] \right\} dx. \end{aligned} \tag{A.18}$$

Using the simple integral relation

$$1 - e^{-\alpha t_a} = \alpha \int_0^{t_a} e^{-\alpha t}\,dt \tag{A.19}$$

and the relation in (A.4), we may rewrite (A.18) as

$$\begin{aligned} A(t_a) &= f_s \frac{\Gamma_a}{4} \int_0^{t_a} dt \int_{-\pi}^{\pi} d\varphi \exp\left[-\frac{t}{2}(1 + \cos\varphi) \right] \\ &= f_s \frac{\Gamma_a}{2} \int_0^{t_a} dt \exp\left(-\frac{t}{2} \right) \int_0^{\pi} d\varphi \exp\left(-\frac{t}{2} \cos\varphi \right). \end{aligned} \tag{A.20}$$

Using the definition of the zeroth-order modified Bessel function of imaginary argument,

$$J_0(i\zeta) = I_0(\zeta) = \frac{1}{\pi} \int_0^{\pi} \exp(-\zeta \cos\phi)\,d\phi \tag{A.21}$$

where ζ is real, Eq. (A.20) will be reduced to

$$A(t_a) = f_s \frac{\Gamma_a}{2} \pi \int_0^{t_a} dt \exp\left(-\frac{t}{2} \right) I_0\left(\frac{t}{2} \right). \tag{A.22}$$

It can be shown that

$$\int_0^{\eta} dx \exp(-x) I_0(x) = \eta \exp(-\eta)[I_0(\eta) + I_1(\eta)]. \tag{A.23}$$

Finally, $A(t_a)$ is written as

$$A(t_a) = f_s \Gamma_a \pi \frac{t_a}{2} \exp\left(-\frac{t_a}{2}\right)\left[I_0\left(\frac{t_a}{2}\right) + I_1\left(\frac{t_a}{2}\right)\right]. \tag{A.24}$$

The expression for the area $A(t_a)$ takes the above simple form. Note that $A(t_a)$ depends on neither Γ_a nor Γ_s, which is a major difference between $A(t_a)$ and $\varepsilon(v)$.

Appendix B
Eigenstate Calculations in Combined Interactions

B.1
Electric Quatrupole Perturbation

The secular determinant equation (2.62) of the combined interactions is

$$\lambda^4 + p\lambda^2 + q\lambda + r = 0, \tag{B.1}$$

where p, q, and r are given in (2.63). According to Ref. [60] in Chapter 2, the roots of this equation are four sums

$$\pm\sqrt{y_1} \pm \sqrt{y_2} \mp \sqrt{y_3} \tag{B.2}$$

with the signs of the square roots chosen so that

$$\sqrt{y_1}\sqrt{y_2}\sqrt{y_3} = -\frac{q}{8} \tag{B.3}$$

where y_1, y_2, and y_3 are the roots of the cubic equation

$$y^3 + \frac{p}{2}y^2 + \frac{p^2 - 4r}{16}y - \frac{q}{64} = 0. \tag{B.4}$$

In the case of pure magnetic interaction (i.e., $R = 0$), $p = -10$, $q = 0$, and $r = 9$ (see (2.63)); hence the roots of (B.4) are

$$y_1 = 0, \quad y_2 = 4, \quad \text{and} \quad y_3 = 1, \tag{B.5}$$

which satisfy Eq. (B.3).

When the magnetic hyperfine interaction is dominant and electric quadrupole interaction is present as a perturbation, i.e., a case with $R \ll 1$, we can expand (2.65) into a Taylor series:

$$y_i(R) = y_i(0) + y_i'(0)R + \frac{y_i''(0)}{2!}R^2 + \cdots \tag{B.6}$$

The zeroth approximation term is given by (B.5) and the first derivative $y'_i(0)$ can be shown to be zero. Letting $\frac{y''_i(0)}{2!} = k_i^2$, we obtain the following solutions to the second-order approximation:

$$\begin{aligned} y_1 &= 0 + k_1^2 R^2, \\ y_2 &= 4 + k_2^2 R^2, \\ y_3 &= 1 + k_3^2 R^2. \end{aligned} \tag{B.7}$$

Because $k_2^2 R^2 \ll 4$ and $k_3^2 R^2 \ll 1$, which we will neglect, such solutions y_i immediately satisfy Eq. (B.3), provided

$$k_1 = \frac{1}{2}(3\cos^2\theta - 1 + \eta \sin^2\theta \cos 2\phi). \tag{B.8}$$

Finally, we obtain the reduced eigenvalues

$$\begin{aligned} \lambda_1 &= -3 + k_1 R, \\ \lambda_2 &= -1 - k_1 R, \\ \lambda_3 &= +1 - k_1 R, \\ \lambda_4 &= +3 + k_1 R. \end{aligned} \tag{B.9}$$

B.2 The Coefficients a_{i,m_g} and b_{j,m_e}

For the eigenstates with $I = 3/2$, the coefficients b_{j,m_e} in the $|Im\rangle$ representation (see Eq. (2.60)) are found relatively easily by a standard procedure, namely by solving

$$\sum_m [\langle Im'_e | \mathscr{H}_{QM} | Im_e \rangle - \delta_{m'_e m_e} E_e \lambda_j] b_{j,m_e} = 0. \tag{B.10}$$

Using the matrix elements of (2.57), the above equation can be explicitly written as four simultaneous linear equations:

$$(R - 3\cos\theta - \lambda_j) b_{j,3/2} - \sqrt{3}\sin\theta e^{-i\phi} b_{j,1/2} + \frac{\eta R}{\sqrt{3}} b_{j,-1/2} = 0, \tag{B.11a}$$

$$\begin{aligned} &-\sqrt{3}\sin\theta e^{i\phi} b_{j,3/2} - (R + \cos\theta + \lambda_j) b_{j,1/2} \\ &\qquad - 2\sin\theta e^{-i\phi} b_{j,-1/2} + \frac{\eta R}{\sqrt{3}} b_{j,-3/2} = 0, \end{aligned} \tag{B.11b}$$

$$\frac{\eta R}{\sqrt{3}} b_{j,3/2} - 2\sin\theta e^{i\phi} b_{j,1/2} - (R - \cos\theta + \lambda_j) b_{j,-1/2}$$
$$- \sqrt{3}\sin\theta e^{-i\phi} b_{j,-3/2} = 0, \tag{B.11c}$$

$$\frac{\eta R}{\sqrt{3}} b_{j,1/2} - \sqrt{3}\sin\theta e^{i\phi} b_{j,-1/2} + (R + 3\cos\theta - \lambda_j) b_{j,-3/2} = 0. \tag{B.11d}$$

From (B.11d) and (B.11a) it follows that

$$b_{j,-3/2} = \frac{\sqrt{3}\sin\theta e^{i\phi} b_{j,-1/2} - \frac{\eta R}{\sqrt{3}} b_{j,1/2}}{R + 3\cos\theta - \lambda_j}, \tag{B.12}$$

$$b_{j,3/2} = \frac{\sqrt{3}\sin\theta e^{-i\phi} b_{j,1/2} - \frac{\eta R}{\sqrt{3}} b_{j,-1/2}}{R - 3\cos\theta - \lambda_j}. \tag{B.13}$$

Substituting (B.12) and (B.13) in (B.11c), we find

$$b_{j,1/2} = \frac{1}{3} \frac{\eta^2 R^2 (R + 3\cos\theta - \lambda_j) - 3\gamma + 9\sin^2\theta (R - 3\cos\theta - \lambda_j)}{2\eta R (R - \lambda)\sin\theta e^{-i\phi} - 2\sin\theta e^{i\phi} (R + 3\cos\theta - \lambda_j)(R - 3\cos\theta - \lambda_j)} b_{j,-1/2} \tag{B.14}$$

where $\gamma = (R + 3\cos\theta - \lambda_j)(R - 3\cos\theta - \lambda_j)(-R + \cos\theta - \lambda_j)$. Recall the normalizing equation

$$\sum_{m_e=-3/2}^{3/2} |b_{j,m_e}|^2 = 1, \quad j = 1, 2, 3, 4, \ldots \tag{B.15}$$

Using the last four equations, we can find four normalized coefficients b_{j,m_e} for each index j.

Similarly, for the eigenstates with $I = 1/2$, the normalized coefficients a_{i,m_g} in (2.61) can be found for $E(\frac{1}{2}, 1) = -E_g$ and $E(\frac{1}{2}, 2) = E_g$, respectively:

$$a_{1,1/2} = \cos\left(\frac{\theta}{2}\right), \quad a_{1,-1/2} = \sin\left(\frac{\theta}{2}\right) e^{i\phi},$$
$$a_{2,1/2} = -\sin\left(\frac{\theta}{2}\right), \quad a_{2,-1/2} = \cos\left(\frac{\theta}{2}\right) e^{i\phi}. \tag{B.16}$$

Appendix C
Force Constant Matrices ($-\Phi$) in fcc and bcc Lattices

First nearest neighbors:

Lattice	Atom position	General matrix for force constant ($-\Phi$)	Central force matrix
fcc	$\pm\frac{a}{2}(1\,1\,0)$	$\begin{bmatrix} \alpha_1 & \gamma_1 & 0 \\ \gamma_1 & \alpha_1 & 0 \\ 0 & 0 & \beta_1 \end{bmatrix}$	$\frac{A}{2}\begin{bmatrix} 1 & 1 & 0 \\ 1 & 1 & 0 \\ 0 & 0 & 0 \end{bmatrix}$
	$\pm\frac{a}{2}(\bar{1}\,1\,0)$	$\begin{bmatrix} \alpha_1 & -\gamma_1 & 0 \\ -\gamma_1 & \alpha_1 & 0 \\ 0 & 0 & \beta_1 \end{bmatrix}$	$\frac{A}{2}\begin{bmatrix} 1 & -1 & 0 \\ -1 & 1 & 0 \\ 0 & 0 & 0 \end{bmatrix}$
	$\pm\frac{a}{2}(1\,0\,1)$	$\begin{bmatrix} \alpha_1 & 0 & \gamma_1 \\ 0 & \beta_1 & 0 \\ \gamma_1 & 0 & \alpha_1 \end{bmatrix}$	$\frac{A}{2}\begin{bmatrix} 1 & 0 & 1 \\ 0 & 0 & 0 \\ 1 & 0 & 1 \end{bmatrix}$
	$\pm\frac{a}{2}(\bar{1}\,0\,1)$	$\begin{bmatrix} \alpha_1 & 0 & -\gamma_1 \\ 0 & \beta_1 & 0 \\ -\gamma_1 & 0 & \alpha_1 \end{bmatrix}$	$\frac{A}{2}\begin{bmatrix} 1 & 0 & -1 \\ 0 & 0 & 0 \\ -1 & 0 & 1 \end{bmatrix}$
	$\pm\frac{a}{2}(0\,1\,1)$	$\begin{bmatrix} \beta_1 & 0 & 0 \\ 0 & \alpha_1 & \gamma_1 \\ 0 & \gamma_1 & \alpha_1 \end{bmatrix}$	$\frac{A}{2}\begin{bmatrix} 0 & 0 & 0 \\ 0 & 1 & 1 \\ 0 & 1 & 1 \end{bmatrix}$
	$\pm\frac{a}{2}(0\,\bar{1}\,1)$	$\begin{bmatrix} \beta_1 & 0 & 0 \\ 0 & \alpha_1 & -\gamma_1 \\ 0 & -\gamma_1 & \alpha_1 \end{bmatrix}$	$\frac{A}{2}\begin{bmatrix} 0 & 0 & 0 \\ 0 & 1 & -1 \\ 0 & -1 & 1 \end{bmatrix}$
bcc	$\pm\frac{a}{2}(1\,1\,1)$	$\begin{bmatrix} \alpha_1 & \beta_1 & \beta_1 \\ \beta_1 & \alpha_1 & \beta_1 \\ \beta_1 & \beta_1 & \alpha_1 \end{bmatrix}$	$\frac{A}{3}\begin{bmatrix} 1 & 1 & 1 \\ 1 & 1 & 1 \\ 1 & 1 & 1 \end{bmatrix}$
	$\pm\frac{a}{2}(\bar{1}\,1\,1)$	$\begin{bmatrix} \alpha_1 & -\beta_1 & -\beta_1 \\ -\beta_1 & \alpha_1 & \beta_1 \\ -\beta_1 & \beta_1 & \alpha_1 \end{bmatrix}$	$\frac{A}{3}\begin{bmatrix} 1 & -1 & -1 \\ -1 & 1 & 1 \\ -1 & 1 & 1 \end{bmatrix}$
	$\pm\frac{a}{2}(1\,\bar{1}\,1)$	$\begin{bmatrix} \alpha_1 & -\beta_1 & \beta_1 \\ -\beta_1 & \alpha_1 & -\beta_1 \\ \beta_1 & -\beta_1 & \alpha_1 \end{bmatrix}$	$\frac{A}{3}\begin{bmatrix} 1 & -1 & 1 \\ -1 & 1 & -1 \\ 1 & -1 & 1 \end{bmatrix}$
	$\pm\frac{a}{2}(1\,1\,\bar{1})$	$\begin{bmatrix} \alpha_1 & \beta_1 & -\beta_1 \\ \beta_1 & \alpha_1 & -\beta_1 \\ -\beta_1 & -\beta_1 & \alpha_1 \end{bmatrix}$	$\frac{A}{3}\begin{bmatrix} 1 & 1 & -1 \\ 1 & 1 & -1 \\ -1 & -1 & 1 \end{bmatrix}$

Second nearest neighbors:

Lattice	Atom position	General matrix for force constant ($-\mathbf{\Phi}$)	Central force matrix
fcc and bcc	$\pm\frac{a}{2}(2\ 0\ 0)$	$\begin{bmatrix} \alpha_2 & 0 & 0 \\ 0 & \beta_2 & 0 \\ 0 & 0 & \beta_2 \end{bmatrix}$	$A\begin{bmatrix} 1 & 0 & 0 \\ 0 & 0 & 0 \\ 0 & 0 & 0 \end{bmatrix}$
	$\pm\frac{a}{2}(0\ 2\ 0)$	$\begin{bmatrix} \beta_2 & 0 & 0 \\ 0 & \alpha_2 & 0 \\ 0 & 0 & \beta_2 \end{bmatrix}$	$A\begin{bmatrix} 0 & 0 & 0 \\ 0 & 1 & 0 \\ 0 & 0 & 0 \end{bmatrix}$
	$\pm\frac{a}{2}(0\ 0\ 2)$	$\begin{bmatrix} \beta_2 & 0 & 0 \\ 0 & \beta_2 & 0 \\ 0 & 0 & \alpha_2 \end{bmatrix}$	$A\begin{bmatrix} 0 & 0 & 0 \\ 0 & 0 & 0 \\ 0 & 0 & 1 \end{bmatrix}$

The atom positions up to the fifth neighbors can be found in Appendix E. The $\mathbf{\Phi}$-matrix is basically determined by the particular symmetry of the lattice. For example, the $\mathbf{\Phi}$-matrices for atoms at $(a/2)\{2\ 2\ 0\}$ and $(a/2)\{1\ 1\ 0\}$ share exactly the same form. The $\mathbf{\Phi}$-matrices for the third nearest neighbors in fcc and the fourth nearest neighbors in bcc lattices are described by four independent constants, α, β, γ, and δ.

Appendix D
Nearest Neighbors Around a Substitutional Impurity

Assuming that a substitutional impurity atom is located at the lattice coordinate origin (0 0 0) in an fcc or a bcc lattice, its nearest neighbor atom positions are given in the following table.

Lattice	Nearest neighbors	Atom positions $(a/2)(x\ y\ z)$	Distance from impurity
fcc	First nearest neighbors (12)	$\pm(1\,1\,0), \pm(\bar{1}\,1\,0), \pm(1\,0\,1),$ $\pm(\bar{1}\,0\,1), \pm(0\,1\,1), \pm(0\,\bar{1}\,1)$	$\frac{\sqrt{2}}{2}a$
	Second nearest neighbors (6)	$\pm(2\,0\,0), \pm(0\,2\,0), \pm(0\,0\,2)$	a
	Third nearest neighbors (24)	$\pm(2\,1\,1), \pm(1\,2\,1), \pm(1\,1\,2),$ $\pm(\bar{2}\,1\,1), \pm(\bar{1}\,2\,1), \pm(\bar{1}\,1\,2),$ $\pm(2\,\bar{1}\,1), \pm(1\,\bar{2}\,1), \pm(1\,\bar{1}\,2),$ $\pm(2\,1\,\bar{1}), \pm(1\,2\,\bar{1}), \pm(1\,1\,\bar{2})$	$\frac{\sqrt{6}}{2}a$
	Fourth nearest neighbors (12)	$\pm(2\,2\,0), \pm(\bar{2}\,2\,0), \pm(2\,0\,2),$ $\pm(\bar{2}\,0\,2), \pm(0\,2\,2), \pm(0\,\bar{2}\,2)$	$\sqrt{2}a$
	Fifth nearest neighbors (24)	$\pm(0\,1\,3), \pm(0\,3\,1), \pm(1\,0\,3),$ $\pm(1\,3\,0), \pm(3\,0\,1), \pm(3\,1\,0),$ $\pm(0\,\bar{1}\,3), \pm(0\,3\,\bar{1}), \pm(\bar{1}\,0\,3),$ $\pm(\bar{1}\,3\,0), \pm(3\,0\,\bar{1}), \pm(3\,\bar{1}\,0)$	$\frac{\sqrt{10}}{2}a$
bcc	First nearest neighbors (8)	$\pm(1\,1\,1), \pm(\bar{1}\,1\,1),$ $\pm(1\,\bar{1}\,1), \pm(1\,1\,\bar{1})$	$\frac{\sqrt{3}}{2}a$
	Second nearest neighbors (6)	$\pm(2\,0\,0), \pm(0\,2\,0), \pm(0\,0\,2)$	a
	Third nearest neighbors (12)	$\pm(2\,2\,0), \pm(\bar{2}\,2\,0), \pm(2\,0\,2),$ $\pm(\bar{2}\,0\,2), \pm(0\,2\,2), \pm(0\,\bar{2}\,2)$	$\sqrt{2}a$
	Fourth nearest neighbors (24)	$\pm(3\,1\,1), \pm(1\,3\,1), \pm(1\,1\,3),$ $\pm(\bar{3}\,1\,1), \pm(\bar{1}\,3\,1), \pm(\bar{1}\,1\,3),$ $\pm(3\,\bar{1}\,1), \pm(1\,\bar{3}\,1), \pm(1\,\bar{1}\,3),$ $\pm(3\,1\,\bar{1}), \pm(1\,3\,\bar{1}), \pm(1\,1\,\bar{3})$	$\frac{\sqrt{11}}{2}a$
	Fifth nearest neighbors (8)	$\pm(2\,2\,2), \pm(\bar{2}\,2\,2),$ $\pm(2\,\bar{2}\,2), \pm(2\,2\,\bar{2})$	$\sqrt{3}a$

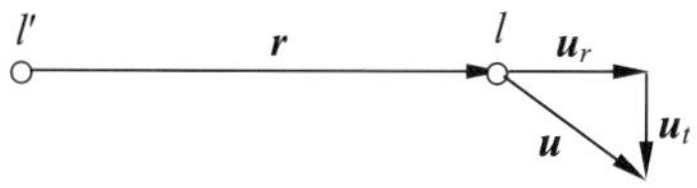

Fig. E.1 Two atoms l' and l are separated by a distance r. Atom l' is fixed while atom l is displaced from its equilibrium position by $\boldsymbol{u}$.

Appendix E
Force Constants for Central Forces

Let two atoms l' and l in a solid be separated by a distance r. One, say atom l', is fixed at (0 0 0), and the other is displaced from its equilibrium position by $\boldsymbol{u}$ (Fig. E.1).

Assume that the force $\boldsymbol{F}$ between this pair of atoms is axially symmetric. The force constant matrix involves two constants, namely a bond-stretching constant A and a bond-bending constant B. In terms of the radial and tangential components, $\boldsymbol{u}_\mathrm{r}$ and $\boldsymbol{u}_\mathrm{t}$, of the displacement of atom l, the force between l and l' is

$$\boldsymbol{F} = A\boldsymbol{u}_\mathrm{r} + B\boldsymbol{u}_\mathrm{t} = \frac{A}{r^2}\boldsymbol{r}(\boldsymbol{r} \cdot \boldsymbol{u}) - \frac{B}{r^2}\boldsymbol{r} \times (\boldsymbol{r} \times \boldsymbol{u}) \tag{E.1}$$

where $\boldsymbol{r}$ is the radius vector from the atom l' to atom l. The constants A and B are related to the spherically symmetric potential $V(r)$ between the two atoms

$$A = -\frac{\partial^2 V}{\partial r^2}, \quad B = -\frac{1}{r}\frac{\partial V}{\partial r}. \tag{E.2}$$

In order to deduce the force constant matrix, it is necessary to write (E.1) also in the matrix notation. The vector $\boldsymbol{r}$ consists of three components, r_1, r_2, and r_3; hence

$$\boldsymbol{r} = \begin{bmatrix} r_1 \\ r_2 \\ r_3 \end{bmatrix},$$

$$\boldsymbol{r} \quad \boldsymbol{u} = \boldsymbol{r}^\mathrm{T}\boldsymbol{u},$$

$$\boldsymbol{r} \times \boldsymbol{u} = \boldsymbol{R}\boldsymbol{u}$$

where

$$\boldsymbol{R} = \begin{bmatrix} 0 & -r_3 & r_2 \\ r_3 & 0 & -r_1 \\ -r_2 & r_1 & 0 \end{bmatrix}.$$

In the matrix form, (E.1) is written as

$$\boldsymbol{F} = \frac{A}{r^2}\boldsymbol{r}\boldsymbol{r}^{\mathrm{T}}\boldsymbol{u} + \frac{B}{r^2}\boldsymbol{R}\boldsymbol{R}^{\mathrm{T}}\boldsymbol{u}.$$

According to Eq. (4.16), the force constant matrix is

$$\boldsymbol{\Phi}(l,0) = -\frac{1}{r^2}(A\boldsymbol{r}\boldsymbol{r}^{\mathrm{T}} + B\boldsymbol{R}\boldsymbol{R}^{\mathrm{T}}) \tag{E.3}$$

and explicitly

$$\boldsymbol{\Phi}(l,0) = \frac{V'' - V'/r}{r^2}\begin{bmatrix} r_1^2 & r_1 r_2 & r_1 r_3 \\ r_2 r_1 & r_2^2 & r_2 r_3 \\ r_3 r_1 & r_3 r_2 & r_3^2 \end{bmatrix} + \frac{V'}{r}\begin{bmatrix} 1 & 0 & 0 \\ 0 & 1 & 0 \\ 0 & 0 & 1 \end{bmatrix}. \tag{E.4}$$

As an example, we take the fcc lattice. Suppose the atom l is at position (1 1 0) and atom l' at (0 0 0), then $r = (\sqrt{2}/2)a$, $r_1 = r_2 = a/2$, and $r_3 = 0$, and we have

$$\boldsymbol{\Phi}(110,0) = \frac{1}{2}\begin{bmatrix} V''(r) + V'(r)/r & V''(r) - V'(r)/r & 0 \\ V''(r) - V'(r)/r & V''(r) + V'(r)/r & 0 \\ 0 & 0 & 2V'(r)/r \end{bmatrix}. \tag{E.5}$$

A special case of $B = 0$ is the so-called central force approximation, which leaves the force along the bond direction. In this case

$$\boldsymbol{\Phi}(110,0) = \frac{V''(r)}{2}\begin{bmatrix} 1 & 1 & 0 \\ 1 & 1 & 0 \\ 0 & 0 & 0 \end{bmatrix}. \tag{E.6}$$

Because of

$$\Phi_{xx}(0,0) = -\sum_{l\neq 0}\Phi_{xx}(l,0) = -4V''(r), \tag{E.7}$$

we obtain

$$\Phi_{xx}(0,0) = -8\Phi_{xx}(110,0) = -8\Phi_{xy}(110,0). \tag{E.8}$$

Substitution of atom l' by an impurity causes some changes in the force constant Φ. The following relation is obviously valid:

$$\sum_{l'}\Delta\Phi_{\alpha\beta}(l,l') = 0 \tag{E.9}$$

which leads to

$$\Delta\Phi_{\alpha\beta}(110, 110) = -\sum_{l' \neq 110} \Delta\Phi_{\alpha\beta}(110, l') = -\Delta\Phi_{\alpha\beta}(110, 0). \qquad (E.10)$$

Here, we assume that, among the nearest neighbors of atom l at (110), only the **Φ**-matrix for this particular pair of atoms (l and l') changes. Inserting (E.10) into (8.11) gives

$$U_{\alpha\beta}(110, 110) = -U_{\alpha\beta}(110, 0). \qquad (E.11)$$

Based on (E.8), the changes in the force constants along the x-axis will be

$$U_{xx}(110, 0) = \frac{1}{2}[V''(r)_{\text{pure}} - V''(r)_{\text{def.}}] = \frac{1}{8}[\Phi'_{xx}(0, 0) - \Phi_{xx}(0, 0)]. \qquad (E.12)$$

Therefore, Eq. (8.11) can be written as

$$8U_{xx}(110, 0) = 8U_{xy}(110, 0) = U_{xx}(0, 0) - \eta M_0\omega^2. \qquad (E.13)$$

The relative change in the **Φ**-matrix is represented by the following parameter λ:

$$\lambda = \frac{\Phi_{xx}(0, 0) - \Phi'_{xx}(0, 0)}{\Phi_{xx}(0, 0)} = 1 - \frac{\Phi'_{xx}(0, 0)}{\Phi_{xx}(0, 0)}. \qquad (E.14)$$

Appendix F
Lattice Green's Function

F.1
Definition of Green's Function

In the harmonic approximation, the Hamiltonian for an ordered crystal can be written as

$$\mathscr{H}_0 = \frac{1}{2}\sum_{l,\alpha} \frac{p_\alpha^2(l)}{M_0} + \frac{1}{2}\sum_{l,\alpha,l',\beta} \Phi_{\alpha\beta}(l\, l')u_\alpha(l)u_\beta(l'). \qquad (F.1)$$

To evaluate the mean square displacement $\langle \mathbf{u}^2 \rangle$, we use the displacement–displacement double-time Green's function defined by

$$G_{\alpha\beta}(ll', \omega) = \int_{-\infty}^{\infty} dt e^{i\omega t} \langle\langle u_\alpha(l, t); u_\beta(l', 0)\rangle\rangle \qquad (F.2)$$

where

$$\langle\langle \boldsymbol{A}(t); \boldsymbol{B}(0)\rangle\rangle = \mp\frac{i}{\hbar}\theta(\pm t)\langle[\boldsymbol{A}(t), \boldsymbol{B}(0)]\rangle \tag{F.3}$$

and

$$\boldsymbol{A}(t) = \mathrm{e}^{(i/\hbar)\mathcal{H}_0 t}\boldsymbol{A}\mathrm{e}^{-(i/\hbar)\mathcal{H}_0 t}. \tag{F.4}$$

Here the upper and lower signs give, respectively, the retarded and advanced Green's function with $\theta(t)$ being the Heaviside step function. $\boldsymbol{A}$ and $\boldsymbol{B}$ are arbitrary operators. Due to (F.3), ω may be replaced in (F.2) by a complex frequency $z = \omega + i\varepsilon$ with Im $z > 0$ (Im $z < 0$) for the retarded (advanced) Green's function, i.e., the retarded (advanced) Green's function is analytic in the upper (lower) half of the complex frequency plane. For applications in physics, the Green's function is always taken as the retarded one. From both these functions a new Green's function can be constructed which is analytic in the entire complex frequency plane, except on the real axis. This new Green's function is related to a displacement–displacement correlation function

$$G_{\alpha\beta}(l\,l', z) = \frac{1}{2\pi\hbar}\int_{-\infty}^{\infty} \mathrm{d}\omega' \frac{1 - \mathrm{e}^{-\beta\hbar\omega'} J_{\alpha\beta}(l\,l', \omega')}{z - \omega'} \tag{F.5}$$

where

$$J_{\alpha\beta}(l\,l', \omega) = \int_{-\infty}^{\infty} \langle u_\alpha(l, t), u_\beta(l', 0)\rangle \mathrm{e}^{i\omega t}\,\mathrm{d}t. \tag{F.6}$$

Now we derive a Green's function equation of motion to determine the Green's function itself. Differentiating Eq. (F.3) with respect to time and taking into account the relation

$$i\hbar\frac{\mathrm{d}\boldsymbol{A}(t)}{\mathrm{d}t} = [\boldsymbol{A}(t), \mathcal{H}_0], \tag{F.7}$$

we arrive at

$$i\hbar\frac{\mathrm{d}}{\mathrm{d}t}\langle\langle \boldsymbol{A}(t); \boldsymbol{B}(0)\rangle\rangle = \delta(t)\langle[\boldsymbol{A}, \boldsymbol{B}]\rangle + \langle\langle[\boldsymbol{A}, \mathcal{H}_0]; \boldsymbol{B}\rangle\rangle. \tag{F.8}$$

It is evident that the Green's function in the time domain can be expressed in the form of Eq. (F.3):

$$G_{\alpha\beta}(l\,l',t) = \mp\frac{i}{\hbar}\theta(\pm t)\langle[u_\alpha(l,t),u_\beta(l',0)]\rangle. \tag{F.9}$$

Based on (F.8) and (F.9), we obtain

$$i\pi\frac{\mathrm{d}}{\mathrm{d}t}G_{\alpha\beta}(l\,l',t) = \delta(t)\langle[u_\alpha(l,t),u_\beta(l',0)]\rangle + \theta(t)\left\langle\left[\frac{\mathrm{d}u_\alpha(l,t)}{\mathrm{d}t},u_\beta(l',0)\right]\right\rangle, \tag{F.10}$$

where the first term on the right-hand side is zero because the quantity in the square brackets becomes zero at $t=0$, and the second term is

$$\theta(t)\langle[p_\alpha(l,t),u_\beta(l',0)]\rangle. \tag{F.11}$$

Differentiating again produces an equation entirely in G, namely

$$-M_0\frac{\mathrm{d}^2}{\mathrm{d}t^2}G_{\alpha\beta}(l\,l',t) = \delta_{\alpha\beta}\delta_{ll'}\delta(t) + \sum_{l'',\gamma}\Phi_{\alpha\gamma}(l,l'')G_{\gamma\beta}(l''l,t). \tag{F.12}$$

Equations (F.10) and (F.12) have been deduced by using the following four relations:

$$\dot{u}_\alpha(l) = -\frac{i}{\hbar}[u_\alpha(l),\mathscr{H}_0] = \frac{p_\alpha(l)}{M_0},$$

$$\dot{p}_\alpha(l) = -\frac{i}{\hbar}[p_\alpha(l),\mathscr{H}_0] = \sum_{l',\beta}\Phi_{\alpha\beta}(l,l')u_\beta(l',t),$$

$$[u_\alpha(l),p_\beta(l')] = i\hbar\delta_{ll'}\delta_{\alpha\beta},$$

$$[u_\alpha(l),u_\beta(l')] = [p_\alpha(l),p_\beta(l')] = 0.$$

Fourier transforming (F.12) gives the equation

$$\sum_{l'',\gamma}[M_0\omega^2\delta_{\alpha\gamma}\delta_{ll''} - \Phi_{\alpha\gamma}(l,l'')]G_{\gamma\beta}(l''l',\omega) = \delta_{\alpha\beta}\delta_{ll'}, \tag{F.13}$$

which can be abbreviated in the matrix notion (the Dyson equation) as

$$(\boldsymbol{M}_0\omega^2 - \boldsymbol{\Phi})\boldsymbol{G}(\omega) = \boldsymbol{I}, \tag{F.14}$$

or

$$\boldsymbol{G}(\omega) = (\boldsymbol{M}_0\omega^2 - \boldsymbol{\Phi})^{-1}.$$

The frequency-dependent Green's function satisfying Eq. (F.13) is

$$G_{\alpha\beta}(ll',\omega) = \frac{1}{NM_0}\sum_{k,j}\frac{e_\alpha(\mathbf{k}j)e^*_\beta(\mathbf{k}j)}{\omega^2 - \omega_j^2(\mathbf{k})}\mathrm{e}^{i\mathbf{k}\cdot(\mathbf{l}-\mathbf{l}')}. \tag{F.15}$$

In a compositionally disordered crystal, the atoms are generally assumed to be placed randomly at the sites of a regular lattice. In the harmonic approximation, the Hamiltonian of such a lattice is

$$\mathscr{H} = \sum_{l,\alpha}\frac{p_\alpha^2(l)}{2M_l} + \frac{1}{2}\sum_{l,l',\alpha,\beta}\Phi'_{\alpha\beta}(l\,l')u_\alpha(l)u_\beta(l'). \tag{F.16}$$

There is an equation similar to (F.13) for the Green's function G' in the disordered crystal:

$$\sum_{l'',\gamma}[M_l\omega^2\delta_{\alpha\gamma}\delta_{ll'} - \Phi'_{\alpha\gamma}(l,l'')]G'_{\gamma\beta}(l''l',\omega) = \delta_{\alpha\beta}\delta_{ll'}. \tag{F.17}$$

When the mass differs from site to site, it is more convenient to introduce the dynamical matrix $\boldsymbol{D}'$ with the elements

$$D'_{\alpha\gamma}(l,l'') = \frac{\Phi'_{\alpha\gamma}(l,l'')}{\sqrt{M_l M_{l''}}}. \tag{F.18}$$

Then, in the matrix notation, the equation (F.17) can be written as

$$(\boldsymbol{M}\omega^2 - \boldsymbol{\Phi}')\boldsymbol{G}'(\omega) = (\boldsymbol{I}\omega^2 - \boldsymbol{D}')\boldsymbol{M}^{1/2}\boldsymbol{G}'(\omega)\boldsymbol{M}^{1/2} = \boldsymbol{I}. \tag{F.19}$$

F.2
The Real and Imaginary Parts of *G*

The real part of $\boldsymbol{G}$ is symmetric and the imaginary part is antisymmetric with respect to ω:

$$\begin{aligned}\mathrm{Re}\{\boldsymbol{G}(\omega)\} &= \mathrm{Re}\{\boldsymbol{G}(-\omega)\},\\ \mathrm{Im}\{\boldsymbol{G}(\omega)\} &= -\mathrm{Im}\{\boldsymbol{G}(-\omega)\}.\end{aligned} \tag{F.19}$$

Using the correlation function (F.6) and the identity

$$\frac{1}{x \pm i\varepsilon} = \frac{x}{x^2+\varepsilon^2} - i\frac{\varepsilon}{x^2+\varepsilon} = p\frac{1}{x} \mp i\pi\delta(x) \tag{E.20}$$

(where p indicates the Cauchy principal value and $\varepsilon \to +0$), the discontinuity of the Green's function across the real axis is

$$\begin{aligned}\mathbf{G}(\omega + i\varepsilon) - \mathbf{G}(\omega - i\varepsilon) &= 2i \operatorname{Im} \mathbf{G}(\omega + i\varepsilon)\\ &= \frac{1}{2\pi\hbar}\int_{-\infty}^{\infty}\left[\frac{(1-e^{\beta\hbar\omega'})\mathbf{S}(\omega')}{\omega-\omega'+i\varepsilon} - \frac{(1-e^{\beta\hbar\omega'})\mathbf{S}(\omega')}{\omega-\omega'-i\varepsilon}\right]d\omega'\\ &= -\frac{i}{\hbar}(1-e^{\beta\hbar\omega})\mathbf{S}(\omega), \quad \varepsilon \to +0. \end{aligned} \tag{F.21}$$

Making use of the symmetry relations, we can write

$$\begin{aligned}\operatorname{Re} \mathbf{G}(\omega) &= \frac{1}{2}[\mathbf{G}(\omega+i\varepsilon) + \mathbf{G}(\omega - i\varepsilon)] = \frac{1}{\pi}p\int_{-\infty}^{\infty}\frac{d\omega'}{\omega'-\omega}\operatorname{Im}\mathbf{G}(\omega')\\ &= \frac{1}{\pi}p\int_{0}^{\infty}\frac{\operatorname{Im}\mathbf{G}(\omega')}{\omega'^2-\omega^2}d\omega'^2,\end{aligned} \tag{F.22}$$

and similarly

$$\operatorname{Im}\mathbf{G}(\omega) = -\frac{2\omega}{\pi}p\int_0^{\infty}\frac{\operatorname{Re}\mathbf{G}(\omega')}{\omega'^2-\omega^2}d\omega', \tag{F.23}$$

which are known as the Kramers–Kronig relations.

Finally, we consider the behavior of $\operatorname{Im}\mathbf{G}'(\omega)$ near a local mode ω_L. Replacing ω by $\omega + i\varepsilon$ in (8.27) leads to

$$\operatorname{Im} G'_{\alpha\alpha}(00, \omega + i\varepsilon) = \operatorname{Im}\frac{G_{\alpha\alpha}(00,\omega+i\varepsilon)}{1-\eta M_0(\omega+i\varepsilon)^2 G_{\alpha\alpha}(00,\omega+i\varepsilon)}. \tag{F.24}$$

The denominator may be expanded around $\omega = \omega_L$:

$$\begin{aligned}1-\eta M_0\omega^2 G_{\alpha\alpha}(00,\omega+i\varepsilon) &= [1-\eta M_0\omega_L^2 G_{\alpha\alpha}(00,\omega_L+i\varepsilon)]\\ &+\frac{d}{d\omega}[1-\eta M_0(\omega+i\varepsilon)^2G_{\alpha\alpha}(00,\omega+i\varepsilon)](\omega+i\varepsilon-\omega_L)+\cdots\end{aligned} \tag{F.25}$$

where the first term on the right-hand side is zero. Taking account of (F.21), we rewrite (F.24) as

$$\operatorname{Im} G'_{\alpha\alpha}(00,\omega+i\varepsilon) = \pi\frac{G_{\alpha\alpha}(00,\omega)}{\dfrac{d}{d\omega}[\eta M_0\omega^2 G_{\alpha\alpha}(00,\omega)]}\delta(\omega-\omega_L). \tag{F.26}$$

F.3
Symmetry Properties of the G-Matrices

The expression in (F.14) indicates that, for a perfect crystal, the ***G***-matrices and **Φ**-matrices have the same symmetry properties. We take the fcc lattice again as an example. The ***G***-matrices between the central atom and its first and second nearest neighbors are given in the following table.

Nearest neighbors	Atom positions	Green's function matrices
First nearest neighbors	$\pm\frac{a}{2}(1\ 1\ 0)$	$\begin{bmatrix} g_1 & g_2 & 0 \\ g_2 & g_1 & 0 \\ 0 & 0 & g_3 \end{bmatrix}$
	$\pm\frac{a}{2}(\bar{1}\ 1\ 0)$	$\begin{bmatrix} g_1 & -g_2 & 0 \\ -g_2 & g_1 & 0 \\ 0 & 0 & g_3 \end{bmatrix}$
	$\pm\frac{a}{2}(1\ 0\ 1)$	$\begin{bmatrix} g_1 & 0 & g_2 \\ 0 & g_3 & 0 \\ g_2 & 0 & g_1 \end{bmatrix}$
	$\pm\frac{a}{2}(\bar{1}\ 0\ 1)$	$\begin{bmatrix} g_1 & 0 & -g_2 \\ 0 & g_3 & 0 \\ -g_2 & 0 & g_1 \end{bmatrix}$
	$\pm\frac{a}{2}(0\ 1\ 1)$	$\begin{bmatrix} g_3 & 0 & 0 \\ 0 & g_1 & g_2 \\ 0 & g_2 & g_1 \end{bmatrix}$
	$\pm\frac{a}{2}(0\ \bar{1}\ 1)$	$\begin{bmatrix} g_3 & 0 & 0 \\ 0 & g_1 & -g_2 \\ 0 & -g_2 & g_1 \end{bmatrix}$
Second nearest neighbors	$\pm\frac{a}{2}(2\ 0\ 0)$	$\begin{bmatrix} g_4 & 0 & 0 \\ 0 & g_5 & 0 \\ 0 & 0 & g_5 \end{bmatrix}$
	$\pm\frac{a}{2}(0\ 2\ 0)$	$\begin{bmatrix} g_5 & 0 & 0 \\ 0 & g_4 & 0 \\ 0 & 0 & g_5 \end{bmatrix}$
	$\pm\frac{a}{2}(0\ 0\ 2)$	$\begin{bmatrix} g_5 & 0 & 0 \\ 0 & g_5 & 0 \\ 0 & 0 & g_4 \end{bmatrix}$

Here the elements g_0, g_1, g_2, and g_3 are given in Eq. (8.44). The $\boldsymbol{G}$-matrices for the third, fourth, and fifth nearest neighbors can all be written in a similar manner.

F.4
The Mean Square Displacement $\langle u^2(0)\rangle$ and the Recoilless Fraction f

Fourier transforming (F.6) yields

$$\begin{aligned}\langle u_\alpha(l,t)u_\beta(l',0)\rangle &= \frac{1}{2\pi}\int_{-\infty}^{\infty} J_{\alpha\beta}(l\,l',\omega)\mathrm{e}^{-i\omega t}\,\mathrm{d}\omega \\ &= \frac{1}{2\pi}\int_0^{\infty} J_{\alpha\beta}(l\,l',\omega)\mathrm{e}^{-\beta\hbar\omega}\mathrm{e}^{i\omega t}\,\mathrm{d}\omega \\ &+ \frac{1}{2\pi}\int_0^{\infty} J_{\alpha\beta}(l\,l',\omega)\mathrm{e}^{-i\omega t}\,\mathrm{d}\omega. \end{aligned} \tag{F.27}$$

Therefore, the mean square displacement of an isolated impurity atom in a cubic host crystal can be found by letting $l = l' = 0$ be the impurity site. Since the vibration of this atom is isotropic we need only consider, say, the x-component of $\langle u^2\rangle$, so

$$\begin{aligned}\langle u_\alpha^2(0)\rangle = \lim_{t\to 0}\langle u_\alpha(0,t)u_\alpha(0,0)\rangle &= \frac{1}{2\pi}\int_0^{\infty} J_{\alpha\alpha}(00,\omega)(1+\mathrm{e}^{-\beta\hbar\omega})\,\mathrm{d}\omega \\ &= -\frac{\hbar}{\pi}\int_0^{\infty}\coth\left(\frac{\beta\hbar\omega}{2}\right)\,\mathrm{Im}\,G'_{\alpha\alpha}(00,\omega+i\varepsilon)\,\mathrm{d}\omega. \end{aligned} \tag{F.28}$$

Hence, we get the recoilless fraction f for an impurity (Mössbauer) atom in a cubic host:

$$f = \exp\left[\frac{k^2\hbar}{\pi}\int_0^{\infty}\coth\left(\frac{\beta\hbar\omega}{2}\right)\,\mathrm{Im}\,G'_{\alpha\alpha}(00,\omega+i\varepsilon)\,\mathrm{d}\omega\right]. \tag{F.29}$$

F.5
Relations Between Different Green's Functions $G_{\alpha\beta}(l\,l',w)$

For nearest neighbor forces only, we may rewrite the equation of motion (4.36) in the following alternative form:

$$\frac{1}{M_0}\sum_\beta \Phi_{\alpha\beta}(0,0)e_\beta(\boldsymbol{k}j) + \sum_l \mathrm{e}^{i\boldsymbol{k}\cdot\boldsymbol{l}}\frac{1}{M_0}\sum_\beta \Phi_{\alpha\beta}(0,l)e_\beta(\boldsymbol{k}j) = \omega_j^2(\boldsymbol{k})e_\alpha(\boldsymbol{k}j) \tag{F.30}$$

where $\boldsymbol{l}$ denotes the position vector of a nearest neighbor around the atom at the origin, and $\Phi_{\alpha\beta}(0,0)$ and $\Phi_{\alpha\beta}(0,l)$ represent $\Phi_{\alpha\beta}(000,000)$ and $\Phi_{\alpha\beta}(000,l)$, respectively. Multiplying (F.30) by $e_\alpha^*(\boldsymbol{k}j)$, dividing it by $N[\omega^2-\omega_j^2(\boldsymbol{k})]$, and then summing it over the first Brillouin zone in the $\boldsymbol{k}$-space, we have

$$\begin{aligned}\sum_\beta \Phi_{\alpha\beta}(0,0)\frac{1}{NM_0}\sum_{\boldsymbol{k},j}\frac{e_\alpha^*(\boldsymbol{k}j)e_\beta(\boldsymbol{k}j)}{\omega^2-\omega_j^2(\boldsymbol{k})}\\ +\sum_l\sum_\beta \Phi_{\alpha\alpha}(0,l)\frac{1}{NM_0}\sum_{\boldsymbol{k},j}\frac{e_\alpha^*(\boldsymbol{k}j)e_\beta(\boldsymbol{k}j)}{\omega^2-\omega_j^2(\boldsymbol{k})}\mathrm{e}^{i\boldsymbol{k}\cdot\boldsymbol{l}}\\ =-\frac{1}{N}\sum_{\boldsymbol{k},j}\left(1-\frac{\omega^2}{\omega^2-\omega_j^2(\boldsymbol{k})}\right)|e_\alpha(\boldsymbol{k}j)|^2.\end{aligned} \tag{F.31}$$

Using Eqs. (4.41) and (F.15), this equation can be simplified to

$$\begin{aligned}\Phi_{\alpha\alpha}(0,0)G_{\alpha\alpha}(000)+\sum_l \Phi_{\alpha\alpha}(0,l)G_{\alpha\alpha}(l)+\sum_l\sum_{\beta\neq\alpha}\Phi_{\alpha\beta}(0,l)G_{\alpha\beta}(l)\\ =-1+M_0\omega^2G_{\alpha\alpha}(000).\end{aligned} \tag{F.32}$$

Two more similar relations can be obtained by multiplying (F.30) by $e_a^*(\boldsymbol{k}j)\mathrm{e}^{-i\boldsymbol{k}\cdot\boldsymbol{l}'}$ and $e_\gamma^*(\boldsymbol{k}j)\mathrm{e}^{-i\boldsymbol{k}\cdot\boldsymbol{l}'}$, respectively:

$$\begin{aligned}\Phi_{\alpha\alpha}(0,0)G_{\alpha\alpha}(l')+\sum_l \Phi_{\alpha\alpha}(0,l)G_{\alpha\alpha}(l-l')+\sum_l\sum_{\beta\neq\alpha}\Phi_{\alpha\beta}(0,l)G_{\alpha\beta}(l-l')\\ =M_0\omega^2G_{\alpha\alpha}(l')\end{aligned} \tag{F.33}$$

$$\begin{aligned}\Phi_{\alpha\alpha}(0,0)G_{\alpha\gamma}(-l')+\sum_l \Phi_{\alpha\alpha}(0,l)G_{\alpha\gamma}(l-l')+\sum_l\sum_{\beta\neq\alpha}\Phi_{\alpha\beta}(0,l)G_{\beta\gamma}(l-l')\\ =M_0\omega^2G_{\alpha\gamma}(l')\end{aligned} \tag{F.34}$$

where l' denotes one of the nearest neighbor sites. It is easy to see that the above three relations can be combined into one:

$$\sum_{\beta',l}\Phi_{\alpha\beta'}(0,l)G_{\beta\beta'}(l\,l',\omega)=-\delta_{\alpha\beta}\delta_{0l'}+M_0\omega^2G_{\alpha\beta}(0l',\omega). \tag{F.35}$$

Appendix G
Symmetry Coordinates

The best method of determining the normal vibrations in cases of symmetrical crystals or molecules has proved to be the method of "symmetry coordinates." Consisting of linear combinations of atomic displacements $\boldsymbol{u}_i$, these coordinates serve as basis functions of various irreducible representations of a symmetrical group. The symmetry coordinates are closely connected with the normal coordinates; the latter can be linearly represented by the former and, in some cases, they are even identical to each other. For instance, the symmetrical nonlinear molecule H_2O belonging to the C_{2v} point group has two normal vibrations of irreducible representation A_1 and one of B_2, i.e., the total representation is decomposed into $2A_1 \oplus B_1$. For B_1 there is only one unambiguous normal vibration, and hence the symmetry coordinate is identical to the normal coordinate. For A_1, on the other hand, there is an infinite number of possible symmetry coordinates and the actual normal coordinates are linear combinations of two mutually orthogonal symmetry coordinates. In the case of the linear symmetrical CO_2 molecule, the symmetry coordinates are the same as the normal coordinates.

The monatomic cubic crystals, such as simple cubic (sc), body-centered cubic (bcc), and face-centered cubic (fcc), have the highest symmetry. Suppose a substitutional impurity atom (say, a Mössbauer atom) is at the origin of the coordinates, i.e., at the (000) lattice site, and is coupled to its nearest neighbors to form an impurity space. The irreducible representations for an impurity site having symmetry O_h in this host lattice are

$$\begin{aligned}
\Gamma_{\text{sc}} &= A_{1g} \oplus E_g \oplus F_{1g} \oplus F_{2g} \oplus 3F_{1u} \oplus F_{2u}, \\
\Gamma_{\text{bcc}} &= A_{1g} \oplus E_g \oplus F_{1g} \oplus 2F_{2g} \oplus A_{2u} \oplus E_u \oplus 3F_{1u} \oplus F_{2u}, \qquad \text{(G.1)} \\
\Gamma_{\text{fcc}} &= A_{1g} \oplus A_{2g} \oplus 2E_g \oplus 2F_{1g} \oplus 2F_{2g} \oplus A_{2u} \oplus E_u \oplus 4F_{1u} \oplus 2F_{2u}.
\end{aligned}$$

For a single impurity, we are only interested in the triply degenerate F_{1u} mode in which the impurity atom moves.

Again we take a host fcc lattice as an example to discuss in details how the basis functions – the symmetry coordinates – for the irreducible representation F_{1u} are found. In the impurity space we have a total of 13 atoms and hence 39 degrees of freedom. At first sight, a basis set for $4F_{1u}$ should have 12 symmetry coordinates:

$$S_j(x), S_j(y), S_j(z) \quad (j = 0, 1, 2, 3). \qquad \text{(G.2)}$$

But one can easily show by group theory that only 4 symmetry coordinates, say the 4 x-component symmetry coordinates $S_j(x)$, are needed. We introduce the following notation for the unit displacements of 13 atoms: $u_x(XYZ) = x_i$,

$u_y(XYZ) = y_i$, and $u_z(XYZ) = z_i$, where the site number i and the respective coordinate are

$i =$	0	1	2	3	4	5	6
$(XYZ) =$	$(0\;0\;0)$	$(0\;1\;1)$	$(0\;1\;\bar{1})$	$(0\;\bar{1}\;1)$	$(0\;\bar{1}\;\bar{1})$	$(1\;0\;1)$	$(1\;0\;\bar{1})$
$i =$	7	8	9	10	11	12	
$(XYZ) =$	$(\bar{1}\;0\;1)$	$(\bar{1}\;0\;\bar{1})$	$(1\;1\;0)$	$(1\;\bar{1}\;0)$	$(\bar{1}\;1\;0)$	$(\bar{1}\;\bar{1}\;0)$	

The symmetry operations R and their respective matrices $\Gamma(R)$ for the irreducible representation F_{1u} are given as follows:

$$E: \begin{bmatrix} 1 & 0 & 0 \\ 0 & 1 & 0 \\ 0 & 0 & 1 \end{bmatrix} \qquad C_2^x: \begin{bmatrix} 1 & 0 & 0 \\ 0 & -1 & 0 \\ 0 & 0 & -1 \end{bmatrix} \qquad C_2^y: \begin{bmatrix} -1 & 0 & 0 \\ 0 & 1 & 0 \\ 0 & 0 & -1 \end{bmatrix}$$

$$C_2^z: \begin{bmatrix} -1 & 0 & 0 \\ 0 & -1 & 0 \\ 0 & 0 & 1 \end{bmatrix} \qquad C_3^{xyz}: \begin{bmatrix} 0 & 0 & 1 \\ 1 & 0 & 0 \\ 0 & 1 & 0 \end{bmatrix} \qquad C_3^{x\bar{y}\bar{z}}: \begin{bmatrix} 0 & 0 & -1 \\ -1 & 0 & 0 \\ 0 & 1 & 0 \end{bmatrix}$$

$$C_3^{\bar{x}y\bar{z}}: \begin{bmatrix} 0 & 0 & 1 \\ -1 & 0 & 0 \\ 0 & -1 & 0 \end{bmatrix} \qquad C_3^{\bar{x}\bar{y}z}: \begin{bmatrix} 0 & 0 & -1 \\ 1 & 0 & 0 \\ 0 & -1 & 0 \end{bmatrix} \qquad \bar{C}_3^{xyz}: \begin{bmatrix} 0 & 1 & 0 \\ 0 & 0 & 1 \\ 1 & 0 & 0 \end{bmatrix}$$

$$\bar{C}_3^{x\bar{y}\bar{z}}: \begin{bmatrix} 0 & -1 & 0 \\ 0 & 0 & 1 \\ -1 & 0 & 0 \end{bmatrix} \qquad \bar{C}_3^{\bar{x}y\bar{z}}: \begin{bmatrix} 0 & -1 & 0 \\ 0 & 0 & -1 \\ 1 & 0 & 0 \end{bmatrix} \qquad \bar{C}_3^{\bar{x}\bar{y}z}: \begin{bmatrix} 0 & 1 & 0 \\ 0 & 0 & -1 \\ -1 & 0 & 0 \end{bmatrix}$$

$$C_4^z: \begin{bmatrix} 0 & -1 & 0 \\ 1 & 0 & 0 \\ 0 & 0 & 1 \end{bmatrix} \qquad C_2^{xy}: \begin{bmatrix} 0 & 1 & 0 \\ 1 & 0 & 0 \\ 0 & 0 & -1 \end{bmatrix} \qquad C_2^{x\bar{y}}: \begin{bmatrix} 0 & -1 & 0 \\ -1 & 0 & 0 \\ 0 & 0 & -1 \end{bmatrix}$$

$$\bar{C}_4^z: \begin{bmatrix} 0 & 1 & 0 \\ -1 & 0 & 0 \\ 0 & 0 & 1 \end{bmatrix} \qquad C_2^{yz}: \begin{bmatrix} -1 & 0 & 0 \\ 0 & 0 & 1 \\ 0 & 1 & 0 \end{bmatrix} \qquad C_4^x: \begin{bmatrix} 1 & 0 & 0 \\ 0 & 0 & -1 \\ 0 & 1 & 0 \end{bmatrix}$$

$$\bar{C}_4^x: \begin{bmatrix} 1 & 0 & 0 \\ 0 & 0 & 1 \\ 0 & -1 & 0 \end{bmatrix} \qquad C_2^{y\bar{z}}: \begin{bmatrix} -1 & 0 & 0 \\ 0 & 0 & -1 \\ 0 & -1 & 0 \end{bmatrix} \qquad \bar{C}_4^y: \begin{bmatrix} 0 & 0 & -1 \\ 0 & 1 & 0 \\ 1 & 0 & 0 \end{bmatrix}$$

$$C_2^{z\bar{x}}: \begin{bmatrix} 0 & 0 & -1 \\ 0 & -1 & 0 \\ -1 & 0 & 0 \end{bmatrix} \qquad C_2^{zx}: \begin{bmatrix} 0 & 0 & 1 \\ 0 & -1 & 0 \\ 1 & 0 & 0 \end{bmatrix} \qquad C_4^y: \begin{bmatrix} 0 & 0 & 1 \\ 0 & 1 & 0 \\ -1 & 0 & 0 \end{bmatrix}$$

Now we start to derive the symmetry coordinates by the standard projection operator which in our case is

$$\begin{aligned}\hat{P}_{11} &= \sum \Gamma^*(R)_{11}\hat{R} \\ &= N\{(1)E + (1)C_2^x + (-1)C_2^y + (-1)C_2^z + (-1)C_2^{yz} + (1)C_4^x \\ &\quad + (1)\bar{C}_4^x + (-1)C_2^{y\bar{z}} + (-i)[(1)E + (1)C_2^x + (-1)C_2^y + (-1)C_2^z \\ &\quad + (-1)C_2^{yz} + (1)C_4^x + (1)\bar{C}_4^x + (-1)C_2^{y\bar{z}}]\}\end{aligned} \tag{G.3}$$

where N is the normalization factor. First, we choose x_0 as a generator. The projecting operation gives

$$\hat{P}_{11}x_0 = 2N(x_0 + x_0 + x_0 + x_0 + x_0 + x_0 + x_0 + x_0) = 16Nx_0;$$

hence we have

$$S_0 \propto x_0. \tag{G.4}$$

To proceed to find the remaining symmetry coordinates, we let x_5 be another generator. The projecting operation gives

$$\hat{P}_{11}x_5 = 2N(x_5 + x_6 + x_7 + x_8 + x_9 + x_{10} + x_{11} + x_{12}),$$

and similarly

$$\hat{P}_{11}x_6 = \hat{P}_{11}x_7 = \hat{P}_{11}x_8 = \hat{P}_{11}x_9 = \hat{P}_{11}x_{10} = \hat{P}_{11}x_{11} = \hat{P}_{11}x_{12} = \hat{P}_{11}x_5.$$

Combining the above eight identical results, we obtain

$$\begin{aligned}&\hat{P}_{11}(x_5 + x_6 + x_7 + x_8 + x_9 + x_{10} + x_{11} + x_{12}) \\ &\quad = 16N(x_5 + x_6 + x_7 + x_8 + x_9 + x_{10} + x_{11} + x_{12})\end{aligned}$$

and therefore we find the second symmetry coordinate:

$$S_1 \propto x_5 + x_6 + x_7 + x_8 + x_9 + x_{10} + x_{11} + x_{12}. \tag{G.5}$$

The next step is taking z_5 as another generator and we have

$$\hat{P}_{11}z_5 = 2N(z_5 - z_6 - z_7 + z_8 + y_9 - y_{10} - y_{11} + y_{12})$$

and also

$$-\hat{P}_{11}z_6 = -\hat{P}_{11}z_7 = \hat{P}_{11}z_8 = \hat{P}_{11}y_9 = -\hat{P}_{11}y_{10} = -\hat{P}_{11}y_{11} = \hat{P}_{11}y_{12} = \hat{P}_{11}z_5.$$

Combining these results gives

$$\hat{P}_{11}(z_5 - z_6 - z_7 + z_8 + y_9 - y_{10} - y_{11} + y_{12})$$
$$= 16N(z_5 - z_6 - z_7 + z_8 + y_9 - y_{10} - y_{11} + y_{12}).$$

Therefore

$$S_2 \propto z_5 - z_6 - z_7 + z_8 + y_9 - y_{10} - y_{11} + y_{12}. \tag{G.6}$$

Finally, the fourth symmetry coordinate can be calculated in a similar way:

$$S_3 \propto x_1 + x_2 + x_3 + x_4. \tag{G.7}$$

From (G.4) through (G.7), the four orthonormal symmetry coordinates, transforming as the first row of the representation F_{1u} of O_h, are calculated as follows:

$$\begin{aligned} S_0 &= x_0, \\ S_1 &= \frac{1}{2\sqrt{2}}(x_5 + x_6 + x_7 + x_8 + x_9 + x_{10} + x_{11} + x_{12}), \\ S_2 &= \frac{1}{2\sqrt{2}}(z_5 - z_6 - z_7 + z_8 + y_9 - y_{10} - y_{11} + y_{12}), \\ S_3 &= \frac{1}{2}(x_1 + x_2 + x_3 + x_4). \end{aligned} \tag{G.8}$$

Appendix H
Mass Absorption Coefficients

The following table lists the mass absorption coefficients ($cm^2\ g^{-1}$) for elements from $Z = 1$ to $Z = 94$. [G.J. Long, T.E. Cranshaw, and G. Longworth. The ideal Mössbauer effect absorber thickness. *Mössbauer Effect Ref. Data J.* 6, 42–49 (1983).]

Absorber			γ-ray energy (keV)												
Element		Atomic mass (u)	6.0	14.41 ^{57}Fe	21.53 ^{151}Eu	23.87 ^{119}Sn	25.66 ^{161}Dy	27.77 ^{129}I	35.46 ^{125}Te	37.15 ^{121}Sb	50	60	80	100	150
H	1	1.008	0.404	0.387	0.367	0.363	0.362	0.360	0.351	0.348	0.335	0.326	0.309	0.296	0.265
He	2	4.003	0.423	0.213	0.194	0.191	0.189	0.187	0.180	0.178	0.170	0.166	0.156	0.149	0.134
Li	3	6.941	0.945	0.227	0.182	0.174	0.171	0.167	0.158	0.156	0.149	0.144	0.135	0.129	0.155
Be	4	9.012	2.51	0.32	0.215	0.195	0.190	0.185	0.175	0.170	0.156	0.150	0.140	0.133	0.119
B	5	10.81	5.38	0.51	0.27	0.25	0.235	0.22	0.19	0.185	0.167	0.159	0.147	0.139	0.124
C	6	12.011	10.6	0.87	0.40	0.35	0.32	0.29	0.22	0.21	0.188	0.175	0.161	0.151	0.135
N	7	14.007	17.9	1.40	0.52	0.43	0.39	0.34	0.26	0.25	0.199	0.182	0.164	0.153	0.135
O	8	15.999	27.6	2.2	0.74	0.58	0.52	0.44	0.30	0.29	0.21	0.19	0.17	0.155	0.136
F	9	18.998	37.4	2.7	0.88	0.66	0.62	0.52	0.33	0.31	0.223	0.193	0.165	0.150	0.130
Ne	10	20.179	54.0	4.0	1.36	1.05	0.87	0.70	0.42	0.38	0.260	0.218	0.179	0.161	0.137
Na	11	22.990	69.2	5.2	1.70	1.24	1.07	0.88	0.51	0.46	0.283	0.229	0.181	0.159	0.134
Mg	12	24.305	92.4	6.8	2.20	1.64	1.18	1.12	0.64	0.56	0.329	0.258	0.195	0.169	0.139
Al	13	26.982	116	9.0	2.74	2.15	1.75	1.40	0.74	0.69	0.368	0.279	0.202	0.171	0.138
Si	14	28.086	145	12.4	3.65	2.8	2.2	1.84	0.96	0.84	0.433	0.317	0.223	0.184	0.145
P	15	30.974	172	14.2	4.15	3.2	2.8	2.2	1.08	0.95	0.49	0.35	0.23	0.19	0.143
S	16	32.06	210	17.0	5.6	4.0	3.6	2.8	1.4	1.20	0.580	0.404	0.259	0.203	0.151
Cl	17	35.453	240	20.0	6.1	4.5	3.7	2.9	1.53	1.36	0.642	0.438	0.270	0.206	0.148
Ar	18	39.948	268	21.0	7.0	5.2	4.2	3.4	1.7	1.5	0.69	0.463	0.276	0.205	0.143
K	19	39.098	326	27.0	8.8	6.2	5.3	4.6	2.2	1.8	0.857	0.565	0.326	0.236	0.159
Ca	20	40.08	383	32.5	10.5	7.7	6.5	5.0	2.5	2.4	1.00	0.649	0.364	0.257	0.168
Sc	21	44.956	407	33	11.2	8.3	7.1	5.6	2.7	2.4	1.07	0.683	0.373	0.257	0.162
Ti	22	47.90	446	42.0	12.4	9.0	7.6	6.2	3.0	2.5	1.19	0.755	0.402	0.272	0.165
V	23	50.941	493	43	14.1	10.6	8.8	6.8	3.3	2.9	1.31	0.827	0.433	0.286	0.168

Absorber			γ-ray energy (keV)												
Element		Atomic mass (u)	6.0	14.41 ^{57}Fe	21.53 ^{151}Eu	23.87 ^{119}Sn	25.66 ^{161}Dy	27.77 ^{129}I	35.46 ^{125}Te	37.15 ^{121}Sb	50	60	80	100	150
Cr	24	51.996	570	48	16.8	12.2	10.4	8.0	4.0	3.4	1.52	0.948	0.486	0.315	0.179
Mn	25	54.938	75.9	53	18.4	13.4	11.3	8.5	4.6	3.9	1.67	1.04	0.523	0.334	0.184
Fe	26	55.847	87.1	64	22.0	15.5	12.4	10.0	5.0	4.4	1.92	1.19	0.59	0.37	0.196
Co	27	58.933	96.0	66	21.4	16.0	13.6	10.6	5.4	4.8	2.08	1.28	0.629	0.390	0.201
Ni	28	58.70	110	75	26.0	18.8	15.8	12.4	6.4	5.8	2.42	1.49	0.720	0.440	0.221
Cu	29	63.546	118	82	28.0	20.0	16.8	13.0	6.8	6.1	2.56	1.56	0.750	0.453	0.221
Zn	30	65.38	132	92	31	22	18.4	14.8	7.5	6.8	2.85	1.73	0.824	0.492	0.233
Ga	31	69.72	141	97	32.8	23.2	19.0	15.2	8.0	7.2	2.99	1.81	0.862	0.511	0.238
Ge	32	72.59	152	102	36.0	25.0	20.5	16.2	8.4	7.6	3.21	1.95	0.921	0.543	0.247
As	33	74.922	168	126	38.5	28.0	23.0	18.5	9.0	8.0	3.54	2.15	1.00	0.586	0.259
Se	34	78.96	180	110	40	30.5	25.5	19.5	10.2	9.0	3.78	2.29	1.07	0.616	0.268
Br	35	79.904	198	130	43	32.0	27.2	21.5	11.2	9.8	4.17	2.52	1.17	0.672	0.286
Kr	36	83.80	212	126	46	34.5	28.5	23.5	11.8	10.4	4.43	2.67	1.24	0.708	0.297
Rb	37	85.468	231	21.0	49	37	31.0	24.8	12.8	11.4	4.81	2.91	1.34	0.761	0.315
Sr	38	87.62	251	24.5	52	39.5	32.5	26.5	13.8	12.2	5.20	3.14	1.44	0.818	0.332
Y	39	88.906	274	26.0	58	42	33.5	28.8	15.0	13.2	5.66	3.41	1.56	0.887	0.354
Zr	40	91.22	296	28.0	62	44	38	31	16.2	14.4	6.11	3.70	1.70	0.951	0.374
Nb	41	92.906	323	29.5	65	46.5	40	32.5	17.5	15.6	6.68	4.04	1.85	1.04	0.404
Mo	42	95.94	343	34	69	49	42	34	18.5	16.3	7.03	4.27	1.96	1.10	0.422
Tc	43	98.906	364	35.5	75	54	44	36	19.6	17.4	7.36	4.45	2.03	1.13	0.432
Ru	44	101.07	391	37.5	13.0	60	49	40	20	18.0	7.87	4.76	2.17	1.20	0.456
Rh	45	102.906	418	38	12.5	60	51	39.5	22	19.5	8.42	5.09	2.32	1.29	0.483

Pd	46	106.4	447	41	14.0	10.6	56	42	23	20.4	8.83	5.37	2.45	1.36	0.505
Ag	47	107.868	479	44	15.5	11.4	57	45	24.5	21.8	9.44	5.75	2.64	1.46	0.539
Cd	48	112.41	501	46	16.2	12.2	10.2	47	25.6	22.4	9.81	5.95	2.72	1.51	0.552
In	49	114.82	530	49	17.0	12.8	10.6	9.0	28.4	24.8	10.5	6.35	2.87	1.58	0.577
Sn	50	118.69	563	54	18.0	13.3	11.2	9.3	28	24.0	10.8	6.60	3.00	1.65	0.594
Sb	51	121.75	579	56	18.7	14.3	12.2	10.1	27.5	25.0	11.6	6.98	3.18	1.74	0.623
Te	52	127.60	604	60	20.1	15.3	12.8	10.5	31	27.5	12.0	7.27	3.30	1.81	0.642
I	53	126.905	655	61	21.0	16.0	13.5	11.0	33	28.4	12.7	7.69	3.48	1.91	0.683
Xe	54	131.30	670	67	23.0	17.2	14.5	11.3	36	30.0	13.0	7.89	3.59	1.97	0.702
Cs	55	132.905	725	71	24.6	18.0	15.8	12.3	6.1	35.6	14.0	8.43	3.82	2.09	0.739
Ba	56	137.33	754	75	25.0	18.6	16.2	13.2	7.0	6.1	15.1	8.68	3.94	2.16	0.763
La	57	138.906	672	76	26.5	20.2	17.0	14.0	6.6	6.0	15.1	9.19	4.20	2.30	0.811
Ce	58	140.12	511	77	29	22	18.6	14.8	7.1	6.5	16.3	9.80	4.43	2.41	0.847
Pr	59	140.098	543	81	30	22.8	19.4	16.0	7.6	6.8	17.3	10.4	4.70	2.56	0.893
Nd	60	144.24	201	88	31.2	24.0	20.2	16.4	7.8	7.0	17.8	10.8	4.88	2.66	0.931
Pm	61	145	210	92	32.8	24.8	21.1	17.6	8.2	7.3	18.7	11.3	5.08	2.78	0.967
Sm	62	150.4	221	96	34.5	25.6	22.0	18.8	8.6	7.6	19.4	11.7	5.29	2.88	1.00
Eu	63	151.96	231	100	35.8	26.2	22.6	18.9	8.9	8.0	20.4	12.3	5.55	3.03	1.05
Gd	64	157.25	238	104	37	27.0	23.2	19.0	9.3	8.3	3.94	12.4	5.71	3.13	1.09
Tb	65	158.925	251	107	38	28.4	24.1	19.7	9.8	8.8	4.17	13.2	5.99	3.25	1.13
Dy	66	162.50	261	110	40	29.8	25.0	20.4	10.4	9.2	4.30	13.7	6.19	3.37	1.16
Ho	67	164.930	273	115	42	31.4	26.4	21.5	10.8	9.6	4.53	14.3	6.46	3.52	1.21
Er	68	167.26	284	120	44.5	33.0	27.8	22.6	11.2	10.0	4.72	14.9	6.73	3.67	1.26
Tm	69	168.934	299	127	45.5	34	28.9	23.3	11.8	10.6	4.92	15.6	7.02	3.81	1.32
Yb	70	173.04	310	133	47	35	29.8	24.0	12.4	11.1	5.05	3.18	7.20	3.93	1.36
Lu	71	174.97	324	139	48	36.5	31	24.8	12.8	11.5	5.27	3.32	7.54	4.10	1.41
Hf	72	178.49	336	145	49	38	32	25.6	13.2	11.9	5.46	3.44	7.66	4.18	1.45
Ta	73	180.948	346	150	51.5	39	33	26.8	13.8	12.5	5.69	3.59	7.79	4.29	1.50
W	74	183.85	364	155	54	40	34.5	28.0	14.5	13.0	5.90	3.70	8.06	4.49	1.56
Re	75	186.207	378	160	56.5	42	35	28.5	15.0	13.6	6.14	3.85	8.70	4.72	1.62

Absorber			γ-ray energy (keV)												
Element		Atomic mass (u)	6.0	14.41 ^{57}Fe	21.53 ^{151}Eu	23.87 ^{119}Sn	25.66 ^{161}Dy	27.77 ^{129}I	35.46 ^{125}Te	37.15 ^{121}Sb	50	60	80	100	150
Os	76	190.2	393	165	59	44	36	29.0	15.6	14.1	6.33	3.99	8.96	4.88	1.67
Ir	77	192.22	407	172	61	45.5	38	31	16.4	14.6	6.61	4.17	9.31	5.04	1.73
Pt	78	195.09	423	178	64	47	40	33	17.0	15.2	6.82	4.29	9.51	5.09	1.74
Au	79	196.966	440	191	66.5	49.5	42	34.5	17.6	15.7	7.13	4.49	2.20	5.23	1.84
Hg	80	200.59	456	165	67	52	44	36	18.2	16.2	7.41	4.65	2.27	5.52	1.86
Tl	81	204.37	472	123	68.5	54	45	37	19.0	16.8	7.66	4.80	2.34	5.66	1.93
Pb	82	207.2	491	120	70	56	46.5	38	19.8	17.4	7.88	4.91	2.38	5.58	1.99
Bi	83	208.980	510	123	71.5	58	48	39.5	20.6	18.2	8.30	5.22	2.56	6.17	2.09
Po	84	209	531	126	73	60	49.5	41	21.2	19.0	8.63	5.42	2.65	6.40	2.19
At	85	210	556	140	76	60	52	41.5	21.7	19.8	9.03	5.68	2.77	6.68	2.29
Rn	86	222	551	60	80	59	53	42	22.2	20.6	8.95	5.64	2.74	6.59	2.26
Fr	87	223	573	62	83	60	54	44	22.8	20.8	9.34	5.86	2.86	1.66	2.34
Ra	88	226.025	591	65	87	62	55	46	23.5	21	9.61	6.05	2.93	1.71	2.41
Ac	89	227.028	613	70	89	63	58	47	24.5	22	10.0	6.29	3.05	1.77	2.50
Th	90	232.038	628	75	92	65	62	48	25.6	23	10.2	6.41	3.09	1.79	2.44
Pa	91	231.036	662	80	95	67	61	48.5	26.2	24	10.8	6.75	3.27	1.89	2.67
U	92	238.029	658	85	81	70	60	49	27	24.5	10.9	6.83	3.31	1.92	2.83
Np	93	237.048	701	90	63	72	62	51	28	25	11.4	7.14	3.46	1.99	2.82
Pu	94	239.13	717	95	63	75	63	53	29	26	11.6	7.27	3.51	2.02	2.71

Copyright Acknowledgments

We gratefully acknowledge the following publishers and/or copyright holders who have kindly provided permission to reproduce the figures and tables indicated. We have made every effort to trace the ownership of all copyrighted material and to obtain permission. Sources of figures and tables are also referenced with full citations in the relevant chapters.

Reproduced with kind permission from the American Chemical Society:
Figure 9.16 & Table 9.5: from R.D. Ernst, D.R. Wilson, R.H. Herber. *J. Am. Chem. Soc.* 106, 1646 ©1984.

Reproduced with kind permission from the American Institute of Physics:
Figure 2.13: from R.W. Grant, H. Wiedersich, A.H. Muir Jr., U. Gonser, W.N. Delgass. *J. Chem. Phys.* 45, 1015 ©1966. **Fig. 7.29**: T.S. Toellner, M.Y. Hu, W. Sturhahn, K. Quast, E.E. Alp. *Appl. Phys. Lett.* 71, 2112 ©1997.

Reproduced with kind permission by the American Physical Society:
Figure 2.14: from G.K. Wertheim, H.J. Guggenheim, D.N.E. Buchanan. *Phys. Rev.* 169, 465 ©1968. **Fig. 4.18**: J.B. Clement, E.H. Quinnell. *Phys. Rev.* 92, 258 ©1953. **Figs. 4.20 & 4.21**: A.S. Barker Jr., A.J. Sievers. *Revs. Mod. Phys.* 47, Suppl. 2, S1 ©1975. **Fig. 4.25**: E.C. Svensson, B.N. Brockhouse, J.M. Rowe. *Phys. Rev.* 155, 619 ©1967. **Fig. 4.26**: R.M. Nicklow, G. Gilat, H.G. Smith, L.J. Raubenheimer, M.K. Wilkinson. Phys. Rev. 164, 922 ©1967. **Fig. 5.5**: J.G. Dash, D.P. Johnson, W.M. Visscher. *Phys. Rev.* 168, 1087 ©1968. **Fig. 5.6**: D.P. Johnson, J.G. Dash. *Phys. Rev.* 172, 983 ©1968. **Fig. 5.8**: L.E. Campbell, G.L. Montet, G.J. Perlow. *Phys. Rev. B* 15, 3318 (1977). **Fig. 5.11**: R.V. Pound, G.A. Rebka Jr. *Phys. Rev. Lett.* 4, 274 ©1960. **Figs. 5.12 & 5.13**: R.D. Taylor, P.P. Craig. *Phys. Rev.* 175, 782 ©1968. **Fig. 6.15(a)**: G. Albanese, C. Ghezzi, A. Merlini, S. Pace. *Phys. Rev. B* 5, 1746 ©1972. **Fig. 6.15(b)**: G. Albanese, C. Ghezzi, A. Merlini. *Phys. Rev. B* 7, 65 ©1973. **Fig. 6.19(a) & Table 6.5**: J.T. Day, J.G. Mullen, R.C. Shukla. *Phys. Rev. B* 52, 168 ©1995. **Table 6.4**: B.R. Bullard, J.G. Mullen, G. Schupp. *Phys. Rev. B* 43, 7405 ©1991. **Fig. 7.11**: F.J. Lynch, R.E. Holland, M. Hamermesh. *Phys. Rev.* 120, 513 ©1960. **Fig. 7.12**: C.S. Wu, Y.K. Lee, N. Benczer-Koller, P. Simms. *Phys. Rev. Lett.* 5, 432 ©1960. **Fig. 7.13**: A.L. Chumakov, G.V. Smirnov, A.Q.R. Baron, J. Arthur, D.E. Brown, S.L. Ruby, G.S. Brown, N.N. Salashchenko. *Phys. Rev. Lett.* 71, 2489 ©1993. **Figs. 7.18 & 7.19**: U. van Bürck, R.L. Mössbauer, E. Gerdau, R. Rüffer, R. Hollatz, G.V. Smirnov, J.P. Hannon. *Phys. Rev. Lett.* 59, 355 ©1987. **Figs. 7.23 & 7.24, Table 7.1**: U. Bergmann, S.D. Shastri, D.P. Siddons, B.W. Batterman, J.B. Hastings. *Phys. Rev. B* 50, 5957 ©1994. **Figs. 8.2 & 8.3**: M. Seto, Y. Kobayashi, S. Kitao, R. Haruki, T. Mitsui, Y. Yoda, S. Nasu, S. Kikuta. *Phys. Rev. B* 61, 11420 ©2000. **Fig. 8.5, Tables 8.1, 8.2, 8.4 & 8.5**: J.M. Grow, D.G. Howard, R.H. Nussbaum, M. Takeo. *Phys. Rev. B* 17, 15 ©1978. **Fig. 8.6**: R.H. Nussbaum, D.G. Howard, W. L. Nees, C.F. Steen. *Phys. Rev.* 173, 653 ©1968. **Table 8.7**: J.F. Prince, L.D. Roberts, D.J. Erickson. *Phys. Rev. B* 13, 24 ©1976. **Figs. 9.9 & 9.10**: M. Peter, W. Potzel, M. Steiner, C. Schäfer, H. Karzel, W. Schiessl, G.M. Kalvius, U.

Mössbauer Effect in Lattice Dynamics. Yi-Long Chen and De-Ping Yang

ISBN: 978-3-527-40712-5

Gonser. *Phys. Rev. B* 47, 753 ©1993. **Fig. 9.23**: B. Fultz, C.C. Ahn, E.E. Alp, W. Sturhahn, T.S. Toellner. *Phys. Rev. Lett.* 79, 937 ©1997. **Table 9.2**: M. Köfferlein, W. Potzel, M. Steiner, H. Karzel, W. Schiessl, G.M. Kalvius. *Phys. Rev. B* 52, 13332 ©1995.

Reproduced with kind permission from Cambridge University Press:
Figures 1.3 & 1.6: from T.E. Cranshaw, B.W. Dale, G.O. Longworth, C.E. Johnson. *Mössbauer Spectroscopy and its Applications.* Cambridge Univ. Press, Cambridge ©1985.

Reproduced with kind permission from EDP Sciences, France:
Figure 5.9: from E.R. Bauminger, A. Diamant, I. Felner, I. Nowik, A. Mustachi, S. Ofer. *J. de Physique* (Colloque) 37, C6-49 ©1976. **Fig. 9.3**: W. Potzel, A. Forster, G.M. Kalvius. *J. de Physique* (Colloque) 40, C2-29 ©1979. **Fig. 9.12**: C.L. Chien, R. Hasegawa. *J. de Physique* (Colloque) 37, C6-759 ©1976.

Reproduced with kind permission from Elsevier Ltd.:
Figure 1.10: from J.M. Williams, J.S. Brooks. *Nucl. Instrum. Methods* 128, 363 ©1975. **Fig. 2.3**: R.C. Axtmann, Y. Hazony, J.W. Hurley Jr. *Chem. Phys. Lett.* 2, 673 ©1968. **Fig. 2.4**: M. Pasternak, J.N. Farrell, R.D. Taylor. *Solid State Commun.* 61, 409 ©1987. **Figs. 2.17 & 2.21**: K. Szymanski, L. Dobrzynski, B. Prus, M.J. Cooper. *Nucl. Instrum. Methods B* 119, 438 ©1996. **Fig. 3.8**: W.F. Filter, R.H. Sands, W.R. Dunham. *Nucl. Instrum. Methods B* 119, 565 ©1996. **Fig. 4.11**: G. Burns. *Solid State Physics.* Academic Press, Orlando, ©1985. **Fig. 4.19**: B.N. Brockhouse, H.E. Abou-Helal, E.D. Hallman. *Solid State Commun.* 5, 211 ©1967. **Fig. 4.22**: T. Hattori, K. Ehara, A Mitsuishi, S. Sakuragi, H. Kanzaki. *Solid State Commun.* 12, 545 ©1973. **Fig. 5.3**: K.N. Pathak, B. Deo. *Physica* 35, 167 ©1967. **Fig. 5.9**: H. Armon, E.R. Bauminger, A. Diamant, I. Nowik, S. Ofer. *Phys. Lett. A* 44, 279 ©1973. **Fig. 5.10**: H. Armon, E.R. Bauminger, A. Diamant, I. Nowik, S. Ofer. *Solid State Commun.* 15, 543 ©1974. **Table 5.1**: L. Cianchi, F. Del Giallo, F. Pieralli, M. Mancini, S. Sciortino, G. Spina, N. Ammannati, R. Garrè. *Solid State Commun.* 80, 705–708 (1991). **Fig. 8.9 & Table 8.8**: C.F. Steen, D.G. Howard, R.H. Nussbaum. *Solid State Commun.* 9, 865 ©1971. **Fig. 8.10**: J.A. Sawicki, B.D. Sawicka. *Nucl. Instrum. Methods B* 46, 38 ©1990. **Fig. 9.15**: A. Gupta, P. Shah, N.P. Lalla, B.A. Dasannacharya, T. Harami. Y. Yoda, M. Seto, M, Yabashi, S. Kikuta. *Materials Science and Engineering* A304–306, 731 ©2001. **Fig. 9.22**: T. Ruckert, W. Keune, W. Sturhahn, E.E. Alp. *J. Mag. Mag. Mater.* 240, 562 ©2002.

Reproduced with kind permission from Institute of Physics Publishing Ltd.:
Figure 2.5 & Table 2.1: from A. Svane, E. Antoncik. *J. Phys. C* 20, 2683 ©1987. **Fig. 2.16 & Table 2.4**: L. Häggström, T. Ericsson, R. Wäppling, E. Karlsson. *Physica Scripta* 11, 55 ©1975. **Fig. 4.27**: J. Lazewski, K. Parlinski, B. Hennion, R. Fouret. *J. Phys.: Condens. Matter* 11, 9665 ©1999. **Fig. 5.7**: J.M. Fiddy, I. Hall, F. Grandjean, U. Russo, G.J. Long. *J. Phys.: Condens. Matter* 2, 10109 ©1990. **Fig. 5.16**: P. Hannaford, R.G. Horn. *J. Phys. C.: Solid State Phys.* 6, 2223 ©1973. **Fig. 6.5**: A. Storruste. *Proc. Phys. Soc. A* 63, 1197 ©1950. **Fig. 6.10**: P.J. Black, I.P. Duerdoth. *Proc. Phys. Soc.* 84, 169 ©1964. **Fig. 6.17**: N.M. Butt, D.A. O'Connor. *Proc. Phys. Soc.* 90, 247 ©1967. **Fig. 9.14 & Table 9.4**: R. Yamamoto, K. Haga, T. Mihara, M. Doyama. *J. Phys. F* 10, 1389 ©1980. **Figs. 9.19, 9.20, & 9.21**: R. Röhlsberger. *J. Phys.: Condens. Matter* 13, 7659 ©2001.

Reproduced with kind permission from International Union of Crystallography, Chester, England:
Figures 6.14 & 6.16, Table 6.1: from K. Krec, W. Steiner. *Acta Cryst. A* 40, 459 ©1984. **Fig. 6.18**: G. Albanese, A. Deriu, C. Ghezzi. *Acta Cryst. A* 32, 904 ©1976. **Fig. 6.19(b)**: C.J. Martin, D.A. O'Connor. *Acta Cryst. A* 34, 500–505 ©1978.

Reproduced with kind permission from Macmillan publishers Ltd. (Nature Publishing Group):
Figure 6.6: from P.J. Black and P.B. Moon. *Nature* (London) 188, 481 ©1960.

Reproduced with kind permission from Mössbauer Effect Data Center, University of North Carolina at Asheville:
Appendix H: from G.J. Long, T.E. Cranshaw, and G. Longworth. *Mössbauer Effect Reference and Data Journal* 6, 42–49 ©1983.

Reproduced with kind permission from the Physical Society of Japan:
Figure 9.13 & Table 9.3: from Y. Nishihara, T. Katayama, S. Ogawa. *J. Phys. Soc. Japan* 51, 2487 ©1982.

Reproduced with kind permission from the Royal Society, London:
Figure 6.7: from P.J. Black, D.E. Evans, D.A. O'Connor. *Proc. Royal Soc.* (London) A 270, 168 ©1962.

Reproduced with kind permission of Springer Science and Business Media:
Figure 2.11: from Y.L. Chen, B.F. Xu, J.G. Chen, Y.Y. Ge. *Phys. Chem. Minerals* 19, 255 ©1992. **Figs. 2.20 & 2.22, Table 2.5**: U. Gonser (Ed.) *Mössbauer Spectroscopy II – The Exotic Side of the Method*, Springer-Verlag, New York, ©1981. **Fig. 2.24**: I.J. Gruverman (Ed.) *Mössbauer Effect Methodology*, vol. 6, Plenum Press, New York, ©1971. **Fig. 4.3**: G. Leibfried and N. Breuer. *Point Defects in Metals I*. Springer-Verlag, Berlin, ©1978. **Fig. 4.14**: P. Brüesch. *Phonons: Theory and Experiments I*. Springer-Verlag, Berlin, ©1982. **Fig. 4.16**: J.R. Hardy, A.M. Karo. *The Lattice Dynamics and Statics of Alkali Halide Crystals*. Plenum, New York, ©1979. **Table 4.2**: T. Nakazawa, H. Inoue, T. Shirai. *Hyperfine Interactions* 55, 1145 ©1990. **Fig. 6.11**: G.V. Smirnov. *Hyperfine Interactions* 27, 203 ©1986. **Figs. 7.7, 7.8, & 7.9**: G.V. Smirnov. *Hyperfine Interactions* 125, 91 ©2000. **Figs. 7.14, 7.16, & 7.17**: U. van Bürck. *Hyperfine Interactions* 123/124, 483 ©1999. **Figs. 7.20 & 7.21**: G.V. Smirnov. *Hyperfine Interactions* 123/124, 31 ©1999. **Fig. 7.22**: H. Grünsteudel, V. Rusanov, H. Winkler, W. Meyer-Klaucke, A.X. Trautwein. *Hyperfine Interactions* 122, 345 ©1999. **Figs. 7.25 & 7.26**: R. Rüffer, A.I. Chumakov. *Hyperfine Interactions* 128, 255 ©2000. **Figs. 7.27 & 7.28, Table 7.2**: W. Sturhahn, A. Chumakov. *Hyperfine Interactions* 123/124, 809 ©1999. **Table 7.3**: W. Sturhahn, E.E. Alp, T.S. Toellner, P. Hession, M. Hu, J. Sutter. *Hyperfine Interactions* 113, 47 ©1998. **Fig. 8.7**: S.N. Dickson, J.G. Mullen, R.D. Taylor. *Hyperfine Interactions* 93, 1445 ©1994. **Fig. 8.8**: P.H. Dederichs, R. Zeller. *Point Defects in Metals II*. Springer-Verlag, Berlin, ©1980. **Table 8.3**: J.T. Devreese, L.F. Lemmens, V.E. van Doren, and J. van Royen (Eds.) *Recent Developments in Condensed Matter Physics*, vol. 3. Plenum Press, New York, ©1981. **Figs. 9.4 & 9.5**: Th. Obenhuber, W. Adlassnig, J. Zänkert, U. Närger, W. Potzel, G.M. Kalvius. *Hyperfine Interactions* 33, 69 ©1987. **Fig. 9.6**: E. Galvão da Silva, U. Gonser. *Appl. Phys. A* 27, 89 ©1982. **Figs. 9.7 & 9.8**: M. Peter, W. Potzel, M. Steiner, C. Schäffer, H. Karzel, W. Schiessl, G.M. Kalvius, U. Gonser. *Hyperfine Interactions* 69, 463 ©1991. **Figs. 9.17 & 9.18, Table 9.6**: A.I. Chumakov, R. Rüffer, O. Leupold, I. Sergueev. *Structural Chemistry* 14, 109 ©2003.

Reproduced with kind permission from Taylor & Francis Group LLC and Taylor & Francis Limited (Informa UK):
Figures 2.2, 5.1, & 5.2: from J.W. Robinson (Ed.). *Handbook of Spectroscopy*, vol. III. CRC Press, Boca Raton, ©1981. **Fig. 4.17**: H.P. Myers. *Introductory Solid State Physics*. Taylor & Francis, London, ©1990.

Subject Index

Mössbauer Effect in Lattice Dynamics. Yi-Long Chen and De-Ping Yang

ISBN: 978-3-527-40712-5

density functional theory (DFT) 165, 168, 355
density of states (DOS) 24, 142 ff., 165
- host 333
- impurity 333
- partial 172, 286, 292, 299, 359
- partial vibrational 354, 366
- phonon 190, 253, 259, 285, 291, 293, 295, 298, 311, 366
- vibrational 317, 352, 358, 359, 362
diamond 26, 150, 335 f.
dispersion relation 124, 126, 138, 141
Doppler
- broadening 4
- effect 2, 14, 79
- energy shift 2
- second-order effect 73
- transverse Doppler effect 196 ff.
- velocity 6, 64, 373
Doppler shift, second-order 148, 177, 318, 322, 333, 345, 348
DOS *see* density of states
^{161}Dy 56, 83, 258
$Dy_3Fe_5O_{12}$ 358
dynamic response function 341
dynamical beats (DB) 265, 274, 281, 284
dynamical matrix 123 ff., 165
Dyson equation 308, 317

e
E2-type radiation 195
effective magnetic field 53 ff.
effective thickness t_a 10, 72, 85, 269, 274, 278
effective vibrating mass 177, 198, 355
Einstein model 143, 323, 326
Einstein temperature 144, 177
Einstein–Debye formula 346
Einstein-Debye model 323
electric field gradient (EFG) 31, 43 ff.
electric quadrupole interaction 29, 38 ff.
electronegativity 37
electron–electron interaction 168
enhancement of coherent channel 266 f.
^{166}Er 56
^{151}Eu 56, 83, 258, 294
$Eu_2Ti_2O_7$ 193
$EuBa_2Cu_3O_7$ 205, 206
$EuBa_2Cu_3O_{7-\delta}$ 189
exchange-correlation energy 169
extrapolation method 165

f
face-centered cubic (fcc) 119, 125, 128, 150, 183, 201, 313, 380, 393
- crystal 134
Fe 149, 294
^{57}Fe 8, 33, 39, 43, 51, 82, 180, 200, 283, 327, 334, 356
α-Fe 34, 36, 56, 57, 66, 100, 149, 191, 206, 265, 283, 294, 299, 342, 356
Fe nucleus (^{57}Fe) 3
$Fe(C_5H_5)_2$ 88, 206, 356
$[Fe(C_5H_7O_3)_3]^{3+}$ 355
$[Fe(CN)_6]^{4-}$ 355
$[Fe(H_2O)_6]^{3+}$ 355
$Fe[Co(CN)_6]$ 151
$Fe[Ir(CN)_6]$ 151
$Fe[Rh(CN)_6]$ 151
Fe^{2+} 45, 54, 189
Fe_2O_3 35, 299
α-Fe_2O_3 88, 206, 216, 223
α-$^{57}Fe_2O_3$ 261
Fe_2Tb 294
Fe^{3+} 45, 54
Fe_3Al 294
Fe_3BO_6 227
$Fe_{40}Ni_{40}P_{14}B_6$ 105, 351, 353
$Fe_{67}Tb_{33}$ 294
$Fe_{74}Si_{10}B_{16}$ 351
$Fe_{75}P_{15}C_{10}$ 351
$Fe_{80}B_{20}$ 353
$FeBO_3$ 294, 364
$^{57}FeBO_3$ 261
FeC_2Ni 364
$FeCl_2$ 189
$FeCO_3$ 50
FeF_2 189
FeF_3 51
FeGe powder 60
Fe–Rh–Ni polarizer 64, 70
Fermi contact field 54
Fermi pseudopotential 157
ferrocene 88, 356
ferrous halides 37
FeS_2 206
Fe–Si 64
$FeSO_4{\cdot}7H_2O$ 88
first-principles methods 165 ff., 299
force constant 117, 138, 150, 152, 291
- central forces 383 ff.
- host–host 305, 310
- impurity–host 305, 310, 322
- matrix 166, 380
- ratio 326, 333, 346
- tensor 307
Fourier transform 108, 184, 296, 306
Fourier-logarithm decomposition 296
fractional absorption intensity 12

– longitudinal 138
– normal 122, 129, 131, 140, 318
– resonance modes 311, 317
– resonant 154
– stretching 361
– transverse 138
moments of frequency distribution 146 ff.
Morse pair potentials 139
Mössbauer diffraction 216, 223 ff.
Mössbauer effect 5, 6, 73, 151
– phonon-assisted 297, 361
Mössbauer lattice temperature 356
Mössbauer spectroscopy 6, 72, 253 ff., 295
– conventional 298
– energy domain 253, 280
– spectrometer 79 ff.
– synchrotron 202, 253 ff., 298
– time domain 253, 265, 280

n
Na 149, 151
$Na_2[Fe(CN)_5NO]\cdot 2H_2O$ 88, 206
$Na_4Fe(CN)_6\cdot 10H_2O$ 336
NaCl 246
NaF 143
Nb 329
Nb_3Sn 189, 201
$Nd_2Fe_{14}B$ 56, 70
Ne 139, 150
neon 92
neutron scattering 156 ff.
neutron scattering, inelastic 366
$(NH_4)_2Mg^{57}Fe(CN)_6$ 274
$(NH_4)_2MgFe(CN)_6$ 359
Ni 151, 243, 329, 342
^{61}Ni 56
$[Ni(CN)_4]^{2-}$ 355
normal coordinates 22, 120
normal mode 122
^{237}Np 33, 56, 83
nuclear Bragg scattering 220, 225, 229 ff., 259, 265, 278
nuclear exciton 265 f., 271
nuclear forward scattering 214, 220, 229 ff., 259, 271, 281, 295, 298
nuclear resonant scattering 155, 214
– inelastic 155, 295, 298, 318, 362
– inelastic incoherent 289

o
operator
– annihilation 16, 20
– creation 16, 20
– density 18
– displacement 17, 20
– Hermitian 18
optical branch 152
– longitudinal 131
optimal sample thickness 84
^{191}Os 5

p
Pb 151, 218, 248
Pd 84, 328, 329
Pd_3Sn 201
$Pd_{80}Si_{20}$ 353
phonon annihilation probability 291
phonon creation probability 291
phonon–phonon interaction 182
plane-wave basis 171
polarization 62 ff.
– π-polarization 64, 69, 262
– σ-polarization 64, 69, 262
– vector 63, 122, 164
principal axis system 31, 40, 48, 56
pseudo-classical quantum states 16
pseudoharmonic approximation 185 ff.
pseudopotential approximation 171
Pt 325, 328, 329
Pt absorbers 6
PtB_2 188
$PtCl_2$ 188

q
QB *see* quantum beats
quadrupole splitting 72
quantum beats (QB) 265, 274 ff., 284
quasiharmonic approximation 185

r
Rayleigh scattering 214, 233
reciprocal lattice 125
recoil energy 2, 161
recoilless fraction f 2, 6, 16, 21, 26, 73, 148, 161, 177, 183, 192, 202, 295, 309, 328, 333, 342, 391 f.
reduced rotation matrix 220
refractive index 272
resonance absorption 9
– coefficient 85
resonance fluoroscence 1
resonance modes 311
Rh 84, 329
^{99}Ru 56

s
saturation effect 13, 71 f., 86
Sb 36

Related Titles

Bordo, V. G., Rubahn, H.-G.

Optics and Spectroscopy at Surfaces and Interfaces

281 pages with 144 figures and 1 tables
2005
Softcover
ISBN: 978-3-527-40560-2

Kosevich, A. M.

The Crystal Lattice
Phonons, Solitons, Dislocations, Superlattices

356 pages with 88 figures
2005
Hardcover
ISBN: 978-3-527-40508-4

Gielen, M., Willem, R., Wrackmeyer, B. (eds.)

Unusual Structures and Physical Properties in Organo-metallic Chemistry

approx. 384 pages
Hardcover
ISBN: 978-0-471-49635-9